ra

ile

f marine

$49.50
ocm69332340
01/02/2008

New ed.

The Facts On File

DICTIONARY of MARINE SCIENCE

New Edition

The Facts On File
DICTIONARY of MARINE SCIENCE

New Edition

Barbara Charton

An imprint of Infobase Publishing

The Facts On File Dictionary of Marine Science, New Edition

Copyright © 2008, 2001, 1988 by Barbara Charton

Illustrations © 2008 by Infobase Publishing

Facts On File, Inc.
An imprint of Infobase Publishing
132 West 31st Street
New York NY 10001

ISBN-10: 0-8160-6383-4
ISBN-13: 978-0-8160-6383-3

Library of Congress Cataloging-in-Publication Data

The Facts on File dictionary of marine science / Barbara Charton.— [New ed.]
    p. cm.
    Includes bibliographical references (p.   )
  ISBN 0-8160-6383-4
    1. Marine sciences—Dictionaries. I. Charton, Barbara. II. Facts On File, Inc.
III. Title: Dictionary of marine science.
GC9 .F28 2006
551.4603—dc22   2006045654

Text and cover design by Cathy Rincon
Illustrations by Richard Garratt

Printed in the United States of America

MP KT 10 9 8 7 6 5 4 3 2 1

This book is printed on acid-free paper.

CONTENTS

ACKNOWLEDGMENTS

The people who encouraged and supported my efforts during the creation of the first edition have been forbearing and were so again. I am grateful. This work would not have been finished without the people and the great library at New York University.

I would also like to thank Frank K. Darmstadt, my editor, and Melissa Cullen-DuPont, assistant editor, for being encouraging, supportive, and challenging. The graphics department at Facts On File has been of invaluable help in putting together this new edition.

Years ago I promised Marvin, The Great Mind, that it would be easier the second time around, and it was. It has also been great fun having a partner with whom to collect rocks and minerals, fossils, and endless new information about the expanding array of invertebrates. Fitting all this new information into both ecological and environmental contexts has taught me much, and I look forward to sharing it with others interested in the continuing story of Earth—our biome.

PREFACE

To some it may seem that the ocean is the same today as it was many years ago. The average mean temperature has risen several degrees since 1900, and this does not create excitement for most people. But in reality, the rise of several degrees is far from insignificant. This increased warmth is enough to dissolve coral reefs and destroy ecosystems. Since the appearance of the first edition of this work, a great deal of research has been produced, much of which is now general knowledge. The general population has become much more aware of the ocean's effect on all life, and phenomena such as global warming are now matters of public discussion. The increase in temperature credited to global warming has finally been accepted as a fact, and this is no surprise to paleoclimatologists. It has happened before, and it will get worse, according to computer projections, which depict that by the mid-21st century there will be a clear shipping channel in summer months around the northern perimeter of North America. The discussion of how human activity affects this phenomenon and what humans—the dominant organism likely to be affected the most—must do to prevent catastrophe is now being actively studied.

The water of Earth defines the planet. As soon as astronauts could see the Earth from space, their name for Earth was the Blue Planet—or the Blue Marble. The photographs taken of planet Earth from space are spectacular and classic. Since water is a substance found everywhere, it is easy for humans to take it for granted, but doing so is a terrible mistake. Since life does not exist in the absence of water, water in its greatest concentration in the world's oceans deserves respect, consideration, and study.

Marine science is a composite field that encompasses other disciplines and their relations to the world's oceans. Thus, biochemistry, botany, chemistry, ecology, geodesy, geography, geology, geophysics, hydrology, meteorology, microbiology, minerology, seismology, and zoology may all be marine sciences in some instances, as are their combinations and subdisciplines. Some of the fields that apply information gathered by natural and physical scientists include engineering, pharmacology, medicine, population study, and toxicology.

The instruments used by scientists and others who apply science have undergone both proliferation and vast elaboration. These instruments range from the simplest tools used by hunter-gatherers to boats, navigation devices, weather instruments, energy-producing engines, elaborate computer projections of future events, and robotic devices that can take the place of humans in hostile environments.

Though scientists have amassed a huge amount of knowledge, much remains to be discovered. The Tree of Life project started in the 1990s to collect data about every currently living organism, and it is growing rapidly. Unknown and unclassified vertebrates are still being found. Since vertebrates are a relatively small group of the biota, it is easy to assume that they are all known. However, life is much more complicated than any biologist thought it might be. For example, it was previously believed that the Sargasso Sea contained few living organisms except a few floating algal species. Since the 1980s better collecting devices have been perfected that have discovered living organisms the previous collecting methods destroyed. It is now known that tiny soft-bodied creatures live in the Sargasso Sea in great numbers. Thus, improved methods and instruments have opened up many areas that were once dismissed as completely understood and catalogued. This continues to happen. Research in the Coral Triangle near western Papua has recently produced a number of as-yet-unidentified species and genera.

Unfortunately, much scientific work is done in a race against destruction of habitat. The ocean, often viewed as a limitless resource, has been misused by humans for centuries. But the ocean's biomes are neither boundless nor inexhaustible. They are as fragile as a terrestrial ecosystem, and as in any ecosystem, the biomes exist in connection with others—air, shore, and underlying rock—all of which are also always changing.

People have interacted with the oceans since the very early discovery that some of the organisms in the water were edible. Therefore, much of the early information about the ocean, its phenomena, and its animals was firmly connected to the need for information about edible marine organisms. Once a basic need for food was fulfilled, curiosity led the adventurous to extend their wanderings to other places thus creating a need for the development of navigation.

It took some time to create devices that would move people from one shore of the ocean to another, so early sailing was very much a coastal business. Eventually, however, the combination of astronomy and better shipbuilding led to extensive commerce. The search for more materials of commercial interest in turn led to great voyages of exploration, and early explorers were driven not so much by scientific motives as by profit motives. These voyages began centuries ago; Herodotus writes of the circumnavigation of Africa by Phoenicians in the employ of Necho II, a 26th-dynasty ruler of Egypt in the seventh century B.C.E. Later rulers funded such expeditions for some combination of personal and national glory and riches.

The great expeditions starting in the 15th century C.E. attempted to collect information about all the physical aspects of voyages: wind, tides, currents, depths, and geography. The sailors made notes and amassed specimens of plants, animals, and peoples encountered. All of this data was meant to serve the needs of the sponsoring rulers or commercial companies that intended to profit from the newly discovered plants, animals, minerals, and people.

These voyages of exploration created a need for better maps and instruments for navigation, which made subsequent voyagers more exact in the information they brought back. In the 18th century the determination of longitude was a much-sought-after goal. The British Admiralty awarded prizes for the construction of instruments that would facilitate its measurement. The instrument developed was a type of clock—not an ordinary clock, but one that is still accurate on ships moving both horizontally and vertically on stormy seas. Then and now, every improvement in instrumentation increases the knowledge of our ever-changing environment, and we become more aware of how much is still unknown.

Since the first wholly scientific expedition, that of the HMS *Challenger* in 1872–76, many others have been launched. The *Challenger* voyage attempted to study all aspects of the oceans using the talents of many scientists working in a number of disciplines. Since then there have been other general expeditions of this type and numerous others to study a specific phenomenon, such as air-water interaction or carbon dioxide turnover in seawater. Some have examined a particular biome, such as the Arctic. There is another year starting (2007–09) that considers a number of disciplines as they pertain to this part of the world and examines the effects of what happens in the Arctic on the rest of the world.

If there is anything consistent about the Earth, it is that it is constantly changing. Continents are not where they once were; oceans have appeared in places where they once were not. The dry state of Nebraska, for example, was once a marine environment. Even climate changes similar to the current one have happened in past times.

One factor that is unchanging, however, is that air is produced by living things. Without the original protists, there would be no free oxygen in the air or dissolved in the water. Some of those primordial organisms are still found on Earth, but it has taken much effort and science to find and identify them. We continue to learn more about the Earth and know that there is much more to learn. We use what we know of Earthbound life to look for it elsewhere in the universe. And since we cannot know what we will find, we keep perfecting present instruments, creating new ones, turning present technologies to new uses, rethinking information, exploring the planet, and finding new things to amaze us.

abalone Snails, class Gastropoda, subclass Prosobranchia, genus *Haliotis*. These animals do not look like snails but superficially resemble clamlike mollusks. The shell exhibits a characteristic row of holes.

Abalone are found in the Pacific Ocean and off the Australian and South African coasts. The red abalone (*H. rufescens*), the largest species, is taken commercially along the western coast of North America. This animal usually has a shell about 30 cm (12 inches) long. The large muscle, the foot, is sliced and broiled as abalone "steak." With the reappearance of a sizable number of sea otters off the coasts of Washington, Oregon, and British Columbia, the abalone is now scarcer than it was in the 1930s and 1940s. *See* SEA OTTER.

abiotic environment The physical factors that affect an organism such as light, temperature, water and its dissolved and suspended materials, and nutrients.

absolute humidity The maximum amount of water vapor that a unit volume of air can contain at saturation. Absolute humidity is a function of temperature because of the gaseous state of air: As the temperature rises the vapor pressure of the water rises, which means that more molecules of liquid water have sufficient energy to make the transition from the liquid to the gaseous state. The higher the temperature, the higher the vapor pressure of the water and therefore the higher the absolute humidity.

absorption The movement of water and dissolved materials in it, i.e., nutrients, minerals, and gases across cell membranes into an organism from the external environment.

abyssal environment The environment at the bottom of deep ocean, one of the most constant marine biomes. The temperature is between 0° and 2°C. The salinity is also relatively constant, hovering at about 35‰, as are the concentrations of various ions (phosphate, silicate, hydroxyl, and carbonate) are also fairly uniform. There is sufficient oxygen to support life. The life forms that exist at great depths depend for food on terrestrial or upper water layers; they are for the most part suspension feeders, scavengers, or carnivores, but their greatest difference from animals found at higher levels is their ability to live at a constant high pressure (200 to 700 atm). The number of animals in the abyssal environment is low, and their rate of reproduction is also thought to be low. *See* ABYSSAL FLOOR, HADAL ZONE.

abyssal floor Also called abyssal plain. The flat, relatively featureless bottom of deep ocean, i.e., at a depth greater than 2,000 m. The average depth of the abyssal floor is about 4,000 m (13,000 feet). There are very few animal species that live in the inhospitable cold, dark regions of the ocean bottom. There is little or no seismic activity, and the major geologic event is sedimentation. Particulate matter coming from a continent, or bits of calcareous material of marine origin, will finally filter down onto the ocean floor. However, the ocean floor is not without activity. Storms and strong currents carry sediment about, scouring the bottom. Also, the Atlantic Ocean receives rushes of very dense, cold water from both polar regions, which carries sediment with it. The great currents of very cold water are more prevalent in the Atlantic than in the Pacific Ocean since the shallow Bering Sea and the Aleutian island chain effectively bar the cold Arctic waters from entering the Pacific.

abyssal rocks Are usually large and formed from magma in the deep ocean. They are the result of upwelling of magma at the edges of crustal plates or directly above hot spots on ocean floors. They form the basic structure of the mountains of the mid-ocean rift system. *See* BASALT, MID-OCEAN RIFT, PLATE TECTONICS.

abyssopelagic Referring to that part of the water column at great depths, usually below 4,000 m (13,500 feet). The term refers to any organism or condition found in these hadal regions. *See* OCEAN BOTTOM.

acantharia Marine protists that are a subclass of Actinopoda; relatives of the radiolarians. Acantherians have had a very long history on Earth, dating from the Precambrian era. Although they are a small constituent of pelagic plankton, 1 to 5% of the total plankton population, they are important factors in the food chain of the Caribbean Sea and the Gulf Stream, particularly during the periods of planktonic bloom.

Acantharia have internal rigid support systems made of celestite (strontium sulfate, $SrSO_4$). This compound is soluble at depths greater than 200 m (650 feet), and as a result, the skeletons of Acantharia do not settle to the ocean bottom when they die. These creatures consist of an external capsule of nonliving material that surrounds the cytoplasm, and the interior and exterior parts of these organisms are connected by a structured network of fibers.

Acantharia migrate vertically each day because they harbor algae or dinoflagellates within their cytoplasm; this periodic rise allows these photosynthesizing symbionts to move into the photosphere during daylight hours. At night Acantharia sink once again to lower levels where they find food and, in turn, provide food for other organisms. *See* MIGRATION.

acanthocephala A phylum of small (about 1 cm or 0.4 in. long) spiny-headed, sausage-shaped worms, all of which are parasitic, that resemble nematodes and are named for the impressive, hook-covered proboscis that attaches them to the gut of their host.

The body cavity of the acanthocephalans is filled almost totally with gonads. The life cycle is a complex one in which larval stages infest intermediate hosts. Isopods are frequently the first or intermediate hosts. The *Corynosoma* is a parasite of several species as it works its way up the food chain. Its first host is an amphipod in which it changes into a more adult form. Then, upon ingestion by a fish, it becomes a parasite of that fish. If this fish is eaten by a bird or seal, yet another change of the worm renders it a parasite in that host. In vast numbers, acanthocephalans can starve their hosts or create fatal gut blockages. *See* FOOD CHAIN, Nematoda, PARASITE, PROBOSCIS.

acartia A plentiful species of copepod that lives in estuaries. It is an important

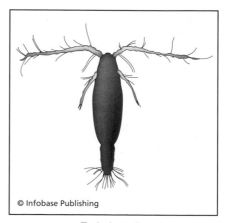

© Infobase Publishing

Typical acartia

source of food for plankton-eating fish, so much so that the fish population of the Black Sea is highly dependent on the population of this organism. Acartia was first classified in 1848. It has been reclassified several times since then, largely because of its unusual number and the appearance of its juvenile forms (nauplia). In most cases, the nauplia of an organism resemble one another, and as they mature, the juveniles begin to resemble the adult. This is not the case with Acartia; several of its juvenile forms have been mistaken and classified as unique genera. *See* COPEPODA, NAUPLIA.

acidity (basicity) A measure of the concentration of acid (base) in water or in a solution. The acidity or basicity of seawater is described by the pH scale. Pure water has a pH of 7; a strong acid would have a pH of 1 or 2 and a strong base would have a pH of 12 or 13; seawater is slightly alkaline at 7.2. The acidity or alkalinity (basicity) of seawater is a function of depth, the quantity and nature of the gases dissolved in the water, local conditions, and the presence (or absence) of plant and animal life. *See* ACIDS AND BASES, BUFFER, MINERALS, pH, WATER.

acids and bases An acid is a chemical entity that can transfer a proton (i.e., a hydrogen ion or a hydrogen atom nucleus) to a base. A base is a chemical entity that can accept a proton. This definition of acids and bases, known as the Brønsted definition, is the most appropriate for an aquatic environment.

Acipenseriformes The order of fish that includes the sturgeons (Acipenseridae). As adults, Acipenseriformes have a cartilaginous skeleton; when immature they have a bony skeleton and have scutes (armor plates) instead of scales. *See* STURGEON.

acorn worm Of the class Enteropneusta (Phylum Hemichordata) whose anterior (head) end resembles an acorn. Not worms at all, these animals burrow in the sand or mud of the sea bottom down to depths exceeding 3,000 m (10,000 feet). Most species are about 5 cm (2 inches) long, but the species *Balanoglossus gigas* is over 2 m (7.5 feet) long.

There is considerable variation between species of acorn worms, with some of the deep-water specimens possibly being free swimming. Feeding modes also differ; some species filter seawater, extracting debris from it, while others feed by trapping food particles in a mucus film and conveying it toward their mouth by cilia. The sexes are separate. Some species produce large numbers of eggs that have almost no food reserve; others have few eggs with a relatively large yolk.

acoustics The study of the production, transmission, and detection of sound. Any disturbance in a medium—gaseous, liquid, or solid—will produce a sound wave, which the medium will disperse.

Sound waves can be used to plumb the depths of the ocean and to map bottom terrain. A ship or buoy can generate a sound wave that will travel through the water at a known rate, strike the bottom, and rebound. Receiving equipment can then measure the interval from the generation to the return of the wave, from which the distance to the bottom may be calculated.

Plankton swarms or large schools of fish also reflect sound waves, because they are so tightly packed that their mass is echo-producing. Commercial fishermen capitalize on this fact and use sonar to find fish. *See* ECHO, LORAN, SONAR, SOUNDING.

Acropora The dominant genus of the family Acroporidae. The species Acropora constitutes almost half of all the true corals and creates the major deposits of calcium carbonate ($CaCO_3$). *See* CORAL, ATOLL, REEF.

Actinopterygii One of the subclasses of the Osteichthyes, the bony fish. They are ray-finned fish with webs of skin on bony or horny spines. These fish are the current dominant aquatic vertebrates and account for about half of the known vertebrate species.

activation analysis The bombardment of samples of rock, water, or fossils by slow (thermal) neutrons, protons, or other nuclear fragments of known energy in order to date them. The radiation coming from the sample is then compared to established standards, which allows the analyst to identify the elements in the sample and determine their ages. The advantage of this type of analysis is that, unlike carbon dating, it does not destroy the sample. *See* DATING, RADIOACTIVITY.

Aden, Gulf of A deepwater area between the Red and Arabian seas and between the African and Arabian coasts. The Gulf of Aden is a part of the East African Rift Valley system. Its most important undersea feature is the Sheba Ridge, a continuation of the ridge system in the Indian Ocean, which is also the site of seismic and volcanic activity.

The deepest part of the gulf is the Alula-Fartak Trench, with a recorded low point of 5,360 m (17,566 feet). The bottom also exhibits a series of fault lines running from northeast to southwest, caused by plate movement as the African and Arabian crustal fragments move apart.

The sediment of the bottom is terriginous and is carried into the water from the surrounding landforms—mainly by the wind. The water of the gulf is warmer than normal ocean water because there is little water entering the system in the form of rain or river runoff, and the gulf is unusually saline because there is considerable evaporation in the hot climate. The surrounding land is largely desert.

The Strait of Bab-el-Mandeb connects the gulf to the Red Sea. There is no large-scale fishing in the gulf, despite the large variety and considerable numbers of fish that feed on the upwelling of nutrients along the coasts. *See* INDIAN OCEAN, RIFT VALLEY.

Adriatic Plate The leading edge of the Eurasian Plate that abuts the Hellenic and African Plates. *See* PLATE TECTONICS.

Adriatic Sea An arm of the Mediterranean separating the eastern Italian coast from the Balkan and Hellenic peninsulas. The sea is about 131,000 km^2 (50,000 square miles). The maximum depth is 1,324 m (4,035 feet) west of Corfu, and the average depth is about 450 m (1,450 feet). The sea has one relatively smooth coast on the west, the Italian one, and a very indented one with many islands, the Balkan coast, on the east. The latter is a continuation of the mountain structure of the East European landmass.

The Adriatic is rather unusual for the Mediterranean because it has a greater tidal range (the difference between the high and low water marks) than that of the larger body itself. The average tide range is less than 30 cm (1 foot) in the Mediterranean, whereas the tide ranges in the Adriatic are three times larger. The nutrient and oxygen loads in this sea are relatively low, like those in the rest of the Mediterranean. *See* MEDITERRANEAN.

Aegean Sea The easternmost embayment of the Mediterranean, bounded by Greece on the west, Turkey on the east, and Crete on the south. The Dardanelles are the strait that connects the Aegean Sea with the Sea of Marmora to the northeast. This is in turn connected to the Black Sea by the Bosporus. The Aegean is dotted with small islands that rise from a relatively deep floor. The deepest point, 3,543 m (11,627 feet), is found near Crete. The total area of the sea is about 214,000 km^2 (83,000 square miles).

Like most of the Mediterranean, the Aegean is relatively tideless except for the strait between continental Greece and the island of Euboea, where an unpredictable and often violent current runs. This anomaly has been the subject of study for millenia; Aristotle commented on it.

The geology of the Aegean is, like that of the islands in it, largely limestone. The water is both nutrient and oxygen poor even when compared to other Mediterranean water. *See* ARCHIPELAGO, MEDITERRANEAN.

aerosol Particulate matter dispersed in air. An aerosol may be either solid or liquid. Solid aerosols are finely divided particles of salt, sand, or other terrigenous solids. The major solid aerosol is salt, which is carried into the air either by evaporation of seawater or by wave action. Liquid aerosols, fogs, are made up of water particles. They are also generated by spray resulting from wave action. The solid particles in some aerosols are windblown pollutants that can be carried for great distances depending on wind velocity and particulate size. The smallest particles travel farthest; therefore, windblown dust is not just a local problem.

One theory for the damaged coral reefs and the withering disease of sea fans in the Caribbean is based on the introduction of dust from the Sahal into American tropical waters. The Sahal is a large region of West Africa adjacent to the Sahara that is plagued by persistent droughts. The dry soil is picked up by the prevailing winds, which blow west toward the Americas, and it falls largely in the Caribbean.

The endangered reef-building corals and related organisms such as sea fans have been studied extensively because they have been exhibiting signs of widespread disease. A living coral is a distinct color, usually greenish brown. When the living organisms die, their tissue sloughs off, leaving the white, bleached skeletal structure. In sea fans, disease is manifested by dark spots that become holes; eventually, only the spines are left. Samples have been scraped from diseased sea fans and cultured in research laboratories. On analysis, the samples were found to contain molds such as *Aspergillis* that are normally found in soil. The supposition is that the mold spores that found their way in soil from Africa to the waters of the Caribbean have been infecting and killing sea fans there. At present, there is no method known to protect the corals from this pollutant. *See* ATMOSPHERE, FOG, POLLUTION, SEA FAN, SEDIMENT.

African Plate The tectonic plate extending under most of the African continent. At its northeastern edge, the Eastern Rift Valley is a subduction zone (the descending plate). Eventually the rift will widen and be invaded by ocean water as the African and Arabian plates continue to separate. *See* PLATE TECTONICS.

Agassiz, Alexander (1835–1910) An authority on coral formations, he was born in Switzerland and came to the United States as a boy in 1849. His professional career was shaped largely by that of his father, the natural historian Jean-Louis Agassiz. Alexander Agassiz's publications ranged over a wide area of interests; he was an authority on echinoderms and had contributed many papers to the literature discussing this phylum and organisms in it. Following an exploratory voyage to South America, he coauthored an account of the trip with his stepmother and fellow voyager, Elizabeth Cady Agassiz.

Agassiz led several scientific expeditions, most of them on the *Albatross*. In 1875 his discovery of a coral reef at a depth of 1,000 m (3,300 feet) led to a contradiction of Darwinian theories of coral reef construction and to Agassiz's continuing study of these formations. Darwin believed that coral atolls and the surrounding lagoons were the result of the subsidence of volcanic cones; Agassiz disagreed. This debate was not finally concluded until the latter half of the 20th century, when core samples proved that Darwin's hypothesis was correct.

Agassiz made numerous trips at the behest of the United States Navy, including explorations of the Caribbean Sea, the path of the Gulf Stream, and the Great Barrier Reef. In 1899 Agassiz sailed from San Francisco to the Marquesas Basin, where he dredged and found many sharks' teeth at great depths, although sharks are not hadal creatures. His explanation for this find (one that later evidence supported) was that in Tertiary times the Marquesas Basin was closer to the photic zone than it is now, "with greater light and warmer water, and thus sufficient prey for sharks." Subsequently the area subsided. Agassiz's last scientific expedition was organized to study the Peru Current (then called the Humboldt Current.)

Alexander Agassiz started his scientific career very much in the shadow of his famous father. For some years, his writings echoed those of the older Agassiz and were firmly antievolutionary in theory. After his father's death Alexander Agassiz incorporated Darwin's theories into his own work, and he became a member of the international scientific community. In his capacity as an expert on echinoderms, he was one of the foreign scholars examining the data brought back by the HMS *Challenger's* scientific studies. One of his many publications on these creatures was "Report on the Echinoidea Dredged by *HMS Challenger* during the Years 1873–1876." Agassiz died unexpectedly aboard ship while returning to Massachusetts from London as he was working on more revisions of the work on the HMS *Challenger* data. *See* AGASSIZ, JEAN-LOUIS; *ALBATROSS;* CARIBBEAN; CORAL; ECHINODERMATA; GULF STREAM; PACIFIC OCEAN.

Agassiz, Jean-Louis Rodolphe (1807–1873)

A natural historian who made major contributions to the study of fish fossils and glaciers. Jean-Louis Agassiz was born in the Swiss canton of Fribourg, was educated in Zurich and Heidelberg, and received a medical degree from the University of Munich in 1830. Agassiz went to France to work with Georges Cuvier, who was studying the classification of fossil fish. While in Paris he met geographer and naturalist Friedrich Humboldt, whose recommendation helped Agassiz obtain his first academic position, as professor of natural history in Neuchâtel (Switzerland).

While in Switzerland Agassiz continued the work he had begun in Paris on the fishes of Brazil. The result was a superb five-volume treatise on fossil fish, published in 1834.

Agassiz belonged to a group, centered in Neuchâtel, that seriously studied natural history, although most of its members were trained in other disciplines. The group made extensive explorations of the mountains around them, and their careful observations were helpful to Agassiz in his own work on glaciers, in which he used existing glaciers to explain phenomena such as scouring scars, transported rocks, and moraines (the mixture of rocks left at the leading edge of a glacier).

Agassiz opposed the Darwinian theory of the evolution of species as the result of natural selection. He believed that changes in and on the Earth created diversification and change in living organisms and fossil remains. Until the work of Mendel on genetics, there was no mechanism for the theory of natural selection, so Agassiz's theory, which was logical and appeared to explain the evidence, found some acceptance.

In 1846 at the suggestion of Charles Lyell, the noted British geologist, Agassiz left Switzerland for an American lecture tour. He was well received, and at the end of the tour accepted a professorship in natural history at Harvard University. He began a new collection of natural history artifacts, which eventually grew to become the Harvard Museum of Comparative Zoology. This institution was later named for Agassiz.

Agnatha

A class of primitive vertebrate fishes. They are survivors of a much larger group that has an extensive fossil record. The name means jawless and as a result, members of the class are all suction or filter feeders. *See* HAGFISH, LAMPREY.

Agulhas Current

A southerly moving oceanic current along the East African

coast. The Agulhas Current begins as an east-to-west flow but is deflected by the coastline. The water temperature varies with depth, although the surface water is warm year-round. The portion of the Agulhas Current between Madagascar and the African continent is called the Mozambique Current.

air A mixture of gases, predominantly nitrogen and oxygen. Air also contains carbon dioxide, water vapor, and trace amounts of some 20 other gases. It is the gaseous vehicle in aerosols and suspensions of larger particles, which it disperses. *See* POLLUTION.

Constituents of Air

Substance	% by volume
Nitrogen (N_2)	78.11
Oxygen (O_2)	20.95
Water vapor (H_2O)	0–7
Argon (A)	0.93
Carbon dioxide (CO_2)	0.01–0.1 (0.03 is average)
Ozone (O_3)	0–0.1 (greatest concentration in the stratosphere)
Neon (Ne)	1.8×10^{-3}
Sulfur dioxide (SO_2)	$0–1 \times 10^{-4}$
Krypton (Kr)	1.1×10^{-4}
Methane (CH_4)	2×10^{-4}
Xenon (Xe)	5×10^{-5}
Nitrogen oxides (NO_x)*	$0–3 \times 10^{-5}$
Hydrogen (H_2)	5×10^{-5} (at higher altitudes)

* This, in turn, is a mixture of several gases, N_2O, NO, NO_2, N_2O_4, and N_2O_6; most of these nitrogen oxides are classified as pollutants. They are produced in gasoline burning engines, where the temperature is high enough to cause normally unreactive atmospheric nitrogen to react with oxygen.

air mass The moving, homogeneous aggregate of air as it pertains to weather. The leading edge, or front, is the site of rapid change as one mass interacts with another. The character of an air mass depends on the land or water it traverses. The meteorologic abbreviation cPK, for example, means a continental and polar or cold, dry air mass, mPK indicates cold polar air of maritime origin. *See* METEOROLOGY.

Alaska Current The final branch of the Aleutian Current moving northeast into the Gulf of Alaska. *See* ALEUTIAN CURRENT.

albatross (bird) A large, web-footed seabird; a member of the genus *Phoebetria* or *Diomedia*. There are about fifteen species of albatrosses. The largest and best known is *Diomedea exulans*, a white bird with a 3.5-m (12-foot) wingspan and a heavy, powerful bill. The usual food of this particular species of albatross is squid.

These large birds live for long periods well beyond the continental shelves in temperate to subpolar latitudes. Like their relatives the petrels, the albatrosses stay on the water for months, coming to land only to nest and raise their young. Most albatrosses are sighted in the Southern Hemisphere. *See* PETREL.

***Albatross* (research vessel)** The first ship to be built as a research vessel. The *Albatross* was designed to explore the environments of commercial fishing grounds, and its construction was underwritten by the U.S. Department of Fisheries. Naturalist Alexander Agassiz used the ship from the 1870s through the 1890s for deep sea sampling. *See* AGASSIZ, ALEXANDER.

albedo The reflecting capability of a body. A white or light body has an albedo of one and a black body's albedo is zero. The Moon's albedo produces a considerable light level, so much so that a full moon on an open ocean produces enough light to trigger planktonic migration. *See* MIGRATION.

Albert I, prince of Monaco (1848–1922) Albert-Honoré Charles Grimaldi was the founder of two outstanding institutions of oceanographic research: the Musée Oceanographique de Monaco and the Institut Oceanographique de Paris.

After service in the French navy during the Franco-Prussian War, Albert devoted his life to oceanography. In 1873 he bought and refitted a 200-ton schooner and renamed it the *Hirondelle*. It made four cruises and was followed by the *Princess Alice* (there were two of that name) and the *Hirondelle II*.

Albert went to sea almost every summer for 40 years and became a self-taught meteorologist. In 1885 his major study of the Gulf Stream appeared. A long study of drift currents in the North Atlantic followed, in conjunction with a mapping of water temperatures and the movements of isothermal bodies of water. He established three observatories in the Azores as meteorological research stations. These weather stations were ultimately linked into a worldwide meteorological net of research stations.

Albert provided financial support for the work of Paul Portier and Charles Richet with poisonous animals, which included experiments with *Physalia* (Portuguese man-of-war)—a coelenterate whose venom results in deep anesthesia. The two physiologists worked aboard the *Hirondelle II*. Their efforts in research produced a large body of evidence for the activity of the toxins produced by coelenterates and the resulting autoimmune response to it. This research ultimately led to the understanding of allergic reactions and garnered a Nobel Prize for Charles Richet.

Albert also sponsored the design of new plankton nets and with them brought to the surface animals from depths exceeding 6,000 m (19,000 feet). He was convinced that many creatures in the ocean were yet to be discovered and that equipment needed frequent redesign.

Albert was an early expounder of the prudent and ecologically harmless use of the sea. He warned of overfishing, particularly whaling in the early years of the 20th century. In addition to the oceanographic institutes in Monaco and Paris, he founded the Musée Anthropologique and the Institut de Paleontologie Humaine for the study of human evolution. His museum and aquarium in Monaco are open to the public. *See* COELENTERATA.

Alcyonacea The order of soft corals. They are cnidarian and live in shallow water. Most of the specimens are tropical and soft-bodied and found in all areas. *See* CNIDARIA.

Alcyonaria An order of cylindrical polyps found in all marine environments. The largest distribution is in tropical waters where many harbor symbiotic zoochlorellae or zooxanthellae algae, which can add color to them. *See* CORAL, ZOOXANTHELLAE.

Aleutian Current An ocean current that moves along the Aleutian island chain from west to east. This subarctic water divides, and part turns north into the Bering Sea and moves along the north coast of some of the Aleutian Islands, another branch forms the Alaska Current. The surface water of the Alaska Current is colder and less saline than the water at the depth of the current. *See* ALASKA CURRENT.

Aleutian Islands Island chain in the North Pacific that was formed by the passage of a crustal plate over a "hot spot," where liquid rock rises and forms a volcanic cone on the sea bottom. These volcanic rises are referred to as either seamounts or guyots if they remain submerged, and as islands if they rise above sea level. *See* ATOLL, GUYOT, HAWAII, ISLAND, SEAMOUNT.

Aleutian Trench An ocean deep that skirts the southern border of the Aleutian Island chain. The deepest portions of the Aleutian Trench are more than 7,300 m (24,000 feet) deep.

algae Large benthic protist and unicellular autotrophe seaweeds that constitute phytoplankton. Algae range from unicellular organisms to large, complex colonies of

Typical algae found in temperate waters.
All of these plants are native to the Pacific coast of North America.

kelp to yellow, green, brown, and red organisms. All species of algae are autotrophic (photosynthesizing). Some algae, such as dinoflagellates and euglenoids, have flagellae and are motile. These life-forms support major marine food chains. All algae, no matter how large, are now categorized in the kingdom Protista. The logic behind this is that algae have a colonial structure no matter how large any kelp blade can grow. Since any one cell could regenerate the entire structure, there is no real differentiation between the cells, and they are officially a colony. *See* AUTOTROPHS, CORALLINE ALGAE, DIATOM, KELP, PHYTOPLANKTON, RHODOPHYTA.

algal bloom The significant increase in the population of algae in a biome. Algae need nitrogen, phosphorus, and carbon dioxide to increase. The upwellings in certain locales provide this, and the end result is the convergence of other organisms that feed on the algae. In areas where there is a concentration of untreated sewage or agricultural runoff containing animal waste and fertilizers, there is a localized overfeeding of the algae population. This is very destructive because the algae, when present in abnormal numbers, form dense mats blocking out sunlight and oxygen. The result of this overfeeding of the algae is a "dead zone," an area in which there is very little left alive. The plants normally present are prevented from getting enough sunlight, and the animals that feed on them cannot get enough oxygen because it is being used by the decomposing bacteria that are feeding on dying plants.

alginate A dried material that is made in industrial quantity from the cell walls of brown algae. This product is used commercially in the paper, paint, food, cosmetic, and pharmaceutical industries because of its colloidal properties.

alkalinity and acidity The hydrogen ion (H^+) concentration of a liquid or solution, usually expressed in logarithmic form as pH. The pH of seawater varies with local conditions. Both alkalinity and acidity are functions of dissolved minerals and gases. *See* ACIDS AND BASES, BUFFER, pH.

allopathic speciation The development of species in isolation. An example of this is the development of different species of finches on the Galapagos Islands—Darwin's great discovery. When a small population is geographically separated, small changes (the mutations that occur) tend to be conserved in that biome and thereby make the population different from the parent stock and from others that have also developed differently from the same parent stock.

alpha ray Alpha rays are produced by radioactive decay. The rate of decay is used in dating marine fossils and other remains or radioactive elements. Alpha rays consist of the nuclei of helium, atoms that are composed of two protons and two neutrons. These particles emanate from a substance as part of its radioactive decay. *See* ATOM, RADIOACTIVITY.

Alps The European portion of the great east-west mountain system stretching from the Pyrenees to Malaysia. The Alps extend from the Gulf of Genoa to the Vienna Woods. The massif is divided into a western section, comprising southeastern France and northwestern Italy; an eastern section comprising Germany, Switzerland, Austria, Slovenia, and Liechtenstein; and a central portion extending from north-central Italy to southern Switzerland.

According to the theory of plate tectonics, the alpine upheaval is the result of the collision of the African and Eurasian plates, each of which has a landmass at its leading edge. The Alps are much deformed and folded, the result of glacial movement. The mountains are relatively young, dating from the Tertiary period. This is notable in light of the number of very sharp, comparatively uneroded peaks, such as the Matterhorn and Mont Blanc, the latter being the highest peak in the range at 4,800 m (15,800 feet) above sea level. Glaciers cover a significant portion of the Alps. Major river systems that rise in the Alps are the Rhine, Rhône, Po, and Danube. *See* FOSSIL, MEDITERRANEAN, PLATE TECTONICS, TERTIARY PERIOD.

Alvarez, Luis Walter (1911–1988) American physicist born in San Francisco. He received a B.A. degree and a Ph.D. degree (1936) from the University of Chicago, and worked with Arthur Compton in early cosmic ray studies, then with Ernest Lawrence in pioneering work with the 60-inch cyclotron. He discovered nuclear beta decay by electron capture (K-capture) and participated in the first measurement of the magnetic moment of the neutron. Alvarez designed the first proton linear accelerator, and he became a leader in the group that developed the liquid hydrogen bubble chamber for studies of muon catalysis and strange particle resonances. This work was the basis for his Nobel Prize in physics, awarded in 1968.

The rock in the area around Gubbio, Italy, has a distinctive layer of clay separating layers of limestone. This clay layer contains compounds of iridium, a metallic element. That is not unusual; however, the proportion of iridium compounds in this clay layer is hundreds of times higher than normal (0.3 ppb). Both Luis and his son Walter Alvarez investigated this anomaly. Their first theory was that this strange deposit was the result of an explosion of a supernova (a giant star). However, this was difficult to support. Their subsequent attempts also involved extraterrestrial phenomenon, this time a meteor strike. Further research linked the crater in the Caribbean, off the Yucatán coast, to the Cretaceous-Tertiary (K-T) boundary. This is the boundary between the Cretaceous and the Tertiary periods in the history of the Earth.

The paleontological records of the history of life on this planet show that the giant dinosaurs and many other animal and plant groups became extinct about 65 million years ago. Subsequent work has shown that this was but one of a series of mass extinctions that have periodically disrupted life on Earth about every 26 million years. The reasons for these major declines in the global biota have been the subject of intensive investigation and debate. In 1978, Alvarez, his son Walter, Frank Asaro, and Helen Michel, colleagues at the University of California, Berkeley, found convincing physical and chemical evidence that showed that the great extinctions which terminated the age of the dinosaurs were related to the high-speed impact on Earth of a great asteroid estimated to have been about 10 km (6 miles) in diameter. The element of catastrophe was thus introduced into the studies of the evolution of life forms. Subsequently, evidence for one or more comet or asteroid impacts has been found at certain other mass extinction boundaries. *See* CRETACEOUS, DINOSAUR, EXTINCTION.

Alvarez, Walter (1940–) Son of Luis Walter Alvarez (see above). With a B.A. degree from Carleton College and a Ph.D. degree in geology from Princeton (1967), he is a professor in the Department of Geology and Geophysics at the University of California, Berkeley. Born in Berkeley, California, his area of specialty is stratigraphic and structural geology and microplate and mountain belt tectonics. He also works in the fields of paleomagnetism, seismology, oceanography, and in archeological geology studies. His work includes major field studies in South America and in Central Asia. He has since concentrated on the structural geology of the Alpine-Mediterranean region from the Alps to North Africa. The work Walter Alvarez did in conjunction with his famous father also made the son famous. He has received many scholarly honors—such as in March 2006 for his work in discovering evidence for the catastrophic meteor strike in the Caribbean. That event led to a series of changes in climate and ultimately resulted in the end of the age of dinosaurs. *See* ALVAREZ, LUIS WALTER, EXTINCTION.

Alvin (**research vessel**) A submersible built in the United States in 1964 and operated by the Woods Hole Oceanographic Institute on Cape Cod, Massachusetts. This lightweight, manned vessel has descended to depths of 3,000 m (10,000 feet) in the Mid-Atlantic Rift Valley, where it has been used to investigate the undersea mountain range. The *Alvin* has also been used to investigate vent communities in the Pacific Ocean and other deep-sea areas. The *Alvin* has been used to explore vent communities in the Atlantic and Caribbean, to check radioactive dump sites, and to explore whale bone deposits. It figured prominently in the location and exploration of the R.M.S. *Titanic*. The vessel continues to be upgraded and used in undersea work alone or with other devices.

More recently, the Alvin has been used to explore the Black Sea. This arm of the Mediterranean was once a large freshwater lake; the bottom sediments contain the remains of freshwater flora and fauna. The entire area was then inundated by seawater, and it is now a saltwater ecosystem albeit less salty than the Mediterranean itself. Some historians have speculated that the earth's movements and the resulting catastrophic inundation of a lakeside ecosystem are the origins of the account of the Mesopotamian flood tales and the biblical flood as well. *See* ARCHIMÈDE, SUBMERSIBLE, VENT COMMUNITIES.

Alvinella pompejana Named after the submersible *Alvin* and commonly called the Pompeii worm, it was first found off the western shore of Costa Rica and is currently the most extreme extremophile known. It is a vent organism and can withstand temperatures of 80° C (176° F) at its head end while its body and tail are in water at 22°C (72° F). This fairly large red tube worm is about 10 cm to 20 cm (4 to 8 inches) long. It is covered by a layer of bacteria that feeds on the mucus produced by the worm. In this possibly commensal relationship, the bacteria—also extremophiles—act as a "thermal protective blanket" for the worm as it protrudes from its burrow on the hot smoker to feed. *See* ALVIN, BLACK SMOKER, COMMENSAL RELATIONSHIPS, EXTREMOPHILE, VENT COMMUNITIES.

Amazon River The world's largest river in volume of water. The Amazon delivers an immense quantity of water into the South Atlantic, draining most of north-central South America.

The river narrows, near Obidos, Brazil, about 320 km (200 miles) from its outlet in the Atlantic Ocean. At this point the river is 2.5 km (1.5 miles) wide and has a channel about 70 m (200 feet) deep. The water surging into the Atlantic creates a 4-m- (13-foot)-high tidal bore called the Pororoca.

The Amazon is 6,275 km (about 3,900 miles) long, making it one of the world's longest rivers. It runs along an almost straight west-to-east line across tropical South America from its headwaters in Peru to its outlet in the Atlantic. Water in the river moves relatively slowly, carrying loads of various sediments. The sediment load is a function of the regions through which the Amazon's tributaries move; some are milky and carry clay; others are black from suspended humus. The increasing deforestation and use of the land for mining and cattle raising has added a toxic mix of cattle waste, debris, and mine tailings to runoff water in almost all reaches of the river.

Navigation upriver is possible on many of the tributaries of the Amazon, as well as on the main river, into Peru.

ambergris A waxy substance found floating in the ocean. It is produced by sperm whales and found in the digestive tracts of these mammals, where it is thought to protect the whale's digestive system against the hard beaks of squid, which form the major part of the sperm whale's diet. When ambergris is fresh, it is a black, greasy, smelly substance. On drying, it becomes grayish-black and smells faintly of musk. Ambergris was at one time used as a pharmaceutical and a fixative, which is the part of a perfume that keeps the volatile aromatic materials dissolved in the perfume medium (usually alcohol). For these uses ambergris has been largely replaced by synthetic materials. *See* SPERM WHALE.

amino acid A chemical component of all proteins, and found in all living organisms. Autotrophic or self-feeding organisms—usually single celled organisms such as bacteria—manufacture amino acids from simpler compounds such as water (H_2O), carbon dioxide (CO_2), and the nitrate ion (NO_3^-). Amino acids are characterized by an acid moiety and an amino group:
There are theoretically thousands of possible amino acids, but only about 20 com-

acid moiety   amino end

monly occur in proteins. These relatively few amino acids, when linked together in specific sequences form the different proteins. The sequences in which they are linked together are species specific. Thus, sharks of one specific genus and species have identical proteins made up of identical sequences of amino acids, and all cod of a certain species have identical proteins, but the shark proteins are slightly different in amino acid composition from the cod proteins. *See* AUTOTROPH, PEPTIDE, PROTEIN.

ammonia (NH_3) A basic chemical compound, the end-product of both protein metabolism and the breakdown of nucleic acids in marine organisms. Bacteria that break down living or formerly living matter liberate ammonia as part of their metabolic processes. Ammonia is usually a gas at normal temperatures, and dissolves in water. *See* NITROGEN CYCLE.

ammonites Fossilized chambered cephalopods, like the nautiloids. Ammonites have been extinct since the Cretaceous period. Ammonites are in the order

Ammonoidea along with clymenias, goniatites, and ceratites. Before their extinction, ammonites were very widely distributed, possibly because their larvae were free swimming or drifting (meroplanktonic). Ammonites were characterized by a tightly coiled planar shell with a bulbous, calcareous protoconch. This protoconch was continuous with the shell, enlarging in a logarithmic spiral. This process by which an organism such as an ammonite moves from a smaller chamber such as the protoconch to a larger one such as the shell is analogous to the molting (ecdysis) of arthropods. The size of the average ammonite was 10 to 30 cm (4 to 12 inches), although adult individuals with a shell size ranging from 1 cm to 3 m have been found. If, like the present nautilus, ammonites contained nitrogen bubbles, they could not have lived at great depths. Their food and method of locomotion are unknown. For unknown reasons also, ammonites managed to survive two near-extinctions and then reevolved many new species, finally dying out at the end of the Cretaceous period. *See* CEPHALOPODA, CRETACEOUS PERIOD, FOSSILS, NAUTILUS.

Ampere Seamount An undersea peak off the North African Coast northeast of the Madeira Islands. *See* ATLANTIS SEAMOUNT.

Amphibia Cold-blooded vertebrates closely related to the fishes from which they evolved. Unlike fish, most amphibians (frogs, toads, salamanders) do not tolerate salt water, although the frog *Rana cancrivora* is an exception. It is a large crab-eating frog inhabiting Indonesia. The general characteristics of amphibians are a moist, absorbing skin, small lungs, four limbs, and gills at some stage of these animals' development. They deposit their eggs, which lack amnions, in water.

Amphidiscophora A subclass of Hexactinellida, the sponges that have siliceous spicules. This is a single surviving order that appears in the fossil record back to the Carboniferous period.

amphidromic point The center of zero tide. The tide crest in any tidal period rotates around this point. *See* TIDE.

amphioxus (lancelet) A member of the subphylum Chordata, *Branchiostoma* and *Asymetron* are the only genera of amphioxus. These animals have a notochord and an elongate, essentially headless, tailless, finless body 4 to 7 cm long. They live in semitropical or temperate coastal waters, where they burrow tail-first into the sand. They are filter feeders with primitive breathing systems. The amphioxus excretory system resembles that of the annelids, and its circulation that of fish. However, the amphioxus has colorless blood and no real heart. The sexes are separate.

Amphipoda An order of about 6,000 described species of small, frequently colorless crustacea. Most species are marine but there are a few freshwater and still fewer terrestrial forms. These small animals are characterized by the lack of a carapace and a laterally flattened body. Amphipoda are for the most part scavengers, although there are some suspension feeding or sucking species.

Amundsen, Roald (1872–1928) The Norwegian explorer best remembered for being the first man to reach the South Pole. Amundsen was a naval officer whose first experience in exploration was with the Belgians on an Antarctic voyage in 1897.

Amundsen led his own expedition in 1901 in the sloop *Gjøa,* to the northeast coast of Greenland. Two years later, again in the *Gjøa,* he determined the location of the North Magnetic Pole, in the Northwest Territories of Canada. He then continued westward and was the first navigator to traverse the Northwest Passage.

Amundsen attempted to be the first person at the North Pole, but after Peary achieved that goal, Amundsen attempted to be first at the South Pole. He used Frijdtof Nansen's ship, the *Fram,* and sailed into the Bay of Whales in early 1911. Amundsen beat the Scott expedition, which was trying for the same goal, arriving at 90° south latitude on December 14, 1911.

Amundsen spent the rest of his life as an explorer and airplane pilot. He felt he had "found his vocation" in aviation. Flying from Spitsbergen, Norway, in 1925, he attempted to be first over the North Pole in an airplane. That attempt was unsuccessful, but he eventually achieved his goal. In 1928, while engaged in an air-rescue operation for a lost dirigible, he and his crew were lost in the Arctic. *See FRAM;* PEARY, ROBERT.

Amur River A river formed in the southeastern part of Russia by confluence of the Argun and Shilka Rivers; it is the river boundary between the easternmost reaches of the Soviet Union and China, then turns north and east and flows into the Tatar Strait. The Chinese name for the river is Heilung Chiang. The Amur forms a large estuary at its mouth in the Tatar Strait, south of the Sea of Okhotsk, delivering large volumes of freshwater into the Strait. The monsoon climate of the river's drainage basin creates spring torrents, which lower the salinity in the Sea of Okhotsk and result in its lack of sediment. The Amur is an important source of freshwater fish and is navigable for much of its 2,800-km (1,700-mile) length. *See* OKHOTSK, SEA OF.

anadromous fish Fish that spend most of their adult lives in the ocean, but move up rivers to spawn. *See* SALMON.

anchovy A herringlike fish (10 to 15 cm or 4 to 6 inches long) of the family Engraulididae. This filter feeder is representative of the entire clupeform (herring) order. Anchovies are found in enormous schools in the Mediterranean Sea and Pacific Ocean, both in northern waters and off the western coast of South America. Anchovies have great economic importance as food and as a raw material in the manufacture of oil, fertilizer, fishmeal, and animal food. They are also an important link in the food chain, since they support a good deal of bird life.

Anchovies are slender, silvery fish that feed on the great plankton populations that bloom in the spring. The Peruvian anchovy, *Engraulis ringens,* and the Atlantic and Mediterranean anchovy, *E. encrasicholus,* produce prodigious numbers of eggs. The female matures at a year old or less, and may spawn as many as 10,000 eggs. The eggs have very little yolk, so the anchovy hatches quickly and must immediately find food. Mortality among the young is therefore high. *See* EL NIÑO, HERRING.

Andaman Sea A marginal tropical sea connected to the Indian Ocean. The Andaman Sea is west of the Malay Peninsula and east of the Malacca Strait. Its average depth is about 900 m (3,000 feet), with its deepest point, 4,180 m (13,800 feet), lying to the east of Nicobar Island. The northern end of the Sea is dominated by the rapidly expanding Irrawaddy Delta. The great volume of water entering the Andaman Sea causes significant changes in its salinity, which falls to 2% in the monsoon months of June to October.

The uplift that produced the Andaman-Nicobar Island probably dates to the Cretaceous Period. Volcanic activity continues on the island and extends south to an area in which numerous seamounts have been found. *See* INDIAN OCEAN.

andesite line A boundary that marks the Pacific edge of the Eurasian Plate. The line is not continuous, and is delineated by a series of deep trenches associated with volcanic and seismic occurrences.

anemone Of the phylum *Cnidaria,* class Zoantharia, and order Actiniaria, this widespread group of benthic invertebrates inhabits almost all waters: They are found from the tropics to polar regions. Most of these solitary animals rest on a basal disk attached to a solid object or sea bottom, although some spe-cies burrow into the bottom sediment (infauna). They all feed on a variety of other marine organisms ranging from copepods to mollusks, annelids, crustaceans, and small fish. They can move slowly, though some species flap their tentacles and propel themselves through the water fairly quickly.

The tentacles look like flower petals, hence the common name of these carnivorous animals. The tentacles are arranged in series of sixes; 6 primary tentacles, 6 secondary ones, 12 tertiary, 24 quaternary and so on around the oral opening. Anemones have highly developed musculature and are either male or female; they are dioecius organisms. The larvae grow to maturity through a series of steps in which they do not resemble their parents (nauplia), although in some species the larval stages can reproduce by budding.

angiosperms Flowering land plants with enclosed or covered seeds. The angiosperms are the dominant terrestrial plant forms and also the most recently evolved land plants. In the context of the marine ecosystem, beach grasses are angiosperms that have evolved from basic terrestrial forms. They can now tolerate a higher than usual salt environment. *See* BEACH, ZOSTERA.

anglerfish (lophiformes, pediculati) Mainly deep-sea creatures, anglerfish are noted for a dorsal fin adaptation attached to the head which functions as a lure. In anglerfish living in very deep ocean waters, the lure contains light-emitting organs called photophores. In some species of anglerfish the males become parasites on the larger females, either permanently or for the breeding season.

Angola Basin A basin located in the South Atlantic Ocean, off the southwest African coast. It lies south of the cone formed by the debris carried into the South Atlantic by the Congo River. *See* ATLANTIC OCEAN.

Anguilliformes Elongated fishes with pelvic fins and an absent or reduced girdle. *See* EEL.

ANGUS The acronym for Acoustically Navigated Underwater Survey System. It is an unmanned sled on which cameras are mounted. This equipment was designed and operated by the Scripps Laboratory for Oceanographic Research at La Jolla, California. In 1977 it made the first observation of vent communities. ANGUS was part of what became a series of unmanned undersea observation stations. Most of these have been designed by the Scripps Institution. *See* FLIP.

Annelida The phylum of segmented worms. These soft-bodied animals have not left a good fossil record but probably arose in the Paleozoic era. The annelids are characterized by a fluid-filled tube within a tube body plan (a coelom), bilateral symmetry, a headlike projection at one end, and a closed circulatory system. The latter included a pumping mechanism and respiratory pigments that contained iron resulting in red blood. Many annelids have hemoglobin. Blood moves forward in the large, dorsal vessel and toward the tail in the ventral one. Tiny vessels bring oxygen, absorbed on the epidermis to the interior. The excretory system, present in each coelomic segment (somite) acts as a filter for nitrogenous wastes, water reabsorption unit, and discharge unit. Aquatic species usually discharge ammonia. The nervous system is highly organized with a ganglion per somite and a continuous double nerve cord that terminates in a large anterior ganglion—not quite a brain, but a control center.

The phylum is classified in a number of ways. One group of taxonomists divides it into three classes: Polychaeta that are mainly free-living and marine and constitute the largest group; Oligochaeta that are free-living, sometimes lack setae and/or parapodia and are for the most part terrestrial; and Hirudinoidea that are aquatic

parasites. Their usual hosts are vertebrates in both marine and freshwater environments. Earthworms are Oligochetes. The classification has also been proposed as Polychaeta and Clittelata with Hirudinea and Branchiobdellida (crayfish parasites) as a subclass.

The polychaetes are marine organisms and for the most part are sexually separate whereas Oligochaetes and the leeches are hermaphrodites. The name Polychaeta indicates many bristles or setae, the chitin structures that protrude from each segment. Polychaetes bear their setae on appendages that extend from the body wall. These are parapodia (almost feet). The parapodia and setae aid in locomotion. They serve to anchor the body while the longitudinal muscle—which elongates the body—moves it along forward; the circular muscle contracts the body and brings it up to the location of the parapodia. This mode of locomotion enables the animal to swim, creep, or otherwise manipulate its environment.

Polychaetes are the most highly diversified of the annelids, both in shape and lifestyle. They occur as sedentary tube worms, primarily filter feeders, or swimmers, or predaceous burrowers, or as ecto-parasites on echinoderms. *See* POLYCHAETEA, TUBE WORM.

Anostraca The order of fairy shrimp. Small aquatic crustaceans lacking a carapace.

Antarctic Circle The 66°30' South Latitude line. *See* ANTARCTIC CONVERGENCE, ARCTIC CIRCLE, LATITUDE, POLAR BIOME, SOLSTICE.

Antarctic Convergence A band encircling Antarctica between 50° and 60° south latitude. In this area, relatively cold surface water of low salinity sinks and spreads northward as Antarctic Intermediate Water. The cold Antarctic water can be detected by the measurement of salinity as far north as 35° north latitude in the Atlantic, but has a much lesser range in other oceans. The Convergence was discovered by the German meteorologist Meinardus during the Gauss expedition of 1901 to 1903, and the Convergence for a time was called Meinardus' Line.

Antarctic Current The principal circumpolar current, known as the West Wind Drift, flows around the Antarctic continent in an easterly direction, while the coastal current, the East Wind Drift, moves west around the land mass. The West Wind Drift is a deep and fast-moving current, which moves a considerable volume of water thought to be greater than that of the Gulf Stream, through the Drake Passage. The currents around Antarctica are global and affect the temperature, salinity, and dissolved gas content of the water in every ocean of the Earth. *See* ANTARCTIC OCEAN, OCEAN CURRENTS, SEAWATER.

Antarctic Ocean An ocean region to the south of South America, Africa, and Australia. The Antarctic serves as the salt, gas, nutrient, and temperature-equilibrating mechanism for the three major oceans. Although the cold water temperature restricts the number of plants and large animals, plankton and krill populations are large and support complex food chains that include birds (petrels, penguins, skuas), large fish, invertebrates, and cetaceans. The high rate of planktonic growth is supported by the combination of nutrient-rich warm water moving south on the ocean's surface and cold, oxygenated water moving north from the continent's edge. *See* ANTARCTIC CURRENT.

Antarctic phytoplankton These microscopic photosynthesizers are the basis of the Antarctic food chain, being the principal food of krill. These organisms can be considered extremophiles since they can thrive in conditions that were once considered too cold for photosynthesis. (All chlorophyll was thought to be temperature-dependent.) The phytoplankton are the

subject of several studies; one based in the Netherlands is examining the iron, cobalt, and zinc availability in seawater as a result of plankton blooms. Another conducted by the Smithsonian Environmental Research Center is examining the effect of the ozone depletion most noticeable around the South Pole. The ozone "hole" has resulted in increased ultraviolet radiation on the water column, which seems to inhibit the growth of phytoplankton. The degree of mixing in the water column is also a contributing factor, as is latitude. Recovery from increased ultraviolet radiation is better in temperate latitudes than it is in polar regions. *See* FOOD WEB, KRILL.

Antarctic Water Several distinct types of water differing in temperature and salinity. The coldest, least salty water is the Antarctic Surface Water, derived from melting ice, deep-water upwellings, and precipitation on land. Antarctic Surface Water reaches its greatest volume in the antarctic summer, December to March. The upwelling of Circumpolar Deep Water, derived from Atlantic, Pacific, and Indian Ocean deep waters, is of higher salinity, is relatively warmer, and does not vary in volume with the seasons. Surface water sinks and continues flowing northward until it encounters the Subantarctic Surface Water, which is warmer and less salty, at the Antarctic Convergence. The mixing of these surface waters produces the Subantarctic Intermediate Water, which continues moving north. Antarctic Bottom Water, formed in the antarctic winter, is cold and highly oxygenated, originating in the salty shelf water and the low-salt Weddell Sea water. This large volume of cold, saline water moves north and also in an easterly direction, away from the Weddell Sea. *See* ANTARCTIC CONVERGENCE, WEDDELL SEA.

anterior The head or leading end of an animal. The term is also used to denote position. Thus, the head of an arthropod is anterior to its thorax. *See* DORSAL, POSTERIOR, VENTRAL.

Anthozoa A class of marine coelenterates (phylum Cnidaria), such as sea anemones and corals. Anthozans are animals with radial symmetry. The dominant form is the polyp, which can be large and quite complex. Most, but not all, anthozoans are colonial. *See* CORAL, SEA ANEMONE.

Anthuridea A family of Isopoda (order). These tiny organisms live in all seas at almost all depths and constitute a major part of the plankton.

anticyclone An upper tropospheric high-pressure system. Anticyclones are generally subtropical high-pressure zones marking either the boundary between easterly trade winds and prevailing westerlies or that between polar and arctic weather systems. *See* METEOROLOGY.

Antilles An island chain that forms the northern and eastern borders of the Caribbean Sea. The Antilles chain is composed of hundreds of islands, most of them tiny. The Greater Antilles are Cuba, Hispaniola (the Dominican Republic and Haiti), Puerto Rico, and Jamaica. The Lesser Antilles are an eastern arc of very small islets, divided into two groups: the Leeward Islands St. Christopher (St. Kitts), Antigua, Guadeloupe, to the north, and the Windward Islands, Barbados, St. Lucia, and Grenada closest to the South American mainland. The islands are rocky, often mountainous outcrops with poor soil and inadequate ground water, particularly on the leeward side. The climate is tropical because the trade winds bring warm weather, making the entire region a center of tourism. Short rainstorms are most frequent in the Antilles summer. It is the occasional summer Atlantic storm that becomes a hurricane. Hurricanes are more likely to invade the northern portion of the region, and most occur in late summer, especially in September. *See* GULF OF MEXICO, HURRICANES, WEST INDIES, WINDS, *NAMES OF INDIVIDUAL ISLANDS.*

Antilles Current A branch of the Equatorial Current that moves surface water toward the West Indies in a northwesterly direction. *See* CARIBBEAN SEA, EQUATORIAL CURRENT.

aphelion The point in the orbit of a planet or comet that is farthest from the Sun. The Earth reaches its aphelion on July 3 in the Northern Hemisphere's summer. The position of the Earth and Sun affects the ocean's tidal ranges. *See* SOLAR TIDE, TIDE, TIDAL RANGE.

apogee The point in the orbit of a moon (or satellite) that is farthest from the planet it revolves around. The path of the moon's orbit affects terrestrial tides; the closer the moon to the Earth, the greater the tidal range. *See* APHELION, PERIGEE, TIDE.

Apostomes An order (although some describe it as a subclass) of Ciliophora. These organisms are parasites or symbionts on invertebrates, usually crustacea.

Appendicularia A class of tunicates found close to the surface in all oceans. This order of tunicates or larvaceans is present in all oceanic environments, though most are found in warm waters. They are among the most common zooplankton in the world's oceans. They are preyed upon by many carnivores, such as medusae, ctenophores, siphonophores, chaetognaths, and small fish. The appendicularians secrete a mucus covering that is also a food source for other small oceanic animals. Thus, they are an essential part of the food web in the ocean.

aquaculture The cultivation of plants and animals in water. Types of aquaculture include growing algae or fish (such as carp, catfish, buffalo fish, and trout) in freshwater pools and raising marine organisms, notably clams, oysters, and mussels, and seaweed, on a commercial scale in protected environments. Aquaculture is an ancient practice: Mollusks have been cultivated in frames that deter predation at least since Roman times. In addition, every medieval monastery and village had fishponds.

While it is possible to raise some animals in captivity, others, notably most ocean fish, present many problems. For reasons not yet understood, the cultured fry of oceanic fish do better once they are liberated. Some shellfish, particularly lobsters, are not amenable to the techniques of aquaculture. Cultivation of some salmon species has not been totally successful either, but techniques are improving. This has been helpful because, in addition to the drop in price, there is less overfishing of the dwindling and endangered wild stock.

Environmentalists have raised serious questions about the results of salmon culture, such as the effect of crowding large numbers of farmed fish. This has led to disease, and the cures applied further pollute local waters, making them inhospitable to the remaining wild stocks. Another involves the possibility of escape and cross-breeding with wild stock, which are leaner and therefore more desirable. At present, attempts are being made to improve the conditions of fish farming and the fish produced. However, pollution of local waters from spilled food, antibiotics, and excrement remains a problem in areas where fish are raised in pens. *See* MARICULTURE.

aqualung An underwater breathing apparatus pioneered by Jacques-Yves Cousteau. The aqualung is a self-contained air delivery system designed to be carried by the diver, thus freeing him or her of the need to be tethered to an air line running to a surface vessel, and providing much greater freedom and range of movement. *See* COUSTEAU, JACQUES-YVES; DIVING; SCUBA; UNDERSEA EXPLORATION.

Arabian Basin A region in the southern part of the Arabian Sea, a part of the Indian Ocean. The Carlsberg Ridge separates the Arabian Basin from the Somali Basin lying to the southwest. The northern boundary of the Arabian Basin is the Gulf of Oman. The deepest point in this basin is 3,658 m (12,657 feet).

The Indus River—one of Asia's largest and most significant waterways—feeds the Arabian Basin, and its cone of sediment is the dominant undersea feature. The continental shelf extends west of India for distances ranging from 120 km (75 miles) to over 300 km (190 miles). *See* ARABIAN SEA, INDIAN OCEAN.

Arabian Plate A small intrusion between the growing African Plate, the Iran Plate (a subduction zone), and the Eurasian Plate. The northward push of the Arabian Plate impels the Turkish Plate westward.

Arabian Sea The northwestern arm of the Indian Ocean, lying west of India and along the coast of the Arabian Peninsula and the east coast of Africa. The Arabian Sea lies west of the Mid-Ocean Ridge and consists of two deep basins, the Arabian and Somali, which are separated by the Carlsberg Ridge. The Persian Gulf, and Gulf of Aden along with the latter's extension, the Red Sea, are all parts of the Arabian Sea. The undersea topography of this area is still the subject of research and conflicting opinion. It is generally accepted by oceanographers that the Arabian Sea developed in the Mesozoic-Cenozoic era, although some features are as recent as the Pliocene.

Currents in the Arabian are the direct result of the monsoon pattern. From November to March the northeast monsoon brings light winds and little rain, and the best weather in the region. The summer monsoon is from the southwest, bringing with it rain and a change in current patterns. The South Equatorial Current turns north at about 5° South Latitude and moves into the sea. At this point it becomes known as the Somali Current. The current continues along the coast and bends east, then south along the Indian subcontinental shore. Areas of upwelling along the African and Arabian coasts are well marked.

The salt concentration of the Arabian Sea is high, particularly in winter. During the rainy season there is a drop of more than 0.1%, from an average of 3.6% to 3.5% or less. *See* INDIAN OCEAN.

Arafura Sea A body of water that separates North Australia from West Irian (formerly Netherlands New Guinea) and Indonesia. A deep trough of 3,000 m (10,000 feet) runs northeast toward West Irian. The area nearest Australia is a shallow shelf 50 to 80 m (160 to 260 feet) deep. The Aru Islands are on this shallow shelf.

aragonite Chemically, calcium carbonate. In its crystalline form, calcium carbonate exists as either calcite or aragonite. Coral reefs are built of aragonite. The health of coral reefs is dependent on the amount of carbon dioxide (CO_2) deposited in the oceans and on the temperature. Warm water dissolves less CO_2, and therefore there is less bicarbonate (HCO_3^-) and carbonate (CO_2^{-2}) ion in the water. This decreases acidity and decreases aragonite deposition by coral organisms. The reefs grow and repair themselves much more slowly. This is another projected effect of global warming. *See* CARBON DIOXIDE, GREENHOUSE EFFECT.

Aral Sea A lake in the Soviet Union east of the Caspian Sea. Both the Aral and the Caspian were once part of the Tethys Sea. Both seas are now river-fed lakes. In earlier times the Aral and Caspian were connected to the Mediterranean Sea. The area of the sea is diminishing, and the level of pollutants is rising. *See* CASPIAN SEA, MEDITERRANEAN SEA, TETHYS SEA.

Archaea Recognized as a group within the Prokaryotes since the 1970s. They are chemically distinct from true bacteria, apparently having separated from those organisms very early in Earth's history, in the Precambrian era. Fossils of Archeobacteria are over 3 billion years old. These simple prokaryotes have no nucleus or mitochondria and no chloroplasts, yet they do utilize carbon and nitrogen, grow and produce filaments that collect sediment, and eventually become structures (stromatolites). These organisms thrive in difficult habitats; they live in very hot, very sulfurous, or salty environments. *See* BACTERIA, CYANOPHYTA, EXTREMOPHILES, PROKARYOTE, STROMATOLITES.

Archaeocyantha A group of extinct, lower- to mid-Cambrian organisms. They are taxonomically difficult to place and may be either in a phylum of their own or related to the hard (stony) corals. They seemed to have ingested food as do the sponges. The Archaeocyantha seems to represent an intermediate stage between sponges and corals. Like both of these phyla, it flourished in warm, tropical, shallow seas. *See* ANTHOZOA, PORIFERA.

Archaeogastropoda Snails exhibiting primitive body plans. The abalones are one of the families in this subclass (Prosobranchia).

Archiannelida A small class of the phylum Annelida. These littoral bottom dwellers are an ancient, relict group of ciliated annelids. *See* ANNELIDA, CILIA, LITTORAL.

Archimède (research vessel) A French submersible used in the 1974 Mid-Atlantic Ridge survey known as FAMOUS (French-American Mid-Ocean Undersea Study). The *Archimède* descended to depths between 2,400 and 3,000 m (8,000 to 10,000 feet) but was not as maneuverable as the smaller American submersible *Alvin*. *See* ALVIN, FAMOUS, SUBMERSIBLE.

archipelago A group of islands. An archipelago may consist of isolated volcanic seamounts or mountain tops whose lower slopes and valleys have been submerged. The Hawaiian Islands are an example of the first type of archipelago; the Caribbean islands are an example of the second. *See* ATOLL, ISLAND ARCS, SEAMOUNT.

Architeuthis dux The world's largest invertebrate, the giant squid. This huge animal was once considered a sailor's tall tale, but when an almost complete body washed ashore on a Massachusetts beach in 1980, it ceased to be a legend. The animal was a young female that weighed 200 kg (440 lbs) and was missing its feeding arms. Had they been there, it was estimated that she would have been 9 m (30 feet) long. It is thought that at maturity the giant squid is twice that size. No juveniles have been found, and little is known of these giants except that they are few in number and probably live at great depths, 200 to 700 m (660 to 2,300 feet). Like all squid, they have complex nervous systems. *See* SQUID.

Arctic Circle The 66°30' north latitude line. *See* ARCTIC OCEAN, LATITUDE, POLAR BIOME, SOLSTICE.

Arctic islands Islands at or above the Arctic Circle. The climate of arctic islands is moderated by their proximity to water and two fairly permanent low pressure areas, one near the Aleutian Islands and the other near Iceland. Some arctic islands are glaciated, most receive a good deal of precipitation, and a few are mountainous. The highest peak on any of these islands is Mt. Forel in Greenland. The dominant groundcover is taiga (grassland). Arctic islands vary greatly in size. The largest is Greenland, with 2,175,000 km² (840,000 square miles), followed by Baffin Island, with 500,000 km² (196,000 square miles). The size then drops off sharply to Victoria Island, Ellesmere Island, the Spitsbergen islands, Novaya Zemlya, and the tiny islands of the Franz Josef Land archipelago.

Arctic Ocean The ocean area north of North America and Eurasia. The Arctic Ocean is never totally navigable and in winter is largely covered by ice. The ice cover on some of the large islands may be more than a mile thick (1,300 m).

The bottom of the Arctic Ocean is a highly varied terrain. The North Pole itself is on the Polar Abyssal Plain and about 4,000 m (14,000 feet) deep. Continental shelves extend toward the pole, most particularly off the Siberian coast, where shallows of less than 15 m (50 feet) exist. The bottom is divided into the Canadian and the Eurasian basins. The Lomonosov Ridge, which runs from the Laptev Sea just to the west of the Pole to the Lincoln Sea on Greenland's north coast, is in the Canadian Basin. The Alpha Ridge is in the Canadian Basin close to the Lomonosov Ridge. The Mid-Atlantic Ridge continues as the Nansen Ridge. It divides the region into the Eurasian Basin and the Canadian Basin. The Eurasian is, on the average, deeper than the Canadian. Water throughout the arctic is fairly constant in temperature and of low salinity.

The plant and animal life of arctic waters is much more varied than that of the Antarctic, particularly with regard to large animal populations. Plankton, algae, and diatoms all exist and provide food for larger organisms, such as crustacea and mollusks, which are frequently larger than their temperate relatives. In the northernmost waters there is less sunlight penetrating the water due to the ice cover, which limits sea life. The current warming trend is obviously melting the polar ice at an unprecedented rate. Ice is a great insulator, and since it is white it reflects sunlight. As the mean temperature increases and the ice melts, the exposed water is no longer insulated from sunlight. Since water is not white and does not reflect the sunlight, it warms, melting the ice faster. At the rate of polar ice melt, it has been estimated that the dream of the early explorers of

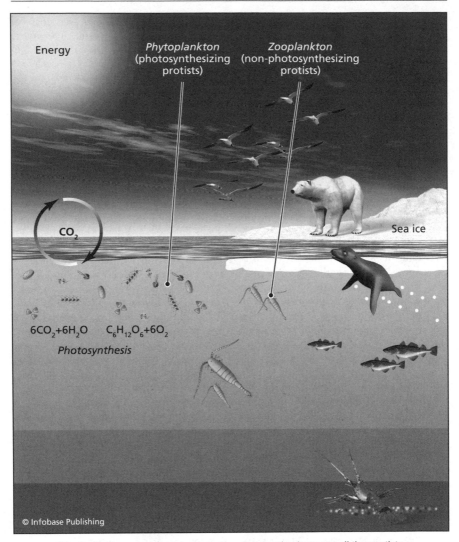

Energy

Phytoplankton (photosynthesizing protists)

Zooplankton (non-photosynthesizing protists)

Sea ice

CO_2

$6CO_2+6H_2O$ $C_6H_{12}O_6+6O_2$
Photosynthesis

© Infobase Publishing

The Arctic biome is unique because the photosynthesizers are all tiny protists.

a Northwest Passage by sea from Europe to Asia will be a reality before the end of the 21st century. *See* ABYSSAL FLOOR, IGY, MID-OCEAN RIDGES.

Argentine Basin A deep, cold-water area lying west of the Mid-Atlantic Ridge off the continental shelf of South America and terminating in the shallower Scotia Basin. The average depth of the Argentine Basin is 5,000 m (16,000 feet).

argon A gas found in seawater at a near-saturation concentration (33.6 ml/1,000 g H_2O at 20°C, 1 atm). The ratio of nitrogen to argon in seawater is fairly stable at 37 to 39. This is used as a base for the determination of the concentration of other gases. Isotopic analysis of the ratio of radioactive nuclei of potassium and argon is used in dating finds that are too old to be reliably dated using carbon 14. *See* DATING, ISOTOPES.

Aristotle (384–322 B.C.E.) A classical Greek philosopher and natural scientist, teacher of Alexander the Great. Aristotle was the son of a physician and as such had early exposure to medicine and the materia medica of his day, which consisted entirely of botanicals. He produced his *Natural History and Zoology* from 344 to 342 B.C.E. The *Historia Animalium* is based on the belief that there are "necessary causes" for structures, and that when these structures are observed in a plant or an animal, their function is apparent.

Aristotle attempted an elaborate classification scheme, known in Latin as the *Scala Naturae*. He ranked animals as more complex and important than plants and man as the most important and complex animal. He put animals into two large categories, blooded and nonblooded. The blooded animals included man, viviparous and oviparous quadrupeds, cetaceans, fish, and birds. These were subdivided according to the degree of perfection of the young at birth. The bloodless animals were mollusks, crustaceans, and insects. Aristotle realized that he had omitted sponges and snakes from his scheme but was not sure where they belonged.

He named about 500 kinds of animals, of about 550 to 600 species. Most of these were known in Greece either as native or available in menageries. His scheme contains about 130 fish and excellent descriptions of many invertebrates, crabs, lobsters, and cephalopods. The sea urchin's mouth-parts, now known as the "lantern of Aristotle," remain today as he described them.

Examining live animals, Aristotle observed sense perception in scallops, fish, and sponges. He saw and commented on the ability of the cuttlefish to adhere to rocks and not be swept away in currents. His statement that sea urchin eggs are largest at the full moon has been confirmed in the 20th century. He opened developing eggs to observe the state of the embryo.

Aristotle was aware of his limitations. He knew that he had not seen everything, nor did he totally understand everything he observed. He noted "suspect entries" and tried to comment on all of the work he did with regard to precision. In addition to his very careful biological work he also observed and commented on the ocean's tides and currents.

The influence of Aristotle's work on European and Arab science endures to this day. Aristotle became the outstanding scientist of his time, and after the decline of science in general, the preservation of his work—however fragmentary—made it the only basis for later workers in science. His methodology was also important: he repeatedly relied on observation—an approach often ignored as philosophy and science became more confused. Not until the idea that observation must be the beginning of science and that philosophy can only follow the facts became firmly established was Aristotle's contribution to man's knowledge of the world really appreciated.

arrow worm *See* CHAETOGNATHA.

artesian water Water from a spring. If artesian water issues forth from the sea-floor with sufficient force, it rises to the surface as fresh and sometimes hot water. Springs drowned by sea risings or collapsed limestone caves may produce artesian water. *See* BLUE HOLES.

Arthropoda An extremely numerous phylum that is also very well represented in the fossil record. Arthropods are segmented; originally, each segment was an appendage, but some segments became fused together, and some appendages were lost. There is always a chitinous exoskeleton which is periodically shed (a process called molting) to allow for growth. The shedding is hormonally controlled. The body cavity is not a true coelom (body cavity), and the circulatory system is an open one with a primary dorsal vessel (a primitive heart). Arthropods have well-developed sense organs. They are distributed worldwide and adapted to all aquatic environments. This is the largest phylum. There are between 1 and 2 million species in four subphyla: Crustacea, Chelicerata, Uniramia (myriapods and insects), and Pentastomida (all parasites). *See* CHELICERATA, CRUSTACEA, CRAB, LOBSTER, SHRIMP.

Aschelminthes A phylum of roundworms. Several rather disparate classes make up this phylum, the taxonomy of which is subject to discussion and revision.

The similarity among the aschelminth classes Nematoda, Rotifera, Gastrotricha, Kinorhyncha, and Priapulida is in their body cavities. Most of these animals are small scavengers.

A disputed phylum, the most commonly recognized aschelminth phyla are

- **Acanthocephala**—spiny-headed parasitic worms; about 1,150 species known
- **Chaetognatha**—arrow worms; about 70 species known
- **Cycliophora**—cycliophorans; 1 species known, microscopic
- **Gastrotricha**—gastrotrichs; about 430 species known, all microscopic
- **Kinorhyncha**—kinorhynchs; about 150 species known, all microscopic
- **Loricifera**—loriciferans; about 10 species known, all microscopic
- **Nematoda**—nematodes or roundworms; about 12,000 species known, but an estimated 200,000 more species extant, mostly microscopic
- **Nematomorpha**—horsehair worms; about 320 species known
- **Priapulida**—priapulid worms; 16 species known, about half microscopic
- **Rotifera**—rotifers or "wheel animalcules"; about 1,500 species known, all microscopic

See individual classes, SCAVENGER, TAXONOMY.

ascidian A class of tunicates. *See* TUNICATA.

Ascothoracica A genus of Cirripedia, parasites of coelenterates and echinoderms.

aseismic ridge Oceanic uplands that are not volcanic sites but may have been built up from volcanic products. The Walvis Ridge and the Emperor Seamount are examples of aseismic ridges. The Lomonosov Ridge, once believed to be an aseismic ridge, may be the result of separated continental rock that moved from its original site by seafloor spreading. *See* ARCTIC OCEAN, MID-OCEAN RIDGES.

Asteroidea A subclass of the stellate echinoderms. They are the familiar starfish or sea stars. The asteroidea are invertebrates with a worldwide distribution, living along coasts and in deep ocean waters. The radially symmetric adult develops from a bilaterally symmetric juvenile. Asteroideans have a flattened central disk surrounded by five arms. The opening that regulates internal water—the madriporite—is on the spiny upper surface. The mouth, anus, gills, and tube feet are on the lower surface. The carnivorous starfish either swallow prey whole and then expel the indigestible parts through their mouth or, more commonly, eject their stomach into or onto their prey and digest it *in situ*. The usual food of sea star is coral or bivalves. They move by retracting and extruding the papillae on their underside, which are part of their internal water canal system, and a distinguishing feature of the Asteroidea. The canal is also the means by which the interior of these animals is aerated. *See* ECHINODERMATA, SAND DOLLAR, SEA STAR.

asthenosphere The semifluid layer in the Earth's mantle. It exists at a depth of 80 to 200 km (50 to 125 miles). Its fluidity makes possible the lateral motion (sidewide slide) of crustal plates. *See* CRUST, MANTLE.

astrobiology A broad discipline that looks for extraterrestrial life in water on Mars, Titan (one of Saturn's moons), and other planets. The underlying assumption is that if life on Earth originated in the oceans, then any water environment in space should be explored for possible life forms or their remains.

astrolabe This instrument for determining the altitude of stars and planets was developed in ancient Greece. Sighting is done along a pivoted rod that points to a circular disk marked off in degrees of a circle. Martin Behaim (1459–1507), a German navigator and geographer, adapted the astrolabe for determining latitude for navigation. It was later replaced by the sextant.

Atlantic Ocean The Earth's largest body of water, accounting for about 20% of the Earth's surface. The Atlantic Ocean is surrounded by more continental land than any other ocean, and receives the most riverine runoff. It is divided into the North and South Atlantic Oceans by the equator. The outstanding geological feature of the Atlantic is the Mid-Atlantic Ridge, which runs from the Arctic Ocean to the Scotia Sea. The ridge breaks at the Romanche Deep (or Fracture Zone) near the equator. There are two transverse ridges, the Walvis Ridge off the South African coast and the Rio Grande Ridge near the Brazilian coast. The large islands of the Atlantic are for the most part of continental origin, while the small ones are volcanic or, in the case of the Bermudas, coral.

The winds and weather in the Atlantic are also divided into northern and southern segments, and the wind-generating currents that cross the ocean move clockwise. The Labrador Current is the cold current in the north that moves south along the North American coast. In the South Atlantic, warm water moves north and south as it approaches the Americas and arches counterclockwise. There is considerable sinking of cold, dense, southern water in winter and thus considerable stirring. The result is water of fairly constant temperature and salinity, the latter about 3.5%. The deep water of the Atlantic is oxygen and nutrient rich, supplying the raw materials needed by plankton, pelagic (open-sea) communities in general, and the predators that feed on them.

Radionuclide studies have shown that the Atlantic Ocean dominates the world's oceans by means of the Antarctic Circumpolar Current, which moves a considerable proportion of the world's ocean water through the Drake Passage. While this thorough mixing makes the global ocean fairly homogeneous, there are temperature and salinity variations that become more intense in shallow or restricted waters. The extent of variation in salinity for the period 1960 to 1981 was only 0.02 parts per million. *See* Arctic, Coriolis force, currents, Drake Passage, Gulf Stream, pollution.

Atlantic-type margin The trailing edge of a continental mass that is subsiding under sediment load. The cooling of the lithosphere (the rock of the Earth) is a factor in creating the subsidence and in its extent. *See* Pacific-type margin, plate tectonics.

Atlantis A legendary island thought to be somewhere in the Atlantic Ocean near Gibraltar. Plato mentioned it in two of his *Dialogues* as a wealthy island with an ideal society, and the legend of Atlantis remained popular in ancient times and the Middle Ages. Some modern archeologists believe that the story might have arisen after the destruction of the eastern Mediterranean island of Thera (Santorini) in a volcanic eruption that occurred about 1450 b.c.e. That catastrophe destroyed the island and inundated all of the eastern Mediterranean. *See* Thera, volcano.

***Atlantis* (research ship)** A U.S. vessel used for research. The first ship of this name was built in 1930, and the second, *Atlantis II*, sailed in 1963 and was used in the exploration of the Red Sea. The Atlantis Deep, named for the *Atlantis II*, is a site of hot water overlying a rich mineral deposit in the Red Sea.

Atlantis Seamount A group of undersea peaks on the eastern slope of the Mid-Atlantic Ridge, south of the Azores islands. *See* Azores, islands, Mid-Atlantic Ridge, seamount.

atmosphere The envelope of gases, vapor, and aerosol particles surrounding the Earth. The major components of the atmosphere vary with altitude, with most of the oxygen and almost all particulates being closest to the Earth. The atmosphere is divided into layers or zones. The layer closest to the Earth, the troposphere, is the most dense and is composed of approximately 78% nitrogen, 21% oxygen, 0.9% argon, and 0.02% carbon dioxide, although these percentages vary with local conditions. In addition, the troposphere contains varying amounts of water vapor as a result of oceanic evaporation. The evaporation of surface water supplies the energy that moves air currents and winds, which in turn contribute to the movement of water. The troposphere is the zone of the greatest circulation of air, as warm air rises from the Earth's surface to the cooler, thinner upper layers beyond, and warm tropical air moves from the equatorial regions toward the poles. The troposphere is thus the layer where almost all weather occurs. The interlocking effect of wind and water and air and water currents demonstrates the unity of the ocean-air fluid system.

The average temperature of the atmosphere falls with increasing altitude from the Earth's surface through the troposphere to the tropopause and then rises in the stratosphere. In the upper stratosphere is the layer containing ozone, which filters the ultraviolet radiation from the Sun. Above the stratosphere is the mesosphere and above that the ionosphere, a zone containing layers of charged particles important for the transmission of radio waves. At even greater altitudes is the exosphere, which merges with the interplanetary medium. *See* CLIMATE, WEATHER.

atoll A ring of land surrounding a tropical lagoon, and which is in turn surrounded by a coral reef. An atoll is frequently formed of basaltic rock of volcanic origin and is capped with limestone covered with coral. The islets that form atolls range in size from those less than a

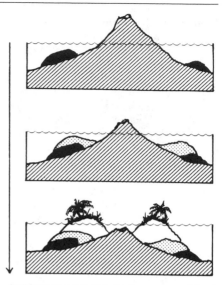

Atoll formation (arrow indicates passage of time)

mile (1 km) in diameter to those over 78 miles (120 km) in diameter.

Darwin proposed a theory of atoll formation that is now generally accepted. He postulated that atolls arose from a volcanic base that subsided and then became the nucleus for a calcareous and coralline buildup that ultimately produced the atoll. *See* CORAL AND CORAL REEF.

atom The smallest complete unit of a chemical element. Atoms are themselves composed of smaller entities, some of which have mass and charge, and others of which have mass but no charge. The major subunits of atoms are protons, electrons, and neutrons. The protons, neutrons, and other subatomic particles excluding the electrons make up the nucleus of the atom. They are electrons distributed in a cloud around the nucleus.

A proton has an atomic mass of one dalton, a unit devised to represent the mass of a proton. It has a positive electrical charge.

The electron, which has a negative electrical charge equal in magnitude to the positive charge of the proton, has a mass of 1/1,800 that of the proton.

Neutrons are uncharged parts of the atomic nucleus that have a mass comparable to that of protons.

The other component parts of atoms are considerably smaller than the three major ones. The entire structure is held together by a series of strong and weak forces.

Most elements are not found as their free atoms; this is true only for a few atmospheric gases: xenon, neon, krypton, and argon, formerly called inert gases. The simplest elements, such as hydrogen (H_2), oxygen (O_2), occur as diatomic molecules. Elemental minerals, such as carbon, (graphite or diamond), gold, and silver, are crystalline aggregates of atoms held together by chemical bonds. *See* ISOTOPE, MOLECULE, RADIOACTIVITY.

ATP (adenosine triphosphate) The energy storage and transferring compound of most biological systems. ATP transfers energy by the reaction:

ATP + a hydroxyl group $\rightleftarrows$ ADP (adenosine diphosphate) + a phosphate ester (found in sugars). ATP is a phosphorylating agent that can transfer a phosphate group to another molecule, releasing energy that is utilized by the living cell in which the transfer occurs. When a phosphate group is added to ADP, ATP is regenerated. The importance of this chemical cycle lies in the role that it plays in metabolism, i.e., the conversion of energy derived from sunlight and fuels (such as carbohydrates) into energy utilized by the cell to do chemical, osmotic, and mechanical work.

auk An arctic marine bird (family Alcidae) varying in length from 18 to 90 cm (7 to 20 inches). The auk is not an adept flyer and roosts on rocks in huge colonies. Auks feed on fish, crustaceans, and plankton. The great auk, which resembled a penguin and was about the size of a goose, became extinct in the 1840s as the result of human activity.

Australia The only continent wholly in the Southern Hemisphere. Australia has a land mass of about 7.5 million km² (almost 3 million square miles). *See* ARAFURA SEA, GREAT BARRIER REEF, PACIFIC OCEAN, SAHUL SHELF.

Australian Plate A large crustal section underlying Australia, New Zealand, and the eastern portions of the Indian Ocean. The eastern boundary of this plate, the Macquairie Ridge, is a subduction zone that causes the seismic activity of New Zealand. *See* MACQUARIE RIDGE, MID-OCEAN RIDGES, PACIFIC OCEAN.

authigenic sediment A fine grained deposit, such as common salt and phosphorites, found on continental shelves. Authigenic sediment is the result of chemical precipitation.

autotrophs Organisms that, together with heterotrophs, constitute the two categories of living organisms. Autotrophs use simple inorganic compounds, found in the environment, to synthesize their own food and protoplasm. The energy source that drives these synthetic reactions is usually sunlight, and those autotrophs that use sunlight are called photosynthesizers. Sulfur bacteria and other autotrophs that use the energy stored in the bonds of simple compounds of sulfur and nitrogen to obtain energy for their metabolic reactions are known as chemosynthesizers. *See* CHLOROPHYLL, HETEROTROPHS, PHOTOSYNTHESIS, SULFUR BACTERIA.

Axinellida An order of sponges that have very varied shapes. They are characterized by spicules enclosed in spongin fibers. Their usual habitat is tidal water down to depths of about 100 m (130 feet). A few species are found at greater depths.

Azores An archipelago in the Atlantic Ocean lying 1,500 to 1,800 km (900 to 1,200 miles) west of Portugal. The islands are the peaks of undersea volcanoes and are arranged in three widely separated groups. The islands have been known, fought over, and used by Europeans since the 14th century. Columbus stopped at the Azores to wait for favorable winds before his first voyage across the Atlantic. The islands had been known to ancient mariners; Carthaginian coins have been found on some of the Azores. *See* MID-ATLANTIC RIDGE, SEAMOUNT.

Azov, Sea of An extension of the Black Sea north and east of the Crimean peninsula and connected to the Black Sea by the Kersh Strait. Both the Sea of Azov and the Kersh Strait are very shallow, the maximum depth in the Sea of Azov being about 13 m (40 feet).

The Sea of Azov is poor in marine fauna. It is less saline than the Black Sea and freezes in winter, which the Black Sea normally does not. However, a large fish population does live in these waters, and sturgeon are taken in commercial quantities. There are many canals, as well as navigable river channels, leading to the Black and Caspian seas. *See* BLACK SEA.

Bache, Alexander Dallas (1806–1867)

An American scientist and superintendent of the U.S. Coast Survey. He helped establish a continuing American interest in marine research. A great-grandson of Benjamin Franklin, Bache was a physicist trained at the U.S. Military Academy at West Point, New York. After serving in the army, he was appointed professor of natural science and chemistry at the University of Pennsylvania. Bache was also in charge of research at the Franklin Institute in Philadelphia.

In 1836 Bache was appointed president of Girard College, and he was also a regent of the Smithsonian Institution while connected with the Coast Survey. As he continued in his family's tradition of public service, Bache was a major figure in 19th-century American science. He lobbied for funding for the Wilkes Expedition and encouraged Matthew Maury's research. Bache, Jean-Louis Agassiz, and others were founding members of the American Association for the Advancement of Science. *See* AGASSIZ, JEAN-LOUIS; MAURY, MATTHEW FONTAINE; WILKES, CHARLES.

backshore The most landward part of a shoreline. It is on the land side of the average spring high water mark and is inundated only during storms and storm-driven high tides. *See* BEACH, COAST, TIDE.

bacteria An order of the kingdom Monera (unicellular organisms with no distinct cell nucleus). Their cells are surrounded by a sugar-based peptide wall, and there may be a capsule exterior to the wall. There are three basic shapes: spherical (coccus), rod (bacillus), or twisted rod (spirillum). Some have flagella and are motile. Reproduction is usually simple cell division, although some species bud. Bacteria range widely in oxygen requirements—from those that must live in air (aerobes) to obligate anaerobes. Some photosynthesize and are thus autotrophs, but not all autotrophs have oxygen-dependent metabolisms. Some bacteria produce methane; others oxidize ammonia to nitrate (nitrifying bacteria), sulfides to sulfates (sulfur bacteria), or ferrous to ferric iron (iron bacteria). Bacteria are important in the nitrogen and carbon cycles; they are found in both marine and terrestrial environments. Some are surface-dwelling heterotrophs (i.e., organisms that use organic material from other organisms as a source of nutrients).

Among the most recently discovered bacterial marine organisms are the bacteria living in thermal upwellings in a number of places in the Pacific and Atlantic oceans. These organisms are dependent on sulfur and/or iron, not oxygen, for their metabolic processes. They are the beginning of the food chain in the rift-deep seabed ecological system. According to current biological theory, these vent bacteria are similar to, if not the descendants of, the organisms first developed on the primordial Earth.

A new way of looking at aquatic bacteria is to assess their importance and population in particular places. They are essential factors in the marine food web. The role of bacteria in the ocean is to be eaten by some organisms or to decompose larger dead organisms. In doing so, they connect two important ecological issues—

sustaining marine resources that may be used as food for humans and being part of the carbon cycle in the ocean, the incorporation of elemental carbon into living tissues of organisms. *See* CHEMOSYNTHESIS, COMMENSAL RELATIONSHIPS, EXTREMOPHILES, FOOD CHAINS, FOOD WEBS, HETEROTROPHS, IRON BACTERIA, METHANOGENESIS, *RIFTIA PACHYPTILA*, SULFUR BACTERIA, VENT COMMUNITIES, VOLCANO.

Baffin, William (1584?–1622) An English explorer of the Arctic and the first English explorer to attempt a scientific determination of longitude. In 1615 Baffin piloted the *Discovery*, which was charged with finding the Northwest Passage through North America. The voyage was inconclusive. The following year Baffin sailed again on the *Discovery* to determine whether the Davis Strait would be a promising route. He went to 78° north latitude, mapped Jones, Lancaster, and Smith sounds, and returned to announce that the project had not been fruitful. He suggested that the London merchants fund a voyage starting in the Far East to sail east, but they declined.

Baffin's next employer was the East India Company. Between 1617 and 1621 he was captain of several ships in which he explored the eastern Mediterranean, the Red Sea, and the Persian Gulf, reaching India. He was killed near Hormuz, in the Persian Gulf, where the East India Company was fighting with the Portuguese on behalf of the shah of Persia.

Baffin Bay The body of water that separates Greenland from Canada. The Davis Strait, to the south, connects Baffin Bay to the Atlantic Ocean. The bay is about 1,100 km (700 miles) long and varies from 112 to 640 km (70 to 400 miles) in width, with depths ranging from 360 to 2,740 m (1,200 to 9,000 feet). The surrounding islands are bleak, cold, and mountainous. The bay itself is ice covered most of the year. The Labrador Current brings icebergs through the bay in summer.

Baffin Island A landmass north of Hudson Strait, the passage leading to Hudson Bay. This mountainous, snow-covered island forms the western boundary of Baffin Bay, an arm of the Arctic Ocean.

Bahamas An Atlantic Ocean island group southeast of Florida and northeast of Cuba. Andros, the largest island of the Bahamas, is about 160 km (100 miles) long and about 60 km (40 miles) wide at its widest point. While most of the Bahamas sit in shallow water on a broad bank, the tongue of ocean east of Andros is more than 4,000 m (13,000 feet) deep. Also to the east of Andros are collapsed caves or blue holes. During the most recent ice age, the one that ended about 10,000 years ago, this area was much higher above sea level. The current islands were mountaintops. *See* BLUE HOLES, CARIBBEAN SEA, GULF STREAM.

Balearic Islands An island group in the Balearic Basin of the Mediterranean. The Balearics lie east of Spain, in relatively shallow water. *See* MEDITERRANEAN.

baleen A horny material that hangs in a fine fringe from the roof of the mouth of the Mysticeti or baleen whales. The triangular plates of baleen are used as food filters. *See* FILTER FEEDERS, KRILL, WHALE.

Baltic Sea An arm of the North Atlantic, separated from it by the Skagerrak, Kattegat, and the Danish landmass. The largest island in the sea is Gotland (Sweden). The Baltic is fairly shallow, particularly on its eastern edges. Although the Gotland Deeps are greater than 450 m (1,500 feet) deep, the Gulf of Riga is about 50 m (165 feet) deep and the shallows near the Danish landmass may be as little as 10 m (33 feet) deep at low tide.

The Baltic is fed by large arctic and subarctic rivers, which accounts for the low salt and nutrient content of its surface-level waters. The salt content is so low in some places that freshwater fish species can make the transition to the sea and back.

With the retreat of the Scandinavian Ice Sheet, a large lake formed, covering what is now the Baltic and extending beyond its present boundaries. By about 8000 B.C.E. the Yoldia Sea extended from the Skagerrak, over southern Sweden to Lake Ladoga, east of the Gulf of Finland. Land rebounding from glacial pressure closed the link and the Yoldia became a lake, known as Lake Ancylus, about 7000 B.C.E. A warm period about 2,000 years later reopened the sea connection (Litorina Sea). The most recent warming has caused the sea level to continue to rise. See SKAGERRAK.

Banda Sea A deepwater area in Indonesia southeast of the Sulawesi Sea (Celebes Sea) and north of the Timor Sea. The numerous islands in the Banda Basin are of volcanic origin. The area is noted for deep, cold water that flows westward to the Indian Ocean. See CURRENTS, INDIAN OCEAN, ISLAND.

Banks, Sir Joseph (1743–1820) A British naturalist who began his career as a scientist while still a student at Oxford. He accompanied James Cook on early voyages to North America, where he collected plants and insects in Labrador and Newfoundland. In 1772 he explored the geology and fauna of the Hebrides. Banks became a member of the Royal Society in 1766 and was selected to accompany James Cook as an unpaid naturalist on the Pacific voyage commissioned by the Royal Society in 1768. He stunned the society with the exotic plants and animals he collected from New Zealand and Australia and was elected its president in 1778—a post he held until he died. Banks continued to collect new specimens as long as he lived and willed his impressive collection to the British Museum. Banks's massive collection of specimens, watercolors, and engravings based on the voyages with Cook are a national treasure in two nations, the United Kingdom and Austra-

lia. His fortune supported this research, and the expenses for the voyages are estimated to have exceeded £10,000. This would be equivalent now to a sum in the millions of pounds—or dollars. See COOK, JAMES; ROYAL SOCIETY.

Barents, Willem (1550?–1597) A Dutch navigator who sailed for Dutch merchants. He sought the Northeast Passage as a quick route to the Indies from Europe. Instead of finding the Passage, he explored Novaya Zemlya, Svalbard (Spitsbergen), and Barents Island. He died during his third voyage.

Barents Island, in the Svalbard archipelago, and the Barents Sea are named for him. See BARENTS SEA.

Barents Sea A part of the Arctic Ocean named for the Dutch Explorer Willem Barents. The Sea lies east of Spitsbergen and is separated from the Mid-Atlantic Ridge by the Barents Abyss. Scandinavia (North Cape) and Siberian Russia (Murmansk) lie to the south, and Novaya Zemlya lies to the east.

The sea bottom of the Barents Sea averages between 100 and 350 m (300 to 1,000 feet) deep and then drops off to about 600 m (2,000 feet) where the Barents meets the Norwegian Sea. The Barents is a boundary region: both the remains of the warm Gulf Stream and very cold arctic water enter it. The air above the water is part of the polar front. The vigorous mixing of the two results in a well-aerated nutrient mixture that supports a planktonic bloom in summer, producing a great deal of oxygen. The phytoplankton population supports a large and varied ecosystem comprising the plankton and many invertebrates. See ARCTIC OCEAN, POLAR BIOME.

barnacle A highly modified crustacean (subclass Cirripedia) of the order Thoracica long thought to be a mollusk because of its "shell." Charles Darwin spent years studying barnacles and compiled information about them in his works *Monograph* on the Subclass Cirripedia (1851–54), *Monograph on the Fossil Lepadidae or Pedunculated Cirripedes of Great Britain* (1851), and *A Monograph on the Fossil Balanidae and Verrucidae of Great Britain* (1854). Barnacles are very successful as a life form: They may have evolved as early as the Silurian period and comprise four orders with more than 800 species.

The free-swimming larvae, nauplia, pass through several molts before assuming the adult sessile pedunculate (stalked) form. Since the juvenile forms frequently look very different from the adult, when Darwin began his exhaustive studies a number of nauplia were named as independent organisms. In some species, the males are tiny and live within the bodies of the females. They were originally identified as parasites.

Some barnacles are enclosed in calcareous plates, while others are "naked." Some are parasites on corals or other crustacean, and some are commensals—independent species that live in communities attached to turtles, whales, or coral. The feathery cirri of the barnacles—the remnants of the arthropod leg—beat food into their stomachs. Respiration is accomplished by the passage of gases through a mantle; there are no real gills. Barnacles are hermaphrodites.

Barnacles live in almost all marine environments, and their negative impact on commercial enterprises is considerable. They are responsible for the extensive fouling of ships, piers, and offshore installations. *See* CIRRIPEDIA, CRUSTACEA.

barracuda A carnivorous fish usually found in tropical waters. There are about twenty species in the family Sphyraenidae. They are usually 1.2 to 1.8 m (4 to 6 feet) long, although some are much larger, notably the great barracuda of Caribbean and West Pacific waters. The family as a whole are aggressive feeders, with large mouths, many teeth, and a heavy underslung jaw. The great barracuda has been reported to attack swimmers.

Barracuda are taken by commercial fishermen and also sought after by sport fishermen, who are attracted by the barracuda's fierce battling ability when hooked.

barrier island A narrow, sandy landform separating open ocean from a lagoon or embayment. The beach usually has a natural canal on its landward side. Barrier islands are separated from one another by shallow inlets. The entire barrier configuration is subject to rapid change as a result of currents and storms. The Virginia barrier islands, for example, have retreated landward in the last 100 years, and some no longer exist as islands. *See* COAST, LONG SHORE CURRENTS.

barrier reef A partially submerged coral outcrop on the seaward side of a lagoon. The barrier reef follows the contour of the land, of which it may be an extension. Water in the lagoon is shallow, permitting coral growth, but the descent is steep on the ocean side. Barrier reefs result from coral growth or a change in the level of the seawater or subsidence of the beach area.

Barrow Strait An arctic passage used by the oil tanker S.S. *Manhattan* on its September 1969 passage through the North Canadian islands from the Atlantic to the Pacific Ocean. The strait is north of the Somerset and Prince of Wales Islands, and leads to Point Barrow, the most northerly point of Alaska and the start of the Alaska pipeline on the shores of the Beaufort Sea.

basalt A dark gray to black igneous rock of volcanic origin. Basalt is a rapidly cooled volcanic material and so exhibits a microcrystalline structure. It consists of about 50% silicon dioxide (SiO_2), roughly 16% aluminum oxide (Al_2O_3), and about 9% calcium oxide (CaO). Other inclusions are well known. Darker basalts contain pyroxene, olivine, or both. Basalt is found as basement rock on land and on the seafloor spreading from the mid-ocean ridges. The basalt slabs called pillows, found on the ocean floor near the mid-ocean ridges, are the youngest rocks on Earth. Basalt on continents indicates that the land was once at the rift edge. An example of this is the line of the Palisades in New Jersey. *See* MID-OCEAN RIDGES, VOLCANO.

base A chemical substance that can attract a proton hydrogen ion, H^+, and remove it from an acid. Bases dissolved in water produce a solution that is alkaline. *See* ACID AND BASES, ACIDITY.

basin A depression in the sea floor delineated by ridges or trenches. It often has seamounts and/or abyssal hills.

basking shark This fish (*Cetorhinus maximus*), of the family Cetorhinidae, is a giant, slow-moving animal found principally in the Atlantic Ocean. Only the whale shark is larger. Basking sharks are usually about 8 m (25 feet) or longer, but there was one reported sighting of an individual about 14 m (46 feet) long. Basking sharks are sluggish plankton eaters who also prey on small fish but are relatively harmless to humans. Their color ranges from gray to brown to black, with lighter undersides. They are hunted for their hides and oil. The name comes from their habit of drifting slowly in the sun just below the surface of the water. *See* SHARK.

bass A valuable food fish of the perch family, found in marine and freshwater environments. Common species are the striped bass, *Roccus lineatus;* black bass, *Micropterus salmoides;* and sea bass, *Centropristis atrarius.*

Bass, George (1771–1803) An English sailor and surgeon. Bass joined the Royal Navy, where his ability in navigation led to his assignment to the HMS *Reliance.* Along with Matthew Flinders and William Martin, Bass explored the Botany Bay region of the Australian coast

and recommended that a settlement be established there.

Bass then continued to explore the coast south of Botany Bay for several additional years. He collected specimens of plants and animals, and explored the region south of what is now Sydney. In 1798 he found the strait separating Australia from Tasmania, which is named for him. In 1799 Bass was made a member of the Linnean Society in recognition of his contributions to botany and zoology. In 1803, while en route to South America from Australia to collect specimens, his ship disappeared. *See* BOTANY BAY; FLINDERS, MATTHEW.

Bass Strait A generally shallow passage separating southeastern Australia from Tasmania, it is named for George Bass, who discovered it in 1798. Water movement is seasonal in the Strait—being eastward in winter and westward in summer. The salinity is lowest in summer, when the snow melt from Tasmania reaches the sea. *See* BASS, GEORGE.

bathyl zone The region of ocean from the edge of the continental shelf to that depth where the water is at a constant temperature of 4°C, usually at about 4,000 m (13,300 feet). The bathyl zone is referred to as the continental slope.

Except for currents or upwellings, the salinity is about 3.4 to 3.6% and the water generally contains less oxygen and nutrients than does shallower water. The bathyl zone has far fewer species and numbers of individuals than does shallower water. *See* CONTINENTAL SLOPE, HADAL ZONE.

bathymetry The charting of the ocean bottom by taking soundings. *See* OCEAN FLOOR, SOUNDING.

bathypelagic zone The portion of the ocean water column 1,000 to 4,000 m (3,300 to 13,500 feet) deep and not in direct contact with the ocean bottom. The bathypelagic region has no sunlight. Plankton are abundant, and cnidarians, arthropods, and fish move freely. *See* ARTHROPODA, CNIDARIA, FISH, PLANKTON.

bathyscaphe A free-moving, three-man, deep-sea research submersible designed by Auguste Piccard in the late 1940s. His first effort was called *FRNS I* (Fonds pour la Recherche Nationale Scientifique). Iron pellets controlled by an electromagnet regulated the vertical motion of the bathyscaphe; a small electrically powered propeller moved it laterally. The first bathyscaphe had limited success. It was improved and the second attempt, the *FRNS II,* dove in 1954 to the 4,050 m (13,365 feet) mark.

Piccard and his Italian coworkers incorporated some of the features of the bathyscaphe into a new submersible, the *Trieste,* which was acquired by the U.S. Navy. In 1958 it dove in the Guam Trench. *See* PICCARD, SUBMERSIBLES.

Batoidea A superorder of widespread, flattened vertebrate but not bony fish of moderate size (0.25–6.5 m; 8 inches–21.5 feet). These are the skates, rays, and guitarfish.

Batrachoidiformes A small order of toadfish. The maximum size of these bottom-dwellers is about 60 cm (20 inches). They are predators, feeding on crustaceans and mollusks. Their habitats include the Atlantic, Pacific, and Indian oceans. Some species are armed with venomous spines.

bay A wide area of the ocean; part of the sea extending landward. For specific bays, see the geographic name.

beach A stretch of sand or rocks along the shore, extending from the highest high-tide point to the lowest low-tide point. A beach is formed by erosion caused by wind and waves and by deposits of sediment brought to the site by ocean waves and rivers. The structure of a beach is subject to considerable variation, depending on the tide, season, storms, and long-term geologic processes.

The areas of a beach have specific names. The foreshore or intertidal zone is the most active part of a beach. It is under water when the tide is in and dry when it is out. The backshore is above the usual high-tide point and is submerged only in storms or exceptionally high tides. This is where the dunes may form. The littoral, which is really under very shallow water, is where water meets sand. It is dominated by wave action and divided into the breakers (where waves become unstable

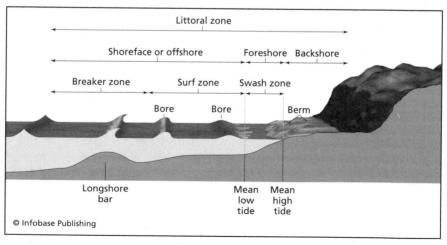

Beach cross section

and crash) and the surf zone (where shallow waves are pushed onto the beach—the swash or splash zone). The sediments of this area are the berm, a ridge at the upper limit of the wave splash and the scarp, a wall that runs upland from the berm. There may be depressions parallel to the shore (runnels) that form as sediment moves in the surf and a beach face, or low-tide terrace, that is the slope below the berm.

Sandy beaches are usually wide and have gentle slopes. They are found on the leeward sides of atolls or on continental shelves. They may be bar- or pier-shaped. Pocket or crescent beaches have sand trapped between cliffs that border the sea. The fauna inhabiting a sandy beach consist of the epifauna, organisms that live on the surface, and infauna, those below the surface. On a worldwide scale, sandy beaches have been lost because of a complicated series of geological events aggravated by human activities such as the damming and restructuring of river channels and the construction of seawalls that, if incorrectly located or not maintained, can increase destruction. *See* BREAKWATER, COAST, EROSION, LITTORAL, LONGSHORE CURRENTS, SAND.

beach grass Members of the loosely defined group of the order Najadales. There are about 10 families of these aquatic or semiaquatic herbs that are, to some degree, tolerant of salt water. The genus *Zostera* is typical. Beach grasses are coarse, reedy plants found on temperate sandy coasts in Europe, North America, and Africa. The stalks are about 1 m (3.3 feet) tall and grow in clusters. Beach grasses have extensive underground stem and root systems and send up new shoots meters away from the original plant. This network of roots and stems acts to stabilize the sand and prevent its erosion by wind.

Beach grass is now being used far from beaches. Since these plants are salt-tolerant, they are being planted in areas where consistent irrigation has oversalted the land. The plants will, over several years,

extract salt from the land and thus restore its fertility. The hay produced from the salty beach grasses is used in cattle fodder.

Beagle, **HMS** A survey ship sent by the British Navy under the command of Captain Robert Fitzroy to chart the South American coast and then circumnavigate the globe. The ship carried naturalist Charles Darwin on board, and his observations during the four-year-long voyage, particularly of the fauna on the Galápagos Islands, led him to develop his theory of the evolution of species. *See* DARWIN, CHARLES; EVOLUTION; WALLACE, ALFRED RUSSELL.

beard worms *See* POGONOPHORA.

Beata Ridge Part of the Caribbean seafloor lying south-southwest of Beata Cape on the island of Hispaniola. The Ridge is between the Colombian and Venezuelan basins, and terminates in the Aruba gap. The Beata Rise is east of the ridge in the Venezuelan Basin. *See* CARIBBEAN SEA.

Beaufort, Francis (1774–1857) A British naval commander and later admiral who was interested in natural science. In 1805 Beaufort designed a wind scale which, although much modified, is still used. The wind scale was used to determine the effect of wind on a man o'war in full sail. Beaufort was elected a member of the Royal Society in 1814.

Beaufort scale A standardized, descriptive scale of wind velocity (see page 421) outlined by Admiral Sir Francis Beaufort of the Royal Navy in 1858. *See* METEOROLOGY, WIND.

Beaufort Sea A branch of the Arctic Ocean lying north of Alaska, with Banks Island on its eastern boundary and the Chukchi Sea to the west. There is a great deal of floating ice in the Beaufort Sea, making its surface the coldest water layer in it. The intermediate layer of the water is warmer and moves north through the Bering Strait. The Alaskan coast of the sea has very little shelf, and the bottom drops abruptly into the Beaufort Deep. The coast is dotted with small islands.

The circulation in the Beaufort Sea has been studied extensively. A research team composed of scientists from the United States and Canada measured current velocities, water temperature salinities, and temperatures over an 11-year period (1986–1997). This area is significant since it slopes sharply down to abyssal depths, which affects conditions in the sea and its boundary currents. See BERING STRAIT, CHUKCHI SEA.

Beebe, Charles William (1877–1962)

An American ornithologist, explorer, and writer, active in a number of scientific fields. Beebe's contribution to marine science involved deep-sea exploration. He descended to a depth of 918 m (3,028 feet) in the Puerto Rico Trench in a submersible vessel of his own design in 1934. The vessel was essentially a hollow steel ball lowered by its mother ship. This bathysphere was able to withstand the pressures of the depths but was not maneuverable. See SUBMERSIBLE.

Beer's Law

Also known as Bouguer's Law or Lambert's Law, it deals with the attenuation of light as it passes through a medium. In the case of the ocean, the water and whatever is in it become the medium. Obviously, the more material in the water, the less light can penetrate it. See IRRADIATION, PHYTOPLANKTON.

Belcher, Edward (1799–1877)

Surveyor of the arctic coasts of Canada for the British Admiralty, Belcher was born in Halifax, Nova Scotia, and devoted his entire career to coastal surveying. In 1825 he mapped the Pacific coast of Canada and the Bering Strait. He then worked along the northern and western coasts of Africa from 1830 to 1833. Between 1836 and 1846 he mapped the Pacific coasts of North and South America, the southern Pacific Ocean, and the coasts of China, Borneo, and the Philippines.

Belcher's surveying career came to an abrupt end as a result of his voyage to find John Franklin, the explorer lost in an arctic expedition in 1848. He was given command of the expedition in 1852, but by May of 1854 he had abandoned his ships, which were frozen in the ice of the Arctic, and was removed from command. Belcher wrote his version of what happened in *The Last of the Arctic Voyages* (1855). He was knighted in 1867 and promoted to admiral in 1872. See FRANKLIN, JOHN; NORTHWEST PASSAGE.

belemnite

Also known as a belemnoid, this extinct cephalopod was characterized by a large internal shell similar to that of its relative, the modern cuttlefish. Like cuttlefish, belemnites were carnivores. They, in turn, were eaten by ichtyosaurs. The belemnites appeared in the Mississippian period and achieved their greatest diversity in the Jurassic and Cretaceous. They disappeared from the fossil record during the Eocene epoch. See CEPHALOPODA.

Belle Isle, Strait of

A strait that connects the Gulf of St. Lawrence to the Atlantic Ocean. Despite its extensive ice cover, due to its chilling by the Labrador Current, the strait is an important commercial waterway leading to the deep-water ports of the Great Lakes. See ST. LAWRENCE RIVER.

Bellingshausen, Fabian Gottlieb von (1778–1852)

A Russian explorer of the Antarctic, Bellingshausen was the leader of the expedition to the Antarctic commissioned by Czar Alexander I in 1819. His two sloops, the *Mirny* and *Vostok,* circumnavigated the Antarctic continent for the first time. The expedition touched the two South Sandwich islands, which Bellingshausen named for Czars Peter I and Alexander I. These islands were once thought to be part of the Antarctic mainland.

On his return to St. Petersburg (Leningrad), Bellingshausen was made an admiral and later governor of the city of Kronstadt, near Leningrad. The Bellingshausen Sea in the Antarctic Ocean is named for him. *See* ANTARCTIC OCEAN.

Belon, Pierre (1517–1564) A French biologist and, with Gesner and Rondelet, one of the three "encyclopedic naturalists." Belon was first an apothecary and then studied with the great botanist of his day, Cordus. Belon's protector, the Bishop of Le Mans, introduced him to the royal court society of France, and he eventually developed a medical practice.

Belon traveled widely to the Middle East and also to England, where he did research at Oxford. He wrote on coniferous trees and, most importantly, on marine subjects. His *Natural History of Strange Marine Fish* (1551) classified fish and included cetaceans, which he recognized as air-breathers that produced milk, yet he still classified these animals as fish.

Belon also wrote *On Aquatic Life* in 1553 and the *History and Nature of Birds* in 1555. He is known as "the father of comparative anatomy" because of his carefully researched work. He was murdered while walking in the Bois de Boulogne in Paris. *See* CETACEA; GESNER, KONRAD; RONDELET, GUILLAUME.

beluga Also known as the white whale (Delphinapterus leucas) of the order Cetacea, the beluga is the only whale that molts. It is an Arctic or sub-Arctic species that is quite large; adult males are 3.5 to 5 m (1.5 to 16.5 feet) long and weigh up to 1600 kg (3,700 lbs). They have a thick fat layer (blubber) that helps insulate them in cold water. The bulging forehead, known as a melon, is characteristic and is thought to be a factor in the whale's use of sound to locate itself. Sound is produced at the blowhole, the whale's nostril at the top of the head. Belugas travel in small groups of about 12 to 15 animals, but this number can rise to several hundred during the southern migrations. *See* ECHOLOCATION.

bends *See* DIVING, NITROGEN.

Bengal, Bay of The northern portion of the Indian Ocean, separating India and Burma. Its area is about 2 million km^2 (800,000 square miles). The major characteristic of the bay is the change in the direction of the current as a function of the monsoon: The water moves clockwise in spring and summer and then reverses its direction in autumn. The salinity of the water in the Bay of Bengal is relatively low, since it is the recipient of torrents of river water from both India and Burma. The rivers also carry tons of sediment into the bay. *See* DELTA, GANGES RIVER, INDIAN OCEAN, IRRAWADDY RIVER, MONSOON.

Benguela Current Wind-driven, cold bottom water of low salinity flowing northward along the west coast of South Africa to about 15° south latitude. The upwelling of this cold water makes the Benguela Current rich in nutrients, resulting in an environment favorable to plankton and other marine organisms. *See* NUTRIENTS, PLANKTON.

benthic ecosystems Ecosystems of the ocean floor whose flora and fauna vary widely from region to region, depending on available light, temperature, salinity, and pressure. The benthic zone, with the greatest diversity of plant and animal life, extends from the high-tide limit down to the edge of the continental shelf. It is characterized by an abundant growth of sessile plants. Subcommunities of plants and animals exist in small, isolated areas such as rock pools or coral reefs. The deeps lie at the edges of continental shelves; their depths range from 400 to about 1,000 m (1,200 to 3,000 feet). They constitute one zone of cold and dark water. Deeper still are the abyssal depths, which are characterized by relatively stable physical conditions of salinity, temperature, and light and by relatively low nutrient levels.

A healthy benthic community, such as an unstressed region of the Chesapeake Bay, would contain several species of clams, worms, oysters, crabs, barnacles, and burrowing anemones. As pollution levels rise

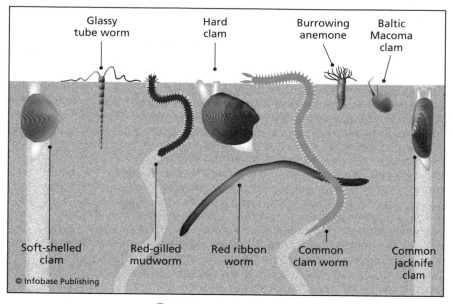

Temperate zone benthic biome

and more pollution-tolerant species survive, the number of species and the biomass of all the organisms decreases. *See* CONTINENTAL SHELF, CORAL REEF.

benthic organisms Organisms that live on, near, or in the bottom sediments of the sea. The organisms of the benthos are classified by their environment—the epifauna live on the bottom; the infauna burrow into it—and by their size: Microbenthic organisms—bacteria, unicellular algae and protists—are less than 0.1 mm (0.004 inches) in diameter. The meiobenthos—copepods, amphipods, small worms, etc.—range from 0.1 to 1 mm. The macrobenthos are greater than 1 mm (0.04 inches) and include the polychaetes, mollusks, snails, sea urchins, and other echinoderms. Megabenthos are large animals: the decapods and large fish. *See* ALGAE, INDIVIDUAL MULTICELLULAR ANIMALS.

Bering, Vitus (1681–1741) A Dane who sailed for Czar Peter the Great of Russia. Although a Russian named Dezhnev had sailed through the Bering Strait in 1648, his record was overlooked and Bering was commissioned to determine whether or not a land bridge between Siberia and North America existed. Bering sailed in 1724 and found the strait later named for him. After Peter's death, Bering continued his explorations for Peter's successor, and mapped Arctic Siberia.

On June 4, 1741, Bering sailed from Kamchatka on a voyage called the Great Northern Expedition. In August 1741 his ship was wrecked in the Gulf of Alaska; Bering died of scurvy and exposure on December 19, and most of his crew also perished. Some survivors straggled back, bringing information about the Alaskan islands and coasts, which led to the Russian interest in Alaska as a source of minerals and furs. *See* DEZHNEV, SEMYON IVANOV.

Bering Sea and Strait The strait that separates Siberia and Alaska, named for Vitus Bering, who explored the Siberian and Alaskan coasts and the Aleutian islands of the North Pacific Ocean. The southeastern region of the Bering Sea near Siberia is the deepest part of the sea, while the northwestern area is a broad, shallow continental shelf. The Bering Strait is narrow and shallow (85 km and 40 m, respectively, or 53 miles and 130 feet), which makes water exchange with the Arctic Ocean difficult. The relatively recent (Pleistocene) submergence of the Siberia-Alaska land bridge divides the two portions of a single continental mass.

The Bering Sea is characterized by tidal variability in its northerly portions, with cool or cold and cloudy weather most of the year. The warmest water temperatures occur from August to October. The large quantity of snowmelt contributes to the low salinity of the Bering Sea water. In the basin, water flow is anticyclonic and comes mainly from the Asian edge. Upwellings of deep water and the nutrients it brings account for the Bering Sea's diversified and abundant fauna.

The Bering Sea is notable for the richness and diversity of its animal life. The region is characterized by its ice pack, and the basis for the diverse animal life is the abundant growth of microalgae on the underside of the ice. This provides a food source for the marine life that in turn feeds the large and relatively unmolested mammals. The ice provides living space for seals and walruses. The dominant predator is the polar bear. *See* CURRENTS.

berm A beach backshore characterized by a cliff or ridge. It is formed by wave action and deposition of sand. *See* BACKSHORE.

Bermuda Rise The best example of a gently sloping, nonseismic elevation in the western Atlantic that is neither continental in origin nor connected with the mid-ocean-ridge system. The islands of Bermuda are on a volcanic foundation that is part of this rise; however, the islands themselves are coral. *See* MID-OCEAN RIDGES.

Bernoulli, Daniel (1700–1782) A Swiss mathematician, physicist, and physician, Bernoulli became an international authority in physics. In his book, *Hydro-*

dynamica, he established the relationship between pressure, velocity, and density in fluids (if velocity is increased, pressure decreases). This work established the principles of movement in fluids, such as air or water currents.

Bernoulli was a member of the Russian Academy of Sciences in St. Petersburg and a lecturer there until 1732. In the years 1725–49 he worked on a number of research topics which were awarded prizes by the Paris Academy of Science. These included such diverse scientific problems as magnetism, ocean currents, tides, and the movement of ships.

Beryciformes A large order of fish (including squirrelfishes and soldierfishes) characterized by an oblong, compressed body. They are marine and found in all but polar waters at depths ranging from surface to below 1,500 m (4,950 feet). Some of them, particularly those in deep waters, are luminescent.

beta rays Emissions of electrons that are part of the disintegration process exhibited by radioactive nuclei. *See* ATOMS, RADIOACTIVITY.

Biafra, Bight of Part of the Gulf of Guinea, extending from the Niger River delta to Cape Lopez (Gabon) on the west African coast. Historically, the Bight of Biafra was the site of much of the west African slave trade, which centered around the Calabar coast (Nigeria). *See* BIGHT.

bight A bay with a headland at each end. A bight is usually fairly shallow. Examples are the Bight of Benin (Africa) and the Great Australian Bight.

Bikini A classic atoll in the Marshall Islands in the Pacific Ocean group. Bikini was the site of United States atomic testing from 1946 to 1958. *See* ATOLL.

Bimini A group of islands in the Straits of Florida and part of the Bahamas group. The Lerner Marine Laboratory is in Alice

Town on North Bimini. The major industry is tourism. The first tourist on record was Ponce de Leon. He believed this to be the site of his "fountain of youth." *See* BAHAMAS, CARIBBEAN SEA.

biofilms Slick, transparent coatings that are aggregates of protein and polysaccharides produced by living organisms. They are a relatively new concept in biology. Biofilms are everywhere; for example, a rock in a pond has a slimy film. That kind of slime occurs in almost all water environments and accounts for much of the carbon cycling through ecosystems. Only with the beginnings of work on organisms that live in extreme environments did biofilms become a studied entity. Then the slime mats began to appear everywhere. They are present in caves in which the atmosphere is a mix of toxic gases; they are present in the open seas where seemingly little life is present.

All sorts of marine life produce slime or mucus. Fish are slimy because the slippery surface cuts down on friction so the fish can move more easily through the water. Since parasites and debris attach to the slime, wrasses, or "cleaner fish," have the never-ending job of picking off these tasty bits that adhere to the fish. Bacterial mats cover ocean bottoms or drift in strands that are meters long. Corals that live in polar waters produce mats of biofilms that attract bacterial populations. These mats are not single individuals or even colonies of organisms. They are communities and, to some extent, food for the coral organisms. Pteropods also produce films that become nets full of plankton and organic debris that the organism feeds on. When the films become too heavy to drift, they sink to the bottom and become food supply for the organisms of deep ocean.

Biofilms are thought to be specific to the organism that produces them, but in general they are nutritious filaments of protein and polysaccharides. The composition and place of biofilms in the food web of the oceans is an area of current research. *See* CARBON CYCLE, CORAL, FOOD WEB, WRASSE.

biogenic reef A ridge or rise above the seafloor made up of organisms or debris composed of the hard parts of organisms, such as bits of shell, teeth, or bone. *See* ATOLL, CORAL, REEF.

biogenic sediment A deposit of once-living organisms or parts of them on the ocean floor. This can be anything from bits of flesh to outgrown shells, or skeletons to plant debris. Biogenic sediments occur in all sizes, and have existed in oceans from the beginning of geologic time, producing a nearly continuous record. *See* SEDIMENT.

biogeochemical cycling An interdisciplinary study of the interactivity between the geology, the geography, and the events of a region on the organisms in that region. It attempts to understand how a particular biome works: What is the driving engine of each particular ecosystem, and how does it handle extraordinary events? For example, does a winter storm have an effect on the plankton population, and how long does that effect last? or What determines the degree of nitrogen fixation in open ocean, and how does that affect the grazing populations? *See* ECOSYSTEMS.

Biogeochemical Ocean Flux Study (BOFS) A data set based on U.S.- and U.K.-sponsored collection cruises that examined the waters of the Atlantic between May 1989 and July 1991. The aim of the study was to quantify data on the biological diversity of the regions, the physical characteristics of the water column, and its air-surface boundary layers.

biological pump Refers to the cycling of materials in a biome and the exchange between its living and nonliving constituents. Sunlight falling on open water that contains nutrients will result in photosynthesis. In this process, phytoplankton use nitrates, phosphates, and carbon to create plant tissue. That process in

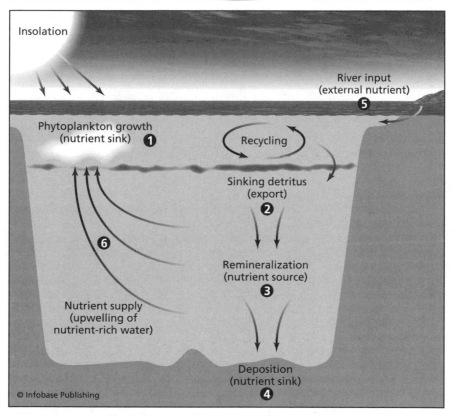

The biological pump is the movement of energy from
the Sun transformed into usable food by phytoplankton.

turn feeds a vast array of organisms and
biological activity. Ultimately, the surface layers of water are demineralized
as debris from once-living organisms,
such as shells and bony parts, sink and
eventually dissolve. Upwelling of deep
water brings minerals and nutrients back
up into the euphotic (photosynthesizing) zone to be reincorporated into living organisms. See CALCIUM CARBONATE,
CARBON CYCLE, CYANOBACTERIA, FOR-
MANIFERA, PHOTOSYNTHESIS, PLANKTON.

bioluminescence The production of
nonthermal light by living organisms,
including species of bacteria, marine
invertebrates, and fish. Bioluminescence
results from a conversion of chemical
energy to light energy. The light may act

as a lure; for example, light-producing
areas (photophores) on the lower and lateral surfaces of bathypelagic (deep-dwell-
ing) fish, such as anglerfish, attract smaller
fish as prey. The bioluminescent glow of
dinoflagellates results from wave action,
which brings them to the surface. Different species have daily or seasonal light
cycles, causing luminescent displays in
some seawater areas, such as Phosphorescent Bay in Puerto Rico. Cnidarians (jellyfish), arthropods (shrimp), annelids, and
mollusks (squid) all have light-producing
genera. There are 16 major oceanic phyla
that produce enough light to be visible.
This is usually accomplished with luciferin, a chemical that is oxidized and from
this reaction releases light rather than
heat energy. This energy then acts on a

fluorescent material that produces light at a characteristic wavelength—longer than the energizing light. Most light produced by marine organisms is in the blue-green wavelength region. This is best transmitted through water. Organisms use bioluminescence to camouflage themselves, to startle prey, or in sexual displays. *See* DINOFLATELLATES, LIGHT.

biomass The amount of living matter, or total sum of the plant and animal organisms of a particular area at a given time. The biomass is measured as milligrams of carbon per unit of volume, which in turn is used as a measure of the productivity of an oceanic system and the growth of phytoplankton, which depends on temperature, salinity, nutrients, carbon dioxide, oxygen, and light. The primary productivity of the ocean varies from 0 milligrams of carbon per square meter per day in the midwinter Arctic Ocean to 3,000 to 4,000 milligrams of carbon per square meter per day for the midsummer Walvis Bay region of South Africa.

biome *See* ECOSYSTEM.

bioturbation The process of structuring and restructuring an environment after it has been disturbed. Bioturbators are organisms that burrow in soft sediment, causing it to move. Some of the sediment will be carried away by bottom currents, will have greater contact with the water, or will be in better contact with dissolved gases. Bioturbators can be crabs or shrimp that are burrowers, or they may be polychaetes or larger plankton.

bird A warm-blooded vertebrate with modified forelimbs and a feather-covered body. Most birds fly, although some have evolved into flightlessness, such as the ratites or running birds (the ostrich, emu, rhea, and kiwi), and the penguins, which are nonflying birds that swim. Birds are land-based almost everywhere on Earth. Individual orders have adapted different body forms, coverings, and behavior to better suit a particular environment. Various orders of the class Aves live in coastal environments. These include the sphenisciforms (penguins), procellariforms (albatrosses and petrels), pelecaniforms (pelicans, boobies, and cormorants), ciconiiforms (herons, ibises, and storks), anseriforms (ducks and geese), grisiforms (cranes and rails), and charadriforms (gulls, terns, and auks).

Although there is considerable variation, the common characteristics of birds include a very short, bony tail; very good hearing and sight; a toothless bill; and a calcareous (calcium-containing) eggshell. The skeleton of birds is a light structure with a flexible neck and a large sternum (breastbone) to which the heavy flight, or pectoral, muscles are attached. The legs and toes of perching birds are also well developed. The circulatory system is a closed one; there is a four-chambered heart and large, nucleated red blood cells. The syrinx, or sound-creating organ of birds, is an enlarged tracheal (throat) structure where the trachea separates into the bronchi. The lungs are small but have connected air sacs that hold air on reserve. Thoracic movement inflates and deflates the lungs. The digestive system includes a crop, or storage vessel, and a gizzard that macerates food by muscle pressure, since birds have no teeth to chew with. (Flesh-eating birds will pick up stones to aid in this process.) The gizzard leads to the intestine and from that to the cloaca. Most birds have almost imperceptible external sex organs.

Birds seem to have evolved in the Jurassic period. The earliest bird fossils show a close resemblance to lizards, though birds are seldom found as fossils, possibly because of the lightness of the skeleton and the lack of teeth. Nevertheless, charting the development of the class is difficult. *Archeopteryx*, a chicken-sized beast with a long bony tail and teeth, along with feathers, has been called the first bird. It was the only such creature for many decades.

Since China has allowed more foreign scientists to join expeditions, a number of transitional forms have been made available to other researchers. These new findings have made the correlation of lizards, dinosaurs, and birds a hotly pursued research topic. Recent fossil finds in China exhibit impressions of feathers on lizardlike creatures and lizards caught and fossilized while napping in birdlike postures. It is now thought that these creatures were incapable of flight, but the concept of bird as contemporary dinosaur has captured the popular mind. *See* FLIGHT, MARSH, MIGRATION, *names of specific birds.*

Biscay, Abyssal Plain and Bay of The Biscay Abyssal Plain is located in the Biscay Rise, a ridge extending west from the northwestern corner of Spain to the Mid-Atlantic Ridge. The Bay of Biscay is the part of the Atlantic Ocean adjoining France, south of the Breton peninsula. *See* ATLANTIC OCEAN.

Bismarck Archipelago and Plate Islands and a crustal plate in the West Pacific Ocean northeast of New Guinea, in an area of considerable seismic activity and many small islands. *See* ISLAND, PACIFIC OCEAN.

bivalve *See* PELECYPODA.

Black Sea The easternmost portion of the Mediterranean. The Black Sea has a large eastern bay, the Sea of Azov. The Crimean Peninsula separates the two. Both seas had more extensive connection with the Caspian and the Mediterranean in earlier geologic time.

The Black Sea is surrounded by Bulgaria and Romania on the west; Russia, Georgia, and Ukraine on the north and east; and Turkey on its southern and eastern coasts. The Black Sea's water is layered. This phenomenon has been reported on by several scientists, the first more than 250 years ago. While there is not much tidal effect in this body of water, there is a surface current and a subsurface current that moves in the opposite direction.

The area is prone to unpredictable and frequent squally storms, but its temperatures are moderate. The Bulgarian and Crimean coasts are resort areas abutting broad shelves of shallow water. The Danube is the major river emptying into the Black Sea. The mountain ranges that come to the shoreline are the Balkans south of the Danube delta, the Crimean Mountains on the peninsula that indents the sea from the north, and the Caucasus and Pontus Mountains on the western and southwestern edges of the sea. The latter two are significant weather-altering ranges of known volcanic activity. These mountains divide the Black Sea from the Caspian Sea, which in the Tertiary period were connected by the remains of the Tethys Sea.

Circulation in the Black Sea is poor, and the interchange with the Mediterranean is limited to the very narrow Bosporus. As a result, there is less biomass than would be expected because of the very low oxygen levels, especially at the center of the sea, where there is almost no oxygen. However, considerable commercial fishing is done, particularly for sturgeon.

The Black Sea and its surrounding area have yielded interesting fossil remains. The fossils closest to the surface are of marine origin, while those further down (and therefore presumed older) are of freshwater organisms. One theory for this finding is that at one time in Earth's history this freshwater area was inundated by seawater after massive earth movement such as earthquake volcanic activity. Robert Ballard, using the *Alvin,* has attempted diving in this region. He believes that the evidence of a massive flood is there and that it might be the event that is recounted in the flood stories of the refugees from that region. *See* ALVIN, MEDITERRANEAN, TETHYS SEA.

Blake Plateau An undersea shelf extending along the east coast of Florida from the Bahamas to Cape Hatteras, with depths of 200 to 1,000 m (660 to 3,300 feet). The plateau drops off sharply at the Blake Escarpment. The Blake Plateau is part of the ancient continental shelf that subsided in the Cretaceous period. *See* CONTINENTAL SHELF, PLATEAU.

Blastoidea *See* CRINOZOA.

blood A fluid present (not always circulated) in complex multicellular organisms. It is the transporting medium for nutrients, waste products, and sometimes gases. The central atom in vertebrate and annelid blood is iron. The blood of crustacea is based on copper; that of tunicates (chordate marine animals) is based on vanadium.

bloom A sudden rise in the population of diatoms evidenced by a change in water color. *See* DIATOM.

blowfish A group of largely tropical marine fish, whose common characteristic is a protective mechanism that allows them to gulp air or water and greatly increase in size. Some species also produce a toxic material or have spiny projections. Blowfish or puffers vary from minnow-sized to meter-long individuals and feed mainly on small crustaceans and molluscs.

blue crab An edible crustacean of the order Decapoda. The name is most often used for the *Callinectes sapidus* or *C. hastatus*—crabs well known on the eastern North American shore. The blue crab's shell is wider than it is long, usually 15 to 20 cm (6 to 6.5 inches) wide and about 7 to 8 cm (3 inches) long. It is a drab greenblue above and has a grayish-white underside. A series of eight short barbs protrude on each side of the carapace.

Crabs are scavengers and live in estuaries, bays, inlets, and muddy shorelines. The females provide minimal care of the young, carrying them around for several weeks after spawning.

Like all other organisms with an exoskeleton, the crab outgrows its shell. It molts and tends to hide until the new one grows and hardens. Between the shedding of one shell and the growth of a larger one, the crab is relatively unprotected, surrounded only by a thick, membranous covering. At this stage it is called a soft-shell crab. *See* ARTHROPODA, CRAB, DECAPODA.

blue-green algae Many organisms that were once called blue-green algae have been reclassified as cyanobacteria. Most true blue-green algae are not marine organisms. Those that are, however, are most abundant in tropical waters. Their remains are found in some of the geologically oldest reefs. These organisms are nonnucleated cells that photosynthesize. In oceans they form a significant part of the plankton population. *See* ALGAE, CYANOBACTERIA, PROKARYOTE.

blue holes Features of the Caribbean's Grand Bahama Bank, blue holes are visible from the surface, where they stand out as dark blue spots against sandy or vegetation-covered bottoms. The hole is an opening into a submerged limestone cavern or sinkhole formed by water moving through limestone and dissolving it away. This occurred during the last ice age, when the water level of the oceans was much lower than it is now. As the ice melted, the water level of the oceans rose and the once-dry caves were flooded. The current that moves water in and out of the holes is powerful and lags behind the tides. *See* ARCHIPELAGO, BAHAMAS, CARIBBEAN SEA, PLEISTOCENE EPOCH.

blue shark Also known as the great blue shark, this large animal, 3 to 4 m (10 to 13 feet) long is known in all of the Earth's oceans, but has its principal habitat in equatorial to warm temperate waters. The blue shark is a great scavenger and a potential danger to man in the sea. *See* SHARK.

blue whale The largest animal on Earth, also called the sulfur-bottomed whale because it may carry along a yellow diatom population. The blue whale is a krill-feeder common to all oceans and, like other krill eaters, is a baleen-type (*Balaenoptera musculus*) whale. Since this huge creature (adults are about 30 to 35 m (95 to 100 feet long) was extensively hunted until recently, even its placement on the endangered species list may not save it from extinction. *See* BALEEN, DIATOMS, KRILL, WHALES.

BOD Biochemical oxygen demand, or the amount of oxygen necessary for the functioning of bacteria engaged in decomposing or breaking down organic matter. Bacteria use oxygen to break down complex compounds, (usually proteins). Thus, the more organic material there is in a given volume of water, the higher the BOD.

Since a great deal of oxygen is necessary to support a population of bacterial decomposers in a region of high industrial effluent or domestic waste (sewage), there is often insufficient oxygen in the water to also support normal flora and fauna. A high BOD, therefore, is an indication of pollution. *See* POLLUTION.

bone *See* VERTEBRATE.

bonito A relative of the tuna and mackerel and like the latter, a member of the family Scombridae. The bonito is a carnivorous fish, found worldwide. It is a sleek fish about 70 cm (30 inches) long, with a sharply forked tail, greenish-gray upper surface, and silvery underside. There are dark lines radiating from the backbone, forming a series of "V"s when the fish is observed from above. The bonito is hunted by both commercial and sport fishermen. Those species of commercial interest are *S. sarda,* found in the Atlantic and the Mediterranean, *S. orientalis,* of the Indian and Pacific Oceans; and *S. chilensis* of the western Pacific. The oceanic bonito is called the skipjack tuna. *See* TUNA.

bony fish Fish with skeletons that contain calcium compounds. They are members of the class Osteichthyes. Bony fish constitute the greatest number of vertebrate organisms now alive.

booby A seabird of the order Pelaniformes which, like the pelican, lives in tropical to temperate waters. The birds are 60 to 80 cm (20 to 30 inches) long. They have long bills and angular, thin wings. They are efficient flyers that skim above the water to hunt squid and fish and dive precipitously into the water to catch their prey.

Common boobies are the red-footed (*Sula sula*) and blue-faced (*S. dactylatra*) boobies. The latter are frequently found on the coast of western South America and in the Galápagos Islands.

The boobies are noted for their elaborate courtship rituals. They nest in large colonies on rocky shores, and the females lay two eggs in messy nests made of stones and debris. The booby is known as a "dumb" bird because it is relatively tame and frequently robbed of its prey by frigate birds. The frigate bird is also the primary predator of booby chicks. *See* FRIGATE BIRDS, PELICAN.

boom A spar (support) holding the foot or bottom of a fore-and-aft sail—the sail is one that is hung either in front of or behind a mast. *See* LATEEN, SAIL.

borers Several unrelated marine organisms that bore holes in the structures of other organisms or into wood and other materials. The shipworm (Teredinidae), a mollusk, and the gribble (Limnoriidae), a crustacean, both bore into wood, and in the case of the gribble into kelps. Boring sponges, the Clionidae, bore into calcareous (calcium-containing) materials. These organisms, however, will not feed on the organism they bore into: they are filter feeders that hollow out mollusk shells, coral, or limestone to make burrows for themselves. These sponges will weaken limestone breakwaters, and the activity of shipworm and gribbles has had a major economic effect. *See* KELP, TEREDINIDAE, TEREDO.

Botany Bay An enclosed arc about 1.5 km (1 mile) wide lying between the La Pérouse and Kurnell Peninsulas in the Australian state of New South Wales. The Bay is where James Cook first landed in Australia in 1770.

The original name given to the Bay by Cook was Stingray Harbor. Because of the great variety of unknown and exotic flora and fauna found there by the Cook expedition's naturalist, Sir Joseph Banks, the name was changed.

The first penal colony in Australia was to be established at Botany Bay in 1778, but unhealthy conditions forced its removal to Port Jackson (now Sydney) to the north. The bay is fed by two rivers, the Georges and the Cook. *See* BANKS, SIR JOSEPH.

bottom *See* OCEAN FLOOR, SHELF, TRENCH.

bottom currents Water movements at the ocean bottoms that are continuous and have a well-defined direction. Bottom currents are present in almost all waters. Early oceanographers attempted to separate the effects of tide, wind, and surface water movement from that of true subsurface currents. The latter are shown to exist in present seas by ripple patterns in bottom sediments. Fossil ripple patterns imply that the effect is not recent. Other indications of the existence of a bottom current are rocky bottoms swept clear of all but the coarsest rocks. Bottom currents are confined to particular areas by sills—elevations such as that which divides the Eastern Atlantic from the Western. The Antarctic Bottom Current is a relatively fast, large current. It moves at about 0.8 km/hour (0.4/mph). Local currents at major passages can achieve much higher velocities. The Mediterranean water leaving through the Straits of Gibraltar moves at 3.5 km/hour (2 mph).

bottom water Deep water that moves toward the equator from a particular polar region. Bottom water is generally high in oxygen content and has a low salinity characteristic which distinguishes it from the ambient ocean water above it as it moves through the ocean. Like any other water mass, bottom water eventually merges into the overall body of ocean water and so loses its identity.

Bougainville, Louis Antoine de (1729–1811) Explorer of the South Pacific and leader of the first French fleet to circumnavigate the globe, in 1766–9. Bougainville described his travels in *Voyage autour du monde* (Voyage around the world), a popular book that helped disseminate Rousseau's vision of the idealized noble, "natural" man. The voyage was underwritten by the French Crown, and its purpose was exploration and the search for new lands. The ship's company included naturalists. The expedition headed west, through the Straits of Magellan and then northwest to Tahiti, Samoa, and the New Hebrides islands. While the Great Barrier Reef was sighted, the party turned north there away from Australia and on toward New Britain. Bougainville spent some time in the Moluccas and then went to Batavia (Jakarta) and on back to France.

boundary current *See* EAST BOUNDARY CURRENT.

Bowditch, Nathaniel (1773–1838) An American mathematician and navigator. Bowditch was instantly recognized and appreciated by the sailing community for his publication in 1799 of an improved edition of J. H. Moore's *The Practical Navigator.* He was also known for his translation and extensive commentary on Laplace's *Celestial Mechanics* (1828), an astronomical work.

He had had considerable experience as a sailor and captain before the publication of his books.

Brachiopoda A phylum of bivalves in which the two valves or shells (dorsal and ventral) are asymmetric, with the ventral shell the larger of the two. The brachiopod shell is also known as a lampshell, since it supposedly looks like a Roman oil lamp. There are now about 300 known species of brachiopod. However, brachiopods appeared in the Cambrian and have had a continuous existence ever since. Over 30,000 species have been identified.

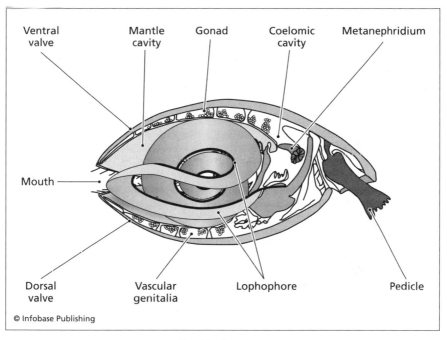

Brachipod anatomy

Their range then and now is worldwide, and they represent a significant proportion of the Antarctic fauna. The present species range in size from 2 to 15 cm (0.75 to 6 inches), while fossils of 35 cm (14 inches) are known. The extant animals have tongue-shaped shells of white, cream, rose pink, brown, or gray.

Brachiopods divide into the classes Articulata, those with shells that are hinged by interlocking "teeth," and Inarticulata, in which the shells are held together by muscles only.

Lingula, a representative species from Hawaii, lives in shallow water (about 40 m, 130 feet) deep, buried in a sand burrow. A footlike extension anchors it in the burrow and also serves in locomotion and digging. The distinguishing characteristic is a lophophore, a filter for extracting food from seawater. The digestive system consists of a digestive gland, stomach, and intestine, but no anus. The mantle that builds the calcareous shell of the animal

has setae or footlike structures on its fringe. *Lingula* is in the class Inarticulata. Since their fossil record is so complete, brachiopods are used as index fossils. The living brachiopods are remnants of what was once a very large and diverse group. They were extremely numerous in the Paleozoic era, and their fossilized shells are now found in the basement rocks of ancient reefs. Their population plummeted during the Permo-Triassic event, the mass extinction that ended the Paleozoic about 250 million years ago. *See* BRYOZOA, LOPHOPHORE, PHORONIDA.

brackish water Water that contains too much salt to be drinkable (potable) but not enough to be seawater. Its average salt content ranges between about 0.5 and 1.7%.

Brahe, Tycho (1546–1601) A Danish astronomer who built a huge wooden quadrant in Augsburg, Germany, with which he observed the motion of Jupiter and Saturn, and realized the inadequacies of the then-best astronomical tables. He was given an island on which to build an observatory, where he pioneered the use of giant astronomical equipment and the notion of reliability and precision in data collection.

His great work, *Mechanisms of the New Astronomy,* was dedicated to the Holy Roman Emperor Rudolph II. Brahe spent his last years in Prague as a guest of the emperor.

Branchiopoda An order of arthropods in the class Crustacea. About 800 species have been identified, and most of them live in freshwater or brackish water environments. Their range is worldwide, and they form a significant portion of the Antarctic fauna. Sizes typically range from 1 to about 10 cm (0.4 to 4 inches), although some fossil forms were larger; one specimen found was 35 cm (14 inches).

Living branchiopods are divided into four groups: Anostraca, the fairy shrimp or sea monkeys that are found in pet stores as fish food; Notostraca, the tadpole shrimp; Cladocera, the water fleas; and Conchostraca, the clam shrimp. The last two orders are exclusively freshwater animals, and water fleas tend to be less than 1 mm (0.04 inches) long.

Feeding habits of branchiopods vary widely; some are suspension feeders, others stir up sediment, and some scrape organic material from rocky bottoms; yet others are predators. There are other differences within the species as well. Many but not all branchiopods are capable of parthenogenesis—asexual reproduction of generations of females. The shells of some groups are calcareous, while others have silicate shells. Some fossil finds were in a deposit that was once a hot spring or geyser pool. The branchiopod species that lived there might have been an extremophile one. This organism is quite remarkable since there is an almost complete fossil record. *See* ARTHROPODA, BRINE SHRIMP.

Branchiura A subclass of the Crustacea in the phylum Arthropoda, branchiurans are small (less than 2 cm or 0.8 inches), disk-shaped parasites. The rear legs of branchiurans are swimming legs. Their prey is bony fish. They attach to the scales of a fish using modified appendages that have become suckers. The animals may detach themselves from one host and move to another. *See* CRUSTACEA, PARASITES.

Brazil Current A current resulting from the South Equatorial Current hitting the South American coast. This warm, salty surface water moves along the Brazilian coast until it is joined by the cold, north-moving Falkland Current. The conjoined stream then moves eastward as the South Atlantic Current. *See* ATLANTIC OCEAN, GUINEA CURRENT.

breakwater Usually a mound of broken rock parallel to the shore, built up on the ocean bottom and extending upward to average wave height. The purpose of a breakwater is to create an area of calm water behind it for safe anchorage, or to protect coasts or offshore towers. The placement and design of a breakwater is critical, lest it create more problems than it solves. Badly designed or misplaced breakwaters can result in dangerous currents in their gaps. These however, are essential for ship passage. In other situations, normal wave action has scoured away the coastline beneath a breakwater, causing it to crumble and thus destroying the area it was meant to protect.

If the breakwater does its job and does indeed protect the coast, the creation of a calm channel between the beach and the breakwater results in the deposition of sand between the breakwater and the beach. Thus, the coast grows out to the breakwater. The deposit must be dredged if the channel is to stay open, since the sand is not removed by wave action. *See* COAST.

Brendan, Saint (485?–577? or 583?) Also called Brendan the Navigator, an Irish monk supposed to be the first European to have traversed the Atlantic. While this is probably untrue, Brendan's voyages were real. In his *Navigatio Sancti Brendan Abbatis* he describes islands which could have been the Canaries, the Azores, Iceland, the Faroes, and the Shetland Islands. Brendan's boats were curraghs, wicker frames covered with caulked hide similar to the fishing boats known and used on the west coast of Ireland for centuries.

breslid shrimp Tiny arthropods found on black smokers. When first discovered in 1989 in a hydrothermal vent field in the Atlantic Ocean, breslid shrimp were the most extreme extremophile, a title that has now gone to the Pompeii worm. The species, *Rimicaris*, was first assumed to be blind, but it was later discovered that these tiny orange shrimp do have characteristic compound eyes that are essentially light-sensitive tissue adapted to receive the low-intensity light available in their habitat. *See* EXTREMOPHILE, VENT COMMUNITIES.

brigantine Or brig. A sailing vessel with two masts carrying square sails.

brine pools Depressions in the floor of the oceans where the saline content of the water is much higher than its average. These pools are biomes, rich in methane. They support interesting fauna that is dependent on the methane bubbling out of the Earth's crust. Dense mussel colonies live at the edges of these pools, very likely in commensal relationships with the methane-metabolizing bacteria that they harbor in their gills. *See* HALOCLINE, SEEPS, VENT COMMUNITIES.

brine shrimp Small crustaceans—most are less than an inch long—in the class Branchiopoda and the order Anostraca. They are related to *Daphnia* and distant relatives of edible shrimp. The single species is *artemia*. These tiny animals can tolerate a wide variety of salinity. They hatch from cysts that can remain dormant for long periods when placed in warm (25°C or 77°F) salt water. These members of the zooplankton are sold as aquarium food. *See* BRACHIPODA, DIAPAUSE.

bristle worms Marine annelids. *See* POLYCHAETA.

brittle star Also called serpent star, one of about 2,000 species of echinoderms of the class Ophiuroidea. Brittle stars have a small central disk and five thin radial arms, which are fairly easily broken but can regenerate. In some brittle stars, the arms are almost exclusively sensory receptors. If the arms of the brittle star are branched, the tangled organism is called a basket star. Brittle stars live on detritus or plankton, although some will catch and devour small fish as well. They are active in darkness and well suited to life on coral reefs. The most common temperate species is the slate blue, luminescent *Amphioholis squamata*. *See* ECHINODERMATA.

bromine Chemical element number 35, a red-brown liquid sufficiently reactive to ensure that it is never found in the elemental state. Bromine, in the form of bromide ion, is present in seawater at a concentration of about 65 parts per million.

The cheapest commercial source of bromine is seawater. The commercial extraction process consists of several steps: The seawater is concentrated, and then air streams are directed over the water to vaporize the bromine. Sulfur dioxide is introduced to react with the bromine vapor, and then water is added to produce hydrogen bromide (HBr) and sulfuric acid (H_2SO_4). The hydrogen bromide, on treatment with chlorine, reacts to produce hydrogen chloride and free liquid bromine.

brown algae Mainly marine algae of the class Phaeophyceae. Brown algae can live in brackish water and are found in salt marshes or estuaries, although some genera, for example *Fucus*, can exist out of water entirely for longer periods and therefore have colonized intertidal zones. Brown algae grow luxuriantly in colder waters and on rocky coasts, albeit in shallow water. Both annual and perennial types are known. The distinguishing feature of these plants is the xanthophyll pigment known

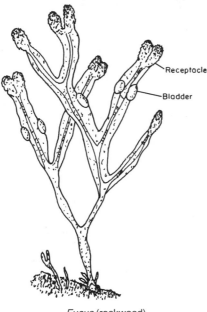

Fucus (rockweed)

as fucoxanthin, which imparts to them their characteristic brown color. Brown algae do not store starches; their carbohydrate reserves are mannitol and laminarin, a glucose-mannitol polysaccharide. Alginic acid, another polysaccharide, is present in the cell walls of brown algae. Derivatives of this, known as alginates, are used industrially as emulsifiers, thickeners, and stabilizers in pharmaceuticals, foods, paint and pigments, and paper products.

The brown algae plants seem to exhibit two characteristics of more highly evolved organisms. They have alternating sexual and asexual generations, and differentiated cells and structures, including "roots" and "stems." Three representative types are filamentous algae (*Ectocarpus*), the kelps (*Laminaria*), which are large-leaved plants, and the rockweeds (*Fucus* and *Sargassum*), which are branched. *See* PLANT.

Bryozoa Aquatic, mainly marine invertebrates, also called ectoprocts (animals with an external anus). Bryozoans are microscopic, sessile colonial organisms.

The colony they form is the result of budding, but begins with a free-swimming larva that lands on a suitable piling or rock or animal shell. Ectoprocts are found most commonly in the littoral (shoreline) zones of the sea at all latitudes, but do occur down to depths of over 6,000 m (20,000 feet). Since they have calcareous shells, they preserve well and many fossil forms are known. The original designation "Bryozoa" was applied to a marine animal with two openings to its digestive tract, as opposed to Anthozoa, with one. There was much controversy among zoologists for more than a century over just what the name Bryozoa, as opposed to Polyzoa, meant. In the older literature, then, there are references to Bryozoa that now refer to the phylum Ectoprocta. *See* ECTOPROCTA.

buffer A term in chemistry describing a solution made up of a dissolved acid and its conjugate base, or a base and its conjugate acid, that is used to maintain a certain pH. Maintaining a particular pH is essential in biological systems, where small deviations in the pH can be lethal to a particular organism.

A common buffer is one containing HCO_3^-, the bicarbonate ion, and CO_3^{2-}, the carbonate ion.

Depending on the concentration of the components, small additions of strong acid (or base) will not materially change the pH of a buffer. *See* ACIDITY, pH.

Buffon, Georges-Louis Le Clerc, comte de (1707–1788) After 1725, George-Louis LeClerc was also known as LeClerc de Buffon. A member of the minor nobility, he studied medicine, botany, and mathematics as a young man.

Buffon published his first scientific work in 1735, a translation of Stephen Hales's *Vegetable Staticks,* a work on plant physiology. In 1740 he translated Newton's *Fluxions* into French. On his appointment to the post of curator of the Royal Gardens, Buffon augmented the inventory of the king's collection. This inventory developed into the multivolume

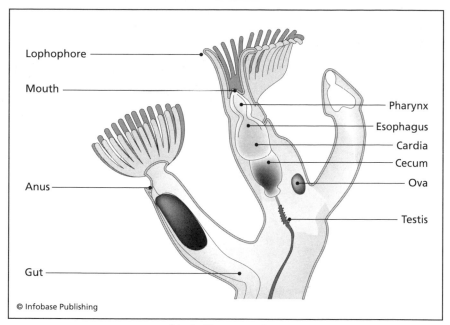

A typical bryozoan colony

Histoire naturelle, generale et particuliere, Buffon's most famous work. The best-known section of this work, *Epoques de la nature,* written in 1778, explained the geologic history of the Earth as a series of stages. Buffon envisioned unknown and extinct species in his works on animals and plants. He began to explore the possibility of evolution but did not continue that investigation. Since he was an appointed government employee and the canonical biblical account of creation was state policy, espousing or even proposing an alternative view of the origin of species would have ended his career. Buffon expanded his duties to include a small menagerie and carried out some experiments on the nature of fire. He was made a member of the French Academy of Sciences in 1753. *See* CATASTROPHISM, EVOLUTION.

buoy A moored or anchored object marking a navigational channel or obstruction. Some buoys collect information about aspects of the environment, such as the large and sophisticated data-collecting buoys FLIP (Floating Instrument Platform) and NOMAD (Navy Oceanographic Meteorological Automatic Device). *See* LIGHTHOUSE, FLIP.

Burgess shale A large, transported rock slab discovered in 1909 by Charles D. Walcott while exploring the Cambrian geology of British Columbia. Walcott found a large and very diverse collection of unknown fauna. He returned yearly through 1917 to continue his excavations of trilobites and other arthropods, worms, echinoderms, brachiopods and primitive mollusks, sponges, and crinoids. Some of his finds were unlike any known living organism or fossil.

The soft-bodied fossils were flattened and preserved as a film which reflects light; they are thus photographically preserved. The shale was probably deposited in deep water off an undersea escarpment or near a deep sea fan. The animals were bottom dwellers, burrowers, or swimmers that were covered by a sudden flow of sediment that was eventually uplifted to its present location in the Canadian Rockies.

Arthropods are the largest group of animals—mostly trilobites—and the next most common species, *Canadaspis perfecta,* is a primitive crustacean. There are more than 140 species in 119 genera present. Some of these are totally unknown anywhere else, and some are truly fantastic. There are a number of theories to explain the evolutionary history of some of these bizarre organisms and their taxonomic placement is under discussion.

byssal threads Threads produced by the byssal gland in the foot of the common bivalves, which serve to attach the animal to a stationary object. The "hairs" or "beard" of mussels are byssal threads.

Cabot, John (1450?–1499) Born Giovanni Caboto, possibly in Genoa, navigator and explorer, attempted to find a westward route to the Orient. He was made a citizen of Venice in 1476. He sailed the eastern Mediterranean for the Venetians and became a skilled navigator.

In 1484 he moved to London, taking his family with him. Like Columbus, Cabot was attempting to amass money and influential sponsors for an expedition to the Orient by a westward route. He received his grant from commercial interests in Bristol, England, in March 1496, with the understanding that any trade ensuing would be the monopoly of the merchants of Bristol.

Cabot's first attempt to sail, in 1496, was aborted because of bad weather and insufficient food; his Mediterranean sailing experience was not sufficient training for the rigors of the North Atlantic. Trying again the next year, he sighted land in the west, probably southern Labrador. Cabot, however, thought he had found Asia. His report of the abundance of fish was most welcome in England, and the area of the Georges Bank and Grand Banks that he had discovered was opened to European fishermen.

Cabot left England again in 1498 on an expedition that may have reached the American coast, but which was lost and its ships assumed to have been sunk.

Cabot, Sebastian (1476–1557) An explorer, probably born in Venice, who came to England as a boy with his father, the explorer John Cabot. Sebastian may have accompanied his father on the expedition of 1497 in search of the westward passage to the Orient. He was a mapmaker for Henry VIII of England, and accepted a commission in the Spanish navy, which he held concurrently with the post of pilot major in the navy of Charles V, the Holy Roman Emperor.

While on an expedition to the Orient, Cabot diverted his ships to the eastern South American coast, where he hoped to find treasure. The diversion resulted in a court-martial that banished him from the Holy Roman Empire, possibly because his attempt was unsuccessful.

Cabot returned to England in 1548 and joined the Merchant Adventurers, a commercial group interested in finding the Northwest Passage. The Adventurers has sponsored several expeditions to find the fabled Passage, all of them unsuccessful. Instead they turned their attention eastward and opened up trade with the Baltic regions. Trade with Russia was one of their successes. *See* CABOT, JOHN; NORTHWEST PASSAGE.

Cabot Strait A strait that connects the Atlantic Ocean with the St. Lawrence River in eastern Canada. The strait, which runs between Newfoundland and Cape Breton Island, is almost 100 km (60 miles) long. *See* ST. LAWRENCE RIVER.

Cabral, Pedro Alvarez (1460 or 1467?–1530) A Portuguese navigator who sailed for Portugal when Vasco da Gama refused to make a second voyage to the Far East. Cabral sailed west in 1500 with a fleet of 13 ships. They sailed southwest and made landfall at a place Cabral called Porto Seguro (now Baia Cabrailia) in Brazil. The fleet then left for India, but that portion of the voyage proved a disaster. A storm drove the flotilla east and a number of ships were lost. The remaining ships required considerable repair when they reached Mozambique. The fleet reached Calicut (Calcutta) in September 1500. It had had only limited trading success because of Arab competition. The seven remaining ships returned to Portugal separately from June to July 1501. Five were loaded with Far Eastern goods, but the other two were empty.

While the goal of the voyage—finding a reasonable jumping-off place from which to make the run around the Cape of Good Hope—was not realized, Cabral is credited with being the first European in Brazil. *See* GAMA, VASCO DA.

caisson A structure that is sunk and used to provide a foundation for a bridge. The pneumatic caisson, which is open at the bottom, is filled with compressed air to prevent water entry. Workers enter through an airlock in order to excavate under it. The excavation is then filled with concrete, and the caisson is moved. *See* DECOMPRESSION.

Calanoida Free-swimming planktonic copepods that have elaborate arrangements of feathery setae (hairlike projections) that function as floatation devices. The calanoids are essential parts of the food web of all oceans. These omnivores constitute a significant part of the zooplankton, eating some organisms and in turn being fed on by others. A typical cananoid is *Canalus finmarchicus,* found in the North Atlantic. The 0.05-mm eggs hatch in spring within days of being produced. An average female will produce about 15 eggs per day through the summer season. That varies with temperature, and Arctic calanoids generally have a shorter reproductive season. When these eggs hatch, a tiny creature emerges that then changes form several times. With each of these nauplid stages,

the organism gets larger, and the resulting adult may be as long as 2 cm (0.8 inches). This process takes some 30 to 40 days depending on the temperature of the water. At all stages, the calanoid is preyed upon by larger sea creatures. As an adult, it is high in fatty acids and particularly sought out as food by herring.

Although the individual calanoid is tiny, the total mass of these creatures is enormous. In the 1960s the summertime *Calanus* mass in the North Sea accounted for about 70 percent of all the zooplankton present. This figure has changed with time; the wind currents have changed somewhat, the volume of cold bottom water in which calanoids hibernate has decreased, and the result is that fewer of the tiny creatures are swept into the North Sea, thereby changing the pattern of herring migration and resulting fishing fleet activity. The movement of calanoids in the Atlantic has been extensively studied, and their population numbers have been correlated with climate oscillations.

The calanoids are a diverse group; numerous species exist in specific locales. *C. glacialis, C. helgolandicus,* and *C. hyperboreus* are Arctic organisms and drift relatively high in the water column.

Another calanoid that is widespread is *C. pacificus,* found extensively in the Pacific Ocean. This organism consumes phytoplankton and bacteria. The phytoplankton exude a waste product composed of complex sugars that form a natural gum that bacteria adhere to. This makes the bacteria available to the calanoid. Without this polysaccharide "glue," the bacteria would be too small for the calanoids to capture. *C. pacificus,* when it is in one of the early immature nauplid stages, is preyed upon by yet another calanoid, *Euchaeta elongata. See* COPEPODA, FLIP, NAUPLIUS, PHYTOPLANKTON.

Calcarea A class of the phylum Porifera, consisting of sponges that have spicules (support structures) composed of calcium carbonate. *See* PORIFERA, SPICULE, SPONGE.

calcichorates Calcite-plated fossils that have no planes of symmetry and are probably Ordovician. They have been variously called echinoderms (of the subphylum Homolozoa) and chordates (of the subphylum Calcichordata).

calcium The fifth most abundant element on Earth. Its compounds form the skeletons of animals, including marine animals. Calcium phosphate forms the bones and teeth of vertebrates, while the shells of bivalves and gastropods are formed from calcium carbonate. See CARBON CYCLE, CALCIUM CARBONATE.

calcium carbonate The calcium salt of carbonic acid. Calcium carbonate ($CaCO_3$) is the structural constituent of many marine organisms. Crystalline calcium carbonate occurs as the minerals calcite and aragonite. Coral, algae, and the planktonic foraminifera secrete calcium carbonate. The carbonate ion [CO_3^{2-}] is in equilibrium with the CO_2 and calcium ion [Ca^{2+}] that are dissolved in ocean water. The ability of water to dissolve calcium carbonate (and therefore the shells of clams, oysters, or coral) increases with lower temperatures and increasing depth (pressure). Thus, in deep ocean the sediment is carbonate-free, although carbonate-containing organisms are present on and near the surface. Their calcareous remains dissolve on sinking below about the 4,000 m (13,000-foot) level. This varies with specific locale and the deep circulation encountered. The depth at which $CaCO_3$ dissolves is known as the carbonate compensation depth.

caldera The bowl-shaped depression at the top of a volcano. In extinct volcanoes on land, these may eventually fill with water. Crater Lake in Oregon is an example. Undersea volcanoes also exhibit caldera, and lava flowing from them forms new seamounts and islands. An exploration conducted by both Woods Hole Oceanographic Institution and Scripps Institution of Oceanography recently reported an active caldera near the Samoan Islands. This caldera was not present in 2001; by 2005 the volcano now named Nafanua had grown 1000 feet high and is about 2,000 feet below the surface. See VOLCANOES.

California, Gulf of A branch of the Pacific Ocean, the gulf separates the Baja California peninsula from the Mexican mainland. The gulf was first named the Vermilion Sea by its circumnavigator, Francesco de Ulloa in 1539. The reddish color of the gulf is the result of planktonic bloom. The gulf is about 1,200 km (700 miles) long and 150 km (100 miles) wide at its widest point. The channel is divided by a constriction into an upper, fairly shallow basin with maximum depths of about 200 m (660 feet) and a more interesting lower basin. The bottom sediments are a combination of Colorado River delta and wind-blown terrigeneous materials. The gulf itself is an arm of the Pacific, not a drowned river valley. It is the junction of the East Pacific Rise and the North American continent, and probably formed as a result of the slide to the northwest of all of California west of the San Andreas fault.

There are several islands in the upper basin; Angel de la Guarda and Tiburon are the largest. Navigation in the gulf is hazardous because of sudden storms and a strong tidal bore. The extremes of temperature of the surrounding region and its aridity are mirrored by conditions in the gulf, where the temperature fluctuates considerably. In winter, cool, nutrient-rich Pacific water enters and wells up. This results in the growth of plankton, which in turn supports a large fish and mollusk population. See EAST PACIFIC RISE, SAN ANDREAS FAULT.

California Current A wind-driven Pacific Ocean current that moves in a southeasterly direction until it meets the

North Equatorial Current. Surface water in the California Current moves northwest near the coast in winter and southeast in summer. Subsurface water flows north all year round. Upwelling contributes to seasonal changes in the temperature, nutrient content, oxygen, and salinity of the water of the current. *See* EQUATORIAL CURRENT.

calorie A unit of heat. It is defined as the quantity of heat required to raise the temperature of 1 gram of water by 1°C at atmospheric pressure (760 mm Hg). Calories in a nutritional context are kilocalories (1,000 calories).

Calvin cycle The part of the photosynthetic pathway that proceeds in the dark. It is sometimes referred to as the dark reaction of photosynthesis. In it, carbon dioxide is added to ribulose 5-phosphate, a five-carbon sugar derivative, to produce a 6-carbon compound that on further reaction is converted to glucose. *See* CARBON DIOXIDE, GLUCOSE, PHOTOSYNTHESIS.

Cambrian period One of the oldest periods dating to about 570 million years ago in the Earth's geologic record, it was preceded by the Precambrian and followed by the Ordovician. It is in the earliest portion of the Paleozoic Era. The Cambrian was named by Adam Sedgwick a British geologist, for Cymru ("Wales," in Welsh), where such formations are found.

The environment was exclusively marine. The plants that existed were algae; some fossil evidence of them has been preserved as impressions on shale and as stromalites. The distinctive zoological development of the Cambrian was the appearance of shelled marine invertebrates. The dominant animals were the trilobites. Brachiopods, stromatolites, and fragments of jawless fish (the last in the very latest Cambrian) were also known.

Geologically, the Atlantic Ocean had begun to form between Laurentia and Gondwanaland, the two supercontinents. Africa, South America, Asia, India, and Antarctica, made up Gondwanaland, while Laurentia comprised the major part of North America, Greenland, Spitsbergen, parts of Scotland, Ireland, and Norway. *See* BRACHIOPOD; GONDWANALAND; LAURASIA; SEDGWICK, ADAM; TRILOBITE; *Appendix: Geologic Timescale.*

camera *See* PHOTOGRAPHY.

camouflage *See* CHROMATOPHORE, COLOR, PREDATOR-PREY RELATIONSHIP.

Canadian Basin Part of the Arctic Ocean nearest to Canada, the basin is fairly shallow near the shore, which in many areas is the result of the submergence of steep valleys. Near Patrick Island it is a drowned (subsided) headland, as is the Baffin Bay area. *See* BAFFIN ISLAND, SUBSIDENCE.

canal A man-made passage through an isthmus. The environmental effect is the introduction of new species to an area.

Canaries Current A branch of the North Atlantic Current that is deflected off Northern Europe and tempers the climate of the Canary Islands, which would otherwise be tropical. On its way south, the cold water of the Canaries Current meets warm wind and water from the Iberian peninsula, producing the fogs common on the western edge of the Pyrenees. *See* CURRENTS.

Canary Islands A group of islands belonging to Spain in the Atlantic Ocean, off the coast of North Africa at about 28° north latitude. The island closest to the continent is about 100 km (65 miles) from the coast. The Canary Islands were known in classical times: Ptolemy used Hierro, the most westerly island, as his line of prime meridian (0° longitude). Arabs later occupied the islands, and they were repeatedly "rediscovered" by Genoese, Portuguese, and

French sailors before the Spanish conquest in 1496. Explorers used the island landfall as a way station for ships before attempting the run around Africa or to America.

Geologically, the western islands in the Canaries group are volcanic mountain peaks, and the eastern ones are on an undersea plateau. The climate is very mild, with almost constant temperature maintained year-round. There is very little rainfall.

Cape Verde Islands A group of islands in the Atlantic Ocean off Cape Verde on the Senegalese coast of Africa. The islands form two chains, the Barlavento (Windward) and Sotavento (Leeward), lying on either side of 10° north latitude. These islands were first noted by Europeans in 1460, and Portuguese colonists arrived in 1462.

The climate is hot and dry year-round. The Cape Verde Islands are volcanic peaks, and some are spectacularly eroded. These islands were visited by the HMS *Beagle* on its outward journey to South America. Darwin explored the islands and noted that there were layers of fossilized sea creatures in the mountains. The beginnings of his theory of atoll formation began with these explorations. *See* ATOLL; DARWIN, CHARLES ROBERT; EVOLUTION.

capillary wave A ripple. Ripples are wind generated and of short wavelength, and can occur on any body of water. A pond is large enough to have ripples on its surface if the wind velocity is sufficient to overcome surface tension. The surface tension of the water restores a smooth surface. Waves of greater wavelength are called gravity waves. *See* WAVES.

Caprellidae A suborder of tiny crustaceans (2–60 mm or 0.08–0.6 inch), also called skeleton shrimp or whale lice. Very little is known of their life cycle, but their worldwide dispersion indicates a possible niche in the food webs of fish and other benthic vertebrates.

Carangidae The carangid fish, the family of pompanos and jacks. Carangids are fished commercially and by sport fishermen. The family consists of about 150 species, most often found in tropical seas. Only a few species live in temperate waters, and a few are freshwater fish.

carapace A relatively large, hard shell covering the back of an animal. Several unrelated animals have carapaces, including turtles, most crabs, all horseshoe crabs, and sea spiders.

caravel A 15th-century Spanish or Portuguese trading vessel, relatively wide for its length, and with a rounded stern and a sharply pointed bow. Caravels had two masts, a main, and a mizzen, the latter behind the former. The sails were either triangular (lateen) on both masts or a triangular sail on the mizzen and a large square one as the mainsail. The poop, or rear castle, extended over the stern. Columbus's *Niña* and *Pinta* were caravels.

The caravel was an important engineering advance in the construction of oceangoing vessels since it made sailing into the wind possible. Thus the Portuguese traders moving along the African coast could go as far as they wished into the southern waters and still get back home by tacking into the prevailing westerlies.

carbohydrate A compound of carbon, oxygen, and hydrogen. The empirical formula for this large group is $C_nH_{2n}O_n$, where n is any number greater than three. The carbohydrate compounds include the glycogens, starch, amylose, and cellulose. Carbohydrates may be polymers made up of monomers that are saccharides. Sugars are saccharides. *See* GLUCOSE, POLYMER, SUGAR.

carbon The element essential to all living structures. Carbon is present on land as the free element, where it exists in one of two forms—amorphous (coal) or

cellulose

$n = 1000\sim2000$

starch

$n = 100\sim1000$

Principal carbon compounds based on photosynthesis

crystalline (graphite and diamond)—or in an array of compounds. Each year, tons of carbon is brought to the world's oceans by rivers. The ultimate distribution of this material is uncertain, though it is known that carbon is moved about by surface currents, deep currents, and seasonal upwellings of cold bottom water.

The compounds of carbon range from the simple gases carbon monoxide and carbon dioxide, present in air and water, to carbonates in minerals and to the complex proteins and carbohydrates of living organisms.

Carbon is a unique element in that it can form an almost infinite number of chemical compounds by bonding carbon atoms to other carbon atoms. All sorts of structures, rings, and branched or linear chains are thus possible. Carbon also bonds to other elements, such as hydrogen, oxygen, sulfur, and nitrogen, to form hundreds of classes of compounds; some are of biological importance. The specific

chemical compounds that are made by an organism are determined by the genetic makeup of the organism that makes and uses them. *See* CARBON CYCLE, CARBON MONOXIDE, CARBON DIOXIDE, HYDRO-CARBON, NITROGEN CYCLE, CARBOHY-DRATE, PROTEIN.

carbonate pump   A part of the recycling system referred to as the biological pump. Photosynthesis removes carbon dioxide from surface water; plankton, both zooplankton and phytoplankton, forms shells that remove both calcium ions and carbonate ions from the water. This changes the acidity of the water.

The absence of shells on the sea floor puzzled early researchers. Their predominant constituent chemical, calcium carbonate, is only sparingly soluble in water. Later research proved that the surface water is more acidic, shells and bones dissolve at or close to the surface, making the carbonate ions more accessible to other creatures that are building shells and bones. *See* ACIDITY, BIOLOGICAL PUMP, CARBON DIOXIDE, PHOTOSYNTHESIS.

carbon cycle A series of interconnected reactions, some of short duration and others that take years. Carbon is circulated

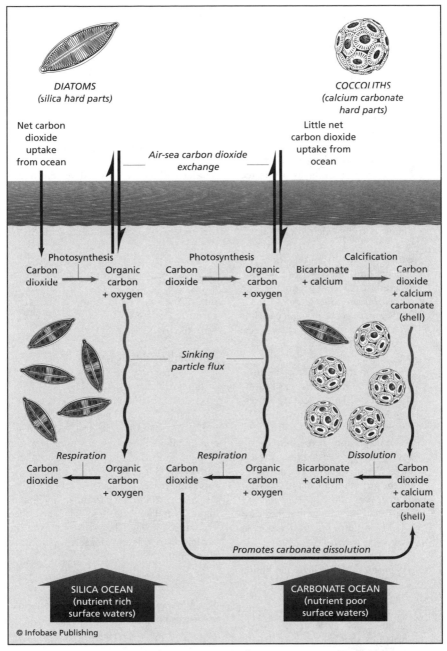

Carbonate interchange in open ocean

through the Earth's biome in a number of both organic and inorganic reactions. These may be categorized as

- *The short-term organic cycle*. This accounts for the interactions of the air with living organisms in the ocean and on land.
- *The long-term organic cycle*. This deals with the creation and weathering of fossils and includes fossil fuels. Relatively little of the Earth's carbon is involved in this cycle.
- *The inorganic cycle*. Carbon dioxide is introduced into the atmosphere in volcanic activity. This gas may become incorporated into the shells of marine organisms if there is calcium available for the formation of shells that are almost entirely calcium carbonate (CaCO$_3$). Shell formation removes carbon dioxide from the atmosphere and from the ocean water where this reaction occurs.
- *The carbonate–silicate cycle*. This is really a part of the inorganic carbon cycle. Silicate ion, SiO$_4$, enters ocean water at plate edges when one plate subsumes another and the limestone is heated in the presence of silicate. The result is the formation of calcium silicate. This reaction releases a molecule of carbon dioxide (CO$_2$), which is less likely to dissolve in the ocean water and escapes into the atmosphere, increasing the CO$_2$ level there. Dissolved silicate is essential for the growth of formanifera, yet another part of the phytoplankton. In general, water containing silicates is richer in nutrients than carbonate-containing water.

The deposition of the shells and buildup of corals is a very lengthy process, as is the weathering of terrestrial limestone when cliffs that were once sea bottom are eroded. The organic carbon cycle mainly consists of the reactions that are initiated by sunlight. Photosynthesis is a series of reactions of carbon dioxide with water in sunlight, catalyzed by chlorophylls at an appropriate temperature. The light energy and chlorophyll split the carbon dioxide molecule, and the resulting carbon atom reacts immediately with water. The end products of photosynthesis are sugars and oxygen. All living organisms require energy; sugars are an energy storage unit. Respiration is the reverse of the photosynthetic reaction. At night, when photosynthesis cannot occur, plants respire; they then use the sugar food supply and react that with oxygen to release energy needed for metabolism. The end product of respiration is carbon dioxide, which is returned to the atmosphere. This is where the ocean becomes part of the biological pump of the Earth. Phytoplankton produce sugars. When they die and decay, their remains drift to the depths of the ocean, feeding other organisms and eventually releasing carbon dioxide to the water and the atmosphere, where it becomes available again for photosynthesis. *See* CALCIUM CARBONATE, CARBON DIOXIDE, CHLOROPHYLL, NITROGEN CYCLE, PHOTOSYNTHESIS, RESPIRATION, SUGAR.

carbon dioxide (CO$_2$) A colorless, odorless gas that now constitutes less than 1% of the Earth's atmosphere. The usual figure given is 200 to 300 ppm (parts per million). In earlier geological times it was much more prevalent. Ancient algae retrieved from sea cores contain alkenones, compounds whose isotopic ratios are affected by atmospheric carbon dioxide. The presence of these isotopic differences is evidence of the CO$_2$ levels in the atmosphere from 45 million to about 25 million years ago. The Earth then was much warmer than it is now, and the carbon dioxide levels ranged from 1500 to 2000 ppm.

The Organic Carbon Cycle

1. When dissolved in water, carbon dioxide forms unstable carbonic acid:

$$CO_2 + H_2O \rightarrow [H_2CO_3] \ (carbonic \ acid),$$

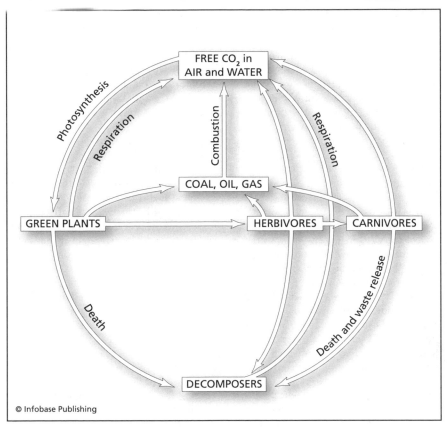

The organic carbon cycle

which is in equilibrium with the bicarbonate ion and hydronium ion:

2. $[H_2CO_3] + H_2O \rightarrow HCO_3^- + H_3O^+$
    (bicarbonate ion) (hydronium ion = water + hydrogen ion).

The bicarbonate ion is in turn in equilibrium with the hydronium ion and carbonate ion:

3. $HCO_3^+ + H_2O \rightarrow H3O^+ + CO_3^=$
    (hydronium ion) (carbonate ion).

Both the carbonate and bicarbonate ions are soluble, much more so than is CO_2. These ions become incorporated into the calcareous shells and skeletons of animals such as coral reefs and planktonic tests, in the form of calcium carbonate ($CaCO_3$). The ions are also incorporated into the blood-buffer systems of numerous species of animals.

Carbon dioxide is produced in the respiratory processes of sea plants and sea animals. Any CO_2 that does not dissolve in the surrounding water is lost to the air. Conversely, CO_2 in the air is taken up and dissolved by seawater to the saturation point. In general, cold and actively moving water dissolves more CO_2 than warmer, calmer water does. This means that with global warming, less CO_2 dissolves in ocean water, and it evaporates into the atmosphere where it becomes a greenhouse gas. The overall turnover time for CO_2 in air to CO_2 in seawater and back to CO_2 in air may be many years. According to some this cycles in about 200 years.

The role of CO_2 in regulating climate has been discussed since the late 19th century. There is considerable disagreement over measuring the capability of the ocean as an absorbing agent and the long-term effects of both human activities and volcanic eruption on the CO_2 content of air and water.

One of the possible techniques for lowering the atmospheric levels of carbon dioxide is sequestration, which involves scrubbing the gas out of power plant smokestacks and storing it underground or undersea, in depleted wells or mines. At present, this technique is expensive and still in an experimental phase. One of the possible negative effects of this process would be the impact of adding CO_2 to sediment that harbor bacteria. The addition of carbon dioxide would lower the pH of the environment, and this would most likely affect the bacteria. The activities of the bacteria are not yet well explored, but their populations are dependent on particular levels of acidity. Even small changes will have an effect on the bacterial populations, and the consequences of those changes are now unknown. See ATMOSPHERE, BIOLOGICAL PUMP, CALCIUM CARBONATE, CALVIN CYCLE, CARBON, CORAL, GREENHOUSE EFFECT, PHOTOSYNTHESIS, PLATE TECTONICS, VOLCANOES.

carbon 14 A radioactive isotope of carbon; a carbon atom with eight neutrons rather than the usual six. This variant of carbon behaves chemically just as C^{12} does and over time will lose the extra neutrons to become C^{12}. The rate at which these neutrons leave is constant, and therefore the amount of C^{14} found in organic remains is an indication of its age. Using these data, a joint British-American team has studied stalagmites found in undersea caves in the Bahamas. They found that the atmospheric radiocarbon trapped in the limestone indicated a C^{14} level much higher than any that at any recent period occurred between 45,000 and 33,000

years ago during the latest ice age. This could have been due to a supernova explosion or other cosmic activity. The conclusion drawn from this observation is that the levels of atmospheric carbon, in the form of carbon dioxide, have fluctuated greatly in geologically recent times. See BLUE HOLES, ICE AGE, RADIOACTIVITY.

Carboniferous period The fifth period of the Paleozoic era. It began about 250 million years ago and lasted roughly 8 million years. Its name is derived from the vast coal beds that were put down in this time. In North America, the early and late Carboniferous period are called the Mississippian and Pennsylvanian epochs.

During the Carboniferous period, as seas advanced and then retreated, the Atlantic Ocean continued to open, separating North America and Europe, or as they still were then, the protocontinents of Laurentia and Gondwanaland. The seas were populated by a variety of invertebrates including sea urchins, crinoids, brachiopods, corals, bryozoans, clams, and trilobites. The last, however, were disappearing in the Carboniferous period. The dominant vertebrates in the seas were fish, which were very diverse, both in number of individuals and in number of species.

carbon monoxide (CO) A colorless, odorless gas that is normally present in air and is produced by living organisms where insufficient oxygen is present. It is also produced by the incomplete combustion of carbon compounds. Carbon monoxide is highly toxic because it combines with hemoglobin in the blood, destroying the oxygen-carrying capacity of hemoglobin that is necessary to support life.

In the presence of sufficient oxygen, carbon monoxide is readily oxidized to form carbon dioxide:

$$2C + O_2 \rightarrow CO$$

$$2CO + \rightarrow 2CO_2$$

Caribbean Current A westward-moving warm-water flow that is the result of the joining of the North Equatorial and Guyana Currents. The Caribbean Current moves past the Yucatán Peninsula of Mexico and past Cuba, then continues on to form the Florida Current and ultimately the Gulf Stream. *See* FLORIDA CURRENT, GULF STREAM.

Caribbean Sea A part of the Atlantic Ocean bounded by South and Central America on the south and west and by the Antilles island chain on the north and east. The Caribbean is a 2.6 million km² (1-million-square-mile) basin.

The underwater terrain is marked by alternating basins and ridges. The greatest average depths are in the Cayman Basin (between the Cayman and Jamaica Ridges) and in the eastern region of the Venezuelan Basin (2,300 and 1,750 m, respectively, or 7,000 and 5,500 feet). Bartlett Deep in the Cayman Trench is about 8,000 m (25,000 feet) below sea level.

The climate is tropical in most of the region. Rainfall depends upon wind direction, and since the prevailing westerlies flow almost all the time, rain is more likely to fall on the windward side of the islands. On mountainous islands, this may produce almost desertlike characteristics on the leeward side.

Water enters the Caribbean from the Atlantic Ocean over the sills marked by the Anegada Passage between the Virgin Islands and the Lesser Antilles, or the passage between Cuba and Haiti. The surface water of the Caribbean is very warm and the bottom water is also warmer than that of the Atlantic, because of the sills, which are high enough to exclude the very cold bottom water of the Atlantic. Tide levels near the Mexican coast are generally higher than they are elsewhere in the Atlantic, since water is driven westward by both tide and wind, and tends to "pile up" on the Mexican and Central American coasts, a phenomenon observed early in the 19th century. The warm, shallow water of the Caribbean is an ideal environment for the luxuriant growth of both plant and animal life. *See* ARCHIPELAGO, CORAL, SHARK, SILL, TURTLE.

Carlsberg Ridge Part of the mid-ocean ridge system, this segment of that undersea mountain chain is in the Indian Ocean. It lies roughly northeast to southwest, with its northern end pointed toward the Red Sea. *See* INDIAN OCEAN, MID-OCEAN RIDGES.

Carnivora The order of Mammalia distinguished by their canine teeth. Most of the genera are terrestrial. The marine order is sometimes divided into two sub-orders and one of them, Caniformia, includes the Phocidae and Otariidae (seals and sea lions) among its seven families. Other taxonomists put the Phocidae and Otariidae into a suborder of Carnivora, called the Pinnipedia. In this grouping, walruses are included in Otariidae. *See* PINNIPEDIA, WALRUS.

carotenes A group of pigments related to the pigment known as visual purple that is found in the mammalian retina, and to vitamin A. These pigments are characteristic of the yellow algae and of diatoms. Their color ranges from pale yellow to red. *See* ALGAE, DIATOM.

carotinoids Fat-soluble pigments found in nature. They are frequently incorporated into protein and found in both plants and animals, where they exhibit a range of colors. The shells or eggs of crustaceans or echinoderms may be blue, green, purple, red, or yellow because of the presence of carotinoid-protein complexes. A cooked crab, shrimp, or lobster turns red because of the carotinoids in the shell. Coral and coralline algal organisms also exhibit red, pink, or purple color because they incorporate carotinoids in their structures. *See* CORAL, CRUSTACEAN, ECHINODERMS, PIGMENTS.

Carpentaria, Gulf of A large bay in northern Australia, between Arnhem Land and the Cape York Peninsula. The gulf was explored by the Dutch in the early 17th century, and was named for one of the explorers, Pieter Carpenter (or Carpentier).

The rectilinear outline of the gulf suggests that it is a graben—a fault line depression—which probably dates from the Pleistocene Epoch. The gulf is fairly shallow, with average depths between 50 and 70 m (150 to 210 feet). The islands of the western edge, Groote Eylandt and the Pellew Islands, are bottom hills.

The climate of the land surrounding the gulf is hot and dry, with a significant wet season that is monsoonal. The runoff in August is large and drops the salinity of the gulf sharply, which may explain the lack of coral formation in an area that is otherwise suited to coral growth. *See* FAULT, GRABEN.

Carpenter's rule (*Bacillaria paradoxa*)

A widespread marine diatom that looks like a folded wooden ruler. It is fairly large for a diatom, and its unusual shape makes it a collector's item. In the ocean it is part of the food web as prey. *See* DIATOM.

carrack A large European trading vessel in regular use from the 14th to the 17th century. Carracks were rigged similarly to the caravel, but were larger and wider. The carrack was three-masted and square-rigged on its fore and mainmasts, with lateen (triangular) sails on the mizzen (the mast behind the mainmast). This style of rigging was a nautical compromise between the square-rigged ships of northern Europe and the triangular sails of the Mediterranean. Carracks had large castles fore and aft. All subsequent large sailing vessels before the age of steam were based on this design. The carrack evolved into the galleon. *See* GALLEON, CARAVEL.

Cartier, Jacques (1491–1557)

The French master pilot who discovered the St. Lawrence River of Canada. Cartier came from St.-Malo in Brittany, a port with a long association with French shipping. He made three voyages west beginning

in 1534, ostensibly to find the Northwest Passage. He may have traveled west earlier, however, with a fishing fleet bound for the Grand Banks.

Cartier's first voyage in 1534 was financed by the French king, François I, in the hopes of finding gold and other treasure. Instead Cartier encountered ice in the Straits of Belle Isle, and moved south. He rounded the coast of Newfoundland and eventually discovered the mouth of the St. Lawrence. His second voyage began in 1536 and continued his exploration of the Gulf of St. Lawrence. The three-ship squadron sailed upriver as far as the present site of Montreal.

Cartier's third voyage in 1541 was a major effort in which he was to rendezvous on the coast with Jean François de La Roque, sieur de Roberval, a military commander. Together they were to defeat the native nation of Saguenay. The two commanders missed each other but finally met in 1542. However, the nation they had been sent to subdue did not exist. They returned to France with fur and wood.

Cartier's achievement was the establishment of a French presence in the New World. The then-dauphin of France, later Henri II, was uninterested in exploration, but a precedent had been set which was later exploited by Champlain, the colonizer of New France.

cartilaginous fish Fish with a skeleton consisting of cartilage; the sharks, rays, and chimaeras. *See* CHONDRICHTHYES, ELASMOBRANCHI, SHARKS.

cartography The art and science of making maps. The origins of the first maps are ancient. Sea travelers and merchants in the ancient world translated oral routes to written and drawn formats. Major trading nations developed pilot books and portolan charts. Maps improved rapidly with the rise of improving measuring instruments.

Surveying techniques were dependent on the development of good optical instruments and chronometers. The development of ocean mapping has progressed from depth soundings made with a weight and a knotted line to sonar to satellite photography and the current computer-enhanced techniques.

Caspian Sea The largest lake in the world. It is almost a rectangle about 300 km (165 miles) wide and 1,200 km (750 miles) long, and lies about 30 m (85 feet) below sea level. The short southern side is in Iran; the rest of the sea is in Russia. Two rivers feed the Caspian, the Ural and the Volga, the latter supplying the bulk of the water inflow. The Caspian was known in ancient times as Caspium Mare or Hycanium Mare.

Since it is a body that loses more water by evaporation than it receives in river inflow, the Caspian is saltier than the average freshwater lake. The major fishing industry centers on the famous sturgeon and its roe, caviar. The largest port is Baku on the western shore. Like the Aral Sea, the Caspian is a remnant of the Tethys Sea. *See* MEDITERRANEAN, STURGEON, TETHYS SEA.

catadromous fish Fish that live in freshwater and migrate to salt water to breed. *See* EEL.

catastrophism An 18th- and 19th-century concept of the history of the Earth, used to explain geologic features and the existence of fossil forms which either resembled existing biota or which were totally different from any plant or animal known.

Catastrophists believed that all organisms were created at once, at the Creation, and that some were demolished in cataclysmic events. The remnants remained in evidence as the biota currently known. Noah's flood was the most recent of these cataclysmic events.

This theory of Earth's history contrasted with the gradualist idea that land formations took millions of years to come into being. The gradualists were far more likely to adopt evolutionary theories. *See* CUVIER, GEORGES; EVOLUTION; HUTTON, JAMES; LAMARK, JEAN-BAPTISTE PIERRE ANTOINE DE MONET, CHEVALIER DE; LYELL, CHARLES.

cave Undersea caves are formed as they are on land: Calcium carbonates are leached out by water to form hollows. Some are in tidal cliff faces, some totally collapse. *See* BLUE HOLES.

Cavendish, Thomas (also known as Candish) (1560–1592) An English explorer and adventurer, who made the embarrassment of the Spanish the basis of his career. Cavendish's first expedition was an attempt to colonize Virginia. On his second voyage in 1586, he sailed to the Cape Verde Islands and then to the coast of South America. His party went through the Straits of Magellan and up the Pacific Coast of South America. In November of 1587 they intercepted the Spanish Manila galleon. Knowing that he would be hunted by the Spaniards all along the American coast, Cavendish headed west and, after touching land in the Philippines, arrived back in Plymouth, England, in September 1588, having been at sea for 25 months.

Cavendish was well rewarded but spent most of his prize money recklessly and used the remainder to finance a third voyage (1592). This time his stated objective was the exploration of a trade route to China and Japan. However, bad luck, no wind, and numbing cold plagued the voyage. The ships entered the Straits of Magellan in April 1592 but were forced back. Cavendish died on the return trip to England.

cavitation The formation of gas-filled cavities—bubbles—in liquids. The cause is rapid motion of the liquid. This phenomenon is important in the design of ship propellers and hydrofoil equipment. *See* BERNOULLI, DANIEL.

Ceara Abyssal Plain A great deep in the Atlantic Ocean due east of the mouth of the Amazon River and east of the Mid-Atlantic Ridge. *See* MID-OCEAN RIDGES.

Celebes Sea *See* SULAWESI SEA.

celestial navigation *See* NAVIGATION.

cell The basic structural unit of most living organisms. All cells contain the genetic material, known as deoxyribonucleic acid (DNA), which holds the assembly code for all of the protein molecules that make up the cell's contents and structural features. In the cells of bacteria and blue-green algae, this genetic material lies free in the cytoplasm (the content of the cell other than the genetic material), whereas in plant and animal cells the DNA is enclosed by a membrane to form a nucleus.

The cytoplasm of the cell, surrounding the nucleus, consists mainly of water, in which are distributed ions (such as sodium, potassium, and chloride), sugars, amino acids, the energy-supplying molecule known as asenosine triphosphate (ATP), and proteins, lipids, and nucleic acids.

Cells without a nucleus are called prokaryotic cells and those with a nucleus are called eukaryotic cells.

Cells were discovered by English physicist Robert Hooke in 1665, but it was not until 1839, when Matthias Schleiden and Theodor Schwann, two Germans, a botanist and a zoologist, respectively, presented their cell theory, that it was realized that almost all living organisms are cellular.

cellulose A linear polymer composed of glucose units. Cellulose is the major constituent of plant cell walls. *See* GLUCOSE, POLYMER.

Celsius, Anders (1701–1744) A Swedish astronomer who designed the temperature scale named for him in 1742, using the boiling point and freezing point of water as the markers on his scale and dividing the difference by 100. On the Celsius scale (prior to 1948 also called the Centigrade scale), the boiling point of water is 100° and its freezing point is 0°. The Celsius thermometer is now standard laboratory equipment everywhere.

Celsius also published works on the aurora, and worked on means to measure the brightness of stars. *See* TEMPERATURE.

Cenozoic era The third era in the Earth's history. During this era the shape of the landmasses we now recognize was determined. The Cenozoic is divided into the Tertiary and the Quaternary periods, which are further subdivided into epochs.

The major new developments in the Cenozoic took place on land: the rise in importance of mammals and of flowering plants. *See* APPENDIX: GEOLOGIC TIMESCALE.

Census of Ocean Life A 10-year program that began in 2000. It is a network of biologists, chemists, climatologists, geologists, geophysicists, and other sciences from 70 nations that will study the diversity, distribution, and abundance of marine life. The Web site is maintained by the Office of Marine Programs at the University of Rhode Island.

Centrohelida An order in the class Heliozoa characterized by a silaceous or spiny exterior. Very little is known about the five identified families in this grouping.

Cephalocarida A subclass of tiny crustaceans (3–4 mm, 0.12–0.16 in.) discovered in 1954. They are considered the most primitive extant crustaceans. These organisms are hermaphroditic, bottom dwellers, and are found in all oceans in depths varying from intertidal to bathyl (1,550 m, 5,115 feet).

Cephalopoda The most highly evolved class of the phylum Mollusca. The most common species of cephalopods today are the octopuses, cuttlefish, and squids. In Paleozoic and Mesozoic times, nautiloids, ammonoids, and belemnoids were widely prevalent. With the exception of the nautiloids, cephalopods have no shell and move by means of a muscular foot that is divided into several appendages. Cephalopods have eyes (and excellent

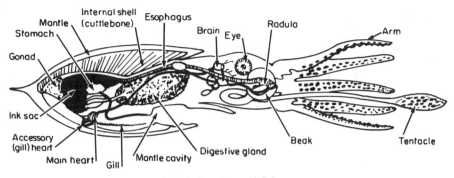

A cephalopod or cuttlefish

eyesight); a mouth with the characteristic molluscan rasping organ (the radula), and a horny beak. The circulatory system is closed and efficient.

The nervous system is complex but not cephalized (not concentrated in a head). An octopus can be taught and learns tasks and responds instantly to stimuli by flight or color change.

Squid and octopuses are voracious feeders who prey on crustaceans or fish, using their sucker-covered arms to catch and hold the prey. In turn, both squid and octopuses are caught and used as a food source by humans and other large mammals. The cephalopods are widely distributed in the ocean, inhabiting water from the equator to the poles and from abyssal depths to tidal pools. They range in size from pigmy squid to giants over 15 m (50 feet) long. *See* CUTTLEFISH, MOLLUSCA, NAUTILUS, OCTOPUS, SQUID.

cephalothorax The combined head and midbody segment of an organism's exoskeleton. Examples of organisms with exoskeletons (skeletons outside their bodies) are spiders and crabs. *See* CRAB, HORSESHOE CRAB, SPIDER.

ceratium A widespread bioluminescent dinoflagellate genus noted for interesting flagella that impede its motion. This organism floats rather than propels itself rapidly through the water. It is an auto-

trophic genus that can tolerate a range of ocean salinity. *See* DINOFLAGELLATE.

ceriatite A primitive cnidarian with modern descendants. The present ceriatites—the order is Ceriatharia, of the subclass Zoantharia—are simple, solitary polyps. *Ceriathus* is a representative genus. *See* CNIDARIA, POLYP.

Cerrantharia An order of Cnidaria that some classified in the subclass Zoantharia. They are also called tube anemones and can grow down into sediment indefinitely. Some individuals are over 1 m (3.4 feet) long. They are hermaphrodites, and the nektonic larvae move until they find a good spot. Their habitat is warm (temperate to tropical) seas at almost any depth but bathyl.

Cetacea The order of aquatic mammals. In general, they are characterized by a broad tail, tiny ears, and hairlessness. They can be found in every oceanic environment. For example, the gray whale, Bering Sea beaked whale, and Dall's porpoise are native to the north Pacific, while the northern bottlenose and North Sea beaked whales are characteristic of the north Atlantic. The snubfin dolphin is localized in Asia, and the bowhead whale and the narwhal are in the Arctic. Pigmy right whales, dusky dolphins, Gray's beaked whale, and other beaked whales are Antarctic dwellers. Though these aquatic mammals are usually found in specific areas, they do not necessarily stay solely in within these regions. The bottlenose whale spends most of the year in Antarctic waters, but it moves into warmer regions in cooler weather. The right whale and longfin pilot whale migrate great distances—toward the Equator in winter for mating and then back toward the poles in summer to feed.

Many species live in temperate waters: the Bryde's, dwarf, spotted, and spinner baleen whales and pigmy sperm whale, as well as many dolphin species. Some of the dolphins—bottlenose, striped, Gray's dolphin, and the false killer whale—will also move into colder waters in summer when food is plentiful there.

The commercially taken whales—the blue, humpback, fin, sei, minke, and sperm whales—are found pole to pole. They are still killed for commerce, and the number of all of these animals is dwindling. *See* DOLPHIN, MAMMALS, PORPOISE, WHALES.

Chaetoceros A genus of diatoms that at a point in early spring (during the North Atlantic upwelling) are the dominant organisms in terms of numbers and biomass. For a period of several weeks, chaetoceros constitute a significant part of the food web. These organisms are so small that they were missed by the usual plankton nets. *See* PLANKTON.

Chaetognatha Carnivorous plankton, also commonly called arrow worms. Although there are only about 70 known species, they are numerous and are a significant part of the food web. They feed on copepods and young fish. The typical arrow worm, shaped like a dart, is about 10 cm (4 inches) long and has a pair of eyes and 4 to 14 spines for grasping its potential food. These free-swimming, hermaphroditic, predatory organisms have many chitinous teeth. Externally they have distinct swim fins and tail. These tail and fins superficially resemble those of a fish's appendages.

The chaetognaths are biologically interesting because of their ecological importance—they are an indicator species because of their effect on the fish population—and their digestive system. The chaetognatha are very primitive examples of coelomites—organisms that have a tube-within-a-tube body form, which is a feature of more complex organisms.

This phylum in turn provides food for larger organisms and is therefore a vital link in the food chains of the continental shelves. Typically, *Sagitta elegans* is found in the phosphate-rich Atlantic and *Sagitta setosa* in the phosphate-poor waters near the coasts. *See* CHITIN, HERMAPHRODISM.

Chain (**research vessel**) A converted tugboat that in 1966 charted the depths of the Red Sea. The expedition observed very hot and very salty areas in the sea. *See* HOT SPRINGS, RED SEA.

Challenger, **HMS** A ship that sailed in 1872 on the first totally scientific circumnavigation of the world. George Nares was in command; C. Wyville Thomson and John Murray were the leading scientists on board. The ship spent almost four years at sea collecting specimens of plant and animal life and taking temperature, salinity, and other measurements of the water from surface to ocean bottom. New organisms were identified from sediments, and typified according to specific locale.

The voyage's findings confirmed earlier work on winds and currents, and the *Challenger* studied the circulation of water, charting the movement of very cold Antarctic water and mapping underwater mountain barriers. The *Challenger* also rediscovered the Mid-Atlantic Ridge. Although sections of this undersea mountain range had been reported earlier, the *Challenger*'s voyage established the single, continuous nature of the ridge.

Because of the large number of experiments done, specially designed instruments used, and far-reaching results achieved, the voyage of the *Challenger* has been called the

Route of the HMS *Challenger's* voyage of exploration; 1872–1875

beginning of the science of oceanography. *See* MID-OCEAN RIDGES; MURRAY, JOHN; NARES, GEORGE; OCEAN FLOOR; OCEANOGRAPHY; SEDIMENT; THOMSON, C. WYVILLE.

Chancellor, Richard (?–1556) A British navigator and the pilot on Sir Hugh Willoughby's expedition to find the Northeast Passage to the Indies. Chancellor left England in 1553 aboard the *Edward Bonaventure,* and was later separated from Willoughby by a storm while attempting to sail around the North Cape of Norway. Chancellor waited for the rest of the group and then continued to the White Sea. He landed and went overland to Moscow, where he met Czar Ivan IV (the Terrible). The result of this visit was the establishment of the Muscovy Company, a trading venture between England and Russia.

On a second voyage, in 1554, Chancellor retrieved Willoughby's body and possessions from Archangel. On the return from an embassy to Russia in 1556, the pilot-ambassador was shipwrecked off Aberdeen and drowned. *See* NORTHEAST PASSAGE; WILLOUGHBY, SIR HUGH.

chart A map for the use of navigators. The first known seaman's chart was drawn by Marinus of Tyre in the first century. This was based on the work of Eratosthenes, who had placed the island of Rhodes at 0° latitude and longitude. From this time to Ptolemy's, two centuries later, there were only minor changes in seamen's charts, and then almost none for centuries. The great developments in charts finally occurred after the Renaissance, with the renewal of interest in exploration. Before the end of the 13th century, sailors relied on the magnetic compass (in use in the Mediterranean since early in the century) and pilots' books as guides to coastlines. By the end of the 13th century portolon or portulan maps appeared in the Italian trading cities of Venice and Genoa. These showed coastlines and distances between points. The *portolano* depicted anchorages, shoals, bars, and other navigational features.

The increasing sophistication of mariners' maps continued and spread from country to country. Aragon's maps were superseded in quality and clarity by those of Catalonia, then Portugal. The best portolans date from the 16th century: By 1530 Pedro Nunes, Portugal's chief hydrographer, realized that meridians of longitude converge at the poles.

The mapmakers of the northern European countries accepted the work of Ptolemy as the basis for their charts. Thus, the mapmaker Ortelius of Antwerp published a chart in 1570 which showed the Canary Islands at 0° longitude, as had Ptolemy. Northern sailors were much more dependent on anecdotal information and the use of the lead line than southerners until the time of Lucas Wagenaer, a native of the Low Countries who combined the ideas of Ortelius and the portolan maps of Spain and Portugal. He published his results in an atlas of sea charts, and attempted to standardize the notation in sailing directions.

In the 17th century, both the English and French surveyed their waters and produced superior series of charts.

By the mid-18th century the British Admiralty was sponsoring surveys of British coasts and nearby island groups, producing an atlas called *The Atlantic Neptune* in 1777. This atlas was the source for sailors in North American waters for over half a century. By 1801 the Admiralty produced the first charts based on its own surveys. The Office of the British Hydrographer of the Navy is now an outstanding producer of navigational charts with a worldwide scope. *See* COMPASS, MAP, NAVIGATION.

Cheilostomata Boxlike colonial zooids. *See* BRYOZOA.

chela A large pincer claw that is the specialized first leg of arthropods. The lobster claw and the fiddle arm of the fiddler crab are examples. *See* ARTHROPODA, LOBSTER.

Chelicerata A subphylum of Arthropoda that dates from the Cambrian. These organisms have no antennae. *See* HORSESHOE CRAB, SEA SPIDER, TRILOBITE.

Chelonioidea The superfamily of sea turtles.

chemautotrophes One of the names of the autotrophes also referred to as chemosynthetic organisms. They are the bacteria that live in darkness in abyssal regions, or deep in the Earth in caves where oxygen is scarce or totally absent. The energy sources of these organisms come from chemical reactions of iron, manganese, sulfur, or other elements. Chemautotrophes are the first step in the food web of their environments. They are thought to be the unchanged descendants of the first living organisms on Earth. *See* ABYSS, CHEMICAL SEEPS, EXTREMOPHILES, HYDROTHERMAL VENTS, VENT COMMUNITIES.

chemical defenses Chemical compounds produced by organisms to protect themselves against predators. Every slow-moving or sessile organism needs some means of protecting itself since every unprotected organism will quickly become another's dinner. A number of these compounds are also of pharmaceutical interest.

Intensive exploration since the 1960s has produced more than 100 chemical defense compounds of pharmaceutical interest. Repellants, both natural and synthetic, particularly shark repellants, are also being investigated. Many other compounds are used as lead compounds in the design of antineoplastics—drugs used to treat cancers.

In the pharmacological exploration for new lead compounds, terpenes have been isolated from soft corals and nudibranchs, carotenoids from carotenoid sponges, and *Nereis* toxin from sea worms. All of these may yield pharmaceuticals. These materials are a toxic mixture of sulfur compounds. It is known that several marine plants and animals produce haloterpenes; some bryozoans yield toxic alkaloids. *See* BRYOZOA, CONE SHELL, CORAL, *Nereidae*, PHARMACEUTICAL RESEARCH.

Chesapeake Bay The largest embayment of the Atlantic Ocean on the North American coast, the bay varies from about 5 to 50 km (3 to 30 miles) in width and separates the Delmarva Peninsula from the rest of Maryland. The greatest deeps are about 100 m (330 feet). A number of rivers, all of them significant in American history, empty into the bay. They include the James, York, Rappahannock, Potomac, and Susquehanna. The very irregular shore extends for more than 5,000 km (3,000 miles). Its relatively mild weather and many inlets have made the Chesapeake Bay area one of the most important sources of shellfish in the world. Crabs and oysters are taken commercially, as are fish—notably, shad and herring. The waterfowl population that feeds on the fish and shellfish is attractive to hunters and birdwatchers alike. This region has a good deal of wild country although it is very close to the large urban centers of Baltimore, Annapolis, and Washington, D.C. Notwithstanding the efforts of both the commercial fisheries and conservation groups, Chesapeake Bay is polluted by industry and by human habitation and encroachment on the area for housing and recreational use.

A new dinoflagellate, *Pfistera,* also a serious problem in other parts of the nation, appeared in the Chesapeake in 1997. The large increase in human population in the area, along with the concentration of large-scale chicken and pig farms, has resulted in an increase in waste in the effluent from human and animal habitation. This adds significant nitrogen and phosphorus to the waters entering the Chesapeake, and the result is an algal bloom. Another complicating factor is the decrease in oyster population due to overfishing. Oysters filter the water and consume algae. When the algal growth is unchecked, there is much less oxygen in the water, resulting in migration of fish out of the area and infection of those that remain. The environmental health of the plants and animals of the region are being seriously studied by NOAA, WHOI, and associated universities. *See* ALGAL BLOOM, COAST, DINOFLA-GELLATE, ESTUARY, PHISTERIA, NOAA, TOXIN, WHOI.

Chile Rise Also known as the Chile Ridge, this is a branch of the East Pacific Rise, running northwest from the Taitao Peninsula of Chile toward Easter Island. Its existence was predicted on the basis of seismic activity before it was charted in 1958. *See* EVOLUTION OF OCEANS, MID-OCEAN RIDGES, NAZCA PLATE, PACIFIC OCEAN.

Chimaera Also known as the rat fish of the class Chondrichthyes, order Holocephalus, this is a deep-bottom-dwelling fish living on mollusks. The Chimaera has a fixed upper jaw and a movable lower one, and most specimens are about 1 m (3 feet) or more long, much of which length is a long, whiplike tail. Chimaeras may have evolved from early sharks. *See* CHONDRICHTHYES, BENTHIC ECOSYSTEM.

chinook A local wind that moves in a southeasterly direction in the Rocky Mountains of North America, carrying a dry air mass that in spring is very warm and creates a great temperature change in a short time span. *See* WEATHER, WIND.

chitin A polymer structurally related to the sugar glucose, chitin is a key component of the exoskeletons of arthropods and the shells of crabs, lobsters, and insects. Chitin is also present in the hard structures of some coelenterates, hard corals, and sea anemones.

chiton Marine mollusk of the order Polyplacophora. Mostly chitons are found in shallow water or intertidal pools. Chitons are very primitive animals, highly adapted to adhering to rock. They are relatively flat, ovoid creatures, 2 to 7 cm (1 to 3 inches) long, although specimens of the tropical species may be more than 30 cm (1 foot) long. Most individuals are dull gray-green or brown. The shell is a construction of overlapping, broad plates, making it possible for chitons to adhere

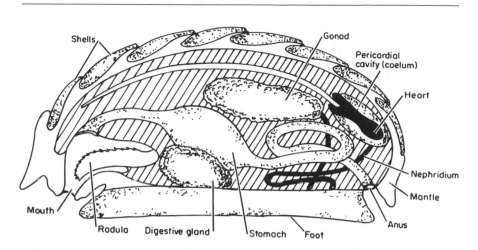

Lateral section of a chiton

tightly to rocks, where they may stay for months at a time. The mantle or girdle and the foot create an excellent suction device. Chitons, like snails, are herbivores. Their nervous system is a primitive network with some regression, as the free-swimming larva loses its eyes upon maturing. *See* MOLLUSCA.

chlorine A chemical element of the halogen family (the other members of the family are fluorine, bromine, and iodine). In its elemental state chlorine is a greenish-yellow gas (Cl_2). In nature, however, it always occurs in combined form as chloride salts. The chloride ion is present in seawater in concentrations of about 19 grams per kilogram of water (about 2%). The chloride ion is the most abundant dissolved negative ion in seawater. About 54% of the total weight of all dissolved inorganic substances in oceanic water is chloride ion. *See* CHLORINITY, SEAWATER.

chlorinity A measure of the concentration of chloride ion in seawater. It is usually expressed as the mass (weight) of chloride (e.g., as NaCl or KCl) in 1 kilogram of seawater. The average chlo-

rinity of seawater is 19 gm/kg, or 2%. *See* CHLORINE.

chlorofluorocarbons (CFCs) A blanket name for a group of compounds that contain chlorine and fluorine. All of the compounds that contain chlorine and/or fluorine are pollutants in the stratosphere. These compounds are small, synthetic molecules that were used universally as propellants in aerosols. While that use has been banned by international consent, the compounds continue to be used in refrigeration and in the production of nonstick polymers: Their emission is regulated.

Because these molecules are so small and relatively unreactive, on release they eventually drift into the stratosphere. There the chlorofluorocarbons are considered to be pollutants because they are part of the mixture of greenhouse gases and, as such, contribute to the global warming conditions that affect oceans worldwide. In the stratosphere these compounds do react with ozone, because that form of oxygen is much more reactive than the more common dioxygen molecule. This reaction depletes the ozone normally present in the stratosphere, thus creating the "ozone hole" first noticed over the

Antarctic in the 1970s. The ozone layer is thinnest at the poles, but it normally covers the entire surface of the Earth. *See* GREENHOUSE EFFECT, OZONE.

Chlorophyceae The phylum of green algae.

chlorophyll The name given to a series of pigments that produce the green color of plants. These compounds absorb blue-violet and red light and therefore reflect green light. They play a basic role in photosynthesis—the use of the energy from sunlight to make carbohydrates from carbon dioxide (CO_2) and water (H_2O)—by absorbing light energy and converting it into chemical energy. The chlorophylls are oxidized by sunlight and in turn reduce some other compound in the leaf of a plant. All chlorophyll molecules are large rings (porphyrins) made up of four smaller rings that contain nitrogen and carbon. The porphyrin ring of chlorophyll has a magnesium atom at its center. There exist different chlorophylls, designated a, b, c, and d. The most common in terrestrial plants is chlorophyll a, where the R = CH_3 (see illustration on page 74). In chlorophyll b, R = CHO. These two types of chlorophyll are present in most plants and algae in the ratio 3:1. Type c chlorophyll, in which R = H, is present in most marine algae. In type d chlorophyll, a CHO replaces the -CH = CH_2- group. All plants produce and use chlorophyll a, but the other variant pigments help capture irradiation in colder and deeper (darker) waters. Protists such as *Rhodophyta*, living at depths once thought too dark for photosynthesis, photosynthesize using the d variant of chlorophyll.

The basic porphyrin ring occurs in a number of biologically significant compounds besides the chlorophylls. Heme, the core of the hemoglobin molecular blood segment, is a porphyrin with an iron atom at its center. The molecules forming the "blood" of arthropods are porphyrins with copper at their center.

chloroplasts The color-containing organelles in cells. *See* CHLOROPHYLL, CYANOBACTERIA, PIGMENT.

Chondrichthyes A primitive class of fish, named for its cartilaginous skeleton. Chondrichthyes first appeared in the Devonian period and are well-represented by fossil remains. The sharks, rays, and chimaeras are chondrichthyes.

While these fish have no true bones, they do have enamel-covered teeth and toothlike processes on their scales, called placoid scales. All chondrichthyes are predators; they have real jaws and paired gills, each in a separate cleft, arranged in groups of five pairs for sharks, five to seven pairs for rays, and three pairs for chimaeras. The group has the usual two-chambered heart of fishes. *See* CHIMAERA, RAY, SHARK.

choreotrichs Among the larger plankton (2–200 μm) that form ocean food webs. They are widely distributed genera of fewer than 10 known species characterized by a circle of cilia around an opening. The cilia beat food particles into a vacuole for digestion. *See* CILIA, PLANKTON, TINTINNIDA.

chromatophore A pigment cell in either an animal's skin or a plant. In plants, the chromoplasts, or color granules within a cell, are most often chloroplasts, containing chlorophyll. In animals they are most likely to contain melanin—a dark brown or black pigment. Other possible pigment colors are reds, blues, yellows, and iridescent colors. In animals, the nervous connections to chromatophores are vital. They allow for fairly rapid (minutes to several hours) color alterations in amphibia, fish, and most strikingly, cephalopods. *See* CEPHALOPODA, FISH, OCTOPUS, SQUID.

Chromophycota A diverse group of photosynthesizing organisms that contain other pigments as well as chlorophyll. *See* ALGAE, CHLOROPHYLL, PLANT.

—CH₃ in chlorophyll *a*
—CHO in chlorophyll *b*

Chlorophyll *a* and *b*

Chrysophyceae All the yellow-green algae having flagella of unequal length.

Chrysophyta A phylum of algae. These plants contain carotene, which colors them yellow because the carotene masks the chlorophyll (green) that is also present in the algal cell. The Chrysophyta are photosynthesizing organisms. *See* ALGAE.

Chukchi Sea A shallow area north of the Bering Strait, its easternmost end reaching Point Barrow, Alaska. Wrangel Island is the western boundary and the 75° north latitude line is the rough bound-

ary of the Chukchi Sea with the Arctic Ocean to the north. Because of its shallow bottom, relatively warm currents entering from the south, and the low salinity typical of arctic waters, the Chukchi is rich in plant and animal life despite the year-round presence of ice. The large Arctic mammals are all represented in it. Typical fish are large cod and char. In summer the bird population consists of several species of duck, goose, and gull.

The area of the Chukchi is of interest to naturalists because of the abundance of wildlife and the fairly recent geological history of the sea as part of the Asia-to-

EFFECT OF OCEAN ON AIR TEMPERATURES			
City	Victoria (Pacific Coast)	Winnipeg (Inland)	St. Johns (Atlantic Coast)
Average maximum temperature in July (°C)	20	22.5	20.5
Average minimum temperature in January (°C)	2	–22.3	–7.5

water. From this work, the researchers learned that some of the deep water now in the southern Pacific Ocean has been in the depths for more than 800 years.

Since the overall effect of the oceans is to act as a heat reservoir, any land mass near a large body of water is protected from great swings of temperature. Comparing the high and low temperatures of three isolateral (same latitude) Canadian cities shows this (see table).

There is evidence that sunspot activity has an effect on climate that, in turn, will affect the mean temperature of deep ocean waters and eventually the composition of the pelagic populations.

Orbital variation is another cause of climate change that in the past has influenced arctic weather. This has produced increased volumes of North Atlantic Deep Water, which eventually surfaced in the South Atlantic: Evidence for this was traced in the microfossil record.

If anything can definitively be said about Earth's climate, it is that it changes. Every major volcanic eruption has contributed hugely to the density of cloud cover, which means that the average annual temperature is lowered. This drop in temperature may persist for years. The retreat of polar glacial ice is happening now and is referred to under the umbrella term *global warming*. There is evidence of this warming and subsequent cooling in past geologic time. The warming and cooling cycles are documented in the abundance or lack of fossils and in the pollen trapped in the ice of polar glaciers. The VOSTOK Ice Project has produced data extending over 400,000 years tracking the history of concentration levels of carbon dioxide and other gases.

Based on evidence of carbon dioxide levels retrieved from glacial ice and fossilized marine organisms, the carbon dioxide levels in the atmosphere in the 21st century are higher than they have ever been. This in turn affects the rate at which this gas dissolves, and that rate ultimately affects the populations in the oceans that depend on it. In turn, the biological populations in the oceans have an effect on the global climate. *See* ATMOSPHERE, CARBON DIOXIDE, CIRCULATION, CORIOLIS, EKMAN SPIRAL, EL NIÑO, GREENHOUSE EFFECT, NITROGEN CYCLE, SULFUR, WEATHER, WIND.

clipper A sailing ship of the late 18th and 19th centuries built to satisfy the demands for speed between American coastal cities. The successful design was then adopted by shipyards worldwide.

There were clipper barks, brigs, and schooners. The usual clipper ship built for cargo was a large vessel with three or more masts rigged with square sails and lateens on the bowsprit.

The great age of the clipper was 1845 to 1860, when these ships carried freight worldwide. Record travel times, such as from San Francisco to New York via Cape Horn in 89 days and from Hong Kong to New York in 74 days, made the clippers the "monarchs of the seas." While clippers continued to be built and used for cargo well into the 20th century, they were displaced as fast transportation by steamships, which, in addition to the advantage

America land bridge. *See* BERING STRAIT, POLAR BIOME.

cilia Short, hairlike extensions of certain cells. Cilia are capable of vibratory motion, and are used by the cell for feeding itself, for locomotion or both. They are frequently based on centriole fibers (tendrils of protoplasm). Nonmotile, ciliated cells in the mammalian respiratory and reproductive tracts, propel mucus and dust out of the former and help propel the egg cell along the fallopian tube in the latter. All motile cells have cilia or the long, hairlike appendages called flagella. The difference between flagella and cilia is one of size and number, their functions being similar. The distribution of cilia on the body of a cell is part of the genetic "package" of the cell. Structurally, cilia contain microtubules—thin, cylindrical, unbranched tubes that consist of similarly arranged bundles of fibers. The cilia of a cell operate in concert: a single beat is like that of the oars of a galley: stroke–recovery–stroke–recovery. The beat of the cilia can move the cell forward or backward or rotate it.

Cilia contain glycoprotein molecules—proteins with carbohydrate chains attached to them—that are used in testing for a particular type of cell, and especially for such single-celled organisms as bacteria and protozoa.

ciliate protozoa A major portion of the microzooplankton that play a significant part of the oceanic food web. Most of the ciliated protozoa are in the class Choreotrichea. They may be best characterized as microheterotrophes and are a significant part of the food supply for copepods. Their populations vary with season and growing conditions: When it is cold or there is a lack of food, many of these organisms encyst and lie dormant until more favorable conditions return. *See* COPEPODA, FOOD WEBS, HETEROTROPHS, PROTISTA.

circulation The movement of masses of water on the surface or in the depths of the oceans. Circulation on ocean surfaces can be described as permanent westerly flow of water near the equator, toward the poles along the western edges of oceans, easterly in mid-latitudes, and toward the equator along the eastern edges of oceans. These large gyres exist in the Atlantic, Pacific, and Indian oceans. There is a subpolar gyre in the Northern Hemisphere, whose closest counterpart is the Antarctic Circumpolar Current. Circulation in the Indian Ocean is seasonal, depending on the monsoon wind direction.

Deep circulation in the oceans is much slower and very much less dependent on winds or seasons than are surface phenomena. Deep water not only is moved laterally over great distances but also rises and sinks to produce considerable interchange between waters of differing regions. *See* BOTTOM WATER, CURRENTS, GYRE, INDIVIDUAL CURRENTS, WEST-FLOWING CURRENTS.

Cirripedia An order of marine crustaceans that use thoracic legs to trap waterborne food. These animals are free-swimming as larvae but are commensal or parasitic as adults when they attach permanently to a hard surface. Their form becomes so modified as it matures that they no longer look like crustaceans. Barnacles are an example. *See* BARNACLE.

cladism (cladistics) A relatively recent methodology used by taxonomists to classify present and fossil organisms. *See* CLADOGRAM, TAXONOMY.

Cladocera Water fleas, a mainly freshwater order of the Crustacea. *Daphnia* is the most widely known of the freshwater species.

cladogram A stylized diagram that shows relationships among organisms. The basis of cladograms is that families of organisms (clades) grow and reproduce. When a viable mutation occurs, one part of the family continues the original line, and the changed ones form a new line shown as a branch on the original. Thus the diagram shows a map with a series of forks in the road, each fork representing a new mutation that changed the original pattern in some way. *See* EVOLUTION.

clam A member of the class Pelecypoda, phylum Mollusca, comprising common marine and freshwater bivalves. The greatest number of species of clams are marine. Clams can range in size at maturity from barely visible to the giant *Tridacna*, which can weigh upwards of 250 kg (550 pounds). Clams usually have two shells of equal size and shape. Like other mollusks, the clams use their siphons to bring in and then expel water, from which they remove food and oxygen. Clams are generally found on or just below water on muddy or sandy bottoms, and use their powerful, muscular "foot" to dig themselves in, to a depth of 50 to 75 m (2 to 3 feet). This ability to "dig in" and the two adductor muscles that close the shell tightly, are the clam's defense against predation by birds and echinoderms. *See* MOLLUSCA.

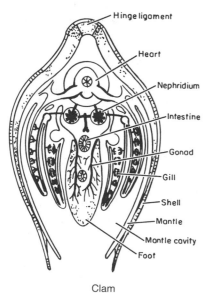

Clam

clastic Made up of rock fragments; unconsolidated sediment or beach deposits. *See* SEDIMENT.

clay Usually aluminum silicates. The particles of clay are larger than silt and smaller than colloidal particles of mud. *See* SEDIMENT.

climate The weather conditions prevalent in an area over a period of time. Weather conditions include temperature, rainfall, sunshine, wind, humidity, and cloudiness. The interaction of atmosphere and water is enormous, and each is affected by the other. The effect of ocean water is based on its heat capacity. In contrast to land, the ocean heats up more slowly and cools more slowly. The atmosphere on an average day absorbs heat that is reradiated from the Earth, raising the atmospheric temperature by about 2°C. This is equivalent to the heat capacity of the uppermost 3 m (11 feet) of ocean. Water also radiates heat to the atmosphere, and as a result, it cools considerably in local areas. This cooler and therefore denser water sinks and from then on is heated only very slowly from above. As a result, the cold bottom water stays on the bottom for very long periods.

Recent research has used coral as a means of explaining climate variation. The exchange rate measures the turnover time for carbon dioxide as it moves between the atmosphere and into the ocean water. The length of time that deep ocean water remains isolated from the surface where the exchange occurs is called the ventilation age. This may have changed over time, causing large variation in climate. To test this, scientists studied corals. Reef-building corals need carbon dioxide to build their shells, and in the process they incorporate radioactive isotopes of uranium. This in turn decays into thorium-230 and protactinium-231, and the rate of decay of these indicates the growth rate of the coral in past time and the ventilation rate of the

of greater speed, required a smaller and less skilled crew.

cloud An accumulation of water droplets or ice crystals in the atmosphere. Clouds were assigned names according to their shapes in the first decade of the 19th century by Luke Howard, an English scientist. "Cirrus" means curl, "cumulus" means heap, and "stratus" means layer.

Most clouds are in the troposphere; few ascend higher than this. Cumulus clouds continually rise in warm regions. This movement, and the rain that falls from the clouds, are the chief causes of air warming as cool air moves toward the equator from the poles.

Precipitation occurs when the droplets of water or ice crystals in a cloud reach a critical size, usually about 0.2 mm (0.01 inch). *See* ATMOSPHERE, OCEAN-ATMOSPHERE, WATER, WATER CYCLE.

Clymenella A genus of the Maldanidae, the family of "bamboo worms." *Clymenella* are polychaete (segmented) animals with a worldwide habitat. They are found at all depths and can form great beds of attached organisms on continental shelves and slopes. They are of very ancient lineage and were a common feature of Paleozoic seas 300 million years ago. *See* POLYCHAETA.

Cnidaria A phylum whose members have tentacles, nematocysts, and radial symmetry. The cnidarians are a varied group. The name is derived from the Greek word for nettle (prickly or stinging plants), and some organisms do appear to resemble plants more than animals. This phylum includes tiny independent organisms and large aggregates—from box jellies to large colonial medusae and siphonophores. What they all structurally have in common are stinging cells. They are, for the most part, marine organisms that include sea anemones, corals, and jellies. Some of the characteristics of this phylum are a central body cavity in which digestion occurs. Although there is

a central gut where digestion occurs, this is not a tube-within-a-tube body plan. This is a two-layer body as is found in sponges. Reproduction is often asexual with an occasional sexual phase. All of these animals have stinging cells. Their habitat is fairly shallow and relatively warm seas.

1) A solitary polyp (sea anemone)

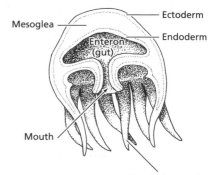

2) A medusa (jellyfish), a typical cnidarian with two-layer body plan.

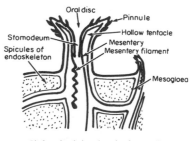

3) A colonial polyp (soft coral)

The greatest numbers of individuals and of different species are concentrated in warm and warm temperate waters. This is not to say that they are found exclusively in warm environments; cnidarians are found in the waters of the North Atlantic as well, but there are fewer of them.

Cnidarians have existed since the Precambrian. Although they consist mostly of soft tissue, enough material has survived or exists in some sort of calcareous or silaceous shell to have preserved a fossil record of them.

Cnidaria is the preferred term but it—in combination with Ctenophora—was formerly used as a synonym for Coelenterata. The Cnidaria have also included the phylum Poriphera (sponges) in an earlier attempt to bring orderly taxonomic classification to these animals. The present state of taxonomic classification of these animals is that cnidarians are present in four major groups:

Anthozoa—the corals, anemones, and sea pens

Cubozoa—box jellies

Hydrozoa—siphonophores, hydroids, fire corals, and medusae (these are the largest number of genera and species)

Scyphozoa—true jellies

Current research now seems to point to the Anthozoa as the most primitive since they never exhibit a medusal form in their life cycles. *See* CNIDOBLASTS, COELENTERATA, CORAL, CTENOPHORA, JELLIES, MEDUSAE, NEMATOCYST.

cnidoblast A stinging cell of a cnidarian. It contains the nematocyst. *See* CNIDARIA, NEMATOCYST.

coast Seashore or land near the shore; the border between land and ocean. The coast is an area of change. Materials brought seaward by rivers and then rearranged by waves and currents form coastlines. In the context of plate tectonics, coasts are either the leading edges or trailing edges of plates. The western U.S. coast, where mountainous cliffs come down to the water, is an example of a leading edge. The Atlantic coast south of Sandy Hook, New Jersey, with its broad, flat beaches is an example of a trailing edge. The trailing-edge coasts in general have more varied forms both onshore and offshore.

A coast may be a primary coast, one shaped by geologic change—erosion, glacial or crustal plate movement, earthquakes, or large transport of sediment by a young river. These coasts are generally "young." A secondary coast is one where wave activity has redeposited sediments; it is generally older than a primary coast. Another means of describing a coast depends on whether it is formed by erosion or deposition. Eroded coasts are generally rocky and are produced by wave activity, rivers, or glaciers. Deposited coasts are those that have accumulated sediment brought in by wave activity, rivers, or glaciers. They are characterized by river deltas, mangroves, salt marshes, barrier islands, and sand dunes.

If a primary coast was formed by a sediment-carrying river or glacier, before or during the last ice age, then the river valley was flooded when sea levels rose and produced a fjord. A river will drop sediment annually, forming a delta. The delta of a large river, such as the Mississippi, changes. If the tidal range is low and the current weak, the river, its current, and sediment load dominate the delta. If the tide is strong, it will extend upriver, and salt water will travel upriver, redistributing the sediment. If the river current and the wave action are both strong, a countercurrent will result, and this forms sandbars and barrier islands.

A current flowing parallel to the beach, the littoral drift, carries beach sand with it. The reshaping of Fire Island, New York, is an example of such drift, as is that of Cape Cod, Massachusetts. Fire Island has, in about a century, grown considerably longer through extension at its western end. Coastal engineers have done

Plants and animals that may be found off the coast of California,
where air temperature is warm and water temperature is cold

considerable experimenting with materials to hold sand in place. These trials include artificial seaweed, old Christmas trees, floating breakwaters, and, in Texas, artificial sand dunes. The overall effect is still uncertain.

The organisms found in a coastal community are dependent on the type of local physical habitat and the temperature. On rocky coasts the type of rock best suited for maintaining an ecosystem is a hard one. To such rocks the plants (algae) and animals can attach firmly. Crumbly shores of shale or sandstones are unlikely foundations for ecosystems. Rocky pools on coasts either near the poles or near the equator have small populations, the former because they are scraped away by the movement of ice in winter and spring, the latter because of drastic temperature changes between tides. The middle latitudes support an array of animals, some on exposed rocks, such as limpets, and others in very shallow water, including barnacles, snails, mussels, echinoderms such as seastars and urchins. They are capable of withstanding tidal surges or live just below them. Crabs, snails, some segmented worms, and mollusks are capable of living temporarily on land, either on or in the damp sand of a beach. The temperature of tropical beaches makes this transition difficult, particularly in the daytime. Most intertidal animals are therefore nocturnal. *See* BARRIER ISLANDS, BEACH, BREAKWATER, CURRENTS, DELTA, DUNE, FJORD, LITTORAL.

Coast and Geodetic Survey, U.S.
An outgrowth of the U.S. Navy's Department of Maps and Charts, now part of NOAA, the National Oceanic and Atmospheric Administration of the National Ocean Survey. *See* NOAA.

cobalt A hard, silver-white magnetic metal present in seawater in much lower concentration than in continental rock. It is assayed at about 1 ppm (part per million) in rock and 0.5 ppm in seawater. Cobalt is an essential micronutrient for plants. It is present in deep-sea sediment and in the ferro-manganese micronodules present on the seafloor. Its origin is uncertain, but it may have precipitated from the seawater or been brought up from subcrustal layers by volcanic activity.

coccolith A crystal of calcite (calcium carbonate, $CaCO_3$) that is deposited as a protective plate by the Coccolithophoridae, a group of unicellular, flagellated, tiny brown algae. The coccoliths are simply bipyramidal or hexagonal crystals formed by a single group of organisms, the Homococcoliths. Heterococcoliths, a category that includes several genera, form complex crystals that resemble cut glass. While each grain or cocclith is tiny (about 500–700 nanometers), their aggregates form massive structures. These formations account for about 75 percent of the deposited calcium carbonate, a major source of dissolved carbon dioxide. The fall of coccoliths to the deep ocean, where they either dissolve or become covered by sediments and remain, may be a factor in the exchange rate of carbon dioxide in the ocean-atmosphere system.

Most of the Coccolithophoridae are photosynthesizers, and they are usually found in tropical and warm-temperate seas. Satellite images show that they have blooms that extent for hundreds of kilometers. Both haploid generations, with half of the animals' usual full number of chromosomes, and diploid generations, with the full complement of chromosomes, may be found in the same heterococcolith. Many simple Coccolithophoridae have a viable gamete (haploid) generation, which usually does not resemble the diploid form.

The deposition of fossil coccoliths occurred worldwide. These phytoplankton are found in limestone deposits of marine origin such as the Dover Cliffs of Britain and in undersea sediments. Their continuous fossil record reaches back to the Jurassic period and is used as a means of dating sediment. *See* CARBON DIOXIDE, CALCITE, CLIMATE.

cockle A clam of almost universal marine distribution. This hermaphroditic bivalve is commonly called the jumping or leaping clam. Cockles are found in waters ranging from the tropics to the polar regions and from intertidal zones to ocean depths of 1500 m (about 1 mile).

The European cockle, *Protothaca staminea,* or rock cockle, is eaten by humans. Many other species are so muscular that they are too tough to be chewed. The average cockle has a shell 2 to 15 cm (0.4 to 6 inches) long, usually gray or gray-white, and ribbed. Several very beautiful species of cockle are collected by hobbyists. *See* BIVALVE, CLAM, MOLLUSCA.

Cocos Islands A group of islands in the Indian Ocean, south of Malaysia and east of Christmas Island. They are the peaks of mountains of the Cocos Rise.

Cocos Plate One of the smaller plates forming the Earth's surface. It is northeast of the Galápagos Islands and is moving northeast, pushing under the plate bearing Central America.

cod A gadiform fish. The major North Atlantic species is *Gadus morhua,* an extremely important fish commercially. This species is usually found in cold water on continental shelves, mainly in the Northern Hemisphere. It is best known around the Grand Banks and St. George's Bank, both areas off the Canadian coast that have been international fishing grounds for centuries.

The cod is a bottom dweller, feeding on other fish and invertebrates. It is usually dark green or brown and has a lighter underside. It has a characteristic chin barb that looks like a sharp tooth. The cod usually landed by a commercial boat are 5

to 8 kg (12 to 20 pounds) in weight, but specimens of well over 50 kg (110 pounds) exist. Traditionally, cod were taken in small boats on individual lines, but the advent of fish factory vessels has changed commercial fishing and possibly the cod population on a long-term—if not permanent—basis. Cod are now endangered.

The economic importance of the cod is underscored by the fact that many of the 16th-century voyages of exploration aimed toward North America were at least in part attempts by European nations to establish fishing rights. Dry salt cod is still a formidable ingredient of the Mediterranean diet, particularly in Portugal.

coelacanth A primitive teleost fish. About 25 genera of fossil coelacanths are known. They first appeared in the Devonian Period; the most recent specimens date from the Cretaceous. The group was assumed to be extinct. The name *coelacanth* is derived from the hollow spine in the dorsal fin.

A living specimen was first caught in 1938 and named *Latimeria chalumnae.* It was about 2 m (6.5 feet) long and weighed 40 kg (90 pounds). Several more have since been caught, and others photographed in their normal habitats. They are nocturnal carnivores and caught very infrequently, but nearly always off the South African coast. The body is purple-brown with pink and blue splotches.

The significance of the coelacanth, besides its great age, is its place as a precursor of land vertebrates. The fossil specimens were often small, and are thought to have had fat-filled, lunglike structures that may have been flotation devices, as well as fins that were protolegs. The original habitat of the coelacanth may have been in fresh water. *See* CROSSOPTERYGII, FISH.

Coelenterata A phylum currently undergoing change in scope. At one point cnidarians and ctenophores were grouped together as coelenterates. The current thinking is that coelenterates are organisms that have a tube-within-a-tube body structure. Cnidarians do not exhibit this very basic body plan, but ctenophores do. This is an active area of taxonomic investigation. *See* CNIDAIRA, CTENOPHORA.

coelom The central body cavity of an animal. It contains the digestive, excretory, reproductive, and circulatory organs.

cold-blooded animals A misnomer, since the blood of these animals is actually slightly above ambient temperature. However, their body temperature is not constant, depending instead on the environment. *See* HOMEOTHERMS, POIKILOTHERMS, TEMPERATURE IN ANIMALS.

cold seeps Slow movements of cool lava at undersea crustal plate edges. Several such sites are in the Monterey Canyon near the California coast, in polar waters, and in the Caribbean. They are less dramatic than the hydrothermal vents, but like those they support diverse communities that are based on the relatively warmer water and minerals, mostly sulfur and gases such as methane, entering the water. These materials slowly enter the water at the plate edges. The introduction of metals, gases, and heat supports the life of the chemosynthetic bacteria, the primary producers in this recently discovered ecosystem. *See* CHEMOSYNTHETIC BACTERIA, PLATE TECTONICS, VENT COMMUNITIES.

colloblast The adhesive cell on the tentacle of a cnidarian or ctenophore. This cell is sticky and enables the animal to adhere to its prey. *See* CNIDARIA, CTENOPHORA, TENTACLE.

colonial animals Animals that live in groups of attached and closely related individuals. They range from autonomous individuals to highly specialized interactive groups. The corals are representative of the former, the Portuguese man-of-war of the latter. *See* CORAL, PORTUGUESE MAN-OF-WAR.

color (of organisms) The characteristic pigments found in living organisms. They are chlorophylls (greens), carotenoids (yellows), phycocyanins and allophycocyanin (blues), and phycoerythrins (reds).

The color of marine organisms is a function of sunlight and of their habitat, mineral content, and distribution. Many invertebrates are transparent or nearly so; they disappear into the background. In tropical, clear waters where light penetrates well, color is useful in camouflaging animals, as a warning system to would-be predators, or as a factor in species recognition. Some deepwater species are red when seen at the surface, but in their deepwater habitats, where red wavelengths of light do not penetrate, they look black. The most brightly colored corals, fish, and invertebrates are found in warm and relatively shallow water. Fish and many invertebrates, notably the cephalopods (squids and octopi), can rapidly alter their color as a defense mechanism. This involves an extensive albeit decentralized nervous system, instant response to stimuli, and the ability of specialized cells to produce different colors on the animals' body surface. Squid are the outstanding color-change masters. In addition to their pigment cells, some squid have bioluminescent cells or harbor colonies of light-emitting bacteria. *See* BIOLUMINENESCENCE, CAMOUFLAGE, CHROMATOPHORE, INDIVIDUAL PIGMENTS.

color (of water) The color of water is a function of its depth, its content (dissolved and/or suspended materials that are present in one area), and the condition of the sky. *See* BEER'S LAW.

comb jellies Members of the phylum Ctenophora; they are no longer considered part of the Cnidaria. They are, however, closely related to cnidarians but lack the stinging nematocysts or the polyp phase of the cnidarian jellies. About 90 species of these comb jellies are widely distributed in the world's oceans. They are among the bigger planktonic organisms, but in the past the populations were very underestimated; most of these fragile animals were destroyed by trawls and nets, and the population estimates were based on few survivors. They feed on smaller plankton such as arrow worms and are in turn fed upon by fish such as herring.

Comb jellies are spherical or teardrop shaped and motile. The outstanding characteristic of the comb jellies are rows of hairlike projections, called cilia, that are fused together and that beat synchronously to propel the animal along. These fused cilia are arranged in eight longitudinal plates, outlined on the gelatinous, almost transparent body of a comb jelly that give the animal the appearance of a minute gooseberry. The internal organs are visible through the body wall, and in dim light these digestive canals that lie between the rows of cilia exhibit bioluminenscence. The comb jelly has two retractable tentacles that are lined with colloblasts—specialized sticky cells that entrap prey. *See* BIOLUMINESCENCE, CNIDARIA, COLLOBLAST, CTENOPHORA.

commensal relationship A relationship between two or more organisms where one derives some benefit from the association. The other(s) may be unaffected or improved by the association. It seems obvious that such a relationship exists between deep sea corals, such as those in Arctic waters, and bacteria. The corals produce thick mucus, which protects the coral from some predation. The mucus is a home and a food supply for bacteria. Large bacterial colonies inhabit the coral's mucus covering. When ocean currents detach it from the coral, this mucus mat, with its bacterial population, floats in the water, providing food for detritus feeders. It is an essential part of the food web in Arctic waters. *See* BACTERIA, CORAL, FOOD WEB.

compensation depth The point in a column of water at which the oxygen generated by photosynthesis equals the oxygen demand of the flora. Below this point plants cannot live. In nearshore areas where, because of high turbidity, light penetrates the water to a lesser degree than it does in the open ocean, the compensation depth is closer to the surface than it is in the open sea. *See* BOD, PHOTOSYNTHESIS.

conch A tropical marine snail characterized by a heavy, coiled shell and a large, flaring, wide lip. The common Caribbean conch (*Strombus gigas*) has a pink, smooth, pearly interior and whitish-gray, grainy surface. Conchs are taken for food, and their shells are sometimes collected.

Average specimens are about 30 cm (1 foot) long. Conchs of the Earth's more temperate zones range in size from 7 to 25 cm (3 to 10 inches). The largest conch is the Triton's trumpet (*Charonia*) of the western Pacific. It may exceed 45 cm (18 inches) in length and 15 to 20 cm in diameter. This shell is a prized collector's item. *See* GASTROPODA.

condensation The change of state by which a gas becomes a liquid. This physical change is always accompanied by a loss of energy. When the gas becoming liquid is water vapor, it changes into liquid water. If condensation occurs around a nucleus of dust or other very fine particulate matter in the atmosphere, fog or clouds form. If condensation occurs very high in the atmosphere, the condensate is ice rather than water. *See* CLOUD, FOG, WATER.

cone shell A marine snail, of the family Conidae, named for its colorful, heavy shell. The shell is anywhere from 1 to 20 cm (0.5 to 8 inches) in length and has been found in a variety of patterns. It may be banded, spotted, or of a single color with dark lines. The colors range from gray to dark brown; the lip is usually white.

Cone shells are carnivores. There are about 450 species of these organisms, which are most often found in shallow, tropical waters. They feed on worms, fish, and other mollusks, which they kill by injection of a neurotoxic venom. All cone shell venoms are dangerous, and those of some species are fatal to humans.

Some cone shells are quite rare, and the shells of some Indo-Pacific species are quite costly. Because of this there is a trade in the shells despite the danger in their collection. Some species are now endangered. *See* GASTROPODA.

conger eel One of fourteen eel species found worldwide. They are black or gray on the dorsal surface and white or gray-white on the ventral. The usual size of the adult that is caught is about 1.5 m (4.5 feet), but some grow to 2.5 m (7.5 feet). The average specimen weighs 2.5 kg (5.5 pounds). Conger eels are caught commercially along continental shores. As food fish, they are more popular in Europe than in America. The species *Conger oceania* breeds in the Sargasso Sea. *See* EEL.

conodont This tiny vertebrate organism was once described as a "tail with teeth" and was used in petroleum prospecting as an indicator for fossils. Its tiny skeleton is composed of calcium phosphate. These remains range in age of origin from the Cambrian to the Cretaceous periods. Their pointed shapes led to the name "conical teeth," but what they were and from what organism they came was long unclear. Geologists and paleontologists have arranged them in groups that are associated with specific geological formations. For this reason, conodonts are index fossils.

In 2002 an entire conodont organism was identified as a minute, ancient relative of the hagfish.

continent A large landmass; that part of the Earth that stands above the water of the oceans. About 30% of the Earth is covered by the continents. They are Eurasia (Europe and Asia), North America, South America, Africa, Australia, and Antarctica. Most of the landmass of the

Earth is in the northern part of the globe; only Australia and Antarctica are wholly below the equator.

Evidence from geological and seismic studies has indicated that at one point in geological history all the landmass of the Earth was concentrated in one supercontinent (Pangaea) and has since drifted to its present positions. This drift still occurs. The granitic continents float on a denser base of the Earth's crust, made largely of basalt.

The material of the continental crust is very old. Each continent has an area called a shield that is the oldest rock, 2 to 4 billion years old. The rock of the ocean floor is youngest. Mountain-making is the result of the upthrusting of relatively young rock at the edges of the floating platforms that carry the continents. As plates of the Earth's crust collide with one another, buckling and folding occurs. The Himalayan Mountains, for example, are the result of the collision of the Indian landmass with the Eurasian one. *See* BASALT, CONTINENTAL DRIFT, CRUST, EVOLUTION OF OCEANS, GRANITE, MAGMA, OCEAN FLOOR, PLATE TECTONICS.

continental drift The name given by Alfred Wegener (1880–1930) to his theory that there was originally only one continent (called Pangaea) which broke up, and that the resulting, newly formed continents then moved away from one another to form the present continents. Wegener also believed that mid-ocean sediments were the oldest rocks on Earth, compacted from continental sediment. The continents themselves float on a semifluid mantle or viscous basement rock. While Wegener's idea concerning the age of the oceanic bottom is wrong, the idea of continents moving away from each other and away from the ancient single continent Pangaea is now supported by considerable evidence. *See* EVOLUTION OF OCEANS, OCEANS; OCEAN FLOOR; PANGAEA; PLATE TECTONICS; SEDIMENTS; WEGENER, ALFRED LOTHER.

A. 225 million years ago

B. 135 million years ago

C. 65 million years ago

D. Present

The progress of continental drift

continental margin The part of a continent that extends from or borders on a steep cliff that falls off toward the ocean floor. The margin is considered to be that portion from the lowest low-tide line to the cliff's edge. The cliff and its component parts—the rise, slope, terrace, and shelf—are extensions of the nearby continent and receive materials from it. The dominant feature of the continental margin is terrigenous sediment covering a plain that may be broken by canyons. *See* CONTINENTAL SHELF, SEDIMENT.

continental rise A name proposed in 1959 for the cliff leading from the continental slope in a seaward direction to the plain of the ocean floor. The Atlantic gradient (east of North America), for example, is steeper near the continent and becomes a more gentle slope as it descends to the ocean floor. Near South Carolina,

this submerged coast resembles a broad staircase. In other areas the slope is a more precipitous one. *See* BLAKE PLATEAU, OCEAN FLOOR.

continental shelf and slope The shallow sea area closest to a continent. This area is about 70 km (45 miles) wide on the average, and deepens as it extends seaward to an average depth of about 130 m (400 feet). Although most shelves are geologically rather featureless, they are very important biologically and ecologically. The edge of the shelf, or shelf break, is the sharply descending cliff or continental slope. Geologically it is less well-explored than the shelf and is frequently deeply cut by canyons. The slope also seems to be the area of greatest faunal change, with new species arising near the shore and spreading out across the shelf. The new species are abruptly stopped from spreading by the drop down to the ocean floor. Continental slopes generally extend from the shelf break to the continental rise.

Shelves in areas that were once glaciated, such as the Grand Banks and Georges Bank, off the coast of Newfoundland, are highly irregular. Continental slopes in tropical seas may be covered with corals and their attendant ecosystems, as in the Bahamas and the Great Barrier Reef. See CORAL, MID-OCEAN RIDGES, PLATEAU.

Cook, James (1728–1779) A British navigator and one of the great 18th-century explorers. Cook made three voyages to the Pacific Ocean and explored the New Zealand and east Australian coasts. He also circumnavigated the Antarctic landmass to the limits of the floating ice, going as far south as 71°10' South Latitude. Cook's meticulous charts made his reputation when he explored the North American coasts. He had charted the St. Lawrence, its great gulf, and Labrador and Newfoundland (1759–67). The Admiralty selected him for his first Pacific voyage (1768–71). The official aim of this expedition was to observe the predicted transit of Venus from a base in Tahiti. The secret aim of the Admiralty was to have accurate maps of the fabled "great southern continent" that was assumed to be between 40° and 60° South Latitude. Rumors of such a continent existed, and the Royal Navy wanted to know if it was of use to the British Empire. Cook verified the existence of Australia as an island. He drew the first useful charts of New Zealand and the eastern coast of Australia on the way home.

Collection of unknown plant and animal specimens and water sampling were integral parts of all of Cook's voyages of exploration. Joseph Banks was on board for some of the North American exploratory expeditions and on the first of the Pacific voyages. The Forsters, father and son, were the naturalists on board for the second Pacific exploration.

On the third Pacific voyage, Cook again went to North America, where he explored the Bering Strait, Alaska, and the coast of British Columbia. This was another attempt to find the fabled Northwest Passage to Asia. Moving on to the Hawaiian Islands for the winter, Cook was killed by the local people on Hawaii in a dispute about a longboat. He had previously discovered the island group and named it in honor of the Earl of Sandwich, first lord of the Admiralty. The voyages brought back a wealth of specimens and information to the Royal Society. Cook contributed to the society's store of knowledge by solving the problem of scurvy, a vitamin-deficiency disease that was frequent on long voyages and was a prime cause of his rapidly ending the first Pacific exploration. Cook was a pioneer in the use of Hunter's chronometer in the determination of longitude. See BANKS, SIR JOSEPH; EXPLORERS AND EXPLORATIONS; LONGITUDE.

Cook Inlet An extension of the Gulf of Alaska, the inlet separates the Kenai Peninsula from the mainland. Captain James Cook explored the area in 1778 while attempting to find an Arctic Passage. Anchorage is the significant city at the head of the 240-km (150-mile)-long inlet. The area is known for its sudden and often violent storms.

Cook Islands A group of islands in the South Pacific about 2,600 km (1,600 miles) northeast of New Zealand. The islands were discovered by Captain James Cook in his voyage of 1773–77. The total landmass of the islands is less than 250 km (93 square miles). They consist of two separate volcanic groups. Rarotonga is the largest island, and the capital, Avarua, is also on this island. See COOK, JAMES; PACIFIC OCEAN.

Cook Strait A body of water that separates the two main islands of New Zealand. It ranges from 25 to 145 km (16 to

90 miles) wide. The strait was found by Captain James Cook, who sailed through it in 1769. *See* COOK, JAMES.

Copepoda A crustacean class in the order Calanoida, Copepoda are the most numerous crustaceans and one of the most abundant animals on Earth, with more than 6,000 species found in fresh- and marine water. Most copepods are truly benthic organisms that never touch the ocean bottom. Some copepods are parasitic, living on fish, but most are free-living. Those that are free-living are only millimeters long (0.5 to about 10 mm, or less than 0.5 inch). They have varied body shapes but are recognizable as arthropods (although there is no distinct head-plus-thorax, or cephalothorax, body plan). Most copepods have distinct and elaborate antennae. Larval copepods, as they develop into adults, undergo several growth stages called instars. Their appearance changes drastically from one instar (intermediate juvenile form) to the next.

Calanoid copepods, which are representative of the free-living suborders, occur in both fresh- and salt water. Copepods are an important link in many food chains; they feed on smaller microorganisms and are in turn food for larger animals, who are attracted to them as a food supply because of their high lipid content. Some are herbivores that eat algae; others are carnivores that eat other zooplankton. They are filter feeders that screen large quantities of water for food, although many species capture individual food particles. They expend great energy in propelling themselves through water and orient themselves by the gravitational field of the Earth, as do decapods such as shrimp. The other suborders of copepods present in the greatest numbers of both individuals and species are the Cyclopoids, planktonic organisms, and Harpacticoids, which are long and slim bottom dwellers. The suborders differ in the shape of the thorax and in their flexibility. *See* ARTHROPODA, ECOSYSTEM, FILTER FEEDERS, FOOD CHAINS.

Copernicus, Nicolaus (1473–1543) A Polish mathematician, best known by his Latinized name, who displaced the Earth from the center of human beings' conceptual universe and made it orbit a stationary sun. Niklas Koppernigk was the original name of this churchman, who specialized in canon law but was also a skillful astronomer. He began what is now considered the first modern scientific revolution when he set about to refine some of the observations in Ptolemy's astronomical work, the *Almagest*. In his search for perfect motion he rejected the Ptolemaic geocentric cosmos, which nevertheless explained astronomical phenomena and predicted events such as eclipses. Copernicus's rejection of this model was based on Aristotle's ideal of perfect motion of perfect heavenly bodies. This idea was central to Copernicus's work: Uneven and therefore imperfect motion was not possible for heavenly bodies.

Copernicus knew of the ancient Greek philosophical view of the cosmos that placed the Sun at the center of the universe. Using that concept as a mathematical starting point, he proposed it in his work *De Revolutionibus Orbium Coelestium* (The revolution of celestial spheres), which did not appear in print until the year of his death. This book profoundly rocked astronomy. Although its practical effect on navigation and on the sciences that became oceanography were minimal in the short run, in the long view the "new astronomy" changed everything. *See* GALILEI, GALILEO; PTOLEMY.

copper A metallic element, and a micronutrient necessary for many enzyme reactions in living systems. Copper is the central metal atom in the blood of some crustaceans, where it is the analog of the iron atom in the hemoglobin of mammalian blood or the magnesium atom in chlorophyll.

Free copper in excess is toxic to many marine organisms, including those in which it is part of enzyme systems. The giant kelp is an example of a plant that requires copper in some quantities and is poisoned by an excess of this element. *See* CHLOROPHYLL, CRUSTACEA, KELP.

coral and coral reefs Cnidarians of the class Anthozoa; they are polyps, predominantly with sixfold or eightfold symmetry, and are either solitary or colonial. The coralline coelenterates construct calcium-containing shells of characteristic shapes. Colonies of coral and their associated foraminifera and symbiotic algae live best in tropical and semitropical environments. The coral reef belt is within the boundaries of 30° north latitude and 30° south latitude. Once coral organisms were recognized as animals, it was assumed that tropical waters were the only habitat for coral organisms. Subsequently, others have been discovered. Some slow-growing, non–reef-building coral organisms are found in Arctic waters. They are not dominant organisms in their habitat and are in commensal relationships with particular bacteria that live in and on the abundant mucus produced by the coral. This is true of deep-sea corals in all latitudes. The mucus is thought to be a protection for the coral against both drying and colonization by other animals. It and the bacteria in it are integral parts of the food web.

Maximum growth of corals occurs in relatively warm water at depths less than 50 m (160 feet) and is optimum at a level of about 7 to 20 m (20 to 60 feet) down. However, corals may be found at greater depths. The dominant corals found on seamounts are gorgonians—sea fans—and this is a diverse group. The sea fans found on seamounts are genetically different from those on other seamounts.

The colonial organisms utilize the skeletons of dead coral, undersea boulders, the continental shelf, and submerged peaks to build coral reefs. Since coral growth is determined both by the quantity of food supply and the availability of calcium carbonate that is necessary to construct the shell, the growth of corals in reefs produces a record similar to that found in tree rings. Dating reefs by the coral growth is

possible and is an indication of the availability of raw materials in ancient seas. Tracking coral growth and correlating it with weather conditions such as El Niño events is important, as is the study of weather and long-term phenomena such as global warming.

There are three main geomorphological categories of reef: fringing reefs, barrier reefs, and atolls. Fringing reefs are formed close to shore on rocky coastlines. Barrier reefs are separated from land by lagoons or channels produced as a result of subsidence. Atolls are found around subsiding volcanoes.

The location of barrier and fringing reefs on the eastern edges of continents is the result of both current flow and terrestrial spin. Warm surface currents move close to the eastern side of landforms that lie away from the equator. This maximizes the growth of algae. Deep polar upwellings move toward the equator along the western edges of continents.

The commensal relationships between coral and photosynthesizing algae and coralline algae, which also form hard structures, are part of a complex ecosystem. In addition to these organisms, there are others that live on a coral reef, using it as support, shelter, or hunting ground. Its collection of fish, echinoderms, mollusks, algae, and arthropods constitutes a large ecosystem unique to each reef.

Fish present in and around coral reefs feed on sea grasses and planktonic organisms. In turn they excrete nitrogen- and phosphorus-containing materials that are recycled and used by the coral and other organisms living in and on the reef.

Reefs have characteristic populations that are part of the unique nature of each reef. There is usually a single species, called a keystone species, that controls the population of its reef. They are predators and prevent the possible intrusion of another species. Damselfish, for example, eat algae, exposing coral, but in turn they aggressively maintain territory and keep other predaceous populations, such as parrotfish, in check. This ensures their use of the algae

REEF TYPES			
	Fringing	*Barrier*	*Atoll*
Locale	Continental shelf	Ocean shallows	Oceanic rise
Shape	Long bar, parallel to continent	Multiple bars separated by deep channels	Circular, surrounding a lagoon
Growth Pattern	Outward from the continent to the edge of the shelf	Rapid in shallows, shifts with opening of new channels by wave action	Enlargement and eventual rise above sea level to form islands
example	Florida Keys	Great Barrier (Australia)	Bikini

REPRESENTATIVE CORALS		
Class Anthozoa	*Representative*	*Common name*
Order: Coenothecalia	*Heliopora*	blue coral
Alcyonacea	*Alcyonium*	soft coral
Gorgonacea	*Gorgonia, Corralium*	sea fan
Rugosa	*Petraia*	compound coral
Scleractinia	*Fungia, Astrangia*	stony (true) coral
Tabulata	*Favositea*	honeycomb coral

population of a given reef. The resulting oversupply of algae feeds other organisms.

The coral organisms are most vulnerable to wave action and storm surges, which destroy the platforms that support them. There are also organisms that prey on live coral, such as the crown-of-thorns sea star and coral-eating parrotfish.

The reproduction of corals is usually asexual or gamete-producing in the typical mode of the phylum. There are, however, highly specialized variations. Acropora is an organism that produces gametes that are externally fertilized at brief periods during the year. Spawning is triggered seasonally by the sea temperature and by specific dark periods. In other species the majority of the gametes are produced in a one- or two-day period. Fertilization is not random, since many species produce gametes at the same time. Eggs are both attractants for compatible sperm and toxic to the sperm of other species. The reproductive advantage of this mass spawning is that fertilization by gametes from another colony of the same species is possible. The soft corals, like the sea pens (gorgonians), are in the subclass Octocorralia; they are eight-sided. The true or stony corals are in the subclass Hexacorralia; they have a six-sided structure. The gorgonians are most prevalent in the Atlantic and Pacific Oceans; the soft-bodied corals are found most often in the Indian and East Pacific oceans. *See* CLIMATE, COMMENSAL RELATIONSHIPS, EL NIÑO, GORGONIANS, ISLAND, REEF, SEAMOUNT, SEA FAN, SEA PEN, ZOOXANTHELLAE.

Coral Sea An area of the Pacific Ocean bounded by Australia's Great Barrier Reef to the west, Papua to the north, and the New Hebrides and New Caledonian island groups to the East. The coral of the Great Barrier Reef gives the sea its name. There are two other major coral reefs in this sea. It is also characterized by trenches and rises that indicate a relationship to Queensland, the northeastern part of Australia. A considerable portion of the Coral Sea plateau may be thought of as related to Tasmanian formations which once were terrestrial.

The Coral Sea was the site of a definitive naval battle in May 1942, which marked the limit of Japanese naval movement and prevented an invasion of Australia. *See* CORAL, GREAT BARRIER REEF, PACIFIC OCEAN, PLATE TECTONICS, TRENCH.

coralline algae Red, pink, or pinkish-gray algae encased in transparent calcareous (calcium-containing) shells that belong to the family Corallinacae and the phylum Rhodophyta. Coralline algae are colonial, and like coral polyps build up hard structures that become reefs. They are found in practically every oceanic environment. Some grow on any kind of rock (epiphytes), others grow on animals (epizoists), and some are parasites on other algae. In 1984 a coralline alga, a photosynthesizing plant, was found at a depth of 280 m (884 feet) in the Bahamas, a depth at which photosynthesis had previously been assumed to be impossible. *See* CORAL, PHOTOSYNTHESIS, POLYP, RHODOPHYTA.

Coriolis force A concept used to describe the motion of an object in a rotating system. In geophysics the term is used to indicate lateral displacement in what would otherwise be a north-south migration of air or water masses. For example, air or water moving from north to south over the surface of the Earth would be seen as moving in a straight line by an observer on the Earth. However, to an observer in space, the path of the moving air or water would seem to be curved because the Earth rotates. Looking in the direction of motion of the air (wind) or water, they would be seen to veer off to the east, or to the right, in the Northern

Hemisphere, and to the west, or left, in the Southern Hemisphere.

In addition, the fluid mass of the air or water is accelerating. The cause of this acceleration is the rotation of the Earth. As the entire Earth moves about its axis eastward, the fluids on it are pushed along at an increasing rate. This effect is called Coriolis acceleration. *See* WIND.

cormorant An aquatic, ducklike bird, 50 to 100 cm (20 to 40 inches) long belonging to the same order as pelicans. Cormorants have long necks and slender bills, and are usually black. The 26 species of cormorants both swim and fly, with the exception of the flightless Galápagos variety. Cormorants have a worldwide distribution and usually feed on fish, with some crustaceans and mollusks for variety. *See* DUCK, PELICAN.

Cortés, Hernán (1485–1547) Spanish explorer adventurer, conqueror of Mexico. At the beginning of his career he went to Cuba to make his fortune. The governor of Cuba sent Cortés to the Yucatán Peninsula of Mexico in 1519, but then reversed this order almost immediately. However, Cortés proceeded with the conquest of Mexico City and its surrounding areas and later claimed not to have received notice of the change.

Cortés made several forays into other parts of Central America, going overland into what is now Honduras. He crossed Mexico and had two primitive ships built on the Pacific side. He used these to explore the region around Baja California. He was the first European to see California (1536). The voyage, looking for spectacular treasure, the fabled lands of the Amazons, or the Strait of Anian (Northwest Passage), was unsuccessful. The politics of New Spain made it unsafe for him to continue his exploration. He returned to Spain in 1540.

cotidal points Those points along a coast that exhibit high tide at the same time. *See* TIDE.

Cousteau, Jacques-Yves (1910–1997)
A naval officer, undersea diver, and marine explorer who, together with Emile Gagnon, developed SCUBA (self-contained underwater breathing apparatus) diving in 1943. After World War II, Cousteau and others created the Undersea Research Group at Toulon, France, which became a center for research in undersea study and techniques.

Through his books and films, Cousteau has promoted interest in life in the sea and the great need for careful use of the marine environment. In 1968 he was asked to make a television series. For the next eight years, *The Undersea World of Jacques Cousteau* introduced the public to a world of sharks, whales, dolphins, sunken treasure, and coral reefs. In 1974 Cousteau started the Cousteau Society to protect ocean life. The membership of this non-profit group has grown to include more than 300,000 members worldwide. Cousteau was awarded the Medal of Freedom by President Ronald Reagan in 1985, and in 1989 he was honored by France with membership in the French Academy. His works have repeatedly stressed the need for intelligent use of the seas, control of pollution and overfishing, the use of mariculture, and the preservation of coastal areas. He was the director of the oceanographic institution founded by Albert I of Monaco. Cousteau died on June 25, 1997. *See* DIVING, MARICULTURE, SCUBA.

cowrie A marine snail of the family Cypraeidae. There are about 300 species of these beautiful animals. They are usually found in warm seas; the greatest number of species are from the eastern Pacific. Cowries have been traded extensively for many years; their fairly heavy, glossy shells have been used as decoration and coinage, including African decorations that have used Asiatic cowries. *See* GASTROPODA.

crab A crustacean with a broad, flat body, a tiny abdomen usually tucked under the cephalothorax, and five pairs of walk-ing legs. The crabs range in size from 1.5 mm (0.1 inch) to 3.5 m (12 feet) in diameter. True crabs, of which there are about 4,500 species, are distributed worldwide and have a range from mudflats to abyssal depths of 3,500 m or 12,000 feet. Some freshwater and terrestrial forms also exist, although most species are marine. They feed on a range of organisms from plankton to live prey or detritus.

True crabs have up to 19 pairs of appendages, such as antennae, antennules, mouthparts, gill ventilators (mandibles and maxillipeds), walking legs (pereopods and chelipeds), and swimmerets (pleopods), which are copulatory organs on the male and egg-carrying devices on the female. In the true crab only, the skeletal plate above the mouth is fused to the chitinous shell, or carapace.

The classification of crabs is subject to differing opinions. One scheme is as follows:

Phylum: Arthropoda
Class: Crustacea
Subclass: Malacostraca
Order: Decapoda
Suborder: Repantia
    Macrura—crayfish and lobsters
    Anomura—hermit crabs
    Brachyura—true crabs

The true crabs are then subdivided into the following groups:

1. Gymnopleura, primitive crabs of which there are about 30 species. They have long bodies and undeveloped chelipeds.
2. Dromiacea, also primitive but numbering about 250 species. On these crabs the fifth pair of walking legs is on the back. These legs hold a piece of camouflage, such as a leaf, clump of algae, or sponge.
3. Oxystomata, burrowing crabs constituting about 500 species. Some of these crabs also hold onto pieces of camouflage.
4. Brachygnatha, comprising about 80% of all crabs, and numbering about 3,600 species. These crabs have a

square mouthframe and have lost the first pair of swimmerets. This group is further divided into the (a) Oxyrhyncha, or Spider crabs, and other quick-moving crabs with hard shells, which have a triangular body shape, narrow head, and a body covered with spines and some camouflage, and the (b) Brachyrhyncha, the best known group of crabs, which range from pea crabs that live in the mantles of oysters and clams, feeding on whatever the oyster dropped, to the commercially important edible crabs, stone crab, blue crab, and Dungeness crab. The tropical, terrestrial Gecarcinidae return to the sea only to breed; the Jamaican tree crab and the Japanese spider crab are part of this large group.

See ARTHROPODA, CEPHALOTHORAX, CHELA, DECAPODA, INDIVIDUAL NAMES.

crabeater seal *Lobodon carinophagus,* is a misnamed Antarctic animal. This seal feeds on krill, not crabs, and has teeth that act collectively to filter these tiny animals from the seawater. *See* KRILL, SEAL.

Cretaceous period A period in the history of life on Earth, the Cretaceous began about 136 million years ago and lasted for about 70 million years. It is the most recent period of the Mesozoic era.

Geologically and geographically, the Cretaceous was a very eventful segment in the Earth's history. The southern continent, Gondwanaland, was still essentially intact, but the Indian subcontinent had detached and started drifting northward. By the mid-Cretaceous, South America had moved away from Africa. North America began to move west, away from Europe, opening the Atlantic, while Africa moved north toward Europe, closing the Tethys Sea and beginning the folding that resulted in the Alps.

The folding of the North American continent that eventually produced the Rocky Mountains also began in the Cretaceous. By the end of the Cretaceous, New Zealand had separated from Australia and Australia itself was moving away from Antarctica.

The climate during most of the Cretaceous was warm. Shallow seas extended over large areas of all continents. The dominant sea animals were the ammonites. Belemnites, brachiopods, echinoderms, and mollusks were widely dispersed. Seagoing reptiles such as the ichthyosaur and the plesiosaur were alive. On land, the dinosaurs reached their zenith during this period, and by its end were extinct. The end of the Cretaceous was marked by the extinction of large numbers of species. *See* EXTINCTION, *Appendix: Geologic Timescale.*

Crinoidea Class of filter-feeding echinoderms especially abundant in the western Pacific Ocean that was once thought to be extinct; the first description of this class of these animals was by Agricola (1546), and they were later depicted in Gesner's encyclopedia (1565). A living feather star was described in 1592 by Fabio Columna, who thought it was a variety of sea star (starfish). It was not until 1761 that a living stalked crinoid was found. This group of very ancient sea creatures includes the stalked, sessile (stationary) sea lily and the mobile feather stars. In fact, the unstalked feather stars account for most of the 690 living crinoid species. The Mesozoic era was the high point in crinoid development. The greatest number of species—more than 5,000 have been identified—and individuals lived then.

Crinoids are covered by unique exterior calcareous plates held together by collagen ligaments. Broken body parts regenerate. The rate of regeneration has recently been studied using the evidence of such rebuilding of body parts in the Paleozoic. Studies of crinoids from the Ordovician to Pennsylvanian periods—approximately 490 million to 290 million years ago—have shown that regeneration rates increased markedly after the appearance of predatory shell-crushing fish. This occurred in the mid-Paleozoic Era, around 380 million years ago, and is referred to as the Paleozoic marine revolution. Ordinarily, the crinoids are avoided by predators; it is thought that the animals that live on the crinoids were the intended targets, and the crinoids were broken in these encounters. The ultimate result was that the crinoids that could best regenerate were the ones that survived. This is an example of predator-prey–driven evolution. *See* ECHINODERMATA, PREDATOR-PREY RELATIONSHIPS.

crinozoa A subphylum of Echinodermata.

Cromwell Current An equatorial, subsurface flow that appears in three major oceans. The Cromwell Current is the equatorial eastward-flowing undercurrent embedded in the westward moving surface current. It is about 300 m (900 feet) wide and about 20 m (70 feet) deep, and moves along at a rate of 2 to 3 knots. *See* CURRENTS.

Crossopterygii An ancient subclass of the bony fishes. These animals are related to lungfish and are sometimes called lobefish. The coelacanth is the only survivor. The Crossopterygii are important because they were the probable progenitors of amphibia. The dorsal fins have fleshy pedicels from which the fin rays emerge. These may have been the precursors of limbs. *See* COELACANTH.

crown of thorns A large sea star, (*Acanthaster planci*) averaging 40 to 45 cm (16 to 18 inches) in diameter, with 12 to 20 arms and covered with red, thick, strong spines. This echinoderm feeds on living coral. A large increase in the crown of thorns population in the 1960s in Australia resulted in the destruction of large areas of coral on the Great Barrier Reef. One possible reason for the population boom may have been the removal of many triton snails by shell collectors. More normal reef conditions have brought the predator population into better control. The crown of thorns overpopulation had also affected other areas in the Pacific where the coral is now recovering. *See* CORAL, ECHINODERMATA, SEA STARS.

crust The topmost layer of the Earth. The ocean floor is basaltic; the continental

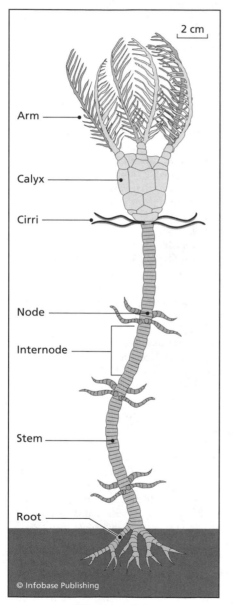

2 cm

Arm

Calyx

Cirri

Node

Internode

Stem

Root

© Infobase Publishing

Crinoid (sea lily)

crust, which is much thicker than that of the ocean, is largely granitic. *See* DISCONTINUITY LAYER, MOHO, OCEAN FLOOR.

Crustacea A worldwide class of arthropods that includes crabs, shrimps, lobsters, and barnacles. There are eight subclasses of the Crustacea and over 25,000 species. They are present in every type of marine biome and are the aquatic analogs of insects. Both fill their respective ecological niches with incredible numbers of species and numbers of individuals. Most crustaceans are aquatic and breathe through brachioles leading from gills. They have five pairs of cephalic appendages: antennae or antennules (or both), mandibles, and maxillae or maxillules or both. The body is divided into a head, thorax, and abdomen and is covered by a chitinous exoskeleton. The last body segment (somite) is the telson or tail, which lacks any appendages. The crustaceans have the typical arthropod circulatory system, with an open heart.

The excretory system consists of a primitive network leading to a pair of glands under the antennules. There is one simple median eye and two compound lateral eyes. Generally, the antennae are more effective in testing the environment than are the eyes. Most crustaceans are dieceous (having distinct male and female forms). Fertilization of the eggs is internal. The female is the usual brooder, carrying the eggs about on her abdominal appendages, which affords some protection for the eggs until they hatch into a larval form. The larvae are called nauplia (singular, nauplius). Successive molts and growth spurts produce the adult. The rate of growth varies with different species, and depends on the availability of food and the water temperature.

While the larvae of most crustaceans are pelagic filter feeders, the adults may develop into an astonishing variety of diggers, borers, or raptors while some remain filter feeders. The adult forms include, in addition to the usual motile animals, parasites (copepods) and sessile (stationary) forms (barnacles). Most crustaceans are grazing and scavenging animals, although the larger, more advanced decapods (lobsters, crabs, shrimp) and stomatopods are more likely to be predators.

The crustacea and other arthropods seem to share an ancestor with annelids. They were certainly present and well-differentiated in the Cambrian Period. Shells from this and other periods are well preserved, and the resulting fossil record is quite complete. *See* ANNELIDA, ANTHROPODA, CRAB, CHITIN, LOBSTER, SHRIMP.

Ctenophora A phylum of marine plankton abundant on the surface, but they may also found in deep water. Seven orders and 19 families have been described. Ctenophora are macroplanktonic, voracious, carnivorous eaters, using their radial arms, which have coeloblasts (adhesive cells), to trap food. Unlike the cnidarians that they resemble, they do not have stinging cells.

Ctenophora have biradial symmetry and a medusa-like three-layer body, the exterior and interior of which are composed of epithelial cells. The intermediate layer, the mesoglea, is gelatinous. Ctenophores are characterized by their body shapes. They can be rather large: The girdle of Venus is a ribbon-like organism of transparent blue and green that grows to over 1 m (3 feet) in length. *See* CNIDARIA, COMB JELLIES, PLANKTON.

Cubozoa A class of the Cnidaria. These animals are in intermediate form that falls between the Hydrozoa and the Scyphozoa. The eggs hatch into four-tentacled polyps that grow fairly rapidly into the adult form. When adult, they are a four-tentacled medusa with a "cube-shaped" body. The Cubozoa are among the most toxic of marine organisms. *See* ANTHOZOA, CNIDARIA, SCYPHOZOA, TOXINS.

currents Parts of a fluid body (air or water) moving continuously in a certain direction. In oceans, currents are responsible for the circulation of vast quantities of waters. They are caused by several factors. The primary one is wind, which creates a friction effect by pushing the mass of water. Another factor is temperature, because cold water, like cold air, flows toward the equator and sinks, since (like cold air) it is denser than warm water.

Local currents are produced by the activity of volcanoes, a sporadic phenomenon; by river systems, whose effect is permanent; and by tidal currents. The last are periodic. In general, oceanic circulation is horizontal. The movement of water is anticyclonic (clockwise) in the Northern Hemisphere. Water moves eastward at higher latitudes (i.e., the Gulf Stream) and returns (moves west) closer to the equator. The east-to-west movement in the Southern Hemisphere is cyclonic (counterclockwise). This gyre is typical of the Atlantic and Pacific Oceans. It reverses itself semiannually in the Indian Ocean because of the monsoon winds. The Antarctic region has a single current: the Antarctic Circumpolar Current. On the whole, less water is transported by gyre in the Southern Hemisphere than in the Northern one. The speed of currents in open ocean is relatively slow, less than 1 knot (0.5 m/second), as opposed to the Gulf Stream, where it is 2 to 4 knots, or 1 to 2 m/second. The Pacific counterpart of the Gulf Stream is the Kuroshio. This current moves northeast out of the South China Sea to warm the Japanese islands and then moves east to the Aleutians. It, too, moves rapidly in comparison to the speed of the average ocean gyre.

In addition to surface currents, subsurface equatorial currents such as the Cromwell are known. Deepwater circulation is slower than surface currents. It is the result of polar water sinking, a phenomenon observed in the Norwegian and Weddell Seas. The cold deepwater mass then moves toward the equator.

While this is the overall ocean pattern, it is very much influenced by conditions in particular locales. The conditions of tide, temperature of the abutting landmass, and the size of the land formation all have profound influences on deepwater currents. The arrangement of continents in the Northern Hemisphere makes it easier to see the circulatory patterns in the ocean, the gyres. Currents are responsible for the deposition of sediments on secondary coasts—those that are being eroded. Sediment can be moved along a coast by wave action in the surf zone, where there are breakers. This is longshore drift. It moves the sand on a beach and may build up sand spits or bars at bay openings or barrier islands. This is the natural evolution of a coast. *See* ANTARCTIC CURRENT, COASTS, EKMAN SPIRAL, EQUATORIAL CURRENT, GYRE, LONGSHORE CURRENT, RIP CURRENT, WIND.

cutter A small ship built for speed. Historically the cutter was a deep, narrow sailing vessel carrying a single mast rigged fore and aft, and had a long bowsprit. Today the U.S. Coast Guard uses diesel-powered steel-hulled cutters that are 25 m (83 feet) or less long.

cuttlefish A cephalopod, not a fish, of the order Sepioidea, related to the octopus and squid. Its origins are in the Miocene. Cuttlefish comprise about 100 species of warm-water bottom dwellers. The cuttlebone is really a shell that has become internalized. It acts as a flexible support rod. Like all other mollusks, cuttlefish produce a calcareous shell by secretions from their mantle.

The various species of cuttlefish range in size from 2.5 to 9 cm (1 to 36 inches) long. They have eight arms, two longer tentacles, and a pair of lateral fringing fins. They eat small fish, crustaceans, and their own young, and are in turn fed upon by large fish and mammals. *See* MOLLUSCA, OCTOPUS.

Cuvier, Georges Léopold Dagobert, Baron (1769–1832) The French founder of the science of paleontology. Cuvier's interest in natural science was evident early in his life; he studied comparative anatomy at Stuttgart, graduating in 1788. For the next seven years he taught privately and studied the Mollusca. This work led to a post at the Museum of Natural History in Paris. He remained at that post for the rest of his long career.

Cuvier's early published works were popular successes. The *Tableaux élémentaire de l'histoire naturelle des animaux* (1797) was followed in 1800–1805 by *Leçons d'anatomie comparée*. In this work Cuvier correlated the structures of a body and showed how they were related and interdependent, concluding that the body plan of an animal was determined by the particular ecological niche it inhabited.

Cuvier created four categories of animals: vertebrates, articulates, mollusks, and radiates. Each, according to him, had distinctive features that were common within the group and different from the distinctive features of the other groups. He devised a table of organization, placing all known animals, living and extinct, on it. This work was a major improvement over existing attempts at taxonomic charting.

In explaining the origins of animals, Cuvier was a catastrophist, holding that species were all created in the form in which he knew them. Thus, fossils were remains of species that had been wiped out in some catastrophic event in the past.

While Cuvier's theory would not have been able to compete with evolutionary theory as an explanation of interrelationships between organisms, his work on comparative anatomy is still basic to that science. His attention to minute features of the structure of an unknown bone could lead him to describe the organism it came from and the habitat it lived in.

In addition to his scientific work and publications, Cuvier was instrumental in establishing universities in the French provinces. He also achieved a general standardizing and improving of the level of public instruction. *See* BUFFON, GEORGES-LOUIS LE CLERC, COMTE DE; EVOLUTION.

Cyanobacteria These organisms, whose name is derived from the Greek word for "blue" may be the largely unchanged descendents of the original life on Earth. Their taxonomy is subject to debate. The cyanobacteria separated from the Archaea about 3 billion years ago. The oldest fossil forms found are about 2.8 billion years old. The difference between the cyanophyta and the archaeobacteria is that the cyanophyta can photosynthesize in an atmosphere that contains oxygen, and that activity creates more oxygen. This was the fundamental shift in atmospheric chemistry, and these organisms caused it, setting the stage for fundamental changes in life on Earth and the very structure of the Earth itself.

The cyanobacteria are prokaryotes; they have a cell wall composed of murein surrounding the protoplasm. There is a central body where most of the chromatin and the ribosomes are found, and pigment rods or granules are scattered throughout the protoplast. These pigments are chlorophyll—essential for photosynthesis and carotenoids. Cyanophytes may also have other pigments; phycocyanins (blue pigments) and phycoerythrins (red pigments)—and are therefore not all blue-green; their colors range from blue-green to red, purple, yellow, brown, or almost black. They reproduce as many bacteria do, by separating the protoplast into two new individuals.

While blue-green algae were known before 1970, they were associated with freshwater and not found in ocean water until the development of better collecting methods. These organisms are an essential part of the picoplankton and the basis of oceanic food webs. They are essential in both the carbon cycle, in which bacterial action uses CO_2 to produce organic matter, and in the nitrogen cycle, where atmospheric N_2 is incorporated in cellular structure.

These versatile organisms can live wherever there is moisture and sufficient light for photosynthesis. This means they can inhabit almost every environment, freshwater or salt, clean or polluted, and very hot or very cold biomes. Some form commensal relationships with algae and plants.

Large aggregates of cyanobacteria are the dark mats commonly found in tidal pools at the ocean's shore. *See* ATMOSPHERE, BACTERIA, CARBON CYCLE, CARBON DIOXIDE, CHLOROPHYLL, COMMENSAL RELATIONSHIP, MUREIN, NITROGEN CYCLE, PHOTOSYNTHESIS, PIGMENT, PLANKTON, PROKARYOTE.

Cyclades A group of more than 50 islands in the Greek archipelago. They are of volcanic origin and have quite different aspects. The probable cause of the tsunami, or tidal wave, that shook the ancient world—and inundated most of it—was a volcanic eruption on Thera (Santorin), an island in this group. Other islands are Siros, Delos, Naxos, and Paros (where Parian marble comes from).

The beginnings of long-distance sailing in the Mediterranean probably developed here in the third century B.C.E *See* AEGEAN, ATLANTIS, ISLANDS, THERA, TSUNAMI.

cyclone A Pacific or Indian Ocean storm that has wind velocities above 75 knots per hour. *See* HURRICANE, STORM.

Cyclostomata A class of jawless chordates, the lampreys and hagfish. The other agnathans or jawless fish are all extinct. *See* AGNATHA, CONODONT, HAGFISH, LAMPREY.

cypris The larval form of the barnacle. It is freely moving. *See* BARNACLE.

Dampier, William (1652–1715) An English buccaneer, navigator, and surveyor, Dampier sailed to Newfoundland and then to Java on commercial vessels before serving the English crown in the Dutch Wars. He joined the pirates of Campeche (Mexico) and with them sailed to the South Pacific in 1683. It was an adventurous trip that took him to China, the New Hebrides, the Australian mainland, and New Guinea. Dampier was shipwrecked and skillfully navigated a canoe to Sumatra, finally returning to England in 1692. His account of this voyage was published in 1697.

After the publication of his memoir of the voyage, *A New Voyage Round the World*, the Admiralty sent Dampier out as captain of the *Roebuck,* with the charge of exploring New Holland (Australia). The unsuccessful voyage included a mutiny and the loss of the ship. Dampier, who seemed to have a charmed life, limped back to England and embarked on yet another unsuccessful voyage in 1703. The real objective of this expedition was piracy. It was on this trip that Alexander Selkirk, the real-life Robinson Crusoe, was put on an uninhabited island. Dampier's last voyage in 1711 was a financial success, and also brought Selkirk back to England. *See* JUAN FERNANDEZ ISLANDS.

damselfish The Pomacentridae, a colorful family of small tropical bony fish, usually found on coral reefs. Some species have defensive spines. Their maximum length is about 35 cm (13 inches).

Daphnia A freshwater crustacean (Cladocera) also called a water flea.

dark reaction, photosynthesis *See* CALVIN CYCLE.

Darwin, Charles Robert (1809–1882) An English naturalist who first formulated the theory of evolution by natural selection. He also produced major works on barnacles, orchids, and climbing plants. Darwin came from a distinguished family: His grandfather, Erasmus Darwin, was a physician and botanist; his father was a physician; and his mother was a daughter of Josiah Wedgwood, the potter, chemist, and ceramicist. After abandoning a medical education, Darwin intended to be a clergyman when J. S. Henslow, a prominent natural scientist, suggested him for the post of accompanying naturalist on the HMS *Beagle,* a position for which he was accepted by the captain, Robert Fitzroy. Darwin's voyage on the *Beagle* turned out to be one of the most fruitful "field trips" ever undertaken in terms of data acquired and the far-reaching conclusions that could be drawn.

Darwin's later work was based largely on observations made during the long voyage on the *Beagle,* 1831–36. He also proposed a theory of coral reef and island formation which is in large part accepted today. He believed that the coral atoll began as a fringing reef surrounding a high island. The reef accreted, and the island eventually subsided. While the sequence may not follow Darwinian mechanics, both the processes of accretion and subsidence do occur. Darwin did not know of tectonic plates and ridge-building mechanisms.

The Darwinian explanation of the differentiation of species, the evolution of current forms from earlier ones, and the process of change was based on his studies of fossils in South America, isolated bird communities on the Galapagos Islands, and careful observation of organisms in tide pools. His work, *On the Origin of Species*, finally appeared in 1859, at the urging of friends. *See Beagle*, HMS; CORAL; ENIWETOK; EVOLUTION; FOSSIL; ISLANDS; WALLACE, ALFRED RUSSEL.

Dasycladales An order of Chlorophycota and the only member of that class present in the fossil record extending back to the Carboniferous. These organisms are characterized by the presence of a thallus (an anchoring base) that is usually calcified. Attached to the thallus is an axis covered with whorls of "branches." The *Acetabularia* (mermaid's wineglass) is the best-known genus. It is found in the Mediterranean and was of great interest to early naturalists.

dating The placement of a specimen in time. The dating of marine specimens relies on the differences in the ratio of the concentration of radioactive to stable isotopes of certain elements in surface samples and core samples. This, in turn, is based on several assumptions about the constancy of radionuclide decay and its inclusion into the living systems of ocean waters. One assumes that a radioactive nuclear sample will decay (break down into smaller nuclei and emit alpha, beta, and gamma rays) at a constant rate. This is a "clock." One also assumes that if a certain element is incorporated into a living plant or animal, that too happens at a constant rate. The latter is subject to much more change. Plant and/or animal populations move. They also mutate. Both of these factors can alter the uptake of a particular element considerably.

The comparison of various ocean waters and the loads of radioactive species of different elements in these waters is a clue to the movement of water from one oceanic area to another and to the rate of water turnover in any particular locale. *See* ISOTOPE, RADIOACTIVITY.

Davidson Current A deep countercurrent (200 m or 650 feet deep) in the temperate Pacific. The current runs north, parallel to the California coast. It appears on the ocean surface north of 35° north Latitude in winter. *See* BOTTOM WATERS, CALIFORNIA CURRENT, CURRENTS.

Davis, John (1543–1605) An English pilot, master mariner, and arctic explorer. Davis's voyages contributed considerably to the knowledge of the arctic.

His first voyage to the far north began in June 1585 and had the stated purpose of finding the Northwest Passage. Instead, he charted many features of the arctic, including the strait that is named for him. He returned to England and made two similar voyages in the two succeeding years.

Davis took part in the fight against the Spanish Armada in 1588. His next exploratory voyage was with Cavendish in 1591. After Cavendish turned back to Europe, Davis continued on into polar (southern) waters and was also driven back by weather. On his return trip to England, he discovered the Falkland Islands.

After his return, Davis found employment in the Far East, where he sailed for both Dutch or English commercial venturers. He was killed by pirates near Sumatra.

Davis was a superb pilot and a skilled navigator. His permanent contribution to seamanship included two books on navigation and development of the backstaff. The French called this navigational aid the English quadrant, the English called it the Davis quadrant. With it one could sight the Sun and measure elevation above the horizon, and therefore calculate latitude. *See* DAVIS QUADRANT, NAVIGATION, SEXTANT.

Davis quadrant Also called the backstaff. It is an instrument developed in the 16th century by John Davis as a means for measuring the altitude of the Sun. There were two versions: One measured altitudes of angles less than 45°, the other angles greater than 45°. The latter is the simpler instrument. It consists of a horizontal vane and a graduated staff held horizontally. With the viewer's back to the Sun, the shadow of the horizon vane was aligned with the graduations on the staff.

The 90° quadrant, which was a later development, consists of two half-transoms, one straight, the other an arc. The straight part is held vertical and perpendicular to the staff and slides along the staff. The arc is fixed to the lower end of the staff; it, too, functions as a horizon vane. As in the simpler instrument, the shadow of the horizon vane is compared to the graduations on the vertical staff. *See* ASTROLABE; DAVIS, JOHN; LATITUDE; SEXTANT.

Davis Strait Part of the fabled Northwest Passage, the strait is the boundary between Greenland and Baffin Island. The strait is navigable for about one month after the summer solstice. *See* DAVIS, JOHN; EXPLORERS AND EXPLORATIONS; NORTHWEST PASSAGE.

dead reckoning *See* NAVIGATION.

Dead Sea A landlocked lake on boundary between Israel and Jordan, part of the African Rift Valley, once a part of the Red Sea but cut off from the Red Sea by steep, barren hills. As a result of its high rate of evaporation, the water in this lake is the densest on Earth. Minerals are extracted commercially on the shore of the Dead Sea. *See* AFRICAN PLATE, EVAPORATES, MINERALS, RIFT VALLEY.

dead zone An area of coastal water in which an excess of nutrients from agriculture, municipal, and industrial waste has triggered algal bloom and led to a net loss of aquatic life. *See* POLLUTION.

Dease Strait A strait in the Canadian Northwest Territories, lying northwest of Hudson Bay and forming part of the "southern route" of the passage from the Atlantic to the Pacific Ocean. While the passage is possible, it is not practical or commercially viable, since the route is icebound for most of the year. *See* NORTHWEST PASSAGE.

Decapoda The best-known order of the Crustacea, comprising the crustaceans most often consumed by humans: the lobsters, crabs, and shrimp. The decapoda have a fused cephalothorax covered by a chitinous carapace, and they have five pairs of legs. The arrangement of the appendages into swimming and walking legs varies with the individual species. *See* CRAB, LOBSTER, SHRIMP.

decomposers Some multicellular animals, fungi, and bacteria that utilize the tissues of dead animals and plants for their own metabolic needs. They eventually break down complex proteins into simple materials that are essential for plant growth. *See* ECOSYSTEM.

decompression A term that refers to the controlled or uncontrolled release from pressure that a diver experiences during and after rising to the surface from ocean depths. If an organism is raised from a depth, the dissolved compressed gases in its tissues will rapidly come to pressure equilibrium with the atmosphere and form bubbles as the gases expand. Thus, if a diver rises rapidly, the gas dissolved in his body will accommodate to the ambient pressure, and bubbles will form in his blood or whatever organ contains the gas. Symptoms of decompression disease, known as "the bends," arise hours after decompression and may be fatal.

By 1907 the British physiologist J. S. Haldane had noted that slow rising in stages will eliminate the bubble formation and thus eliminate decompression sickness.

Most organisms would appear to be subject to decompression distress. The hadal (abyssal) bacteria will die at surface pressure. Whales and dolphins both exhibit deep-diving and surfacing behavior, but for reasons that are not well understood they do not develop decompression sickness. It is theorized that they may have no "nuclei" around which gas bubbles might form, as well as a circulatory system that operates so smoothly that no eddies form in the blood which might give rise to gas bubbles.

An alternative mode of decompression, one that would enable a diver to rise rapidly if this should become necessary, involves the use of a hyperbaric (greater than normal pressure) chamber. If a diver, immediately after ascending, is put into a pressurized environment in which the pressure is gradually dropped to the normal 1 atm (atmosphere), there is no decompression distress. See DIVING.

deep-scattering layers Ocean layers at depths greater than 200 m (660 feet) that produce a sonic effect detectable by echo-sounding equipment. There may be one or more deep scattering layers at any given location, and these layers move up in daylight or bright moonlight. They then descend at night or on heavily overcast days. The movement corresponds to that of schools of small fish, cnidaria (jellyfish), and possibly squid and larger crustaceans. The sonic difference is attributed to the echo produced by gas bubbles within the animals. See PLANKTON, SOUNDING, VERTICAL MIGRATION.

deep sea A sea zone at depths below approximately 3,000 m (6,600 feet), or the 4°C isotherm. The sea zone at depths below 6,000 meters (19,800 feet) is hadal. The depth of these deep sea zones is set by international agreement. However, there is usually no plant life at depths where there is no light. The bacteria, protists, and animals in deep seas are adapted to both pressure and darkness. They represent all major invertebrate phyla and many fish. The deeper an organism's environment, the greater the likelihood that it is a detritus feeder, living on pieces of dead plants and animals, or a deposit feeder—a mudeater. Some species that are carnivores near the surface may have deposit-feeding hadal relatives; for example, mud-eating sea stars live below 4,000 m. See ABYSSAL ENVIRONMENT, OCEAN FLOOR, SEAMOUNT.

deep-sea fauna See BENTHIC ORGANISMS.

defense mechanisms Means by which organisms defend themselves against predators. Some organisms use camouflage or move away rapidly, some "clam up," and others, particularly slow or nonmotile organisms, use chemical defenses. For example, some brown algae use chemical defenses to make them unpalatable to predators, several species of sole produce shark repellants, and some sessile (stationary) organisms produce toxic substances. See CHEMICAL DEFENSES, PREDATOR-PREY RELATIONSHIP.

Delaware Bay A bay of the North Atlantic at 39° north latitude, separating the state of New Jersey from the eastern shore of the Delmarva peninsula (Delaware/Maryland/Virginia). The Delaware River is the major source of terrestrial water to Delaware Bay. The bay, in spite of its location as the water-catchment basin for water draining past the heavily industrialized areas of Marcus Hook, Pennsylvania, and Wilmington, Delaware, has a wetland and swamp region surrounding it. These wild areas serve as breeding grounds for many species of plant and estuarine animal life. See CHESAPEAKE BAY, COAST, ESTUARY.

delta A fan-shaped plain at the mouth of a river, created by sediment deposited by the river. In general, the foundation of a delta depends on the rate of sedimentation and the currents of the sea at the point at which the river discharges its deposit. Thus, a strong current can prevent the formation of a delta or cause the erosion of an existing one. The movement of sediment in a delta and its rearrangement by local currents causes continuous change in the local coastline. See BREAKWATER, COAST, FAN, SEDIMENT.

Demerara Abyssal Plain A part of the Guyana Basin off the northeastern coast of South America. One side of this plain is the abyssal cone of accumulated sediment brought downriver by the Amazon. The plain is also fed by the waters of the Orinoco and other river systems of northeastern South America. *See* AMAZON RIVER, ATLANTIC OCEAN, ORINOCO RIVER.

Demospongia A class of the phylum Porifera. This group of sponges is a fairly large one, with about 9,500 species. They are classified by the type of spicule (support structure) they have: Some have silicate spicules and others have spicules made of spongin, a rigid protein, while still others have no spicules. They range in size from thin layers a few centimeters across to cake-shaped masses more than 2 m (7 feet) wide. Their mode of sexual reproduction is oviparous or viviparous; larval cells are then rearranged to their adult—sessile—position. Over 90% of all existing sponges are in this class. The group has evolved from Cambrian sponges and has a 500 million–year history. *See* PORIFERA, SPICULE, SPONGE.

Dendroceratida An order of sponges. These organisms have no hard skeleton but one of spongin (a protein containing sulfur) fibers which are difficult to tear or cut. They protect themselves from predators by chemical means—they produce an array of unpalatable terpenoid compounds. The three families of this order are abundant in tropical and subtropical shallow water, and most individuals found are between 20 and 50 m (65–170 feet). The deepest an individual was found was at 750 m (2,475 feet). This is the commercially significant sponge.

density The mass of a substance per unit volume. The density of seawater is a function of salinity, temperature, and pressure. The vertical change in density is greater at lower latitudes, (i.e. near the equator), since the temperature of the water varies so much with depth. In colder regions the density of a vertical column of seawater is fairly constant. An increased salt content increases the density of seawater. Thus, the introduction of large quantities of river water or glacial water (icebergs) will decrease the density of the water locally.

In areas where the temperature is high and humidity low there is considerable water loss due to evaporation. This increases the density of the water locally. The locale in which this occurs can be quite large—the Mediterranean Sea is one such high-density, high-salt area.

Increased pressure will also increase the density of seawater. However, liquids are not very compressible, and the increase in density due to pressure is therefore small and negligible compared to that caused by salt and temperature effects.

deposits *See* MINERALS, SEDIMENTS.

Deryugin, Konstantin Mikhailovitch (1878–1938) An outstanding Russian oceanographer and expert on the oceanography of the White Sea. In 1899, while still a student, he made his first trip to the White Sea. He took his doctorate at the University in St. Petersburg, and remained there as a professor for the rest of his career. Starting in 1920 he was also the manager of the Oceanic Division of the State Hydrological Institute in Leningrad.

Deryugin's first major scientific reputation rested on his work on the classification of aquatic arctic biota. He went on to study the entire ecosystem of the Barents Sea: its biota, shoreline, bottom, geology, hydrology, and chemistry. In 1921 he reestablished the water-sampling station on the Kola meridian, at 33°30' east latitude, 75° north latitude. This station sampled the North Cape Current, whose direction, temperature, and water volume have a profound effect on the quality and quantity of fish in the Barents Sea.

By 1922, Deryugin transferred his scientific interests to the White Sea. This body of water is unique both hydrologically and biologically. The principal reason for its singular composition is the Gorlo Strait, which lies between the White and Barents seas and acts as a biological barrier. The strait does not have a high sill: The barrier is instead tidal.

Deryugin was in charge of the organization of the Pacific Ocean Expedition of 1932–33. The explorations of the Sea of Japan, the Bering, and the Chukchi Sea brought a "new world of organisms" (Deryugin's comment) to scientific scrutiny.

This able organizer was also instrumental in the direction of more than 50 expeditions and was a pioneer in the methodology of oceanography. *See* BARENTS SEA, BERING SEA, CHUKCHI SEA, WHITE SEA.

desalination The process of removing salt and other dissolved minerals from seawater to make it useful for agriculture, industry, and human use. This is an energy-intensive procedure and therefore an expensive one.

The process used in desalination is either distillation (controlled boiling) or reversed osmosis. In distillation, the seawater is boiled and the resulting steam is forced to leave the boiler through a conduit, which is cooled by surrounding air or by water. The dissolved or suspended minerals remain behind in the boiler, while the cooled, condensed water is collected at the delivery end of the conduit as potable water.

A simple distillation apparatus, which is both affordable and portable, uses solar energy to raise the temperature of the water in a plastic boiler. The use of plastic cuts down on maintenance costs, which are considerable if metal is used, since the dissolved materials in seawater are frequently corrosive. The water does not necessarily have to boil as long as the temperature is high, since evaporation will still occur and the distillation apparatus will still work, albeit more slowly. In regions of high insolation there is sufficient light and heat to power a still. The removal of the mineral residue of seawater and its separation is another industrial process. It is also possible to freeze out fresh water, which freezes at a higher temperature than salt water. This is less successful than the evaporation process.

In the process of reversed osmosis, saline water is forced through a membrane while under pressure. This concentrates the minerals and releases potable water. The membrane simulates a cell membrane, which ordinarily allows water flow from regions of high salt and other solute concentration to regions of lower concentration. Commercial membranes are made of cellulose acetate or nylon. Pressure applied to the flow of water pushes desalinated water out of the membrane container. This process may also be solar-powered and may be used for collecting minerals.

The electrolytic separation of water as a means for purifying it is economical enough to be used in large-scale operations. By passing a current through a sample of water it can be separated into its component elements, hydrogen and oxygen. *See* WATER.

detritus Loose, unconsolidated material that is either finely divided rock or the finely divided remains of animal or plant tissue or both. *See* ECOSYSTEM, *MARINE SNOW*.

deuterium An isotope of the element hydrogen, also referred to as heavy hydrogen. The atomic mass of this naturally occurring isotope is 2 daltons, as the result of an extra neutron in the nucleus of the deuterium atom, whereas the usual form of hydrogen has a mass of 1 dalton, from the single proton that constitutes the nucleus of its atom. Deuterium forms the same compounds as does the more common isotope of hydrogen. *See* DATING, HYDROGEN.

Devonian period A period in the Earth's history from 400 million to 345 million years ago, a part of the Paleozoic era that preceded the Carboniferous period. The English geologists Adam Sedgwick and Roderick Murchison worked on the classification of Devonian rocks and named the period for the distinctive rocks and fossils in Devon, England.

The Devonian period is called the age of the sea, since more of the Earth was under water then than now. During the Devonian, the large landmasses of Europe and North America were in contact. The southern continents were all grouped at the equator. The Tethys Sea was the major divider of the landmasses. The face of the Earth changed dramatically in the middle Devonian as the northern Appalachian buildup occurred. Major changes in South America, Asia, and Australia resulted from violent volcanic and seismic activity. The rise of mountains trapped seawater in North American and Siberian basins, while coastal erosion produced much of the red sediment now found in basement ocean floors.

The biota of the Earth was also rapidly diversifying during the Devonian. While trilobites and grapholites declined and disappeared, conodonts, brachiopods, and cephalopods became numerous and varied. Corals appeared in many new species, taking advantage of the warm, shallow marine environments. There was spectacular diversification of fish in the mid to late Devonian, which explains why the Devonian is perhaps best known as the age of fish. Marine forms increased in number of species and number of individuals. Some moved into areas that were occasionally out of water and eventually developed as lungfish, while some of them ultimately became amphibians. Others moved into brackish and then freshwater habitats.

The end of the Devonian was a sharp break in evolutionary development. Catastrophic extinctions occurred. Several theories, including one of meteorite bombardment, have been advanced as causes for this. Whatever the reason(s), many corals and attendant reef animals and plants ceased to exist during the end of the Devonian, as did many cephalopods and other invertebrates. See CEPHALOPODA, *Appendix: Geologic Timescale.*

dewpoint The temperature at which the air is completely saturated with water vapor. If cooling occurs, the water condenses as dew on land or as fog near coasts and over cold, open oceans.

The dewpoint can be read directly using a smooth metal plate and a thermometer. The temperature at which there is a film of water condensed on the plate is the dewpoint. *See* AIR, CONDENSATION, FOG, HUMIDITY.

Dezhnev, Semyon Ivanov (also spelled Dezhnyov) (1605?–1673) A Russian (Siberian) explorer who made the first transit of the Bering Strait. Cape Dezhnev, the easternmost point of the Chukchi Peninsula in Siberia, which, in turn, is the easternmost point of Asia, is named for him. In English atlases it is called the East Cape. Dezhnev, sailing from the Lena River in a series of voyages, rounded the East Cape in 1648–49, sailing with F. A. Popov. Their account of this voyage was lost and not found until after Bering's transit of the same strait. *See* BERING, VITUS.

dhow A sailing vessel used on the Red Sea and in Arabian waters. The dhow is designed to beat against the wind, particularly in regions of very gentle winds. The usual dhow has one mast, although there may be two, and the masts are lateen rigged, carrying triangular sails. The mainsail is very large and intended to catch whatever wind is available.

Diadematoida The order of deepwater echinoderms that are extremely photosensitive.

diapause A period of cessation or suspension of metabolic activity. Insects exhibit this ability to lie dormant for long periods, as do some coastal copepods. Spore formation in yeasts may "come to life" years after formation. In 2005 bacteria thousands of years old were revived from ice from an Alaskan glacier. Ocean bottom sediments that are very old have been brought to the surface, and organisms were found that are still living albeit in diapause. These diverse organisms retrieved from a sediment core at a depth of 5,900 m (19,600 ft) were cultured. The diverse sample containing bacteria, fungi, and radiolarians was brought up from layers of Earth that are between 0.43 to 0.18 million years old. It is now known that there are diverse but sparse communities living beneath the sediment, subsisting on the basement rock. *See* PROTEOBACTERIA, SEDIMENT.

Dias, Bartholomeu (or Diaz) (1450?–1500)

A Portuguese navigator who first sailed around the Cape of Good Hope. The Portuguese empire had established trading positions on the Guinea coast of Africa when the king, Alfonso V, assigned further exploration to his son and heir, Prince John, later crowned King John II. It was then thought that Africa and India were connected by land, and that an overland trek from the West African coast would eventually end at Abyssinia and India. Somewhere between the latter two places, one might find the legendary Christian kingdom of Prester John.

Prince (later King) John (or Infante João) sent Alfonso Paiva, a military commander, overland and Dias by a sea route. Part of Dias's charge was to explore the west coast of Africa south of 15° south latitude. The flotilla of three ships left Portugal in August 1487. They rounded the Cape of Good Hope in a storm, probably in the first week of February 1488, but did not actually see the Cape. They did sight it on the return, having established the possibility of a sea route to India. The name originally given the Cape by Dias was Cape of Storms. It was possibly King John II who settled on the present name.

In 1497 Vasco da Gama led a large flotilla to India with Dias as captain of one of the ships. The armada sailed too far west and actually sighted the Brazilian coast before rounding the Cape of Good Hope. Dias was drowned near the cape.

diatom A group of unicellular algae essential to the support of most marine food webs. These organisms contain a yellow-brown pigment in addition to the essential chlorophyll and possibly xanthophylls or carotenes (yellows and reds) needed for photosynthesis. The outstanding feature of the group is the hinged, two-part, transparent, siliceous cell wall. These cell walls are called tests. Diatom tests may be circular (Centrales) or elongated (Pennales). Fossil diatoms are well known, since the cell walls of these animals are virtually indestructible.

Recent research has demonstrated the vital role that diatoms play in the regulation of the planet's atmosphere, since about 25% of the photosynthesis on Earth is performed by these organisms and their commensals. They were much more prevalent about 40–60 million years ago, when the carbon dioxide level in the atmosphere plummeted. However, diatoms do need CO_2; the porous, siliceous test concentrates this vital ingredient within the test and therefore keeps the level of CO_2 in the water low.

The reproduction of diatoms is complicated, since it involves the unhinging of the valves of the cell wall. One half of the wall, with one valve, goes with each new daughter cell. It has been discovered very recently that the bacteria that form symbiotic associations with mats of diatoms add nutrients to the ocean by nitrogen fixation. Previously, it had been assumed that only land plants supported nitrogen-fixing bacteria.

The intricate, beautiful tests of the diatoms have interested scientists since microscopy first made them visible. It is now thought that the elaborate construction of the tests makes maximal contact with water possible and thus increases surface for photosynthesis. *See* BROWN ALGAE, CARBON DIOXIDE, COMMENSALS, FOOD CHAIN, PLANKTON, NITROGEN FIXATION, OOZE, TEST.

diffraction The bending or redistribution of light around an obstacle. It is a different phenomenon from reflection or refraction. Diffraction occurs not only with light waves but also with water waves and sound waves. *See* REFLECTION, REFRACTION.

diffusion The transfer of matter or energy from a region of high concentration to one of lower concentration. This is the mode of transmission of light through the photic zone of the ocean.

dinoflagellates Single-celled organisms, less than 1 mm in size, that were formerly classified as plants or animals since some, but not all, contain chlorophyll. These protists have been found in fossils that date back to the Cretaceous period and may have existed before that era. These organisms constitute a part of the phytoplankton—microscopic autotrophs that are the basis of all sea life. They have two nonidentical flagella that provide them with locomotion. The organism uses one flagellum to turn and the other to move vertically in response to light. All of these organisms are covered by a unique exterior—the amphiesma, which is a continuous inner and outer membrane with flattened vesicles between the membranes. The dinoflagellates are found in two main groups: armored and naked. The former, the Peridinales, are found inshore. The armor is made of cellulose plates in the flattened vesicles of the exterior. The latter, the Gymnodinales, are found in the open ocean, floating in warm seas.

The lifestyles of dinoflagellates vary. In some the exterior is ridged or crested, and these indentations in the amphiesma provide niches for cyanobacteria, which form commensal relationships with the dinoflagellate and provide it with nitrogenous compounds necessary for protein building. Some forms are fish parasites; others live symbiotically with corals, specifically the Alcyonacea. Other species are photosynthetic and contain the photosynthesizing mechanism in a separate organelle. Some have light-detecting organelles using carotenoid compounds as the light-detecting mechanism, and a few species have distinct lenslike structures, ocelli. Depending on locale, the organisms produce light displays in warm waters. Species that live in tropical or semitropical regions are present year-round; others in colder waters and/or higher latitudes will encyst and drop into the bottom sediment when there is less available warmth and daylight in winter.

The nuclei of dinoflagellates are interesting since they possess more DNA in their chromosomes than do any other eukaryotes, which results in large nuclei. The process of mitosis is also unique to these organisms, since it is unlike other eurkaryotic reproduction.

A characteristic of many dinoflagellates is bioluminescence, or phosphorescence, and their abundance in certain places has given names to specific locales, such as Phosphorescent Bay (Bahia Fosforescente) in Puerto Rico.

Dinoflagellates "bloom," or greatly increase in number, periodically, sometimes with disastrous results, such as the toxic red tides. These dinoflagellate blooms are toxic to shellfish, causing gastroenteritis in the shellfish and in fish, birds, and other predators that eat them, including humans. Thus, while an oyster may have indigestion, the bird (or human) that eats several oysters may receive a fatal dose of the toxin. *See* BIOLUMINESCENCE, FLAGELLA, MIGRATION, PROTISTA, RED TIDES, TOXIN.

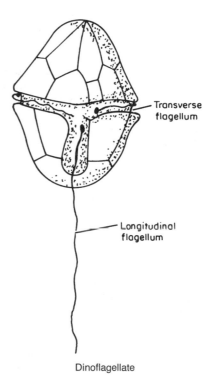

Transverse flagellum

Longitudinal flagellum

Dinoflagellate

dinosaur A popular name for the usually very large herbivorous or carnivorous reptiles found on the Earth and in its oceans in the Jurassic, Triassic, and Cretaceous periods of 195 to 65 million years ago. Most of the best known and most spectacular dinosaurs were terrestrial animals. *See* ICHTHYOSAUR.

***Discovery*, HMS** The name of several British ships used in explorations. The first *Discovery* went to Hudson Strait, in a voyage led by George Weymouth (1602); Hudson used the ship to sail into the same waters on his last voyage in 1610.

Another *Discovery* started life as a collier (coal carrier) named the *Bloodhound*. This ship and the *Resolution* were under the command of James Cook when he sailed on his last voyage in 1788.

The third significant *Discovery* was part of George Nares's polar expedition. The ship had a reinforced bottom designed to withstand ice pressure. It was the model for Scott's ship on his 1901–04 Antarctic expedition. The same ship (*Discovery*) was used as the base for antarctic surveying in 1925–27, and again to examine the Australian coast. The *Discovery II* (a rather confused designation since it was the fourth vessel of that name) charted antarctic waters immediately after World War II. Much of the work that led to the mapping of the Antarctic Currents and the Antarctic Bottom Water was done by the exploration teams on these ships.

Part of the charge of the "*Discovery* Expedition" of the 1920s was an attempt to explain the serious population decline of whales in the South Atlantic. It was this expedition that foretold of the possible extinction of several whale species because of overfishing. This problem has worsened considerably since then.

Another significant biological discovery of this research effort was the realization that the Antarctic Convergence is a "biological barrier." Many organisms such as plankton, birds, and fish are present either on both sides of the convergence line, or exclusively on one side or the other. The sinking of the Antarctic Water and the different layers, different temperatures, oxygen, nutrient, and salt content produce very different growing conditions and therefore very divergent populations on the two sides of the line. *See* ANTARCTIC CONVERGENCE; COOK, JAMES; NARES, GEORGE.

diurnal oscillation A term referring to variations in the planktonic population in the course of a day. Plankton move up toward the surface of the ocean in daylight and down at night. The reason for this migration varies with different species. Some animals migrate for reproductive purposes, others follow their food supply, still others, which harbor photosynthesizing protists, migrate to give the symbionts their daily exposure to the Sun. *See* DEEP-SCATTERING LAYERS, PLANKTON, TIDES, VERTICAL MIGRATION.

diurnal tide A tide that exhibits one high and one low tide during each tidal or lunar day. *See* TIDE.

divergence *See* CONVERGENCE.

diving Descending beneath the ocean's surface. Dives of more than one to three minutes (the length of time a person can hold his breath) require the use of breathing apparatus. The use of the hard-hat diving apparatus, made reasonably safe by the early 20th century, tethers the diver to an air pump at the surface. While it does provide security and an air supply, the hard-hat apparatus encumbers the diver with bulky, heavy equipment and restricts his or her movements. Scuba (or SCUBA) equipment, developed in the 1940s, allows a diver to carry his or her own compressed air supply in portable tanks, thus providing free range down to about 80 m (250 feet). Saturation diving is descent in a submersible, which is intended to remain submerged for some extended period.

The physiological effects of diving result from pressure on tissues and from gases. Without an air supply, a diver cannot go below a point determined by his particular physiology—usually about 100 m (300 feet), where the water pressure forces air out of the lungs. Given an air supply, the pressure on the lungs exerted by the inhaled gases increases, and this is injurious to lung tissue. Increased pressure within the lung causes collapse. The tiny air sacs, the alveoli, burst as an overfilled balloon would. These "balloons" are enmeshed in a tracery of tiny blood vessels and these too can rupture. An increased nitrogen content dissolved in the blood at about the 100-m mark produces intoxication by obstructing oxygen transport by the blood. This impairs judgment, reflex action, and accurate performance and is potentially highly dangerous. The exact mechanism of nitrogen narcosis is unknown.

The collection of symptoms that characterize decompression sickness are exhibited by a diver emerging from compression too quickly. The nitrogen dissolved in the blood bubbles out and symptoms can occur at any time—up to about 18 hours after decompression. The presence of nitrogen bubbles in joints or muscles causes localized pain, hence the common name for decompression sickness—*the bends*. Over 80% of all decompression sickness involves localized pain. Other symptoms depend on the location of the bubbles. In the lungs, they cause labored rapid breathing. In the brain a range of symptoms from headache to vertigo, to unconsciousness or paralysis or death have been documented.

The replacement of some of the nitrogen in the breathing mixture by helium eliminates many of the problems encountered in using nitrogen. Helium is less soluble in human tissues, and therefore makes decompression after a dive easier but not problem-free. Helium does, however, extend the practical work limit of a diver to depths of about 175 m (580 feet).

The pressure of both oxygen and carbon dioxide increases in the body at increased depths, and is potentially dangerous. Increased oxygen in the tissues is toxic, and can lead to convulsions that may be fatal. The most common adverse effect of this is labored breathing. Tingling sensations in the arms and legs, vertigo, nausea, and confusion may also occur. Unfortunately, such symptoms do not occur until several days after the pressure increase itself. Carbon dioxide is retained in greater than normal quantities in tissues at great depths. This produces an effect similar to that of nitrogen. *See* COUSTEAU, JACQUES-YVES; MARINE ARCHEOLOGY; NITROGEN NARCOSIS; SCUBA; SUBMERSIBLES.

DNA (deoxyribonucleic acid) The
genetic encoding molecule of all living organisms. Each plant or animal has its own unique DNA. The constituents of DNA, which is a long chain molecule, or polymer, are the sugar deoxyribose, phosphate groups, and the four nitrogen-containing organic bases adenine (A), guanine (G), cytosine (C), and thymine (T). (*See* diagram.)

The DNA molecule exists in the form of a double helix—a coiled ladder in which the "sides" are the sugar-phosphate strings and each "rung" is a base pair. Each sugar is attached to its respective base by a strong chemical bond, whereas the bonds that hold together the base pairs forming the rungs of the ladder are weak chemical bonds (hydrogen bonds).

The genetic information is carried in the linear sequence of the four types of bases along the DNA chain. DNA replication, which duplicates the genetic information and allows it to be passed on during cell division, occurs as the weak chemical bonds in the base pairs forming the rungs of the DNA ladder come apart and a new DNA chain is polymerized onto each of the two old chains that formed the sides of the original ladder (replication).

By an analogous process, a segment of the genetic information contained in the linear sequence of the bases along the DNA chain is transcribed, or copied, into a complementary strand of messenger RNA (ribonucleic acid). This is then translated into protein in a reaction catalyzed by a large complex aggregate of RNA known

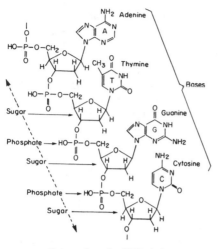

1) A section of a DNA chain

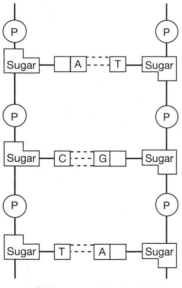

2) DNA structure (double helix)

3) Base pairs between strands
of the double helix

codon, specifies a particular amino acid, and in this fashion the "genetic code" in the DNA molecule is translated into the proteins that make up the skin of an eel or the feathers of a penguin.

Recent research has examined the structure of deepwater corals, organisms that were unknown until the 1990s. This work has been concentrated on the analysis of DNA to establish lines of descent of the coral colonies of the same species and the genetic drift as one colony might have changed and become another species. Since corals reproduce by both sexual reproduction and dispersal of a colony's fragments, this work will lead to a better understanding of the dispersal of corals to new habitats. This information is useful in the determination of the life cycle of both coral reefs and seamounts, places where corals are expected to be found. These areas are also subject to destruction by natural or man-made catastrophes, and their rebuilding is ecologically important.

as a ribosome. The amino acids used for protein synthesis are brought one at a time to the messenger RNA chain by transfer RNA molecules, each of which recognizes a specific sequence of three bases. The sequence of bases in the messenger RNA is "read" from one end to the other in sets of three. Each such triplet, called a

Mitochondrial DNA is usually used for this analysis; it mutates rapidly and is easier to read. The DNA analysis will yield information about the biological, oceanographic, and genetic factors that lead to isolation of a colony; it is isolation that triggers the small genetic differences that drive to speciation. Darwin's finches are a prime example of this. *See* CORAL, EVOLUTION, PEPTIDE FORMATION, RNA.

dogfish A member of one of the most familiar groups of sharks (for example, the Squalidae and Carcharhinidae). The dogfish, which at maturity is about 1 m (3 feet) long, is a well-known biological laboratory specimen. It is fished commercially as it swims in schools following herring. Its oil is used industrially.

The dogfish is ovoviviparous. Its eggs are fertilized internally and hatched about a year and a half later. It is a long-lived animal, specimens having been tagged and then recorded more than 25 years later. *See* MARINE OIL, SHARK.

Dogger Bank A submerged sandbank, roughly in the center of the North Sea. It is about 250 km (160 miles) long, quite shallow (13 m or 42 feet), and composed of glacial moraines and outwash. There are submerged peat bogs in the Dogger Bank. This area is one vast commercial fishing ground used by the vessels of several nations. Historically, it is remembered as the site of the largest World War I naval encounter, the battle of Jutland. *See* NORTH SEA, PLEISTOCENE EPOCH.

doldrums *See* INTERTROPICAL CONVERGENCE.

dolphin A relative of the whale, belonging to the family Delphinidae. Dolphins are slender and have a beaklike snout. They are the most agile cetaceans. The different species vary from 1.25 to 25 m (4–85 feet) long and range up to 230 kg (500 pounds) in weight. They move in herds that follow and feed on schools of herring or sardine. Tagged specimens have shown

that in the wild dolphins live more than 25 years, possibly 35.

The most common characteristic of the dolphin is that it has both fins and flippers. Like other cetaceans, it breathes in great gulps at half-hour or longer intervals, descends rapidly to great depths without compression or decompression effects, and is noted for its familial and community adherence.

The most frequently seen dolphins are the common dolphin and the bottle-nosed dolphin. Both are highly intelligent, and communicate with other members of their group and with humans using a series of high-pitched honks, barks, and squeaks. Dolphins appear to like humans, and seafarers' tales of dolphins saving shipwreck victims seem to have some validity.

Dolphins swim and share food populations with tuna, and if they become enmeshed in the large seines used to catch tuna, they will drown. In response to this, dolphin-proof tuna nets have been designed. *See* CETACEAN, ECHOLOCATION, WHALE.

Doppler effect A change in the frequency of sound, light, or other waves. In acoustical Doppler, a moving source emitting sound at a constant frequency seems to emit sound at a higher pitch as the distance between the source of the sound and the listener increases. Radar used in navigation is based on this phenomenon. A Doppler radar system measures the velocity of the echo. The Doppler shift in frequency in the target echo is proportional to the radial component of the target's velocity.

dorsal Relating to the back or upper surface of a fish or other animal.

dory A small rowboat-sized craft with high, curving sides. Dories are dropped by fishing boats and used to set lines. *See* FISHING INDUSTRY.

dory fish *See* JOHN DORY.

Dover, Strait of A body of water that separates England from the European continent and constitutes the southern boundary of the North Sea and its connection to the English Channel. The strait is about 33 km (21 miles) wide at its narrowest point, and about as long. Since it is in the shallowest part of the North Sea perimeter (60 m or 180 feet deep at most), relatively little Atlantic water enters the North Sea through this strait. It is thought that this area was once exposed during the Pleistocene. The strait itself was carved out by the now-drowned ancient Thames river system. The strait was undercut by the Anglo-French venture which built a tunnel to connect the road and rail systems of the two nations. *See* ATLANTIC OCEAN, NORTH SEA.

Drake, Sir Francis (1540–1596) English navigator, explorer, and buccaneer; the first Englishman to circumnavigate the globe. Drake first went to sea in 1566 and 1567–68 with John (Jack) Hawkins, a buccaneer, sailing on slave ships to the West Indies. In 1570, 1571, and 1572 he sailed again on voyages to raid Spanish treasure ships. The last of these was very successful: Drake captured the Spanish treasure convoy in Panama, which made him an international figure.

A syndicate of Protestant gentlemen financed Drake's voyage of circumnavigation. The stated objective was to explore and exploit the western coast of South America. It also served to promote the cause of the English Protestants and to get Drake out of England at a time when the political climate made it important to Queen Elizabeth I to attempt to retain diplomatic relations with Spain even while her privateers were harassing Spanish shipping.

Six ships left Plymouth in December 1577 (the number was quickly reduced to five). The flagship was the *Pelican,* which was later renamed the *Golden Hind* in honor of Sir Christopher Hatton, one of Drake's patrons, whose family crest included a golden deer. The expedition first moved south along the West African coast and then across the Atlantic at its narrowest point to South America.

At this point every narrative about Drake becomes vague. It is known, however, that Drake sailed through the Straits of Magellan in the winter of 1578. His fleet was reduced to three ships, the others being abandoned because they were unseaworthy. Drake then moved up the western coast of South America, raiding Spanish ports and ships as he went. On April 16, 1579, he left Spanish America (Mexico) and made landfall on June 5 in what is now Oregon. Drake named the West Coast of the United States New Albion—the first New England.

He sailed south and commented (unfavorably) on the brown uninviting hills frequently covered with fogs. This seems to have been his sighting of San Francisco Bay.

On July 24, 1579, the *Golden Hind* headed west. Sixty-five days later it made land, probably on the Palau island group in the Pacific, and then went on to Mindanao. The ship then rounded the coast of Sulawesi (Celebes) and struck out into ocean waters from Java. Sailing around the Cape of Good Hope without stopping, it finally anchored in June 1580 in what is now Sierra Leone, on the west coast of Africa. The long run at sea was intended to avoid Spanish or Portuguese vessels. The *Golden Hind* sailed into Plymouth harbor on September 26, 1580. Drake brought back a ship in good condition, and a healthy crew of 58 men, having lost 17 to illness, fights, and desertion.

Although memorable, the voyage did not bring many new discoveries but was a great example of seamanship.

Drake became a fabled hero in his own lifetime. He continued to harass the Spaniards and died in the West Indies of yellow fever on January 28, 1596.

Drake Passage (Drake Strait) The narrows of the Southern Ocean, probably opened in the late Oligocene epoch; it connects the Atlantic and Pacific and separates the tip of South America at about 50° south latitude from the South Shetland Islands and the Antarctic Peninsula. It is about 1,000 km (450 miles) wide and has an average depth of about 3,500 m (12,000 feet). Sir Francis Drake, for whom the passage is named, discovered it inadvertently in the Antarctic in the spring of 1578, when blown off-course by a storm.

The Antarctic Convergence bisects the Passage along an east-west line. Water from the Antarctic continent moving north sinks at about 50° south latitude and then spreads north and rises to the Subtropical Convergence. There it sinks again and moves further north as Antarctic Intermediate Water. It is still recognizable as cold, low-salt, and low-oxygen-content water. Intermediate Water from the middle latitudes moves south to replace what is lost. Since the Bottom Water comes up from great depths, it is responsible for the great fertility of the Antarctic waters.

Because it is so constricted, the Drake Passage is the Earth's greatest carrier of water moving in a west-to-east direction. The total volume of water pouring through this passage into the Atlantic has been variously estimated at anywhere from 75 to 150 million cubic meters per second (20 to 40 billion gallons per second)—a volume hundreds of times greater than the volume of water moved by the Mississippi River, for example. *See* ANTARCTIC CIRCUMPOLAR CURRENT; ANTARCTIC CONVERGENCE; ATLANTIC OCEAN; DRAKE, SIR FRANCIS.

dredge A floating device used for removing underwater material, usually held in place by supports that descend to the bottom.

The removal of sediment from canals, rivers, and harbors is vital for navigation and sanitation. The fill has traditionally been used to create new land or build dams, dikes, and breakwaters. Dredges are also used in salvage or to remove commercially significant materials from the sea bottom.

Dredging techniques have been used for centuries. One of the earliest examples of this work is the reshaped landscape of the Netherlands. Dredges and mudmills were common features of the medieval Dutch landscape. Earlier still, dredging was part of the general maintenance work done on

the canals of ancient Babylon. When the irrigation canals silted up, the life of the city ended.

Basic modern dredge design involves some combination of mechanical and hydraulic techniques. Dredges are either dippers which operate at the end of a flexible arm—a type of dredge that can operate at almost any depth—or buckets that grab bottom material and are brought up by cables or a continuous feed. The continuous feed or ladder is a line of buckets that digs into the mud and brings it to the surface. Hydraulic dredges liquefy bottom silt and pump it to the surface. *See* COAST.

drowned river valley An ice age V-shaped river channel that was submerged by rising ocean levels as the ice melted. The Chesapeake Bay is an example. *See* FIORD.

dry dock A shipyard structure used for major overhaul and repair to ships' bottoms. The dry dock is a basin cut into a harbor and separated from the rest of it by a wall. A ship can be positioned over supports and the basin emptied, leaving the ship standing on the supports. After repair and cleaning, the water is allowed to flow back into the basin and the ship is refloated.

Floating dry docks have at least one section that can be pumped out to contain and service a ship. These containment basins can be complex and large enough to float a naval vessel of battleship size. The dry dock is a direct descendant of the primitive earthen walls with which the ancient Greeks held back the sea while standing a vessel on the beach in order to repair it.

duck One of about 90 species of aquatic birds of the family Anatidae, order Anseriformes. Ducks generally have broad bills, short necks, and legs located on the tailside of their centers of gravity. This results in a waddling type of gait. The plumage of the male is markedly different from that of the female. Ducks undergo two molts each year, and true ducks have overlapping scales on the exposed skin of their legs. Ducks do not mate for life, as do geese and swans.

Ducks that live in brackish coastal waters or on open coasts, such as the eiders, are likely to be divers who pick their food out of the water or off the bottom. Among sea ducks that winter at the seacoast and nest inland are scoters and mergansers. Almost all of these ducks live in the Northern Hemisphere. See BIRD.

dugong A herbivorous Asiatic mammal living in coastal, tropical waters, where it does little more than eat. These slow, fat animals reach a size of about 3.5 m (12 feet) and 270 kg (600 pounds). They have gray-to-brown exteriors, front flippers, no rear appendages, and a broad, powerful bilobed tail. The dugongs have been hunted almost to extinction. The only surviving species, *Dugong dugon,* is in the same family as the manatees, Sirenia. See SIRENIA.

Dumont d'Urville, Jules-Sébastien César (1790–1842) A French explorer of the South Pacific and Antarctica. He also made contributions to archeology because, while charting the eastern Mediterranean Sea for France, he was part of the venture that brought the Venus de Milo to France (1820). He also participated in a French voyage that circumnavigated the world in 1822–25.

Dumont d'Urville was commissioned to find the remains of Jean François Galoup, comte de La Pérouse, who disappeared in the Pacific Ocean in 1788. On this reconnoitering mission, he stopped at and charted parts of New Zealand, New Guinea, the Loyalty Islands, Van Diemen's Land (now Tasmania), the Carolines, and Sulawesi (Celebes). He found what he believed to be parts of La Pérouse's wrecked ships in the Santa Cruz islands and returned to France in 1829.

The careful maps made on this voyage led to the designation of the charted island groups into Micronesia, Polynesia, Melanesia, and Malaysia (Fiji Islands,

Loyalty Islands, Maritius, etc.). The expedition was also a scientific success since it returned with samples of plants, animals, minerals, and ethnological artifacts, as well as dictionaries of native languages.

Dumont d'Urville was promoted to captain and again commissioned to explore the Pacific for France. He departed in 1837 for a voyage to the antarctic, hoping to better the record of the English explorer James Weddell, who had gone as far south as 74°15', in 1823. Dumont d'Urville charted the Straits of Magellan and found ice at 63°29' south latitude, and then drifted east from the Straits to the South Orkney and South Shetland islands. After this he headed west again to Fiji, New Guinea, and Borneo. After taking on supplies the expedition again returned to the antarctic, hoping to locate the South Magnetic Pole. It sighted the Adelie coast of Antarctica in 1840, which it named for the captain's wife. The French expedition and the American one led by Wilkes from 1838 to 1842 each claimed to be the first to have sighted Antarctica. Dumont d'Urville returned to France late in 1841, and died in a train wreck the following year. See EXPLORATIONS; WILKES, CHARLES.

dune A mound of windblown sand or a mix of soil and sand found downwind of large lakes, ocean beaches, or dry riverbeds in areas where the prevailing wind is slowed or stopped.

Dungeness crab A large, commercially important, edible crab, *Cancer magister,* found in western North American waters. The Dungeness is a decapod found from Alaska to Baja California. It is usually 18 to 25 cm (7 to 9 inches) wide and 10 to 15 cm (4 to 6 inches) long. The upper surface is dark, the belly is yellow. The Dungeness is a relative of the Atlantic rock crab *C. irroratus,* the Pacific rock crab, *C. antennarius,* the Jonah, *C. borealis,* and the common crab of North European waters, *C. pagurus.* All of these are edible; the last is usually even larger than the Dungeness of the Pacific coast. See CRAB, DECAPODA.

dust The wind-borne product of continents present over open ocean and deposited on polar or glacial ice. Both erosion and volcanic activity propel dust into the atmosphere. The dust is composed largely of iron compounds. Iron is present in ocean water in very dilute concentrations, but physiologically it is important. It is present in a number of essential catalysts, particularly those influencing the growth of phytoplankton, which thrive in relatively iron-rich water. In turn, the phytoplankton are essential in absorbing the carbon in the world's oceans as carbon dioxide and thereby making it biologically available.

earthquake A rapid movement of rock caused by the buildup of strain created by the movement of one crustal plate against another. When this occurs in water, the displacement is spread both vertically and horizontally in all directions. Since water is a medium that disperses shock poorly, tsunamis (incorrectly called tidal waves) are created.

Earthquakes are associated most closely with specific geographical areas. The rim of the Pacific Basin is a series of deep trenches parallel to the South American coast, the Aleutian and Japanese island chains, and New Zealand. At these trenches, crustal plates are moving together, and as one plate moves below another one (i.e., as it is being subducted and moves into the molten interior of the earth), volcanoes are sometimes formed behind the trenches and eruptions and earthquakes result.

Earthquakes are categorized as either shallow-focus, occurring at a depth of about 400 km (250 miles), or plutonic or deep-focus, occurring at depths of about 750 km (470 miles) depths. Seismic waves of the two basic types spread in different ways.

P waves (primus) are compressional, s (secondus) have more shear quality. They arrive at different times at a distant detector. The time difference depends on the composition of the rock traveled through. *See* CRUST; TRENCH; TSUNAMI; VOLCANO; WILKES, CHARLES.

East Australian Current The western portion of the gyre (the circulatory route

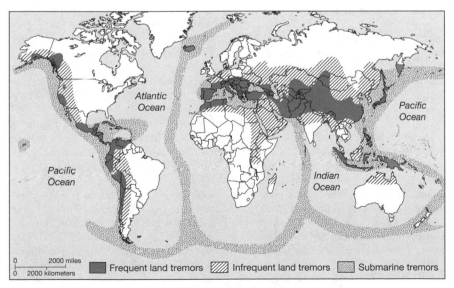

Earthquake zones

of surface water in an oceanic region) in the South Pacific Ocean. It forms between the Great Barrier and Chesterfield reefs and is augmented by warm equatorial water moving west-southwest, pushed by the spring (January–March) monsoon. From April to December this warm water comes from the Coral Sea. At its strongest point the East Australian Current moves at an average speed of about 50 km/day (30 miles/day). There is an offshore countercurrent in winter.

South of about 32° south latitude, the Current broadens to about 200 km (150 miles) wide and disappears into an eddy system. These warm-core eddies are a notable feature of the East Australian Current, and last about a year each. The eddies can coalesce to form new systems.

This current has been called the Southern Hemisphere's Gulf Stream. However, unlike the Gulf Stream or the Kuroshio Current, it disappears rapidly when it moves away from the coast. The Australian continent ends at about 44° south latitude. South of the continent the water moving west to east is swept along by the Antarctic Circumpolar Current, which, because of its size and volume, incorporates any other moving water in the area. *See* ANTARCTIC WATER, GYRE, MONSOON, OCEAN CURRENTS.

east boundary currents Currents that flow toward the equator on the eastern side of an ocean. These currents include the Canaries, California, Peru, West Australia, and Benguela Currents. The east boundary currents are surface currents with water temperatures lower than currents further west at the same latitude. The currents are relatively slow. Upwellings, when they occur, are on the eastern edges of the currents, bringing nutrients to the surface. The result is a large number of both plant and animal species and individuals. *See* BEN-GUELA CURRENT, CALIFORNIA CURRENT, CANARY CURRENT, CURRENTS, PERU CURRENT, WEST AUSTRALIAN CURRENTS.

East Cape The easternmost part of the Eurasian continent. It is on the Chukchi Peninsula and is the eastern boundary of the Bering Strait. The Soviet name for East Cape, Cape Dezhnev, after its discoverer, is becoming internationally recognized. *See* DEZHNEV, SEMYON IVANOV.

East Greenland Current Part of the water of the Greenland Sea. This body of cold, polar water of low salinity moves south along the east coast of Greenland, picking up icemelt and water of terrestrial origin. This further decreases the salinity of the Greenland Sea to below 3% in summer. Polar Intermediate Water joins the East Greenland Current near Spitsbergen to form the Greenland Current, a flow of south-moving surface water that pushes a belt of ice before it. *See* ARCTIC OCEAN, FRAM STRAIT, ICE.

East Pacific Rise Or East Pacific Ridge. A broad, relatively continuous formation that runs roughly north-south. It rises 2 to 3 km (1 to 1.5 miles) above the ocean floor and is 2 to 4 km (1 to 2 miles) wide. The average depth of the rise is about 2.7 km (or about 1.7 mile). The East Pacific Rise disappears under the North American Plate at the Gulf of California.

Recent studies have revealed considerable volcanic activity in this region of seabed spread. Chimneylike vents pour hot sulfides into the surrounding water. An ecosystem based on bacteria that survive using sulfur instead of solar energy makes possible a wide variety of organisms that can survive in hydrothermal environments.

Geologic studies of the northern edge of the East Pacific Rise near the Mexican coast reveal recent lava flows, sulfide-rich areas, and significant manganese and iron accumulations. It is thought that similar finds would be made in other parts of the Rise. *See* EVOLUTION OF OCEANS, PACIFIC PLATE, SULFUR BACTERIA, VENT COMMUNITIES.

East Siberian Sea A broad continental shelf north of Siberia. It lies between the New Siberian Islands to the west and the Wrangell Islands on the east. The

Wrangell Islands are the traditional division between the East Siberian Sea and the Chukchi Sea. North of the East Siberian Sea is the Wrangell Abyssal Zone. The area was explored by the Russian Baron Wrangell in the 1820s.

The East Siberian Sea is quite shallow. The depths at the eastern or deeper end are frequently in the 30- to 40-m (100–130-foot) range. The temperature and salinity of the water depend on river input. Since the sea is so shallow, there is thick ice on its surface through most of the year. This results in a relative diminution of large animal life. However, there are large fish communities, such as char and salmon, and the birds that feed on them, including cormorants, terns, and gulls. Seals and walruses are found at river mouths and on terrestrial bars in the sea. See JEANETTE; WRANGELL, BARON FERDINAND PETROVICH VON.

Easter Island A small (130 km² or 50 square miles), volcanic, South Pacific Island, lying about 3,300 km (2,300 miles) west of Chile. It is a part of the Easter Island Ridge, which is the Pacific Ocean equivalent of the Mid-Atlantic Ridge. The ridge system of which Easter Island is a part is somewhat unlike the Atlantic Ocean counterpart in that it is asymmetrical, sloping inward toward the South American continent. Easter Island is best known for the *moi*, huge and unusual statues found on it. See HEYERDAHL, THOR; MID-OCEAN RIDGES.

Echinodermata An exclusively marine phylum of great age and worldwide distribution. There are about 6,000 living species. The name comes from Greek and means "spiny skin." While this is true for most of the phylum, which tends to exhibit calcareous spines, some species have warts rather than spines.

The echinoderms are classed as follows:

Crinoidea: sea lilies
Holothuroidea: sea cucumbers
Asteroidea: sea stars
Echinodea: sea urchins and sand dollars

Ophiuroidea: serpent stars and brittle stars
Ophiocistioidea: extinct class of small domed animal, usually 7–8 cm in diameter

The extensive fossil record of these animals dates from the Ordovician to the Devonian period. The common features of the Echinodermata are the absence of a head or end of some sort and radial symmetry, with arms in a pattern of five or a multiple of five. The crinoids are sessile (stationary), and their original ancestor may also have been stationary. The rest of the phylum is generally bottom dwelling and moves by creeping along, some using the arms for locomotion, although some sea cucumbers are pelagic.

Echinoderms have a water-vascular system, in which water enters the body through a valve called the madreporite; in asteroids, this lies on the aboral (upper) surface of the animal. Water is circulated through the highly calcified ($CaCO_3$) "stone canals" to the "tube feet"—suction devices on the legs that adhere to the shells of mollusks and pry them open. The stone or calcite in the canals is deposited in ossicles. These are characteristic structures and account for the difference in the rigidity of the interior skeleton. In some, such as the sea urchins, they have little flexibility, while the crinoids can bend and the sea cucumbers have almost no interior stony structure. Most echinoids, with the exception of sea stars, are filter feeders. See CRINOIDEA, DIADEMATOIDA, SEA STAR, SEA URCHIN, STARFISH.

Echinoidea The class of Echinodermata that represents the sea urchins and sand dollars. These creatures are characterized by a rigid covering called a test. There are about 950 living species and over 5,000 described fossil species. See ECHINODERMATA, SAND DOLLAR, SEA URCHIN.

Echiura The spoon worms, a small phylum of about 140 species of plump marine burrowers of worldwide range and long history. U-shaped burrows such as the ones modern echiurans create are present in Cambrian rocks; the oldest fossil dates from the Pennsylvanian period. The burrows of echiurans are shared by commensals—small crabs, polychetes, and fish—leading to the other common name for this phylum, the innkeeper worms. Most of the individuals are small, though they range in size from about 1 to 60 cm (0.4 to 24 inches), and most are free-living. They share some characteristics with the annelids, and until recently these animals were classified as annelids. Fertilization is external, and the young are nektonic, swimming freely on their own. Unlike the annelids, the majority of echiuroidean species are dimorphic, with distinct male and female members.

These animals have some unique characteristics, the most notable of which is an elaborate, grooved, ciliated proboscis (food-gathering) organ that amasses food particles, either detritus or plankton. The proboscis may be extended 90 to 100 cm (30 to 40 inches) long, although the body of the average worm is only about 8 cm (3 inches). Individuals may be dull brown, gray, red, rose, yellow, or transparent, and in some the color of the animal is dependent on its food supply. Thus, Bonellia tasmanica is green because of a porphyrin pigment derived from the chlorophyll a of the algae it feeds on. The species of echiuroidean, Bonellia, is remarkable because the males are tiny and parasitic on the females. They attach to the female's proboscis while still larvae and move down into the body. Fertilization in this species is anomalous because it is internal. See ANNELIDA, NEKTON, PLANKTON, PROBOSCIS.

echolocation The ability to determine the shape and size of an object and its distance by the echo it makes. The best-known terrestrial animals who find their way about by echolocation are the bats. In the sea, the range of species using echolocation is much greater. All toothed whales, the Odontoceti, use echolocation, as do some baleen types, such as the blue and

gray minkes. Some pinnipeds, including the California sea lion, the Weddell seal, and the walrus also echolocate. Echolocation is useful to these mammals because although they have good vision, they frequently find it obscured by a high sediment load in the water; alternatively, they may be beneath ice or below the photic zone, where there is no light.

The sound used for echolocation is produced by several structures in the animal's head. In the whales it may (or may not) be amplified by the blowhole. The sound that is used is specific to a given species, and of a frequency range that is usually greater than humans can detect. The pattern of sounds is also species-related, and different sounds mean different things. There have been considerable attempts made to investigate the meaning of the sounds used for echolocation.

The sound receiver for echolocation is not always the ear, even in species that have external ears. The entire skull may be a receiver, and in some whales the lower jaw is an auditory receiver. The mechanism by which the sound that is received is processed into information is conjectural. See ECHO SOUNDING, WHALE.

echo sounding A technique used to map undersea terrain. Sound is carried well by water, a phenomenon observed by the ancient Greeks. In seawater it travels approximately 500 m/sec (or 1,600 feet per second). Not until the 20th century, however, was there a reliable method for both making a sound and timing its rebound from the bottom.

By timing the echo from the bottom, the shape of the bottom surface can be plotted, using a number of soundings. Computer enhancement of this plot then gives a good, fairly reliable picture of the bottom to an experienced observer. It is essential to use repeated soundings and to check the results using lead lines and other methods, since plankton and schools of fish will also produce echo, called a false bottom. The echoes from schools of fish are used by commercial fishermen to locate such schools. *See* SONAR, SOUNDING.

ecliptic The plane of the orbit of the Earth as it revolves about the Sun. The line through the Earth from the North Pole to the South Pole makes an angle of 25°30' with the ecliptic. This is the declination of the Earth. *See* DECLINATION, EQUINOX, SOLSTICE, TIDE.

ecosystem A term used to describe the interrelationships among all organisms in a given area and their relationships to the nonliving materials that make life possible. The specific environments in the oceans are divided according to physical conditions of depth, light, temperature, and salinity. In many parts of the ocean, the density of living organisms is low, but excluding insects in their incredible numbers of species and individuals, the oceans contain a wider variety of organisms and are more densely populated than land areas.

All aquatic life is dependent on the abiotic (nonliving or inorganic) materials present in water. Dissolved calcium salts (carbonates) as well as nitrates, silicates, phosphates, and sulfates must be present, and also a large variety of micronutrients, including nickel, manganese, and copper, as well as gases such as nitrogen, carbon dioxide, and oxygen. Ultimately, almost all life depends upon solar energy, which is used to make sugar molecules by autotrophic (photosynthesizing) organisms. The greater part of this process, but by no means all of it, is carried out by algae, diatoms, dinoflagellates, and nanoplankton.

The living things in aquatic environments are divided into three arbitrary groups: plankton, nekton, and benthos. Plankton are either animal-like (the zooplankton), or plantlike (the phytoplankton). This plantlike photosynthesizing plankton includes Monera, microscopic, cellular, photosynthesizing autotrophs. Zooplankton includes animal eggs, larvae, protists such as radiolarians and foraminiferans, tiny crustaceans, copepods, mollusks, and other small animals. The zooplankton is subdivided according to the size of the organism. Nekton are swimming organisms, including fish, cetaceans, crustaceans, squid, and octopus. Benthic organisms are bottom dwellers. It does not matter whether the bottom is a tidal pool or an abyssal depth. These organisms are the grasses and large algae, either infauna, organisms that live in sediment, or epifauna, those that cling to rocks, and animals that move such as crabs, fish, oysters,

The ecosystems and organisms of the ocean can be divided into their various zones:	
Habitat	*Organisms*
Open ocean	phytoplankton: bacteria, diatoms, coccolithophores zooplankton: formainferans, radiolarians, copepods, krill nekton: fish both cartilaginous and bony, birds, reptiles. large mollusks, sea mammals
Deep-sea	crabs, sea stars, urchins, worms, mollusks, sea fans, sea anemones
Vents	chemosynthetic bacteria, giant clams and worms
Rocky shore	periwinkles, snails, barnacles, crabs, mussels
Tide pools	sea anemones, fish, crabs
Sandy shore	clams, snails, sand dollars, crabs, birds, sea otters, seals
Muddy shore	clams, worms, snails

sea anemones, echinoderms, and corals. Benthic organisms feed on plankton and sediment coming down on them from the surface water and on each other. Animals that as adults are benthic may be planktonic or nektonic as larvae. Oysters and barnacles are examples. Immature oysters are planktonic and drift with the current; developing barnacles are swimmers. The juveniles float in the water currents, while the adults are sessile.

The marine ecosystem is further divided according to depth. The two subdivisions are neritic and oceanic, or crudely stated, green water and blue water. The neritic inshore water is green because it contains more phytoplankton and other plants and is nutrient rich. Nutrients are either stirred up from the bottom or brought by terrestrial runoff. This green water zone is further divided into the littoral, which is sometimes exposed between tides, and the shallows, where there is always a water covering of varying depth depending on the tide. The oceanic region is graded according to depth: epipelagic, mesopelagic, bathypelagic, and abyssopelagic (or hadal).

The microscopic organisms in each environment, in addition to serving as photosynthesizers, include the bacteria and detritus feeders, the tiny scavengers that recycle all other living and once-living material and release from it the raw materials that, in turn, are necessary for the photosynthesizers and other autotrophs. Not until the Viking Lander searched for microbial life on Mars was similar examination applied to Earth. Before 2003 about 99% of the Earth's oceanic microbes were largely overlooked. Bacteria are present in huge numbers in the ocean and account for most of its biomass and metabolism. The ultraplankton—the bacteria in the oceans—may account for between 50 and 70% of all the photosynthesis on Earth.

The relationship among competing populations is being explored both by commercial interests such as the fishing industry and by environmentalists, whose interest is in the interrelationships of varying oceanic populations. Any species in a given area depends on the food supply; the physical conditions of temperature, light, salinity, and mineral content; and local predators, as well as the presence of other species competing for food. For some species, competing species are also predators.

Oceanic ecosystems

Opposite page: An oceanic ecosystem is based on the plankton in it. All other organisms are dependent on this population.

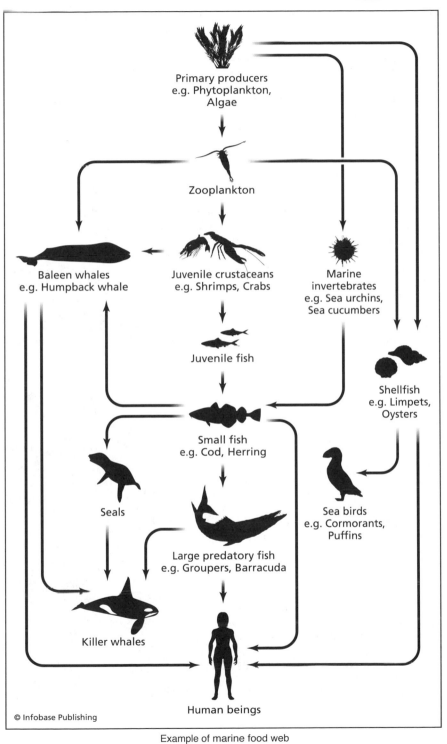

Example of marine food web

Herring and mackerel, for example, use each others' larvae as food. Species may also compete for space. Fishing for a dominant population may reduce its numbers to a point at which a subordinate species becomes dominant. This interrelationship of species and their physical requirements is not yet well understood for many commercially fished organisms.

Heterotrophs consist of all the nonautotrophic organisms in an ecosystem. Ultimately, all of them feed on the autotrophs. Herbivores do so directly. The detritus feeders live on the dead of all species; the carnivores feed on herbivores and on one another. Eventually the detritus feeders and scavengers consume both the carnivores and herbivores.

See ALGAE, BACTERIA, CHEMOSYNTHETIC BACTERIA, ESTUARIES, PELAGIC ENVIRONMENT, VENT COMMUNITIES.

ectoderm The outer layer of cells of an embryonic organism. This layer in the embryo engenders the nervous system and skin.

Ectoprocta A phylum of bryozoans. There are about 60 species. These colonial animals are found as wrinkled, crackling sheets on the shore. The sheets are masses of calcareous (calcium-containing) or chitinous "boxes" that form the individual organism's living space. *See* BRYOZOA, CHITIN, COLONIAL ANIMALS, ENTOPROCTA.

eddy A fluid flow that is markedly different from the general movement of a large mass of that fluid. The number and velocity of eddies increase as the velocity of the surrounding mass of fluid (air or water) increases. More eddies and faster-moving ones spin off in rapidly moving water than in a more sluggish stream. The Gulf Stream is a prime example of an eddy-creating system.

Eddies are a means of energy dissipation. Larger eddies form quickly and then break up into smaller ones, which eventually dissipate.

In the ocean, eddies are spun off by wind activity, upwellings, or waves encountering an obstacle. If the water moves around an obstacle, an eddy may form at some critical velocity (this depends on the nature of the obstacle) on the leeward side of the obstacle because the pressure of the fluid is reduced at that point by the rapid movement of the flow. If the velocity of the general flow is great enough, a vortex or whirlpool will form. Whirlpools are permanent eddies. Normally an eddy moves in a particular path. Cyclonic storms are examples of atmospheric eddies. *See* GULF STREAM, STRAITS OF MESSINA.

Ediacaran fossils Found in 1946 in the Ediacaran Hills, an area northeast of the Great Australian Bight. These fossils were first declared to be the casts of jellyfish, worms, and frondlike structures. They were paleontologists' first view of Precambrian animals. These soft-bodied organ-

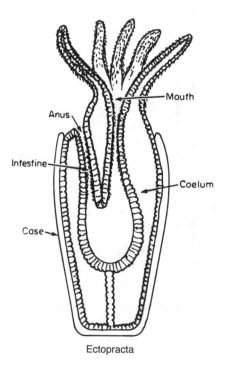

Ectopracta

isms were preserved by rapid burial in sand. Ediacaran remains have since been found on every continent except Antarctica. The largest deposits are in the Northwest Territories and the Yukon. Other substantial finds were made in Newfoundland and British Columbia.

eel The order Anguilliformes; elongated, usually scaleless, bony fishes with an average size of about 1 m (3.3 feet). Eels have been caught and eaten as a delicacy since classical times. There are about 500 species. One major family, the Anguillidea, are found in freshwater streams in North America and in the Caribbean, northern Europe, the Baltic, the eastern Mediterranean, and the Far East. These eels breed in the Sargasso Sea.

Another family, the Moringuidae, consists of small burrowing and nocturnal sea eels that are sometimes found in brackish water. The morays, or Muraenidae, comprise about 120 species of tropical eels that are often brilliantly colored and marked with spots or bars of black. They have strong jaws and poison glands, live in rock crevices, and hunt in darkness. The morays are the largest of all eels; one Indo-Pacific species exceeds 3 m (10 feet) in length.

A third family, the Congidae, or conger eels, are also toothed. They are generally under 3 m (10 feet) in length and live in all of the world's warm seas except the eastern Pacific. They live on fish and squid, pouncing on them from rock crevices and other dark hiding places. Congers die upon spawning.

The Ophichthyidae or snake eels live on coral reefs. They bore holes in the sand, tail first, and feed on fish, crustaceans, and copepods. The Nemichthyidae or snipe eels are long-nosed deep-sea dwellers found at depths ranging from 400 to 4,000 m (1,300 to 13,000 feet). The suborder Saccopharyngoidei are deepsea organisms. The pelican fish is one. Its whiplike tail is longer than its body. It has many arched teeth and distends its mouth to swallow prey whole. Gulpers are classed as eels by some taxonomists and not by others. These slender, black, 50-cm (1.5-foot)-long fish live on plankton or small crustaceans. See SARGASSO SEA.

eelgrass Also known as wrack, this term comprises a group of angiosperms (flowering plants) of worldwide distribution found in protected inshore locales. Eelgrass belongs to the family Zosteracea. It grows in areas where there is abundant sunlight and serves as both food and shelter for littoral communities. It is unusual because it lives and flowers while submerged. See ANGIO-SPERMS, ZOSTERA.

egret A long-legged, long-necked heron. Egrets are usually white and have plumes called aigrettes, which play a part in their courtship display. These feathers were popular as hat decoration in the 19th and early 20th centuries. The egret was almost driven to extinction by the feather trade.

The common egret is found in warm to temperate regions worldwide. It is a thin, rangy bird, between 50 and 100 cm (20 to 40 inches tall). The bill is also long and slender, designed for picking at the small fish, frogs, worms, and crustaceans that are the egret's food.

Egret communities usually nest in trees, and they may be found almost anywhere near water. Some species live in saltwater marshes; others range far inland. Cattle egrets are most often found in grasslands, where they are commensal with cattle, picking insects off them. See HERON.

eider duck A sea duck related to the scoter. While there are some Pacific species, most eiders are found in the North Atlantic. They are large birds, about 70 cm (28 to 29 inches) long, and have large wings. The head is large and the bill is distinctively shaped and forked. The male has a white back and a black underside. The female is red-brown with black bands.

Eiders nest in large colonies. Each breeding pair builds a grass-and-seaweed nest on the ground, and the female lines the nest with down plucked from her breast. When the female leaves the nest to fish, she covers the five-egg clutch with these remarkably light, resilient insulating feathers. The down is prized and used commercially, collection taking place after the birds are fledged. Those areas of Norway, Iceland, and Greenland where the eider duck nests are protected. *See* DUCK.

Ekman, Vagn Walfrid (1874–1954)

A Swedish oceanographer best known for his work in ocean currents. He was a student of physics at Uppsala University and went from there to the International Laboratory for Oceanographic Research in Oslo, Norway. There he worked on several projects, among them the design of devices for sampling and measuring ocean water. Ekman also did the work that explained the "dead water" phenomenon observed by Frijdtof Nansen. This occurrence in fiords and other enclosed areas, such as narrow bays, rendered slow-moving vessels motionless. Ekman explained that this was caused by nearly fresh meltwater spreading over the saltier ocean water and setting up internal waves.

In 1910 Ekman went to the University of Lund, Sweden, as a professor of mechanical and mathematical physics. While there he did the work on currents that made him well known to oceanographers. This was the culmination of a lifelong interest.

Ekman joined the German-sponsored *Meteor* expedition, and spent five years, 1920–25, collecting data. The final report of the expedition was not published until after World War II. Ekman, the consummately careful researcher, had been separated from his data and did not work on the project until he had free access to his notes again. *See* METEOR RESEARCH EXPEDITION; NANSEN, FRIJDTOF.

Ekman spiral

A mathematically constructed spiral describing the effect of wind on the ocean. It is named for the Swedish oceanographer V. Walfrid Ekman, who formulated it in 1905. This effect is a composite of the force of wind and the force related to the Earth's rotation, the Coriolis force, which deflects movement of air or water to the right or east in the Northern Hemisphere and to the left or west in the Southern. At the same time, this deflection is diminished by friction. The Coriolis deflection causes circular motion, the friction creates a spiral. At a given depth, which depends on the force of the wind, the Coriolis and frictional forces both approach zero. The layer of water above that depth is the *Ekman layer.* It moves as an entity above the water further down. The details of how this layer moves and what angles it makes with the prevailing wind are unclear. The *Ekman transport*—the net effect of water movement in the upper part of the layer if the water moves seaward—causes coastal cold-water upwelling, as evidenced off the coasts of California and Peru. *See* CIRCULATION; CORIOLIS FORCE; CURRENTS; EKMAN, VAGN WALFRID; EL NIÑO; NANSEN, FRIJDTOF.

Ekman transport

The total transport or displacement of surface water as it moves with the wind. Excluding the friction factor, the transport is at a 90° angle to the right of the prevailing wind in the Northern Hemisphere and at 90° to the left of the wind in the Southern Hemisphere. *See* CORIOLIS FORCE, CURRENT.

El Niño

A southeast current in the southeastern Pacific Ocean bringing warm water toward the coast of South America in December or January. Because of its proximity to Christmas and the tropical fauna it carries along, it is called El Niño, the Christ Child. Another notation for this often-observed phenomenon is ENSO (El Niño Southern Oscillation). When used, this notation is also dated; the most recent studied major occurrence is ENSO-97/98.

At irregular intervals of as few as two or as many as 10 years, this benevolent warm current becomes a very warm current and brings catastrophe. This is part of an overall phenomenon involving a change in wind direction, the Southern Oscillation, and changes in air and water temperature (SST, or sea and surface temperature) as well as oceanic circulation. An El Niño event begins to build about a year and a half before it is evident in South America. The wind in the western Pacific drops, the sea level falls, the water temperature rises, and coral begins to die. The change in the height of the water level moves warm air and water into coastal areas and keeps it there through the winter, deflecting the South Equatorial Current. This means that cooler water, instead of moving west across the Pacific, goes north as far as San Diego, California.

This bottling up of water of higher temperature along the South American coast results in the dying off of plankton, squid, and fish populations that are adapted to cold water. With the death of the anchovy and sardine populations, hake, herring, and mackerel disappear or die. The seabirds and seals that feed on these fish abandon nests and lose their young. The decaying fish accumulate on beaches, producing huge quantities of sulfurous gas.

The climatic changes bring torrential rain and floods to the normally dry South American regions while producing drought in the western Pacific. These climate changes have effects on the weather in North America as well. The severe winter and spring flooding experienced in North America has its origins in the anomalous wind pattern associated with El Niño.

This phenomenon of destruction has been chronicled in the West since 1726. However, the height of the flood of the Nile has been measured since 715 CE using structures built specifically to measure the river's height at crest. Before that, others did take measurement; the remains of the Roman measures are available. The Nile flood correlates directly with El Niño appearances. The present Nilometer on Roda Island near Cairo is the most recent construction that monitors the state of the river.

Some El Niño years in the 20th century have been 1925, 1930, 1941, 1953, 1957–58, 1969–70, 1976–77, 1982–83, 1993–94, 1997–98, 2006–07. All of these were major events and were associated with worldwide climate anomalies. *See* CURRENTS, KELVIN WAVES, LA NIÑA, PERU CURRENT.

Elasmobranchii The largest subclass of Chondrichthyes (jawed, cartilaginous fish): the sharks, the rays, and the chimaera. They are medium to large in size (0.15 to 21.1 m or 6 inches to 70 feet). These fish have 5 to 7 gill slits, platelike gills, numerous, replaceable teeth, and placoid scales. Their buoyancy is controlled by oil storage. They are hunted for oil, particularly the squalene-manufacturing sharks. *See* MARINE OIL, RAYS, SCALES, SHARKS.

electric eel A freshwater elongate fish, *Electrophorus electricus,* related to the carp and not a true eel. It is capable of giving a strong electric shock with its electric organs.

electromagnetic spectrum The range of energy emitted by a star or other energy source as waves or particles. The wavelengths in the spectrum vary from about 10 km (6 miles) or more for very long radio waves to less than 1 billionth of a kilometer for cosmic rays. A wave is represented as a curve about a nodal line. If the wavelength is short, the frequency is great; if the wavelength is long, the frequency is small. Visible light is only a small part of the electromagnetic spectrum.

element A chemical substance that cannot be decomposed into simpler substances by chemical reactions. All atoms

Electromagnetic spectrum

of the element have the same atomic number and thus the same number of electrons outside the nucleus. *See* ATOM.

Emiliania huxleyi A tiny, photosynthesizing coccolithophore also known as *Ehux*. It is the most numerous of the coccolithophores and is found in almost all oceanic water except polar. During its periods of bloom, this phytoplankton can turn 100,000 km^2 areas into turquoise islands in a blue sea. The organisms secrete plates of calcium carbonate that detach from the living organism, presumably during asexual division or upon the death of the organism, and drift down into the bottom sediment.

The tiny protist is named for Thomas Henry Huxley, who first used the term coccolith when he detected the plates in ocean-bottom sediment. The organism was not observed until the development of the electron microscope.

Ehux is significant because it exists in such vast numbers that it turns the water "white"—actually it is opaque turquoise—and this is observable from space. This change in the color of the water means that sunlight is reflected back into space so that the ocean water is cooler than it would ordinarily be. In addition, the carbon system of the water and its carbon dioxide load is greatly affected by the creation of the plates. This in turn affects global climate. The continual deposition of these plates forms oceanic bottom that on solidification becomes large limestone deposits. One such notable and massive formation is the Dover cliffs of southern England. *See* CARBON CYCLE, CARBON DIOXIDE, CLIMATE, COCCOLITHOPHORE, PHYTOPLANKTON.

Emperor Seamounts A north-trending line of seamounts (undersea mountain peaks) in the Pacific Ocean. *See* HAWAIIAN ISLANDS, HOT SPOTS, SEAMOUNT.

endangered species A variety of marine organisms are endangered; some because of an increase in the number of natural predators or diseases, and some because of human activity. Untreated sewage discharges into embayments will foul the water and render it uninhabitable by the usual organisms. Overfishing has seriously depleted stocks of important commercial fish such as cod, herring, and pollock. Several species of whales and sea turtles are endangered.

Many species of sharks are now endangered. These creatures suffer from humans on two levels; they are demonized as indiscriminant killers and therefore considered a pest group worthy of extermination, and they are also sought as a source of food, oil, leather, and sport. The body of scientific information about sharks is insufficient for marine biologists to make more than general statements about the number of individuals needed to form viable populations.

Coral reefs are extremely complex ecosystems and many are endangered. Several stretches of reef off the Florida coast have recently been declared sanctuaries. Some recovery has been noted. The coral reefs are such complex communities that it is difficult to establish just what factors influence their growth or decline. It is known that terrestrial runoff is a major polluting factor, as is dust that has been windblown from continental masses that experience drought; that lack of water may be the result of El Niño activity. *See* CLIMATE, INDIVIDUAL ORGANISMS, MARINE RESERVES.

endocrine systems Aggregates of glands and/or tissues that produce hormones. Hormones are specific proteins, peptides, or steroids that produce an effect in some other interconnected organ or structure. Annelids, mollusks, and arthropods produce hormones; one highly specialized group of these are the neurohumors. Neurohumors are secreted by the nervous system. Some of them control color change. Crustaceans have sinus glands located near their eyes that produce pigment-concentrating and pigment-dispersing hormones. The receptors are the chromatophores. Others are retinal hormones that adapt the crustacean eyes and control the amount of light entering the photosensitive eye cells.

Almost all multicellular organisms produce reproductive hormones that stimulate spermatogenesis and/or egg production. The secondary sex characteristics of crustacea are also evident in animals with functional androgenic glands. Molting, a process that is essential in the growth of animals with exoskeletons, is initiated by a group of growth and maturation hormones. The hormone systems of vertebrates are well known. Those that are specialized for marine organisms are those that control salt and water balance. *See* CHROMATOPHORE, EYE.

endoderm The innermost layer of cells of an embryonic animal. In the embryo it develops into the digestive and excretory systems.

energy The capacity for doing work. This transformation is never complete—there is always a loss of energy (entropy). The forms of energy are mechanical, electrical, chemical, thermal, nuclear, and electromagnetic. These are rather arbitrary names because there is some overlap among them. Ocean thermal energy is the energy derived from the circulation of layers of water of different temperatures.

The energy source that is the basis for most but not all terrestrial life is sunlight. The energy requirements of the aphotic zone of the ocean—the deeps where sunlight never penetrates—are met by the activity of an energy web that is supported by the chemosynthesizing bacteria and the heat of the molten magma that seeps or occasionally explodes through crustal fissures. The bacteria use pyrite (FeS_2), chalcopyrite ($FeCuS_2$), and sulfur (S) as energy sources. These materials are present in abundance at

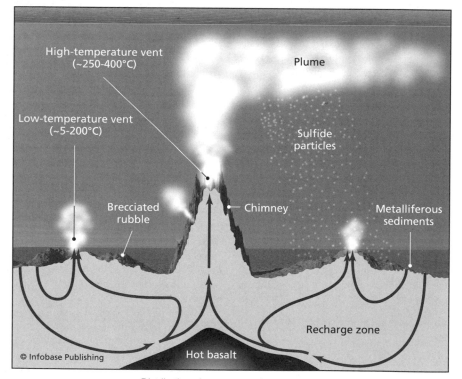

Distribution of energy at a deep sea vent

the openings in the Earth's crust. *See* ELEC-TRICITY, ELECTROMAGNETIC RADIATION, HEAT, OCEAN THERMAL ENERGY, SUN, VENT COMMUNITIES, WIND.

English Channel An arm of the Atlantic Ocean separating England from France. It slopes steeply on the English side to a maximum depth of about 120 m (400 feet). Tidal currents in the channel travel at an average of 1.5 to 3 m/second (5 to 10 feet/second) but at twice this speed in some areas, moving warm water from the Gulf Stream into the colder North Sea, along with a considerable quantity of sand. This is very much a surface-only phenomenon, since there is a high sill blocking entry to the North Sea. The usual wind in the channel is westerly and strong. *See* DOVER, STRAITS OF; NORTH SEA.

Eniwetok A group of coral islands in the west central Pacific, part of the Marshall islands group. The main island is a volcano whose top subsided. The coral growing on it has produced a limestone cap about 1,400 m (4,000 feet) deep. The structure of Eniwetok has proven Darwin's hypothesis of the origin of coral islands. Darwin postulated the origin of atolls as undersea mountains that subside and form the base for coral growths. The oldest coral on Eniwetok dates from the late Eocene epoch.

The island was used as an American atomic test site in the 1940s. *See* ATOLL, CORAL, SEAMOUNT, VOLCANO.

Enteropneusta A class of Hemichordata commonly known as acorn worms. These vermiform animals have soft, cylindrical bodies with no appendages. There are about 70 species in 4 families of these bottom burrowers of the littoral. Their movement resembles that of earthworms, and they perform a similar function of digesting organic material in the sediment. They range in size from 2.5 cm to 2.5 m (1 inch to 1.5 feet), although some 2.5 m (8 feet) specimens have been found. They range in color from white to purplish black. The sexes are indistinguishable but

separate. The taxonomy of these creatures changes: The larval state resembles the larval state of echinoderms, and DNA evidence supports these observations. *See* ACORN WORM, GRAPHOLITE, HEMICHORDATA PTEROBRANCHIA.

Entoprocta A phylum with similarities to the Rotifera and some to the Bryozoa. Entoprocta larvae sometimes resemble annelids. The adults are small polyps that use tentacles to collect diatoms and other plankton. They are either solitary or colonial. Some species of the class Ascidia (sea squirts), siphynculids, and porifera (sponges) are associated with entoprocts, but the nature of the relationship is unclear. *See* ANNELIDA, BRYOZOA, POLYP, ROTIFERS, SEA SQUIRTS.

environment All of the factors—physical, chemical, and biological—that touch the life of an organism or a community. *See* ECOSYSTEM.

environmental stress The reaction of groups of organisms to changes in the environment. Dinoflagellates, organisms that exhibit periodic "blooms," are a prime example of an organism that experiences sharp increases and decreases in population. They are most obvious in temperate waters; neither tropical nor polar waters experience the rapid increase, or bloom, in population, which is the culmination of several factors: the increase in sunlight in spring, the upwelling of colder water, wind, and the relative decrease in diatom population. Diatoms are the main food competitors of dinoflagellates, and when the diatom population increases, dinoflagellates either die—possibly in a programmed cell death event—or form cysts. Cysts are yet another response of a population to stress. These organisms are then in diapause and will be carried as inert plankton until favorable environmental conditions allow them to emerge and grow again. *See* DIAPAUSE, DIATOM, DINOFLAGELLATE.

Eocene epoch The longest epoch of the Tertiary period of the Cenozoic era. This epoch followed the Paleocene and preceded the Oligocene. It lasted about 20 million years, ending roughly 50 million years ago. The name was devised by Sir Charles Lyell, who meant it to represent the "dawn of the recent time," because fossils from it are recognizably like presently found mollusks.

Eocene marine sediment is present in the Alps, the Caucasus, the Himalayas, and the Rocky Mountains. The sands of the Gulf Coast of North America in which oil deposits are now being exploited were laid down at this time. At the same time, California and the North Pacific coast were undersea, as was much of the Atlantic coast.

Relatively warm seas supported an extensive biota, which continued most of the organisms of the Paleocene epoch. Birds, teleost (bony) fishes, and siphonate gastropods (snails) thrived, as did echinoderms and nummulites. Plankton also flourished. As the sea level rose, Australia was cut off from other landforms, establishing the "Wallace line." *See* CENOZOIC ERA; LYELL, CHARLES; NUMMULITES; WALLACE LINE; *Appendix: Geologic Timescale*.

epibenthic A term referring to the region of the ocean just above the ocean bottom and extending to a height of about 3 m (10 feet) and to the organisms in it. *See* ALGAE, BENTHOS, CORAL, ECOSYSTEM.

Epicaridea A family of tiny isopods (2–4.5 mm, 0.03–0.1 inch) that are crustacean parasites. Their characteristic feature is that the female is larger than the male.

epicenter The point on the surface of the Earth directly above the focus of an earthquake. *See* EARTHQUAKE.

epifauna Animals that live on the ocean bottom, comprising both motile and sessile (relatively stationary) species. *See* EPIBENTHOS, SESSILE.

epipelagic A term referring to the uppermost layer of the ocean water column, down to a depth of about 200 m (700 feet). *See* PELAGIC ENVIRONMENT.

Equatorial Currents Currents at the equator that move east to west, pushed along by the trade winds. These currents are fairly shallow and move at 0.5 to 2 knots per hour, or 0.25 to 1 m/second. These currents are separated by a slow, narrower, eastward-flowing countercurrent at about 5° north latitude. There is also a large, eastward-flowing undercurrent beneath the surface system. When the South Equatorial Current is weak, this undercurrent, which moves at a faster rate (2 to 3 knots/hour or 1 to 1.5 m/second), breaks onto the surface of the ocean. *See* CURRENTS.

equinox The date when the hours of sunlight and the hours of darkness are equal. There are two equinoxes each year, occurring when the sun is directly overhead at the equator at noon. For the Northern Hemisphere, the vernal equinox occurs on or about (leap years make a difference) March 21. The autumnal equinox is September 23, with a day or two variability. The dates of the equinoxes are reversed for the Southern Hemisphere. *See* SOLSTICE.

Eratosthenes (276?–194 B.C.E.) Curator of the Alexandrian library, mathematician, geographer, and founder of geodesy, the study of the Earth. He believed, as did Pythagoras, that the Earth was a sphere. He measured the circumference of the Earth by a relatively simple calculation: A vertical hole (really a well) at Syene (now Aswan), Egypt, had the sun directly overhead on the day of the summer solstice, June 21. At Alexandria, which was directly north of Syene, the shadow of an upright stick made an angle of 7°1'. By measuring the distance between the two cities, Eratosthenes could then calculate the circumference of a full 360° for the Earth. His measurement was only about 4% in excess of the modern one, despite several errors.

Eratosthenes' map of the known world put Rhodes at the center. His parallels of latitude and meridians of longitude were unequally spaced. The limits of his map were from the Pillars of Hercules (Gibraltar) to the Ganges, and in the north-south span from Taprobane (possibly Sri Lanka) to Borysthenes, which may be the outflow of the Danube.

Ericsson, Leif (980?–1001)

A Norse explorer who is associated with the Viking colony in North America. He had grown up in Greenland and as a young man went on a trading voyage to Norway. The king, Olaf, was surrounded by a court of adventurers who were full of stories of voyages of discovery. The originator of some of these tales was Biarni Heriulfson, who claimed to have seen lands to the west with trees growing on them. This was significant, since both Iceland and Greenland are treeless, as are many of the other islands of the North Atlantic.

In the summer of 1001 Leif Ericsson sailed west. He found one of the islands described by Heriulfson. It is now known as Baffin Island. His party then continued westward. The Norse sagas based on this voyage describe what might be Belle Isle. The next anchorage was most probably L'Anse aux Meadows, Newfoundland. The "vines" of Vinland could have been red currant bushes, which are found at the site. A settlement was established and from the remains seems to have lasted about 15 years, with continual fighting among the settlers and between them and the "skrallings," or local inhabitants.

Eric the Red (10th century)

A Norse sailor and explorer, immortalized by his red hair, who left Europe after a fight in 984 or 985 in which another man was killed. Sailing west, he was heading for Iceland with a party of settlers and was blown off-course to an unknown land. In a prime example of early advertising, he called it Greenland. Eric assumed that giving the new land a good name would entice settlers and encourage them to stay. When the Norse settlers found the eastern coast of Greenland icebound, they moved around Cape Farewell and established a settlement near what is now Godthaab.

estuary The area near the mouth of a river where seawater and freshwater meet. In the estuary, freshwater draining from the river system dilutes the salt water and tides also push salt water upriver to some extent, depending on the strength of the tide, geologic formations, the offshore currents, and the quantity of freshwater entering the system via the rivers.

All estuaries are geologically "recent." In the last ice age (Pleistocene), when much of the world's water was present as ice, river mouths were considerably seaward of their present positions. The present aspect of many estuaries was very different; they were longer river valleys which are now "drowned." Examples of these are the Hudson and the Thames river valleys.

In 1818 the English naturalist John Fleming published a work detailing the mechanism of mixing of waters in a well-mixed estuary, building on the earlier assumption that an "invisible wall" or dam separated fresh water from salt water.

A true, or positive, estuary is one in which more water flows in from rivers than is brought in by the tide. True, or *positive,* estuaries take several forms. One form, called a salt wedge, occurs when a large volume of fresh water is discharged and flows above the denser ocean water that reaches inshore. There is some but not much mixing, which results in a sharp difference in the salinity at a given depth. This line of demarcation, the break in the salt level, is the halocline. Some salt water is moved by wavelets into the fresh layer, in a process called entrainment. When rivers seasonally bring great quantities of water and sediment down to the sea and there are no strong offshore currents to disperse this load, deltas form, such as the Mississippi and Niger deltas.

Another type of estuary is formed when there is a strong tide and the river load is more or less the same year-round (this is more likely in temperate zones), the frictional drag of the tide on the bottom serves to stir the water, and the halocline is less steep than in the salt-wedge estuary. In the Northern Hemisphere, fresh water flows to the right as it moves downstream, and salt to the left (in the Southern Hemisphere the reverse situation prevails), the result of the Coriolis force. The Chesapeake, the Thames, and the Gironde are all estuaries of this classic type.

In a well-mixed estuary system, the tidal range is large and the water in the estuary is thoroughly mixed at high tide. This leaves salt pools and salt marshes when the tide ebbs.

Fiords are deep, glaciated cuts with much freshwater drainage and a sill separating them from open ocean. Some salt water comes over the sill, which is usually less than 100 m deep (300 feet). The fiord is usually deeper than 600 m, and the salt water in it is completely overlain with freshwater. There is very little mixing of the water systems. This is typical of the Canadian Northwest coast, Norway, and New Zealand.

A bar estuary is one built up on a low coast where there is some sediment coming toward the sea, where coastal currents move it around. The Texas coast is typical of such shoals interspersed with inlets whose size and position are highly variable.

Negative estuaries occur in areas where the rate of water loss due to evaporation exceeds that of water inflow from rivers and precipitation. In such a landform, the water flows in from the sea at the surface and out at the bottom.

Because they are partially enclosed, estuaries can become polluted. The pollution may be biodegradable or not; its likelihood of being swept out to sea or rendered harmless depends on both its quantity and the likelihood of the biodegradables in it being attacked by decomposers, sunlight, or sufficient oxygen to degrade it.

The interaction of plants, animals, and their physical world in an estuary is very complex and not at all well understood. Because rivers bring loads of inorganic essential materials, including nitrates and phosphates, turbulence aerates the water,

and there is adequate insolation, plants thrive in estuaries. Certain species are arranged in accordance with their tolerance of brackish water, but estuaries are a prime locale for algae of all varieties, from tiny phytoplankton to huge kelps. The fauna living off and on this plant community is also varied and selected for its tolerance to salt water and its exposure at ebb tide.

Each estuary has its own collection of living organisms, which will change with an alteration in any of the factors shaping it. This has enormous significance for the local and ultimately the global ecosystem. *See* CURRENTS, INSOLATION, MARSH.

eukaryote   A member of the superfamily of single-celled or multicellular organisms (living and extinct) that possesses an organized nucleus with pairs of DNA-containing chromosomes. The cells also contain the usual extra-nuclear structures such as mitochondria, Golgi bodies, lysosomes, peroxisomes, endoplasmic reticula, and microfilaments. Some have color-containing plastids. *See* PROKARYOTE.

Euphausiacea   An order of swimming crustaceans (1 to 6 cm or 0.25 to 1.4 inches) in size. They are found in the open ocean and throughout depths of over 3,500 m (10,000 feet). They constitute about 100 species and are important links in the oceanic food chain, where they are the principal food of whales. Euphausids look like tiny shrimp; some are transparent, some are red, most are luminescent. This luminescence, which is a product of special photophores, appears on the thoracic appendages, abdominal somites (body segments), or eyes of these animals. Euphausids are filter feeders. *See* CRUSTACEA, FILTER FEEDERS, LUMINESCENCE, SHRIMP.

euphotic zone   The uppermost region of the ocean, the area where photosynthesis occurs. It is the well-lit area of the sea, both its surface and the region directly below the surface. The depth of the euphotic zone is dependent on the clarity of the water, which in turn is dependent on what is in the water. Muddy water may have a euphotic zone that is about 20 meters (65 feet) deep, whereas in clear, clean water, light will penetrate to depths of 4,000 meters (13,000 feet). Temperature is also a factor, and it will vary depending on surface conditions.

Eurasian Deep A part of the Arctic Ocean near the North Pole. It is in the European Basin of the Arctic and adjacent to the Lomonosov Ridge. The water depth averages 4,300 m (13,000 feet). The floor of the Eurasian Deep is a flat, clean plain which tilts slightly upward near the Lomonosov Ridge. The part of the deep closest to Spitsbergen has the warmest water, since warmer Atlantic water enters the Arctic at that point. *See* ARCTIC OCEAN, CURRENTS.

Eurasian Plate The large portion of the Earth underlying Europe and Asia, extending west to the Mid-Atlantic Ridge. On the east the plate ends in the Sea of Japan, where it meets the Pacific Plate and Philippine Plate. The Java Trench is to the south, and the southern edge of the plate abuts on the Iran, Arabian, Turkey, Hellenic, and Adriatic Plates, which mark a line of seismic instability. *See* CONTINENTAL DRIFT, PLATE TECTONICS, *individual plates.*

euryhaline A term referring to the capability of an organism to live in environments of variable salinity. *See* ANADROMOUS FISH, CATADROMOUS FISH.

eurythermal A term referring to the capacity of some organisms to survive in a wide range of temperatures. *See* HOMEOTHERM, POIKILOTHERM, TEMPERATURE OF ANIMALS.

eutrophication An accelerated rate of plant growth, particularly the growth of algae, most often associated with freshwater, but which can occur in fairly constricted bays. Algae can "bloom" if there is comparatively little flushing out of an area and a high nitrogen and phosphorus load.

The increase in nitrogen, phosphorus, or both may be the result of agricultural fertilizer runoff, manufacturing waste, or the draining of domestic sewage. This is most noticed in fresh water where the introduction of fertilizer from agricultural runoff and/or sewage from human habitation or large animal populations in feedlots, pig, fish, or chicken farms encourages the growth of algae to the point where all the available oxygen in the water is exhausted and massive die-off occurs. This can occur in ocean water as well; the causes are the same. The algal bloom is devastating to other organisms in the biome, and the end result is a "dead zone" in the ocean. This is almost always near the mouth of a large river or near a large coastal city. The algae will grow as long as sunlight and temperatures above freezing and below about 25°C are available.

As the algae proliferate, and if there is no large animal population to keep them in check, dense mats of algae form. These choke off the surface and block the water column's access to sunlight. The algae below the surface begin to die for lack of light. The resulting decaying mass deoxygenates the water as it decomposes. The then-poisonous mass makes the water unfit for anything to live in, and there is a rapid die-off of many species.

Eventually the water will become clean again, but unless it is restocked, it will not exhibit the same populations it once had. *See* ALGAE.

evaporation The change of a liquid into a gas. When water evaporates, it becomes water vapor. This change requires energy, the natural source of which is the Sun. The warmer the water is, the greater the rate of evaporation. *See* CONDENSATION, OCEAN-ATMOSPHERE RELATIONS, WATER.

evaporites Deposits of salts left by the evaporation of a body of water. When water containing dissolved minerals stands, the water evaporates, leaving deposits of gypsum ($CaSO_4 2H_2O$), rock salt (NaCl), and a variable mixture of carbonates, sulfates, halides, and nitrates. These salt deposits are formed when the growth of a reef, or wave action, or shifts in terrain isolate an arm of the sea. The discovery of marine fossils in the same area as salt beds establishes the origin of the evaporites and dates them. *See* MINERALS, SEDIMENT.

evolution The process by which life forms change with time and accumulate successful mutations. There have been many theories in the course of human thought to explain the variety of life found on Earth. For example, Linnaeus, whose system of taxonomic nomenclature is used to describe life forms, believed that the species he saw had been present in the same form from the beginning of time. Earlier, it was held by some that any one species could produce almost any other; thus, barnacles with long necks could be the relatives of geese with similarly long necks.

Erasmus Darwin, the grandfather of Charles Darwin, was a naturalist and contemporary of Georges-Louis Leclerc, comte de Buffon. In 1794 he developed the theory of "gradual transformation." Lamarck, curator of the natural history museum in Paris, believed that the creation of a new species as a response to some environmental stimulus was possible, and that the characteristic difference in that new species was heritable. Lamarck's great critic and rival, Georges Cuvier, was a catastrophist. He believed in a scheme of three great epochs, each of which was devastated by some calamity. The first was that of great fish and reptiles, the second that of small mammals, and the third that of the giant animals: the hippopotami, mastodons, elephants, and the like.

The idea of gradual transformation created scandals. The idea of change in species became linked in the respectable (and essentially English-speaking) world with the excesses of the French revolution.

The idea of natural selection was not original with Darwin. Any animal breeder knew how to develop a particular characteristic. However, the theory of natural selection that best explained all the evidence was proposed independently by Wallace and Darwin. It provoked intense controversy and an immense literature. This outpouring still continues, as much effort is devoted to theories of the causes of catastrophic extinctions.

Although Wallace and Darwin explained natural selection, it was not until the work of Gregor Mendel and his concept of the gene was appreciated that the mechanism for this phenomenon was understood. According to this, each individual has a finite possibility of producing a slightly altered chromosome—a change in the DNA of the individual. If this change is not life-threatening, and is viable, the descendants of this individual organism will also have this genetic alteration. Thus, in a given population, many versions of a characteristic are possible.

If the population is geographically isolated, the various separated, related species will continue to change slowly along ever-divergent lines. If, however, the environment changes, only some of the versions may survive the changes. Fairly frequent changes in environment will lead either to extinction or to a rapidly evolving organism, as an individual of one or another genotype (specific genetic makeup) is favored by chance and circumstance.

The aspect of the environment that favors the survival of an individual of a particular genotype can be any one or a combination of a vast array of factors, such as the temperature of air or water or the nutrient value or salinity of water. Alternatively the organism may change color, rendering itself more (or less) noticeable by predators (or prey), or it may develop (or lose) an enzyme system that makes digestion of a particular food possible (or impossible). A migrating species (i.e., one that is new to one area but not so different that it would produce sterile matings) may introduce still greater variety into a gene pool and again increase the variability of a population.

Current evolutionary theory explains extinction in terms of wide divergence. If

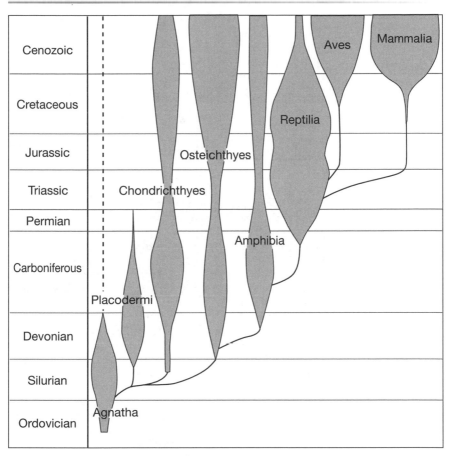

Evolution of vertebrates

a particular group of organisms becomes so specialized that only a narrow band of conditions will support its existence, a small change in the environment can be fatal. Thus, if a small predator population is wholly dependent upon a particular prey, a sharp decline in the prey population (due to any one of many variables) will eliminate the predators because they will be unable to get to enough food for a long enough period to maintain a reproducing population.

Current areas of evolutionary research involve the relationships between organisms, their lines of development and the similarities (or differences) in their structure, and the protein structure and enzyme systems of organisms. *See* AGASSIZ, JEAN-LOUIS RODOLPHE; BUFFON, GEORGES-LOUIS LECLERC, COMTE DE; CATASTROPHISM; CUVIER, GEORGES; DARWIN, CHARLES; FOSSILS; LAMARCK, JEAN-BAPTISTE PIERRE ANTOINE DE MONET, CHEVALIER DE; LINNAEUS.

evolution of the oceans The currently accepted theory of the evolution of the oceans holds that at one point in the Earth's geologic history there was only one supercontinent—actually a consolidation of several fragments—called "Pangaea," which was centered around the equator. As Pangaea broke up, the water of the primordial ocean also split into the separate oceanic areas, the Atlantic Ocean being the oldest. Pangaea broke into a distinct northern segment (Laurasia) and a southern segment (Gondwanaland) during the Jurassic period, about 195 to 135 million years ago. By 150 million years ago, the central Atlantic had spread to about one-third its present width. Europe began to move away from North America, leaving remnants (terranes) on the Carolina mountains. Bits of North America are in Scotland. By the Lower Cretaceous period, the Iberian Peninsula had moved counterclockwise, creating the Bay of Biscay. By 60 million years ago the last of the large Atlantic basins formed. This created the Norwegian Sea and separated Greenland, a North American fragment, from Europe. At the same time, all the continental mass was moving northward and opening the water channels, making possible water circulation between tropical and polar water.

Gondwanaland fragmented about 140 million to 130 million years ago. The Indian landmass split away from Australia and Antarctica somewhat later; the usual figure is 110 million years ago. An island chain formed at the Ninety Degree East ridge (90°) and acted as a barrier between the deep eastern basin of the Proto-Indian Ocean and the shallower western part. The eastern basin is now the Wharton Basin. The Ridge continued to act as a barrier, and by 53 million years ago came close to the Mid-Indian Ocean Ridge, which once passed north of Australia, but as that continent drifted northward, the ridge appeared to its south. This opened the channel between Antarctica and Australia, making polar water and Indian Ocean water freely exchangeable.

The Pacific Ocean dates to the time when India moved north of Antarctica. About 100 million years ago the Pacific Ocean area was composed of four large plates whose borders are now not well-defined. They were the Antarctic, Pacific, Kula, and Farallon Plates, each moving away from the others. Trenches or subduction zones consumed much of this oceanic crust. The ridges that seamed this Proto-Pacific Ocean were connected to those of the Indian Ocean, which was opening at the same time.

Between 80 and 60 million years ago, the Tasman Sea developed, pushing Australia and New Zealand apart. By 55 million years ago Australia began to move north and away from Antarctica. One of the four plates that formed the area, the Kula Plate, which lay between Siberia and the Pacific coast of North America, all but totally vanished into the Aleutian Trench. The Farallon Plate disappeared about 27 million years ago; it fragmented into the remains of the Gorda, Cocos, and Nazca Plates. *See* ATLANTIC-TYPE MARGIN; CONTINENTAL DRIFT; CRETACEOUS PERIOD; JURASSIC PERIOD; MID-OCEAN RIDGES; PLATE TECTONICS; TRENCH; WEGENER, ALFRED LOTHAR; *individual oceans; individual plates.*

Ewing, William Maurice (1906–1974)

An American geophysicist and the first director of the Lamont Doherty Laboratory of Columbia University. He was instrumental in the reexamination of Alfred Wegener's theory of continental drift, and explained it by the idea of seafloor spread, or creation of new bottom at the mid-ocean ridge. Ewing took the first seismic recordings in open seas (1935) and the first deep-sea photographs in 1939. He was one of several scientists who suggested that earthquakes and the undersea rifts associated with oceanic ridges were linked. One of his major publications was *The Floor of the Oceans: The North Atlantic* (1977). *See* CONTINENTAL DRIFT; HEEZEN, BRUCE CHARLES; WEGENER, ALFRED LOTHAR.

explorers and explorations

An explorer is defined as one who travels in search of geographic or scientific information. However, the Europeans of the 15th century and the explorers that followed, until the 18th century, were impelled to travel great distances for what some have called the causes of gold, God, or guns. We know their names, but they were not the first ocean explorers, nor were they unique.

Polynesians had travelled vast distances in open ocean, navigating by stars and returning to particular island groups. Their voyages were most certainly not unidirectional. Chinese sailors navigated successfully as far as Africa, making regular trips. Arab mariners and traders had well-established routes along the African and Indian coasts.

The first known name in connection with exploration is that of Necho, the Pharaoh who financed a Phoenician circumnavigation of Africa. Herodotus, in the 5th century B.C.E., produced a map of the world that stretched from Gibraltar to India. The astronomer-geographer Pytheus was also an explorer. He sailed north to Iceland in 325 B.C.E., using celestial navigation to arrive at a measure of latitude. Eratosthenes measured the Earth and developed the concept of latitude and longitude. Ptolemy, in the 2nd century C.E., continued and expanded on these works.

The Romans were users of geographic works, but were not particularly interested in exploration for the sake of science. After the fall of the Roman Empire, the center of exploration shifted to both the east and west. Brendan and the Norsemen in the west continued to travel and chart new lands; Arabs in the east did the same.

With the Renaissance and the renewed interest in the world in general came the potential for new trade. That and the technological advances of better ships, better navigational instruments, and more accurate maps made long voyages possible. These were undertaken both for national honor and for profit. This was the age of da Gama and Columbus.

These early navigators were followed by traders and, in the 18th century, by the mainly scientific explorers such as Cook, LaPerouse, and Bougainville. Naturalists accompanied the geographers, thus reviving a practice of the ancient Greeks. Banks, a botanist, and Darwin, a natural scientist, were on voyages of exploration as "scientists on board."

By the 19th century it was recognized by all maritime nations that maps and soundings of all parts of the world were important. This led to a renewal in various arctic interests and the voyages of the Rosses. Surveying techniques were developed by Maury and others.

Increasing information led to the need for consolidation and classification of data. The need to know and to provide as much information as possible as a public service to all scientists led to the *Challenger* expedition around the world in 1872–76 and then the polar explorations of Nansen in the early 20th century. The *Meteor* survey was the major work of the 1920s, although its report was not published until the 1950s.

Exploration in the second half of the 20th century has produced images sent to Earth from space, photography from the deeps, and the development of free diving. New genera unseen and unknown before 1950 are now classified. The use of submersibles has greatly extended the researchers' reach. Recent expeditions in 2003 have used *Alvin* in a number of trips to the sea floor in the Gulf of Alaska. The object of this search was the exploration of seamounts such as the Chirikof and the Warwick in an attempt to unravel the volcanic history of these structures and the biota that live near or on them.

One of the more unusual recent finds is cold-water coral. Most corals live in shallow, warm water, but deep-water specimens thrive in the waters of the polar ocean. Like their warm-water counterparts, the slow-growing corals of colder seas create a habitat for other creatures. Yet we still know more about the moon than about the Earth's ocean deeps. There is much left to learn. *See* ALVIN, BACTERIA, HISTORY, MUCUS, SEAMOUNTS.

extinction The natural end, or disappearance, of an organism. Throughout geologic time, many taxonomic families and groups have become extinct. At least four—some researchers say five or more—distinct periods saw large-scale or mass extinctions. The disappearance of a taxonomic family was an ongoing process that lasted about a million years.

Under normal extinction patterns, fewer than eight families per million years disappeared. The major variations in this occurred around the ends of the Ordovician, Permian, Triassic, and Cretaceous periods. At the end of the Cretaceous up to 19.3 extinctions per million years were noted.

When calculating the number of extinctions, allowances have been made for those organisms, both plant and animal, that are rarely preserved. Shells and skeletons become fossilized, but soft tissue usually disappears without leaving a trace.

There have been a number of theories to explain extinctions. It is no coincidence, however, that the major breaks occur at the ends of geologic periods. The original 19th-century geologic timescale was constructed on the evidence of large-scale change, and geologic periods ended with more or less abrupt (give or take a million years) changes.

Salinity changes, the deepening of ocean basins, and meteor impact have all been proposed as causative agents of extinction. Changes in the salt content of oceans would kill off marine organisms if there were suddenly less salt and would poison freshwater species if the salt content increased. If ocean basins became deeper, marsh organisms would be drowned or littoral life-forms would become stranded on dry beaches. Meteor impacts might have showered the Earth with dust and changed the weather much as a large volcanic eruption does now. The anomalously large iridium layer corresponding to the late Cretaceous extinctions may be an example of this phenomenon.

Changes in the polarity of the Earth are still another possible cause of extinctions. Some researchers have also argued that the level of extinctions and emergence of new groups has slowed because the more adaptable—meaning more highly evolved—organisms are better able to withstand environmental shock. Therefore they are more long-lasting or durable than those in past ages that were felled by small but, for them, significant changes. Plants and animals have been driven to extinction by human activities, either from overuse or a loss of habitat.

An attempt to study the ecosystem and establish the consequences of extinction involves the sampling of ocean mud—the top layer of the ocean bottom. The mud holds the remains of creatures that have lived in the ocean for thousands of years. This sediment is disturbed and churned by a large number of animals, such as crabs, clams, sea urchins, brittle stars, and marine worms. As these animals seek their food, they churn up the mud, thus oxygenating both water and mud and making life possible in the mud. Looking at the rate of sediment mixing produces a model of the biota that lives in it and leads to prediction of what would be the result of losing species.

As larger organisms create more disturbance in any biome, they are the first to become extinct. The loss of the large organisms results in less oxygen in the sediment, and that will push smaller organisms to extinction. Thus, some organisms have more impact on the rate of sediment mixing than others, and the extinction of those is critical to the overall health of a particular biome. Since the food web in the sea is intricately linked to what happens in the sediment and how nutrients are generated, the sediment affects every creature that lives in the ocean.

The effects of human activity have disrupted the ocean bottom, in some places beyond recovery. There are "dead zones" in the ocean resulting from the dumping of untreated sewage in some places or the introduction into ocean water of agricultural runoff that contains a high concentration of fertilizers. An example of the latter is the dead zone in the Gulf of Mexico. It is the result of fertilizers delivered to the area by the Mississippi River, and almost nothing lives there. *See* EUTROPHICATION, PFISTERIA.

extremophiles Organisms that live in regions of temperature and/or pressure that until recently would have been called lifeless. The deep sea vents are an example of an extreme environment; they are covered with living organisms that can withstand the incredible pressure of a water column thousands of meters high and temperatures of 200°C or higher. At that pressure, the water does not boil. The organisms range from bacteria to giant worms, clams, crabs, and shrimp.

An ongoing investigation of extremophiles is being conducted at the Hawaii Undersea Research Laboratory (HURL), which has been studying Loihi—the underwater volcano that will produce the next Hawaiian island—since 1988. This project is attempting to sample the organisms present, retrieve them, and maintain them in laboratories at the surface. *See* ALVINELLA, ICE WORM, POMPEII WORM, RIMICALIS, SEEPS, TUBE WORMS, VENT COMMUNITIES.

eye A real eye has a distinct transparent cornea covering a flexible, spherical lens that receives light images, concentrates them, and directs them to an image-receiving retina. This structure is composed of rods that receive size and shape information and cones that are color receptors. This visual information is then transmitted to the brain by way of the optic nerve.

Almost all animal phyla have some photoreceptor system. In vertebrates, the eye is a brain substructure. In invertebrates, the eye is an epidermal structure. The blue "eyes" of scallops are light-sensitive tissue, not real eyes. Jellyfish, flatworms, and echinoderms have ocelli-pigmented photoreceptors on their surfaces. Many cephalopods have both eyes and light-sensitive vesicles near them. The vesicles are thought to have physiological functions such as timing of reproduction. Some polychaete worms have image-forming eyes, and crustaceans as a group have well–developed eyes and optic networks.

Therefore, what an animal "sees" is dependent upon the array of photosensitive tissue—both its quantity and concentration—and whether or not a cornea and lens exist, which determines the extent of motion detection possible. The presence of a lens separates animals that can really see from those that can only detect differences in ambient light. Arthropods tend to have lenses of fixed focal length, and some have an increased visible field because they have eyes on stalks. A compound eye is not a collection of little eyes; instead, it is arrayed so that a number of receptors form an image by concentrating light. It may have advantages in motion detection. The Alciopidae (segmented worms) have the best developed polychaete eyes. They are hydrostatically focused. The lenses of octapod eyes are moved by muscles.

In general, predators' eyes tend to be more complex and capable of image formation than those of related species. Predators with modified eyes include the octopus. The eyes of deep-sea animals, including chimeras and gulpers, are further modified; they are set forward on the head, have temporal rods that have longer light-sensitive areas adapted for dim surroundings, and have a wide binocularity that increases spatial perception. Some animals have tubular eyes—with the usual round lenses—as a space-saving device. Modifications notwithstanding, bottom dwellers and fish in the abyssal depths do not have good vision.

F

Fahrenheit, Gabriel Daniel (1686–1736) A German meteorologist and instrument-maker renowned for his work on the measurement of temperature and the design of a thermometer scale. Fahrenheit studied physics in the Netherlands, Germany, and England and was elected to the Royal Society in 1724.

In devising his thermometer scale, Fahrenheit used ice and salt to establish the lowest temperature he could make. He called this 0°. His second point of reference was that of a normal human body. He then divided the interval by units of 12. On this basis the freezing point of water is 32° and the boiling point is 212°. The Fahrenheit temperature scale is no longer used in most of the world, the United States being the major exception. The world's entire scientific community uses the Celsius scale.

Possibly more important than his design of the thermometer scale was the experimentation Fahrenheit did using liquids other than water as indicators of temperature. His attempts at low-temperature work led to the use of various alcohols as indicator liquids, and alcohol is still used in this manner. *See* CELSIUS, ANDERS; TEMPERATURE.

fairy shrimp The order Anostraca. These small animals are less than 10 cm (4 inches) long and characteristically swim upside-down. They are fairly widespread in shallow waters, where they feed on bacteria, algae, protozoans, and organic debris. At least one species feeds on other anostracans. They are unusual marine organisms in that their eggs can withstand some desiccation and remain viable.

Falkland Islands A group of islands on the continental shelf off the southeastern coast of South America, less than 300 km (about 200 miles) west of the Straits of Magellan. They are windswept, rocky, treeless, small peaks perched on a very old Devonian landmass that is analogous to that of the South American mainland. The islands are inhabited by over a half million sheep and several thousand people, mostly of British descent. The islands are claimed reluctantly by Britain and vigorously by Argentina, the latter claim being based on an older claim by the Spanish crown. This unresolved series of claims and counterclaims led to a war between Great Britain and Argentina in 1982. *See* ANTARCTIC OCEAN, DEVONIAN PERIOD, EXPLORERS AND EXPLORATIONS.

Falklands Current A cold polar current flowing parallel to the eastern South American coast; it meets the Brazil Current at about 35–40° south latitude. *See* ATLANTIC OCEAN, ANTARCTIC WATER, BRAZIL CURRENT, EXPLORERS AND EXPLORATION.

Falklands Plateau An undersea plateau east of the Falkland Islands and at a considerably greater depth than the ocean floor surrounding the islands. It ranges from 2,000 to 3,000 m (or 6,600 to 9,900 feet). Steep escarpments separate it from the trench east of South Georgia Island and the Scotia Ridge to the south. *See* ATLANTIC OCEAN, ANTARCTIC WATER.

FAMOS (Fleet Application of Meteorological Observation) This weather-studying project began in 1962 and was at the very beginning of environmental studies using satellites. It functions as a part of the National Ice Center. *See* NOAA.

FAMOUS (French-American Mid-Ocean Undersea Study) In 1974 this expedition explored the axis of the Mid-Atlantic Ridge. *See* MID-ATLANTIC RIDGE, MID-OCEAN RIDGES, OCEAN FLOOR.

fan A geologic formation at a river or canyon mouth. It is the characteristic spreading pattern of a river-borne load of sediment.

Faroe Islands A group of islands directly north of Great Britain; they are a part of Denmark. There are 17 inhabited islands and many small, uninhabited islets in the Faroe group. The group is one of a series of volcanic peaks with high, steep cliffs and very irregular coastlines. The climate is mild due to the surrounding warm water of the North Atlantic gyre (water circulation). There is almost daily rain and fog. The dominant native animals are birds, notably the puffins and eider ducks. Because of the constant wind, the islands are essentially treeless. The first colonists of the Faroes were Irish monks, who came about 700 C.E. The Vikings arrived about a century later. *See* EIDER DUCKS, GYRE, PUFFINS.

fast ice Sea ice that is frozen onto a shoreline and is, therefore, stationary. *See* FLOE, ICE, ICEBERG.

fathom An old measure of water depth, now defined as 1.83 m (6 feet). The word is of Danish origin and means outstretched (arms). It is supposed to be the length from left middle finger to right middle finger of a large man standing with his arms outstretched.

fault A fracture zone in the crust of the Earth. Displacement of blocks of crust occurs at faults. A fault block is a piece of crust, usually rectilinear. A fault block that is pointed down is called a graben; one that is tilted up is called a horst.

fauna The animal life of a place at a particular time. *See* ECOSYSTEM.

feather star *See* CRINOIDEA.

felucca A small sailing vessel of ancient design. These craft were once quite common in the eastern Mediterranean; they are now almost exclusively Nile River craft. They usually have two masts—infrequently there is only one—and lateen sails. The smaller feluccas had rowers.

fiddler crab A member of the genus *Uca*; a true crab with its first cheliped, or large walking leg, highly developed into a pincer claw. This oversized right claw is a feature only of the male; females have symmetrical legs. The modified leg is used in courtship and mating. *See* BACKSHORE, CHELA, CRAB.

Fiji A group of Southwest Pacific islands at about 15° south latitude and 180° west longitude. Fiji is composed of over 200 islands, of which fewer than half are inhabited. The population is Polynesian and Melanesian. The islands are an important resort area and also produce agricultural products.

Geologically the Fiji islands are interesting because they are not the usual atolls of volcanic and coral construction but have continental basement rock. *See* FIJI PLATE, ISLANDS.

Fiji Plate A small but separate part of the Earth's crust, originally attached to Antarctica. It is therefore "continental," and its various large islands are a composite of continental igneous and sedimentary layers, with adducts of volcanic eruptions and coral encrustation. The smaller islands in the Lau group are truly oceanic and form a tail that gives the entire Fiji Island group a "lambchop" configuration.

The Fiji islands rest on the plate, which is divided into two areas. The North Fiji Basin (Pandora) is about 2,000 m (6,600 feet) deep and circumscribed by the Melanesian Border Plateau, the north-south line of the New Hebrides, and the Hunter Ridge. The South Fiji Basin is a triangularly shaped area with an average depth of about 4,000 m (12,000 feet), pointed at New Zealand. The "tail" of the Fiji "lambchop" is separated from the almost true north-south line of the Tonga-Kermadec Trench by the Tonga-Kermadec Ridge. The slippage of the plate is northwest in the South Fiji Basin and southwest in the Pandora Basin. *See* PACIFIC PLATE.

filter feeders A variety of animals living mostly on the ocean bottom or in the plankton, whose feeding mechanisms are a filter, a means of creating a current running through the filter, a scraping device to move food from the filter, and a water exit.

The filter's function may be accomplished by a mucus net put out by the animal and ingested along with anything it has trapped. The filter itself may be made of cilia, setae (and in crustaceans), antennae, or antennules and may be part of gill (mollusk) or maxilliped (crustacean leg) structures. The cilia, or setae, may beat in concerted waves to create a current that moves detritus or plankton past the net to entrapment. Either the cilia or setae or maxillipeds then deliver the item to the mouth.

The filter will absorb particles of an acceptable size and reject those that are too large. In the sponges and sea cucumbers the entire body acts as the filter, water inlet, and water exit. The baleen whales, which are the largest filter feeders, have blowholes that expel the water ingested along with the krill that is filtered out of the water. *See* BENTHOS, CILIA, DETRITUS, KRILL, PLANKTON, SETAE, SPONGE.

Finland, Gulf of A 400-km (250-mile)-long arm of the Baltic Sea between Finland and Russia. Its maximum width is 130 km (80 miles). The depths range from 8 m (26 feet) to 100 m (290 feet). The Gulf is separated from Lake Ladoga, which was once a part of it, by the Karelian Isthmus. The eastern end of the gulf is quite narrow, being only about 20 km (12 miles) wide.

The bottom of the gulf is sandy and has many bars that, along with the heavy winter ice, make navigation difficult and perilous. The gulf receives the outflow of several rivers, the largest of which is the Neva, the river that runs through St. Petersburg. *See* BALTIC SEA.

fiord (or fjord) A waterway in a rocky or mountainous region that may be currently glaciated or have been covered by moving ice in the past. The coastal cliffs are deeply infiltrated by seawater. Fiords are characteristic of the Alaskan, Chilean, Norwegian, and southern New Zealand coastlines. *See* ESTUARY, GLACIER.

fire worm A polychaete (also called a bristle worm). One of the most widespread families in this order are the Amphinomidae. Bristle worms range from 10 to 20 cm (4–8 inches) in length and usually hide in mud or under rocks. They are part of almost all the oceanic biomes and are carnivores, preying on coral polyps and other small organisms. The segments of this animal bear setae that deliver a toxic material, which produces a painful chemical burn. The setae are also brittle and break off on contact. These animals therefore present a danger to divers. If a bristle worm is touched, the bristles (setae) that touch skin must be removed to prevent secondary infection. *See* CHEMICAL DEFENSES.

fish Aquatic animals of the superclass Pisces, which breathe by means of gills. The jawless fish, the Agnatha or Ostracoderms, are sometimes put into a phylum of their own. Fossil agnathans are the evidence of the earliest vertebrates; they were well established in the Ordovician period (500 million years ago). Agnathans had, in addition to bony skeletons, formidable bony armor plates.

The living species of fish represent a considerable portion of all the members of this superclass. The classes and orders of fish are as follows:

Class Agnatha
    Order Cyclostomata
        Lampreys and hagfish
Class Placodermi (known only as fossils)
    These creatures had both jaws and armor. They have been found in some very ancient rocks.
Class Chondrichthyes (cartilaginous fish)
    Subclass Elasmobranchii
        Order Selachi
            Sharks and dogfish
        Order Batoidea
            Skates and rays
        Order Holocephali
            Chimaeras
Class Osteichtyes (bony fish)
    Subclass Actinopterygii (ray-finned fish)
        Superorder Chonrostei
            Most of these fish are fossils only. The sturgeons, distinguished by their ganoid (tilelike) scales, are living examples of this group.
        Superorder Holostei
            Most of this group is also extinct; the mudfish are living specimens.
        Superorder Teleostei (spiny-finned fish)
            About 95% of all living fish are in this group of about 20,000 species.
    Subclass Choanichthyes (lobe-finned fish)
        Order Crossopterygii (stalked fins. Most species extinct)
        Suborder Rhipidistia
        Suborder Coelacanthini
        Order Dipnoi (lungfish)
            These animals have swim bladders that act as air reservoirs. A major evolutionary advance in these fish is the nose-to-mouth connection. This "minor" improvement makes possible the transition to air-breathing.

Although fish exist in an incredible variety of forms, sizes, colors, and habitats, they share several unifying structures. One is their covering; their skin is covered by colorless dermal scales. The color comes from the dermis above and below the scales. Fossil scales are osmoid, either bony or ganoid. The extant ray-finned fishes also have ganoid scales. These contain ganoin, a hard, translucent material similar to tooth enamel. The arrangement of the scales in ganoid fish resembles floor tile, being edge-to-edge as opposed to overlapping. Sturgeon scales have no ganoin but the pattern is ganoid. Placoid scales are present in the Elasmobranchii; each scale has a rounded element with a spine or dentine projecting toward the tail. Ctenoid scales are thin, translucent, and overlapping, with comblike projections on the free edge; their lines of growth vary with species and season. Cycloid or roughly circular scales are found on some primitive fish; this type of scale is present earlier in geologic time than is the ctenoid scale.

The digestive system of fish is fairly simple. A mouth is the passage for food and for water used in respiration. A short esophagus leads to the stomach and digestive glands. The liver in some species is unusually large—a phenomenon leading to much speculation but as yet no concrete answers. The anus (or cloaca in some species) is anterior to the tail structure.

Respiration is conducted by the gills. These are made up of filamentous epithelial tissue arranged in tiny parallel lamellae in the interior gill. The exterior gill is covered by ectoderm. The elasmobranchii have gill slits as the external opening, the teleost fish have an operculum. This gill cover acts as a one-way valve. Water rushing into the gill is passed through the lamellae, which remove oxygen from it and deposit carbon dioxide and gaseous nitrogenous waste into it, after which it passes out of the mouth. The removal of most of the waste nitrogen from the body by the gills explains why fish have only rudimentary kidneys, if any.

Swim bladders may be respiratory organs but usually arise as an offshoot of the digestive system. They are the equilibrating hydrostatic structures that make change of depth possible without the animal experiencing decompression problems (i.e., the bends). The mechanism by which air is introduced into the swim bladder is not known, nor is there a satisfactory explanation of why fish or other diving organisms without swim bladders can manage rapid changes in pressure without experiencing illness.

The reproductive mechanisms in this vast group of water creatures vary enormously. Most fish are either male or female, but some are hemaphroditic and some change sex during their lifetimes. Most produce vast numbers of externally fertilized eggs with little yolk supply. This means rapid hatching and many fry that do not live to grow up. Other fish, such as many elasmobranchs, have internally fertilized eggs that develop either from a large yolk, (the ovovivaparous fishes) and produce live young after a long development, or have a placenta-like organ that nourishes the embryos (the viviparous fishes).

The circulatory systems of fish are analogous to those of other vertebrates. They have two-chambered ventral hearts, closed circulatory systems, and red blood cells that contain hemoglobin as an oxygen-carrying substance. Oxygen can also diffuse directly into the blood from the gills and through the skin. This may explain why there is not very much hemoglobin in fish blood. It constitutes about 2% of the blood in most teleosts, as compared to about 15% in humans.

The sensory reception of fish is good. While there is probably no sense of taste because the palate is too bony and the tongue lacks papillae, the sense of smell is excellent. The senses of hearing and sight also seem good, since the receptor areas for these senses in the brain in all varieties of fish are well developed. The lateral line of fish acts as a receptor for vibrations.

The generalized fish body plan has an oval shape with fins attached laterally and a tail at the end. The tail in teleost fish is a symmetrical one, its upper and lower halves, being of equal size. In cartilaginous fish the upper lobe of the tail is larger than the lower one. The fins of fish are most often paired appendages of dermal origin (as is the tail). Thus, the pectoral and pelvic fins are paired. The dorsal, anal, and ventral fins may or may not be present in any particular species, but when present they are usually unpaired. *See* COLOR, EYE, LATERAL LINE, PARASITES, SCHOOLING.

fishing The catching of marine or freshwater fish and other organisms for food or sport, individually or on a large scale.

Currently the annual worldwide commercial catch is over 75 million metric tons. With some improvements this can be extended to 100 million metric tons. This is considered to be an upper limit, since many fishing grounds are being used to the maximum. A few nations catch the great bulk of fish. In terms of tonnage taken, Peru ranks first, Japan second, and Russia third, with about 25% of Japan's total. In order to prevent overfishing, a number of recent treaties involving two or more North Atlantic nations have been signed with the aim of maximizing the catch while not depleting the breeding stock.

Fishing methods, whether mechanized or not, follow a few basic patterns. Inshore or coastal fishermen rarely exceed the 5 km (3 mile) limit. They do not trawl, and they take most of the known food fish and shellfish, including shrimp, crabs, lobster, herring and the related pilchards and anchovies, cod and its relatives, the pollack and haddock, mackerels, tuna, salmon, flatfish, hake, and mullet. The fishing techniques they use depend on the size of the operation. Pole and line are still used in fishing from small boats. Small tuna are also taken by hand.

The large factory ships or large fleets set purse nets which encircle fish, particularly schooling fish. Drifting long lines bearing multiple hooks are used for tuna. Bottom dwellers such as cod, flatfish, and rays are taken with bottom lines. The major part of the crustacean catch is brought up by trawlers using conical bag nets.

Finding fish can be an art; they do not always appear in the same places. Schools move in response to water temperature, food supply, salinity, wind and water currents, and other yet unknown factors. They are spotted by the likelihood that they will be in a certain place at a certain time of year, and more scientifically by using echo sounding, bioluminescence, or the observation of accompanying species. The last technique is based on the frequent observation of tuna with accompanying porpoises, for example, or of birds on the surface following herring schools. *See* MARICULTURE, NETS, SEINES.

fish oil Oil obtained from the liver of fish, principally cod and sharks. The main commercial use of fish oil is as a dietary supplement, since it is a good source of vitamins A and D. While this use has declined, the oils are still used in the manufacture of paints and coatings. *See* MARINE OIL, OIL.

Flabellifera A large and widely dispersed family of isopods. Some are economically important as fish parasites; others are wood-boring "gribbles." *See* ISOPODA.

flagella Tail-like organelles on a cell surface, which move rapidly and usually rhythmically. They are structurally similar to cilia, the difference being one of size and number: Flagella are larger, stronger, and present in smaller numbers than cilia. Structurally, both flagella and cilia are bundles of muscle fibrils arranged in distinct packages that divide longitudinally when the cell or organism bearing these organelles divides by fission or reproduces asexually.

Flagellate motion involves a bending backward of the flagellum, which both undulates and rotates about a lateral axis. The body of the cell then rotates and moves forward in response to this motion, with an action analogous to that of a screw propeller. *See* CILIA, DINOFLAGELLATES, PLANKTON.

flagellate Any planktonic organism that has an appendage or two that aid in motion is a flagellate; those with many appendages are ciliates. The role of these organisms in the food web is under study. Flagellates may be naked, or shell-less, and have flagella of different sizes. They, along with the phytoplankton, form the food supply for copepods. *See* CILIOPHORA, COPEPOD, FOOD WEB, PHYTOPLANKTON, PLANKTON.

flatfish Fish with both eyes on the same side of the head, such as sole, flounder, or halibut. The eyes and skin of flatfish, however, differ from species to species. The larval flatfish has one eye on either side of the head, but as the larva matures, one of the eyes migrates so that both are on the same side. Depending on the species, either the left or the right eye dominates and stays in place, while the other eye migrates. There are other changes in flatfish during maturation. The scales on some flatfish may become cycloid (smooth and rounded) on the underside and ctenoid (with sawed edges) on the superior surface. The swim bladder also tends to disappear on maturation. *See* FISH, FLOUNDER, SCALES.

Flinders, Matthew (1774–1814) An English explorer and navigator, he served on a voyage to Australia with George Bass. They became friends and colleagues and surveyed southern Australia together. Flinders circumnavigated Australia in 1802. The voyage was a very difficult one, and his ship, the *Investigator,* almost did not survive it. Most of the crew died, and Flinders was never healthy afterward.

Flinders did very careful work determining the extent of compass deviations due to iron components in the ship. His precise calibrations of compass error and deviations due to nearby ferrous material were vital after the advent of metal ships.

Flinders was to have returned to England after his voyage on the *Investigator,* but he was shipwrecked and captured by the French on Mauritius, where he remained a prisoner until 1811. He finally returned to publish *A Voyage in Terra Australia* in 1814.

FLIP (Floating Instrument Platform) This is a 355–foot long research platform, owned by the U.S. Navy and operated by the Scripps Institution of Oceanography. It was launched in 1962 and carries a scientific complement of 11 and a crew of five. It can float or be moored in water deeper than 2,000 fathoms (3,000 m, 12,000 feet). Some of the studies conducted have yielded results in determining acoustic propagation through water, wave study, and laser communication.

floe A relatively flat piece of floating ice. Ice floes vary in size from quite large (over 1 km in diameter) to smaller than 25 cm (less than 1 foot). *See* GLACIER, ICEBERG, PACK ICE.

flora The plants of a given locale at some defined time. *See* ECOSYSTEM.

Flores Sea A body of water separating the islands of Flores and Sumbawa from Sulawesi (Celebes) to the north. The Makassar Strait is also north of the Flores. The bottom of this small sea is composed of four distinct regions. The west is a submerged plain with depths of 500 to 1,000 m (1,500 to 3,000 feet). The Postiljon and Pasternoster Islands are seamounts that have accreted coral rising up from this plain.

Another feature of the sea is the Flores Trough, parallel to the Sunda Shelf. It has a maximum depth of over 5,000 m (16,000 feet). The ridges south of South Sulawesi Sea are marked by islands, of which Salajar is one.

It is believed that the Flores Sea is the result of downwarping and subsidence of continental areas. The focus of seismic activity in Indonesia, a region where earthquakes are well known, is in the Flores Sea. *See* PACIFIC OCEAN, SEAMOUNT, SEISMIC ACTIVITY, SUBDUCTION.

Florida Bay A body of water at the southern end of the Florida peninsula, its southern boundary is the string of small islands called the Florida Keys. The western end of the bay opens into the Gulf of Mexico and is delimited by the last of the Keys, the Dry Tortugas, which are west of Key West. However, Key West, at about 81° west latitude, is sometimes referred to as the last key.

The bay and the keys are of geological interest. The bay is quite shallow, and calcium carbonate precipitation in it is an ongoing process that has continued throughout geologic time. This is a present-day example of the process that has produced the limestone beds found in various places on Earth and used by paleontologists and geologists interested in the age of such formations. The keys are exposed coral reefs which formed in the last ice age (Pleistocene) and are still growing by coral accretion. *See* CARIBBEAN SEA, CORAL, GULF OF MEXICO, GULF STREAM, LIMESTONE.

Florida Current The large volume of water leaving the Caribbean and moving northeast at an average speed of 2 to 4 knots/hour through the constriction between the Pourtales Terrace and Cuba that forms the Straits of Florida. The straits, which separate the Atlantic and the Gulf of Mexico, are quite narrow and only about 800 m (2,500 feet) deep. As a result, the warm Florida Current flows into the Atlantic Ocean with great force. This current forms part of the Gulf Stream. *See* CARIBBEAN SEA, GULF STREAM.

flounder (Pleuronectiformes) A widely distributed marine flatfish taken for food. Flounder are bottom dwellers whose eyes are on one side of the body. There are two families of flounder: right-eyed (Pleuronectidae) and left-eyed (Bothidae). In a right-eyed fish the right eye migrates. The shape of the head is also deformed. Both have free-swimming young. The right-eyed types, about 100 species, are found in north temperate and polar waters. They include the halibut, plaice, winter flounder (*Pseudopleuronectes americanus*), and starry flounder (*Platichthys stellatus*). All of these are fished commercially.

The Bothidae comprise about 200 species inhabiting warmer water, and most of these are smaller than the right-eyed types. The sand dab, California halibut, turbot, and summer flounder (*Paralichthys dentatus*) are among the species belonging to this group that are fished for food. *See* FISHING, FLATFISH, HALIBUT.

flowmeter An instrument for measuring the velocity of water. A flowmeter usually has a wheel, propeller, or rotor whose revolutions per minute can be transmitted to a monitoring station as a measure of the velocity of the water, which causes the rotor to turn. Flowmeters are designed to be attached to ocean bottoms or to a stationary object, although some float on a buoy. They may be anchored at some height above the bottom when the object is to determine the velocity of ocean currents at varying depths. *See* CURRENTS.

flying fish (Exocoetidae) A small fish (usually less than 45 cm or 18 inches) with large winglike pectoral fins, found in tropical and semitropical waters. Some, such as the California flying fish, which looks like a four-winged flying object, also have enlarged pelvic fins. Flying fish build up speed in the water and break out of it using their fins as gliders, while the still-submerged tail acts as a propeller.

fog A meteorological phenomenon that occurs when the water content of the air is at saturation. Fog usually occurs when warm, moist air contacts cold air or water. The resulting temperature drop forces the equilibrium between water vapor and water in the direction of liquid water. This is condensation. If the water condenses around dust or other particles in the air, fog results.

Fog is an almost constant feature of some regions, notably polar areas of air or water convergence. *See* POLAR CONVERGENCE.

food chains Interlocking dependencies of organisms on each other and on their environment. Algae and other chlorophyll-containing organisms, utilizing solar energy, extract minerals from the water, perform photosynthesis, and store the resulting carbohydrates. They are in turn consumed by other organisms, which are prey to larger animals or, on dying, are consumed by scavengers and bacteria. Intermediate steps in this process involve commensal organisms—those that live together without seeming to affect each other directly—and symbiotic organisms, which do affect each other directly and favorably. Such symbiosis occurs between the sea anemone and the fish that live among its tentacles and lure other fish to the anemone as food.

The predator population, then, is directly affected by the population it feeds on. Great population fluctuations can occur after natural disasters, earthquakes, weather changes, or El Niño or as the result of human activity, such as overfishing or pollution with oil spills or chemical dumping.

A completely different series of interlocking relationships exists on ocean bottoms at the edges of crustal plates, where hot, sulfurous vents create an environment totally dependent on sulfur as the first element in the food chain. *See* CORAL, ECOSYSTEM, EL NIÑO, ESTUARY, SULFUR BACTERIA, VENT COMMUNITIES.

food webs The complex pathways by which marine organisms transport energy through the biome. Food webs have been an object of study only very recently. Everything that is living must have both a food supply and energy if it is to live, grow, and, most important, reproduce. The autotrophs—the primary producers—manufacture organic carbon using inorganic carbon dioxide and derive energy either from sunlight or chemical reactions of the Earth's

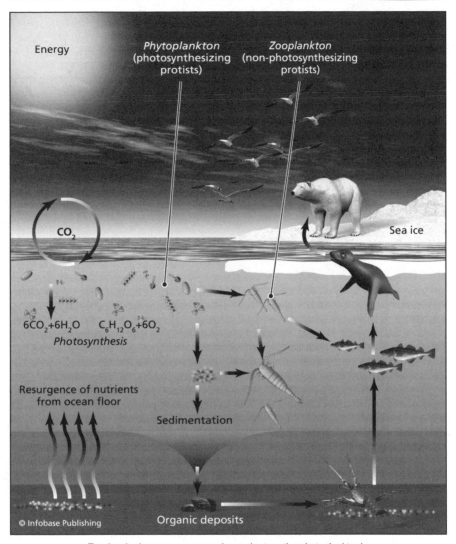

Food webs in open seas are dependent on the phytoplanktonic photosynthesizers, the primary producers.

lithosphere. All the other organisms must consume primary producers as their food supply. The autotrophs are the first level of a food web, the herbivores and omnivores that consume them are the second level, the carnivores that eat the second-level organisms are the third level, and the carnivores that consume carnivores form a fourth level. A reasonable measure of the amount of energy that is passed from one level to another is the "10% rule"; 10% of the energy in the primary level is available to the next level that is supported by it.

In the water column, where sunlight is the energy source, phytoplankton are the producers and are grazed on by filter feeders, which may be other planktonic organisms, such as copepods, or larger animals,

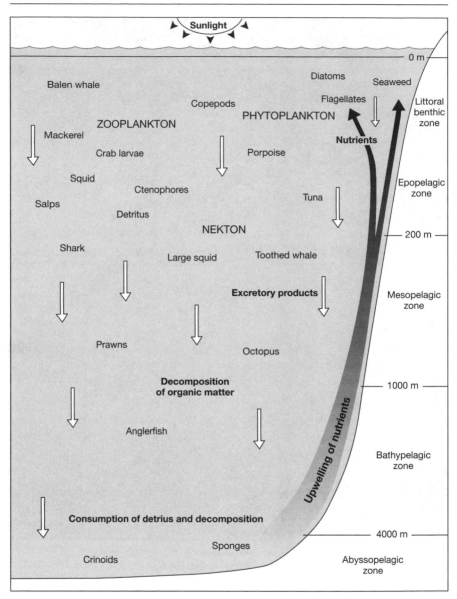

The interaction of photosynthesizing plants, animals, and organic debris and the recycling of resources through the layers of the ocean is dependent on light as the energy source.

live sponges, or mollusks. These in turn feed the fish or other carnivores. Many fish are not dependent on a single food source; juveniles, for example, will feed on all the plankton and fish smaller than themselves.

Nitrogen plays an essential role in the food web; without it, there is no protein synthesis. As on land, nitrogen is present in water as ammonium, nitrite, and nitrate ions. Some cyanobacteria can "fix" nitrogen—convert the unreactive dinitrogen molecule (N_2) to NH_4^+, NO_2^-, or NO_3^-. As organisms decompose, their nitrogen content (proteins) sinks to the ocean bottom as part of the sediment. It remains there until the water column is agitated by either surface winds or upwellings of cold water. The now-available nitrate is used by the phytoplankton, and the increase in the phytoplankton population is a "bloom" that is noticeable in the water because it changes the color of the water. Thus, both light and nutrient content are the controlling factors in the euphotic zone.

Because the interactions of the microbial organisms in the food web were not studied until the last quarter of the 20th century, the interactions among the planktonic organisms is still unclear. The discovery of the bacterial populations living in the mucus produced by deep-sea corals adds yet another factor to the many that seem to determine the complexity of the food webs of the oceans.

In open ocean, the water is stratified, trapping the nutrients at the bottom. Both coastal and Arctic waters are much better mixed and therefore higher in nutrient content and largely dependent on an herbivorous food web. The Sargasso Sea is an excellent example of an area that according to the traditional model of the food web was considered a "marine desert" because the fish population is small. It is, however, extremely rich in plankton—mostly nanoplankton (2–4 mm in diameter). These congregate around colonies of cyanobacteria and consist of bacteria, fungi, diatoms, dinoflagellates, ciliates, juvenile copepods, and other juvenile crustaceans. The debris resulting from all of these organisms eating and dying is nutritious "marine snow." The most significant large animals in this nutrient-poor region are the eels. Both European and American eels migrate to the Sargasso to reproduce. The Benguela upwelling, which would be expected to be rich in fish, is not. The cold, nutrient-rich water is typified by a high rate of nitrogen recycling. It is a complex and long food chain that is not understood: It seems to have all the ingredients necessary for a complex food web that contains larger animals such as fish, yet it does not. See AUTOTROPH, BENGUELA CURRENT, CARBON, COPEPOD, CORAL, ECOSYSTEM, MARINE SNOW, PLANKTON, SARGASSO SEA.

Foraminifera A very abundant order of almost exclusively marine protists. Most of them have a test, or shell, either of chitinous or calcareous material. The tests fall to the bottom as the organisms die.

The tests are either single chambered, the chamber expanding as the animal grows, or multichambered, with a series of larger spaces that the animal moves into when it becomes cramped. The walls between the chambers are perforated. The tests are usually colored: Globigerina is often red-brown but may be pink or dark orange. Other foraminiferan tests are yellow or brown.

There are two types of tests: either a large proloculum (or initial living space) or a megalosphere. The union of two zygotes, in animals whose reproduction is sexual, produces a smaller proloculum. The animal then grows by asexual fission, and there are a number of asexual generations before the production of zygotes.

Foraminiferans have been incorporated into rock since the Cambrian period and appear as chalk or limestone. The Egyptian pyramids are faced with limestone produced by fossil foraminiferal products. The ooze of ocean bottoms, ranging from continental shelves to depths of about 4,000 m (13,000 feet), consists of unconsolidated foraminiferal tests.

The Foraminifera extant in the Eocene and Oligocene eras were large, measuring several inches across, but most of the specimens found today are quite small, with some tropical exceptions. Some foraminiferans have commensal (not mutually beneficial, merely parallel) relationships with algae. The relationship with other species, however, is symbiotic.

The living foraminiferans are bottom dwellers, some crawl around, others are sessile. Their food consists of bacteria, diatoms, and other algae. In spite of the vast number of Foraminifera, their life history is not well understood. It is believed that foraminiferans can withstand adverse conditions by forming spores, but this is not certain.

Since they are so plentiful and ubiquitous (about one-third of all sea bottom sediment consists of discarded foraminiferan tests), the specific species of these animals are used as index fossils. *See* CARBONATE, CHITIN, DIATOM, Globigerina, Protista, SYMBIOSIS, TEST.

Forbes, Edward (1815–1854)

An English naturalist, Forbes was the scientist-on-board the HMS *Beacon,* commanded by Thomas Graves, which charted the Mediterranean. In the 1840s, prior to the voyage of the *Beacon,* Forbes had studied the marine animals of the English coasts and documented the different groups of species that occur at various ranges of depth. His *Beacon* voyage allowed him the chance to extend this work to warmer and deeper waters.

Forbes divided organisms into eight groups with regard to depth and the type of bottom they inhabit. He noted, quite accurately, that not all sea bottoms are capable of supporting the same ecosystem. Some are literally undersea "deserts." He noted that the deeper his sample came from, the less abundant was the array of organisms. He theorized that plant life disappeared at a depth of 100 fathoms (200m or 660 feet). He extrapolated this to a "zero point" for animals too, concluding that any remains found on deep bottoms died higher up and sank.

This idea of an "azoic" zone at about 300 fathoms (600m or 2,000 feet) so enthralled the scientific community that in spite of evidence to the contrary in existence at the time of Forbes's publication of his work in 1844 (e.g., John Ross's sea star [starfish] brought up from a depth of 800 fathoms (1,600m or 5,300 feet) in 1818, and Aimé finding animals at 1,800m or 900 fathoms), the idea of an azoic zone hung on. Even the work of a

Beacon officer named Spratt, who continued Forbes's research, and a refutation by David Forbes, Edward's brother, in 1869 did not completely dispel this comfortable but erroneous idea. *See* BENTHIC ECOSYSTEMS, HADAL ZONE.

foreshore The part of the shore between high and low water—the intertidal zone. *See* BEACH, INTERTIDAL ZONE, LITTORAL.

fossil Evidence of an organism's past history: It may be the original organism, a mineral deposit in the interstices of the organism's skeleton or soft tissues, a mineral replacement of the organism's tissues, or an impression of the entire organism or its tracks. Since living things decompose quickly, as opposed to mineral replacement, which is a slow process, there are very few fossils relative to the numbers of individuals that lived. The fewer hard parts an organism had, the less likely there is to be a fossil of it at all.

Early attempts to explain fossils ranged from scientific to theological. The ancient Greek, Xenophanes, and the Chinese of the first century B.C.E., both recognized fossils as remains of once-living plants or animals. The Chinese by the sixth century were trying to classify the "stone fish" they knew. These ideas were not revived in the West until after the Renaissance.

Although canal builders regularly turned up fossilized remains and wrote accurate descriptions of them, as did the natural historians of the time, this was not widespread information. Leonardo da Vinci described fossils in 1500, as did Bernard Palissy (1510–90) in France later in the century, but the official position of all religious authority in Europe forbade any scientific explanation for them. The official Western religious view of a six-day creation certainly did not allow for any creation to arise and disappear in that period.

In 1706, the governor of Massachusetts, Thomas Dudley, wrote to Cotton Mather describing a mastodon tooth as the remains of a giant killed by Noah's flood. Twenty years later another theologian

ERAS	PERIODS	YEARS X 10^6
Cenozoic	Tertiory	65
Mesozoic	Cretaceous	135
Mesozoic	Jurassic	195
Mesozoic	Triassic	230
Paleozoic	Permian	280
Paleozoic	Carboniferous	355
Paleozoic	Devonian	405
Paleozoic	Silurian	440
Paleozoic	Ordovician	510
Paleozoic	Cambrian	590

Column labels: Angiosperms, Gymnosperms, Pteridophytes, Bryophytes, Bacteria, Algae, Fungi, Protozoa, Porifera, Simple Metazoa, Ammonites, Trilobites, Myriapods, Insects, Molluscs, Echinoderms, Fish, Amphibians, Reptiles, Birds, Mammals

The fossil record

(Johannes Beringer, in Germany) viewed fossils as instruments of God (or Satan) sent to test men. Beringer gave accurate accounts of both real fossils he had found and those his students had "baked" for him as practical jokes.

Eventually, as religious authority lost control of the scientific community, the study of fossils increased. Also, large public works such as the construction of canal networks turned up more and more fossils.

By the 18th century and the work of Carolus Linnaeus and then Georges Cuvier (1769–1832), it was established that these remains of formerly living things could be fitted into an almost continuous pattern of living organisms, with some related to organisms currently existing.

With the growing interest in stratigraphy (the study of the Earth's strata) in the 18th and 19th centuries, fossils became identified with distinct eras in Earth history. The fossilized flora and fauna of each epoch (i.e., the fossil record) gave a clue to successive stages of life's development. Through the work of William Smith (1769–1879), Cuvier, and Charles Lyell (1797–1875), fossil analysis became, and still is, the stratigrapher's tool for identifying unknown strata. Although various 19th-century natural scientists (even Charles Darwin) were skeptical about the fossil record, eventually an extended fossil history for humankind became accepted. See EVOLUTION, INDIVIDUAL ERAS AND PERIODS OF GEOLOGIC TIME, PALAEONTOLOGY.

fracture zone A long, narrow, fresh break in the ocean floor, with sheer drops of more than 1 km. In the Atlantic, fracture zones are more-or-less perpendicular to the Mid-Atlantic Ridge, where the north-south orientation of the line of the Ridge is offset by numerous fractures. The East Pacific Rise has no corresponding single ridge but more spectacular fracture zones. Of these, the Clipperton, Clarion, Murray, Molokai, and Mendocino fractures extend for thousands of kilometers perpendicular to the edge of the North American Plate.

Magnetic anomalies are offset by the fracture zones. This is to be expected from the theory of seabed spreading. Fracture zones are not usually sites of volcanic activity unless at the edge of a plate.

While the fractures in the Atlantic and Pacific oceans usually run east-west, those in the Indian Ocean run north-south or north northeast-south southwest, demonstrating the complicated series of shifts of continental plates in that area. See MAGNETIC ANOMALY, MID-OCEAN RIDGES, OCEAN FLOOR, PLATE TECTONICS.

Fram The ship on which the Norwegian explorers Frijdtof Nansen and Otto Sverdrup drifted across the Arctic Ocean in the late 19th century. The name means forward. This ship was built according to Nansen's specifications; it was a three-masted schooner and had a rounded hull that enabled it to rise onto pack ice and not be crushed by it. Nansen and 12 others were frozen into the ice near the New Siberian Islands on June 14, 1893. They drifted with the ice, hoping to cross the pole, thus proving the nonexistence of an arctic landmass. They missed the pole but proved their point. The *Fram* drifted with the ice for three years before it broke free in 1896 near Spitsbergen. It returned to Norway on April 16, 1896, eight days after Nansen. Nansen and a companion had left the ship on skis, hoping to reach the pole. They got as far as 86°14′ north latitude before turning back.

Later, Roald Amundsen used the *Fram* on his expedition to the South Pole. The ship is now in a museum in Oslo, Norway. See AMUNDSEN, ROALD; EKMAN, VAGN WALFRID; NANSEN, FRIJDTOF.

Fram Strait A body of arctic water separating the east coast of Greenland from Spitsbergen. Cold water of low salinity moves south fairly slowly along the surface of the strait. This comprises the upper 30 to 50 m (90 to 150 feet) of water in the strait and is the exit for a great part of the Eurasian Basin runoff. The very distinct layers of arctic water make mixing almost impossible, particularly with the warmer, more saline intermediate water that moves north.

The Fram Deep is a very small cut in the Fram Strait, with reported depths of over 5,000 m (16,000 feet). It has not been thoroughly explored, however. The bottom of the strait is highly varied, with seamounts rising from it. One seamount comes to within 730 m (2,400 feet) of the surface pack ice. *See* ARCTIC OCEAN, FRAM, PACK ICE.

Franklin, Benjamin (1706–1790)

An American polymath who made contributions to oceanography among many other sciences. Although he did no systematic work in the field, his observations and comments on significant phenomena led to systematic work by others. Thus, Franklin was the first on record to refer to the Gulf Stream as a "river in the ocean." He also observed the phenomenon of internal waves, which are waves created by motion at an interface or boundary where saline water is overlain by fresh water, such as in an estuary of the fiord type. Internal waves have only recently been given extensive study, and are important in the understanding of water movements.

"Pouring oil on troubled waters" is an ancient cliché. Franklin tested the validity of this statement on a pond. Oil on the windward side was spread by the wind over the surface of the pond, and its waves were indeed stilled. If the waves on the pond were wind-created, the layer of oil (it would have to be a very thin film) would cut the aeration of the water that creates the foaming crests of waves. This is due to the surface tension of the oil particles and their lack of adhesion to water particles. The smaller the body of water involved, the more likely it is that the film of oil will completely cover the surface, and the more effective it will be in stilling wave cresting. *See* GULF STREAM, WAVES.

Franklin, Sir John (1786–1847)

An English naval captain and explorer, the nephew of Matthew Flinders, the explorer of Australia. Franklin was in the Royal Navy from boyhood; he had been at the battle of Copenhagen in 1801. Since he showed talent as a surveyor, he was chosen to accompany Buchan on the first of that captain's voyages to the Arctic in 1818, which charted western North America. By 1821 Franklin had been promoted to captain and was introduced as a member of the Royal Society.

A second voyage to North America and the Northwest Territories began in 1825. From 1836 to 1843 Franklin was in the Southern Hemisphere as governor of Tasmania. In May 1845 as commander of the *Erebus* and accompanied by Francis Rawdon Crozier, captain of the *Terror*, he sailed into the Arctic to find the Northwest Passage. The ships were last seen afloat on July 26, 1845. After the ships became icebound, the surviving crews attempted to walk south from King William Island to a settlement. None survived. Captain (later Admiral) Francis Leopold McClintock, on one of the many exploratory expeditions financed thereafter to find Franklin, found the log of the expedition, the last entry dated April 28, 1848. It is unclear whether Franklin made the last entries. The expedition had found the passage but didn't know it.

A small cemetery on Beachey Island holds the remains of three crew members who died in January 1846, when the expedition was still well-equipped and in reasonable health. While the graves were known for years, they were untouched until 1981, when Dr. Owen Beattie of the University of Alberta opened the grave of John Torrington. Beattie believes that Franklin lived at least a year after the foundation of the cemetery. The Inuit (Eskimo) who witnessed the end of the Franklin party thought them mad. They may have been acting in a bizarre manner from lead poisoning brought about by eating food preserved in badly soldered cans.

free diving *See* DIVING, SCUBA.

freshwater Water that is drinkable and of almost no salt content. Freshwater enters the ocean most regularly at river mouths. Some freshwater is introduced into seawater through artesian springs under the sea. Water of relatively low salinity is also produced as glaciers and icebergs melt. The result is a specialized ecosystem in bays and estuaries, where the salinity is very different from that in open ocean. *See* ESTUARY, SALT MARSH.

frigate A class of naval ship. In the days of sail it was a three-masted ship that carried 24 to 38 guns on one gun deck. Frigates were more maneuverable than ships of the line. The name frigate was revived in World War II and is now used to mean a smaller warship of multi-purpose capability.

frigate bird Also known as the pirate bird, a member of the family *Fregatidae,* order Pelicaniformes. The alternate name is the result of the bird's habit of stealing food from other birds while in flight. The frigates are large birds and almost entirely black. The females, which are larger than the males, have white undersides. The frigate has albatross-sized wings with a span of over 2 m (7 feet). It is almost constantly in flight. The birds nest in trees on warm coasts on all continents. Both parents care for the single chick. Frigates live on fish and small crustaceans, which they skim out of the water. The male frigate bird has a spectacular courtship display, inflating a large, bright-red, featherless neck. *See* BIRD, PELICAN.

fringing reef A reef attached to an island or a continent. The seaward side may be submerged and therefore a navigational hazard. *See* ATOLL, CORAL, ISLAND.

Frobisher, Martin (1535?–1594) An English explorer who made three voyages west in search of the Northwest Passage. The first expedition sailed in 1576; it explored the Davis Strait and other points in the Canadian north. Frobisher brought back minerals, sure that he had found gold. He was deceived by mica and pyrites,

but made a good enough case for the commissioning of a second voyage. In 1577 he sailed into Frobisher Strait and the next year returned to the Arctic again. This time he commanded a fleet of 15 ships. They charted the waters they explored but found nothing of commercial interest, and the project was abandoned.

The last voyage marked the end of Frobisher's career as an explorer. He later sailed with Drake (1585) to harass the Spanish and then took part in England's defense against the Spanish Armada. In 1590 and again in 1592 he was in the West Indies, attempting to capture the Spanish treasure fleet. In 1594 he sailed for France to defend the port of Brest against Spanish attackers and was killed in the battle.

front The leading edge of an air mass. Both polar and tropical air masses move into the temperate regions of the Earth from west to east. A warm front is a warm air mass moving into a cold one. A cold front means that cold air is invading a body of warm air. In either collision the warmer air rises above the cold air. Thus, the water vapor in the warm air condenses as it rises into colder regions of the atmosphere and precipitates as rain or snow. *See* METEOROLOGY, POLAR FRONT, WEATHER.

frustule The hard "shell" of a diatom. It is composed of two hinged plates of opal, hydrated and polymerized silicic acid, $Si(OH)_4$. It is both strong and transparent: The two halves are held together by so-called girdle bands that are flexible. These pliable structures allow the organism to grow within the shell. The size, shape, and pattern of the frustule is species-specific. *See* DIATOM.

Fucales An order of brown algae. *Fucus* is one of the genera in the family Fucaceae and is the genus most noticed in temperate North Atlantic waters. *Sargassum* is one of the genera in the family Sargassacoeae and is the most prevalent algae in tropical and subtropical North Atlantic waters.

fucoxanthin One of the pigments contained in phytoplanktonic chloroplasts—the structures within cells that contain pigments. In the phytoplankton one of the pigments must be chlorophyll. The other pigments serve other purposes; it is believed that the xanthins protect the photosynthesizing pigments from excessive sunlight during the periods of dinoflagellate "blooms." *See* BLOOMS, CHLOROPHYLL, DINOFLAGELLATE, PHOTOSYNTHESIS.

Fucus With *Sargassum,* it constitutes the class Phaeophyceae, or brown algae. *Fucus* grows in temperate intertidal zones. Specimens of 2 m (7 to 8 feet) long and more are known. The holdfast cells (anchoring cells), midribs, air receptacles, and conceptacles (reproductive cells) of this alga are examples of cellular differentiation. In *Fucus* a diploid organism produces oogonia or antheridia, which in turn produce eggs or sperm. These, when they unite, produce the zygote that grows into a new plant.

Sargassum is usually a tropical plant. It looks like a vascular plant because it is branched. It is algal and a truly floating species that has pillowlike structures to buoy it up. It grows luxuriantly, sometimes forming dense mats. *Sargassum* in turn supports an ecosystem of crustaceans, mollusks, small fish, and plankton that live on the plants or hide in the branches, feeding on the alga or on each other. *See* ALGAE, PHAEOPHYCEAE.

fulmar A gull-like oceanic bird. The Procellariidae (order Procellariiformes) includes petrels and shearwaters—scavengers who eat everything from dying fish to carrion and garbage. As a result, they are called "stinkers." Fulmars are usually white or grayish white. Their habitat is near the Earth's poles. The great fulmar of the Antarctic, *Macronectes giganteus,* is about 90 cm (3 feet) long and has a wingspan of over 2 m (6.5 feet); most fulmars,

however, are closer to half that size. *See* PETREL, SHEARWATER.

Fulton, Robert (1765–1815) An American inventor best known for his invention of the steamboat. In his first career Fulton was an apprentice jeweller and then an artist; he finally settled on engineering. His first success was a submarine, the *Nautilus,* whose development had been supported by a grant from the French government in 1806. Although it dove, the French, English, and American navies were not interested in it.

Fulton's attempts to power vessels with steam were more interesting to private backers. He displayed a steam-powered craft on the Seine in 1803, and by 1807 he ran his famous steamboat, which used an English engine, up the Hudson River in New York. By 1815 he had constructed an oceangoing paddlewheel steamboat, the *Fulton,* this time for the U.S. Navy.

Fundy, Bay of A large inlet on the North American Atlantic coast, lying in a northeast direction separating the Maine and New Brunswick coast from Nova Scotia. It is about 240 km (140 miles) long and about 75 m (250 feet) deep. At its head end, the bay splits into a forked tail with Chignecto Bay to the north and west of the Minas Basin.

Because of its shape and orientation, the Bay of Fundy, and particularly the Minas Basin, exhibits some of the highest tidal ranges in the world. The standing wave is over 2 m (7 feet), and the tide range is greater than 17 m (56 feet). The tide enters the bay along the south shore and leaves via the north shore, creating an eddy within the bay area itself. Periodically the idea of utilizing the enormous force of the moving water to generate electricity has been explored. Very recently, commercial exploitation of this energy source has begun. *See* RIVER BORE, TIDAL RANGE.

fungi Essentially land plants that have no chlorophyll. They live on dead plant tissue. The origin of fungi is unknown, but one theory is that they developed out of filamentous seaweeds, whose structure and growth pattern they duplicate. *See* ALGAE, HETEROTROPHS.

Fury and Hecla Strait A short, barely navigable stretch of water in the Canadian Arctic, it is part of the boundary of the western end of Baffin Island. Hudson Bay is south and east of the strait. The strait is part of the so-called southern route of the Northwest Passage. *See* NORTHWEST PASSAGE.

Fusulinacea An extinct superfamily of Formanifera of the phylum Protozoa. They are a significant component of the fossil record.

G

Gadiformes An order of actinopterygian fish. They include the economically significant cod, haddock, and grenadier. *See* COD.

Gaia The name that James Lovelock gave to his concept of Earth as a system in equilibrium in the first publication of his book *Gaia: A New Look at Life on Earth* in 1979. This was a popular concept in the 1970s. A number of environmental organizations that are exploring the use of renewable resources now use the name.

Galápagos Islands An island group isolated in the Pacific Ocean about 1,000 km (650 miles) west of Ecuador, of which the Galápagos are a part. The group consists of 14 large and many smaller volcanic outcroppings near the mid-ocean ridge where the Pacific Rise branches off into the California Rise and the Easter Island Rise.

Darwin visited the Galápagos Islands as part of the explorations made while he traveled on the *Beagle*. He was not the only naturalist to observe the highly differentiated biota on the islands. Many have noted the exotic animals: the tortoises, iguanas—both terrestrial and marine—sea lions and penguins.

The islands are unusual since they lack coral. They are temperate because of the cold Peru Current that touches them in spite of their position near the equator.

Recent geologic exploration of the Galápagos has shown that some of them formed under the sea while others have limestone layers indicating coral accretion at some time in the geologic past. Espinola (or Hood) Island has an eroded undercone with pillows of younger lava above. This is evidence of a much greater age for these islands than was previously theorized. The erosion means that at some time in the past the island was above sea level, making its age about 3 million years.

Another instance of the changing face of the sea bottom in this area is the Galápagos Spreading Center, a site of hot springs which bring calcium and magnesium into the seawater. This region is part of the Mid-Pacific Ocean Ridge. *See* DARWIN, CHARLES; MID-OCEAN RIDGES; PERU CURRENT; VOLCANO.

galatheids Crustaceans of the family Galatheidae. They are related to hermit crabs and are commonly referred to as squat lobsters because of their oval bodies. Their very well-developed chelipeds look like lobster claws. These small crustaceans are found on both hard and soft corals as commensal organisms. They are part of reef communities. *See* COMMENSAL RELATIONSHIP, CORAL, CRUSTACEAN, HERMIT CRAB.

Galilei, Galileo (1564–1642) An Italian mathematical physicist who discovered the laws of falling bodies and the parabolic motion of projectiles. Galileo was perhaps the outstanding physical scientist of his time. Although he did not invent the telescope, he made one and used it to observe objects on Earth as well as in the skies, including sunspots and the phases of Venus. His use of the telescope to plot many new stars led to experimental refutation of the "perfect Aristotelean" cosmos with the stationary Earth at its center, and these ideas brought Galileo into conflict with the Church. His publication of the *Dialogues*

on the *Two Chief World Systems—Ptolemaic and Copernican* led to his being
considered a dangerous heretic. His last
published work, *Discourses Concerning
Two New Sciences...*, was published in
1638 in Leiden, the Netherlands.

Galileo spent his last years working on
the use of the pendulum in clockmaking,
thereby laying the foundation of precision
timekeeping.

galleon A redesign of the carrack. John
Hawkins, an English navigator, is credited
with the modification of that large commercial vessel to create the galleon. The
change removed the large forecastle that
had caught the wind, forcing the ship to
leeward (away from the wind). The elimi-
nation of the large bow structure made the
ship more maneuverable and better able to
tack, or sail into the wind. *See* CARRACK.

galley A warship of the Mediterranean;
galleys existed before 3000 B.C.E., and
were finally retired in the 18th century.
The typical Greek galley was about 5.5 m
(18 to 19 feet) wide and about 33 to 34
m (130 feet) long. The oldest examples
had a line of rowers on each side. Later
there were biremes with two banks of oars
and triremes with three. The long oars of
the upper decks of the biremes (two-tiered
galleys) and triremes were manned by
several rowers for each oar. It is thought
that the rowers were freemen or pirates in
the early days of the galley; from classical
times onward they were most often slaves.
Sails were used on the mast and there was
most often only one square-rigged mast.
The oars were the chief means of propulsion, making the vessel very maneuverable. The principal weapon of the galley
was the ram—a metal-sheathed protrusion
from the bow. After the use of gunpowder
became widespread, small cannon were
also mounted on the galley's bow.

Gama, Vasco da (1460?–1524) A
Portuguese navigator who made three
voyages to India and established Portugal
as a world power. He first fought for the

Portuguese king John II, and was involved
in the capture of a French fleet. John was
succeeded by Manuel in 1495. The latter,
like his father, was interested in funding
voyages of exploration that would lead to
the opening of a route to the Indies that
would allow Portugal to bypass Muslim
intermediaries. Da Gama and his four-ship
squadron sailed from Portugal on July 8,
1497. They were accompanied by Dias,
who was en route to the Gold Coast (now
Ghana). They all spent several weeks on
the Cape Verde Islands and finally came
round the Cape of Good Hope in early
1498. Da Gama stopped in Mozambique,
where the expedition found much treasure
and stories of Prester John, the mythic
king of a Christian kingdom. A characteristic stone pillar was left to claim the
land for Portugal. The group left the African coast in April and made a 23-day run
across the Indian Ocean to Calcutta (then
called Calicut).

The trade component of this voyage
was a failure. There were Muslim traders on site in Calcutta whose goods were
much superior to those of the Portuguese,
who had come with what they had used
on the West African coast. The Indians
were unimpressed. Again leaving a pillar,
the Portuguese left for a long and difficult voyage home. Much of the crew died
of scurvy, and the remainder limped into
port in July of 1499.

After Cabral's expedition, a small factory (trading post) had been left in Calcutta, protected by a garrison. They were
overrun and slaughtered. When the news
finally drifted back to Lisbon, da Gama
was sent to India as an avenger. Sailing in
1502, he stopped again in the Cape Verde
Islands and then on the coast of East
Africa, where he captured an Arab trading
vessel, killed the crew, and then proceeded
to India. There he again captured several
vessels and killed their crews. He returned
to Portugal in October 1503.

After this punitive expedition, da
Gama took an inactive role in Portugal's
seafaring ventures. He was the king's advi-
sor on Indian affairs but largely spent his

time as a country gentleman. In 1524 the administration of the reestablished factory in India having fallen into chaos, da Gama was sent to reorganize it. He died, presumably of overwork, in December 1524, in Cochin, India. *See* CABRAL, PEDRO ALVAREZ; DIAS, BARTHOLOMEU.

Gambia Abyssal Plain A long valley perpendicular to the Mid-Atlantic Ridge and to the African coast. The Cape Verde islands are north of it and the Sierra Leone Rise is to the south. *See* ABYSSAL PLAIN, ATLANTIC OCEAN, CAPE VERDE ISLANDS.

gamma rays *See* ATOMS, ELECTROMAGNETIC RADIATION.

Gammaridea Large suborder of Amphipoda. They are generally small (1–15 mm, 0.04–0.6 inches), free-living omnivores that live in almost all benthic environments. Some of the intertidal species, particularly the sandhoppers, make a cricketlike noise. *See* AMPHIPODA.

Ganges River The most important North Indian river (2,500 km or 1,550 miles) long. However, it is not the longest river in India: One of its tributaries, the Brahmaputra (2,900 km or 1,800 miles), is longer. The combined system drains a territory of about 4.5 million hectares or 422,000 square miles. The area consists of clay, sand, and silt interspersed with peat bogs. There is evidence of forestation in earlier times. The abundant river water of the Ganges comes from Himalayan meltwater and the summer monsoon rains.

The mouth of the Ganges at the Bay of Bengal opens into a series of shifting channels. The eastern fringe of the delta is swampland. To the west, the Sundarbans (the word means beautiful forest) is a wildlife preserve. *See* DELTA, INDIAN OCEAN.

gannet A fish-eating seabird found in arctic to temperate waters. It is the largest northern seabird. The gannets are variously classified as *Morus* or with the boobies as *Sula*. They are in the order Pelecaniformes. The adults are white with black wing feathers and "black eyes." They produce a single egg and the chick reaches adulthood in three to four years.

Crowded colonies of gannets are a feature of sea cliffs in Canada, Greenland, and northeastern Europe. *See* BIRD.

gaper A large, commercially significant clam. Its habitat is the Pacific side of the North American coast. *See* HORSE CLAM.

gar A mainly North and Central American freshwater fish, although some species are found in brackish and salt water. Gars breathe air some of the time.

Gars have descended from the Eocene epoch; their fossils are found both in the Americas and in Europe. Their long slender bodies, behind small heads and sharp, strong beaks, are encased in an armor of ganoid scales. The alligator gar of the southern United States grows to lengths of 3 m (10 feet) and more. Technically the gar is edible, but few people try to eat it.

Unrelated species, such as the needlefish and its relatives, are sometimes mistakenly called gars.

GARP (Global Atmospheric Research Program) This international group maps and follows meteorological occurrences and their effects on the world's weather. *See* CYCLONE, EL NIÑO, HURRICANE, WEATHER.

gas One of the three states of matter; the others are solid and liquid. Gas exists in seawater, which dissolves the various components of air. Surface water is saturated with oxygen, nitrogen, carbon dioxide, and the rest of the gases in air because of its contact with air.

As water moves, its gas content is affected by the temperature encountered at different latitudes because the temperature determines the solubility of gases in water. Additionally, subsurface waters are usually less well aerated because less gas can dissolve in water that is under greater pressure.

Respiration and the decomposition of organisms tend to consume more oxygen than the subsurface water layer can generate. This means that subsurface layers of water will have less oxygen dissolved in them than will surface water, regardless of the pressure they are under. This in turn limits the kinds of plants and animals that may be found at a given layer of water. *See* BOD, CARBON DIOXIDE, NITROGEN, OXYGEN.

Gaspé Current   A current that moves St. Lawrence River water around the Gaspé Peninsula in Quebec, Canada. The current skirts the southern shore of the Gulf of St. Lawrence before merging with the Labrador or Cabot Current. The combined flow moves through the Cabot Strait. The Gaspé Current is strong in summer and weak in winter, when the river and its tributaries are icebound. *See* CABOT CURRENT, CABOT STRAIT, LABRADOR CURRENT.

Gastropoda   This class in the phylum Mollusca is that of the snails and slugs, animals characterized by an asymmetric body; if they are shelled, there is only a single shell. They are present in very many species and are second in total numbers of species—insects are first. There are about 38,000 species of gastropods, ranging from some that are so small that they are barely visible to some sea hares that weigh more than 10 kg (22 pounds). Their presence is marked in the fossil record back to the early Cambrian period, or about 600 million years ago.

This class is divided into three subclasses: Prosobranchia, the largest grouping, are composed of the familiar snails, which are mostly marine but some are terrestrial (abalones, conchs, whelks, and limpets); Opisthobranchia are both marine and terrestrial slugs; and the smallest group, the Pulmonata, is composed of terrestrial snails.

The gastropods are usually hermaphrodites. They occupy a vital part in the food web of the ocean. Some species are herbivores, while others are carnivores; the oyster drill is an example of a carnivorous species. Gastropods are also the food supply for other organisms—including humans. *See* FOOD WEB, HERMAPHRODITE, NUDIBRANCH.

Gastrorichs   Small transparent roundworms (about 0.1 to 0.5 mm) found in both freshwater and marine habitats. They are considered by most taxonomists to be a separate phylum of about 400 species, though they resemble nematodes. Their name, *Gastroichs*, means "hairy stomach," and this refers to the dense carpet of cilia that covers the head and ventral surface of the organism. Their normal mode of movement is gliding along algal strands, where they feed on smaller plankton, or moving between sand grains in intertidal regions. *See* CILIA, NEMATODA.

geochemistry   The study of the chemistry of the Earth, and particularly its crust, waters, and atmosphere. The chemistry of the Earth affects the ocean, since a number of ions are present in both sea water and freshwater. Most often the ions in seawater come from terrestrial runoff; others are introduced by undersea volcanic activity or from vents in the mantle at the mid-ocean ridge lines. *See* CRUST, MANGANESE, MID-OCEAN RIDGES, OCEAN FLOOR.

geochronology   The dating of events on Earth by analysis of the radioactive elements in seawater and corals. Several radionuclides are useful in the dating process. Uranium is a terrigenous element brought to the sea by river water. Thus, its incorporation into corals and the tests of foraminiferans dates those structures. The isotope carbon 14 is present in greatest quantity in the oceans and is useful in assigning an age to fossil shells.

Much of the Earth's structure is too old to use only carbon decay for dating, because there are other radioisotopes whose decay period is far longer than of carbon. For example, the potassium-to-argon shift has a half-life of 1×10^9 years. Sulfur and molybdenum isotopes have also been used to date specific mineral formations.

When examining a core sample of ocean bottom and assigning a date to it, allowances must be made for the problem of turbulence. Ocean-bottom sediments may be disturbed by bottom currents, the movement of animals, or shifts in the bottom as a result of volcanic activity. Any and all of these may seriously disturb the bottom sediments. *See* CARBON 14, RADIOACTIVITY.

geodesy A branch of geophysics that concerns itself with the study of the size and shape of the Earth and its gravitational field. *See* GRAVITATION.

geology The study of the history of the Earth and its life as recorded in rocks. Geologic stratifications were well known by the early 19th century. Since marine fossils were found on land, some early naturalists assumed the reverse would be true and that land fossils would be found on the sea bottom. Not until the *Challenger* voyage reports were published, however, was it generally accepted that this was not the case. However much sea levels had changed in the history of the Earth, the ocean basins were never land areas.

Still, the oceans do change with time. Very obvious events in oceans are rapid erosion and land slippages. Both were noticed by the cable engineers who found breaks and burials of cables even in the absence of volcanic activity. The activating force in the movement of the sea bottom is the underlying movement of the Earth beneath the water. This was very poorly understood for a long time. Another misunderstanding of the age of ocean basins and the rapid events that occur in them resulted in the ignoring of Wegener and his arguments for continental drift as a long-term, ongoing event. His theory was not considered valid in spite of the striking similarities of the edges of continental shelves and the relationships between the Earth's landmasses.

Results of fairly recent dredging that revealed the Earth's youngest rock at the mid-ocean ridge and thickest crust at the continents, together with evidence of the movement of the plates and mantle, have totally changed the geological picture of Earth. *See* ABYSSAL FLOOR; CATASTROPHISM; CONTINENTAL MARGIN; ISLAND ARCS; LYELL, CHARLES; MID-OCEAN RIDGES; MURCHISON, RODERICK; PLATE TECTONICS; TRENCHES; WEGENER, ALFRED LOTHAR.

geomagnetism The study of the Earth's magnet properties, including its magnetic field and the changes and reversals of that field. *See* GRAVITY, MAGNETIC POLES.

Gesner, Konrad (1516–1565) A Swiss theologian and one of the "encyclopedic naturalists." Gesner was a protégé of Zwingli, the Protestant reformer. He was attracted to medicine and studied that subject along with theology and ancient languages. He practiced the latter two professionally.

Gesner's five-volume *History of Animals* was begun in 1551. He worked on it for the rest of his life, but the massive collection was not published until 1587, well after his death. Gesner attempted a total reclassification of all animal species. The book was so complete that it achieved instant scientific success and remained a major reference work into the 18th-century Enlightenment. It was considered a foundation work by Cuvier, the French expert on paleontology of the late 18th and early 19th centuries, who sought to classify species by their fossil remains. *See* BELON, PIERRE; BUFFON, GEORGES-LOUIS LE CLERC, COMTE DE; CUVIER, GEORGES-LEOPOLD DAGOBERT, BARON DE; RONDELET, GUILLAUME.

ghost crab *See* SAND CRAB.

giant clam A clam of tropical Pacific and Indian Ocean waters (*Tridacna*) that measures over 1 m (3.3 feet) in diameter and very occasionally closes its shell on divers who manage to step into it.

Because it is so large, the clam is hunted for its shell. One very large specimen is the holy water font in the church of St. Geneviève de la Sorbonne in Paris. *See* CLAM, MOLLUSCA, SEA MONSTER.

giant crab *Macrocheira Kaempferi,* a spider crab found in waters near Japan. The habitat of this creature is usually between 50 and 300 m (150 to 1,000 feet) deep. It is the largest known arthropod, the length of the crab from one claw tip to another frequently measuring more than 4 m (13 feet). It weighs about 18 to 20 kg (40 to 45 pounds). *See* CRAB, SEA MONSTERS.

giant squid The genus *Architeutis* was long thought to be a sea legend. The physical evidence of sperm whales bearing scars of squids' hard suckers and large fragments of what was obviously squid tissue, followed by the recovery of an almost entire giant squid body proved that their existence was no legend. These giants are voracious deep-sea predators that range up to 20 m (68 feet) in length and weigh about a ton. The carnivorous squid will eat and ingest anything it catches. They, in turn, are preyed upon by sperm whales. *See* SQUID.

Gibraltar The limestone top of a granite formation, rising sharply at the southern tip of Spain at the western end of the Mediterranean Sea. The highest point is 450 m (1,350 feet), and the formation is a natural bastion. Because it commands the strait between Europe and Africa and is easily defended, Gibraltar has been a fortified bastion throughout recorded history. *See* MEDITERRANEAN.

Gibraltar, Straits of A body of turbulent water separating North Africa from Europe (Spain) at the western end of the Mediterranean. Only 13 km (8.5 miles) wide at the narrowest point, the straits are very ancient, apparently having existed when the Mediterranean was much larger than it is at present. The depth of the sill separating the Atlantic from the land-locked Mediterranean controls the water circulation in the latter. Cold Atlantic surface water of relatively low salinity spills into the Mediterranean, and warm, relatively high-salt water leaves the Mediterranean and enters the Atlantic. *See* ATLANTIC OCEAN, MEDITERRANEAN, SILL.

Gilbert, Humphrey (1539?–1583) An English soldier, navigator, and relative of Sir Walter Raleigh, Gilbert was the leader of an English movement for the colonization of the Americas, which resulted in the English annexation of Newfoundland.

His interest in the New World began with the proposal for expeditions to find the Northwest Passage. In 1578 he was given a charter entitling him to plant a colony in America, so long as it was not in the dominion of any other Christian prince. He sailed in November 1578 with seven ships but could not control his men; some turned pirate and the others drifted back to England. After a second attempt, which left Plymouth in June of 1583, Gilbert arrived in Newfoundland and claimed it for England. At the end of August, this expedition began moving south, but several ships were wrecked near Sable Island (Nova Scotia), and the rest turned east in a storm. Gilbert's ship, the *Swallow,* went down near the Azores. There were no survivors.

gill A respiratory organ of fish, mollusks, crustaceans, and some worms. A gill is constructed of finely divided, featherlike tissue that is well supplied with blood. The gill is often enclosed in a supportive or protective structure made of bone, shell, or chitin. In teleost fish the gills are covered by an operculum (covering flap).

The finely divided structure of the gill makes possible maximum contact of the blood in the gill with surrounding medium. The blood brings carbon dioxide to the gill, and this gas is passed from the blood to the surrounding water. At the same time, oxygen diffuses from the water onto the gill and into the circulatory system. In many organisms the water is forced over the gill by muscle action; respiration in gilled organisms is not a passive process. *See* CARBON DIOXIDE, CHITIN, RESPIRATION.

glaciers Slow-moving masses of ice accumulated either on mountains or in polar regions. Glaciers are found in areas of heavy precipitation where warm, moist air or warm water meets cold air or water. Ideal sites near oceans are in Alaska, Antarctica, Greenland, and New Zealand. More than 95% of glacial ice is polar. Glaciers reach an equilibrium between accumulation or growth and ablation or decay. In polar regions, this occurs at sea level. Glaciers move slowly; their movement is lubricated by a slush layer under the ice and, barring surges, they will move downhill eventually to sea level, or toward the coast, and into the water as icebergs. This process is called calving. The most rapidly moving glaciers are in western Greenland, and these produce the largest number of Northern Hemisphere icebergs.

Until the late 20th century polar glaciers were considered extremely stable, with little overall change of volume. However, within the last 20 to 30 years, glaciers everywhere are retreating faster than what was once considered normal. They are becoming smaller both in polar regions and on mountains in more temperate latitudes as well. This new and disturbing phenomenon is considered by climatologists to be a direct result of overall global warming.

Glacial surges are more often than not mountain occurrences. They are of uncertain cause but are very impressive. In a surge, a glacier can move up to 15 m (60 feet) in one day. *See* GLOBAL WARMING, ICE, ICEBERG.

glass sponge A sponge with siliceous spicules (support structures) of the classes Hexactinellida and Hyalospongiae phylum Porifera. On drying, the skeleton resembles glass. *See* HEXACTINELLIDA, HYALOSPONGIAE, SPONGE.

Global Ocean Observing System A branch of UNESCO composed of four separate agencies: the Intergovernmental Oceanographic Commission (IOC), the World Meteorological Organization (WMO), the United Nations Environment Programme (UNEP), and the International Council for Science (ICSU). These international groups form a network for observations, computer modeling, analysis of marine conditions, and descriptions of resources and climate.

global warming The phrase used to describe the rise in the Earth's average mean temperature. This rise has been observed since the development of thermometers, and it has been accelerating in the last century. There is fossil evidence of similar temperature cycles in Earth's history. This phenomenon of global warming has been studied intensively for the last 20 years. It is obvious that the Earth's mean temperature is increasing and has been doing so for most of the 20th century. There was steady increase in mean temperature from 1905 to about 1940 followed by relatively steady mean temperatures. From the mid 1970s to the present, mean global temperature has been rising again. There is considerable and ongoing study concerning the cause of this, and the factors include not just carbon dioxide levels but nitrogen oxide levels and the tilt of the Earth on its axis (it wobbles).

Examination of glacial ice cores and coral cores has revealed the cyclic nature of climate on Earth. There are warm periods that alternate with cold ones; we are now living on a warming Earth. There is concern that the rapid warming of polar waters will affect all biomes, not just polar ones, and that sea level worldwide will be higher than at present. This in turn would inundate major stretches of continental coasts and significantly change almost all island communities. Another concern is the rapidity of the polar icemelt. This is a destructive cycle wherein the ice cover of polar water warms and melts, resulting in the exposure of water to sunlight. The ice acts as a solar reflector, bouncing sunlight back into space and keeping the ice frozen. The open water is darker in color and absorbs insolation, warming the water still more and melting more of the ice cover. This process has continued sufficiently to interest shipping companies, which are now looking forward to using the fabled Northwest Passage from Europe to the Far East, a water route north of Canada.

Climate changes are part of global warming; areas that were arid are becoming more so, as in the North American southwest and the plains of Australia. More frequent and more destructive cyclonic storms—hurricanes and typhoons—are also attributed to the global warming climate change phenomena. There is no question that greenhouse gases are related to this pattern; the role of humans and their industries as producers of greenhouse gases and the extent of that contribution to global warming is under study. *See* CARBON DIOXIDE, CLIMATE.

Globigerina   Shelled protozoans. Their shells are the single predominating sediment that covers much of the ocean bottom. There is a continuous layer of this deposit in all oceans north of the Antarctic Convergence.

The tests of *Globigerina* are about 30% calcium carbonate. This compound usually dissolves at depths greater than 2,000 m (6,000 feet). Some anomalous *Globigerina* deposits exist at greater depths, where the waters are oxygen deficient.

John Murray, in the *Challenger* expedition's samplings, brought up *Globigerina*. He correctly surmised that it was a surface-dwelling organism whose test was deposited on the seafloor after the death of the organism. He also correctly explained the absence of deposits of tests at great depths. *See* CHALLENGER, HMS; FORAMINIFERA; MURRAY, JOHN; OCEAN BOTTOM.

Glomar Challenger   A drilling ship built in the United States, in operation since 1968. Although primarily an oil rig, it has more recently been used to take core samples. The use of this vessel for commercial drilling, however, depends on the cost of petroleum. If the price of crude oil is high, the high cost of the *Glomar Challenger*'s operation can be justified. During periods of low prices for crude oil, drilling for oil from deep ocean bottoms is not economical, and the ship is used to take scientific samples. *See* SCRIPPS INSTITUTION.

goby   A small perciform carnivorous fish. The gobies are a group of about 800 species distributed in all waters. The greatest number of species are in tropical oceans. About 85% of all species are in the family *Gobiidae*.

The crystal goby (*Crystallogobius nilsonni*) is found in Europe. It is about 10 cm (4 inches) long and transparent. It is one of the largest species of goby. Gobies may be as tiny as the Philippine goby (*Pandaka pygmea*), which is only 13 mm (0.5 inches) long.

Many gobies are scaleless. Gobies also lack a lateral line but do exhibit prominent spots of color. Some gobies are commensals, living near or on large fish and picking parasites—usually copepods—off their hosts. They attach themselves to the larger animals by means of a suction plate that is a modified pelvic fin. Other species attach themselves to rocks or other features of the ocean bottom with the same plate. *See* BENTHOS, COMMENSAL RELATIONSHIPS.

Gondwanaland   The archaic southern continent that resulted from the breakup of the still older supercontinent Pangaea. Gondwanaland then broke up again to form South America, Africa, Australia, Antarctica, and India. The name is based on the Gondwana region of India. *See* CONTINENTAL DRIFT; PLATE TECTONICS; WEGENER, ALFRED LOTHAR.

goniatites   A group of extinct clams (genus *Goniophora*) that appeared in Silurian seas and continued into Devonian times. Their shells have a distinctive heart-shaped outline and a network of radial growth lines. Modern clams have annular ring–type (once a year) growth lines. *See* CLAM.

Gorda Plate A small segment of the Earth's crust under the Pacific Ocean. This plate abuts the western North American plate north of the San Andreas Fault. It has not yet been forced under the North American Plate by the pressure of the much larger Pacific Plate. *See* CRUST, PACIFIC PLATE.

gorgonians Also known as "horny corals," this group of sea animals includes the sea whips, sea feathers, and sea fans, all of which are related to the soft-bodied corals. Both the soft and horny corals are in the subclass Octocorallia, having eight-part symmetry.

Although gorgonians are found in all seas, they are most prevalent in warm waters at maximum depths of about 1,000 m (3,300 feet) but usually in much shallower waters. These sessile (stationary) animals attach themselves to a firm surface by means of a basal plate or stolons, constructed of a calcareous (calcium-containing) mass of spicules in a horny matrix called gorgonin. These basal plates in turn serve as anchorages for hydroids, brachiopods, and copepods.

Most gorgonians are both graceful and attractively colored: reds, yellows, orange, and various shades of purple are the most common colors. The coral used in jewelry is a gorgonian.

Some gorgonians produce toxic compounds, usually terpenes, as chemical defenses against predators or other animals that would compete with them for the same planktonic food supply. *See* CHEMICAL DEFENSES, CORAL, DEFENSE MECHANISMS.

Gorlo Strait An Arctic waterway separating the Barents and White seas in Russia. *See* BARENTS SEA; DERYUGIN, KONSTANTIN MIKHAILOVITCH; WHITE SEA.

graben A valley formed by rock that is downthrust between two parallel faults in the Earth's crust. Also called a rift valley. *See* CRUSTAL BLOCK, PLATE TECTONICS.

gradient wind A theoretical wind based on the combination of the Coriolis force and atmospheric pressure. This calculation is useful in predicting the wind acceleration. It is vital to meteorology because it provides some basis for predicting the force of wind in a particular storm such as a hurricane. *See* CORIOLIS FORCE, STORM, WIND.

gram-negative bacteria Most pelagic (open ocean) bacteria seem to be coccid (round) and gram-negative. This means that the cell walls of these organisms do not absorb Gram's stain (a watery solution of iodine and potassium iodide). They also seem to be more adaptable to changing conditions and constitute a large proportion of the marine snow that feeds larger pelagic organisms. *See* PELAGIC ENVIRONMENT.

Grand Banks A submerged continental margin east of Nova Scotia and separated from it by the Laurentian Channel. Newfoundland is northwest of the banks, where the average depths are between 40 and 100 m (130 and 330 feet). The banks are roughly triangular in shape, with Newfoundland forming one angle of the triangle. The northernmost point on the banks, Flemish Cap, might be thought of as another angle, but it is shallow and separated from the rest of the banks by much deeper waters (over 1,000 m, or 3,300 feet).

A part of the remains of the Gulf Stream reaches as far north as the Tail of the Banks, where the cold, low-salt Labrador Current moves south along the coast and into the area.

The Grand Banks are a prime fishing ground and the source of much of the North Atlantic's cod and herring catch. Despite fluctuations in the fish population, overfishing, worries about pollution, international competition, and fluctuation in the cost-to-profit ratio of the industry, this region is still paramount in its tonnage of several commercially caught fish species. *See* COD, HERRING, FISHING INDUSTRY, SAINT LAWRENCE, GULF OF.

granite Igneous rock that is largely feldspar and quartz. It is both light in color and less dense than basalt. The continental masses are largely granitic at their lowest levels.

grapholite An extinct marine colonial organism. They first appeared in the Cambrian period and were most prevalent in the Carboniferous. Their remains indicate their existence for more than 250 million years, and they are found in limestone deposits worldwide. These abundant sticklike fossils are also found in Paleozoic shales, and were common in the Ordovician and Silurian. The "sticks" may be straight, curved, spiral, single, bifid, or multibranched. When flattened in the shale, one or both edges look like tiny circular saws or gear-blades. The graptolites are another type of indicator fossil. Taxonomically, they have been placed with a number of organisms and were once thought to be related to the corals. The current placement is as a class of Hemichordata—the Graptolithina. *See* CHITIN, COLONIAL ANIMALS, HEMICHORDATA.

gravimeter An extremely sensitive spring balance that will deflect with minute changes in gravitational force.

After World War I, V. Meinesz made several voyages during which, using gravimeters, he charted belts of gravitational anomalies. The combination of his charts with the seismic readings of the areas gives a picture of the shapes, geologic features, and foldings of the seafloor. Trenches show negative anomalies; island arcs show positive ones. *See* MAGNETIC FIELD; MEINESZ, VENING; TRENCH.

gravity wave A wave in which gravity is the force that restores equilibrium. These waves are of relatively long wavelength, being greater than 1.75 cm (0.5 inches). *See* WAVE.

gray whale A relatively small (about 15 m or 50 foot long), slender baleen whale. It is usually gray to gray-black and splotched with white. The grays are found most often closer to shore, as opposed to out in the open ocean. Their habitat is the northern Pacific. *See* WHALE.

grazing The consumption of bacteria or other producers by larger organisms. *See* FOOD CHAIN, FOOD WEB.

Great Australian Bight The wide (1,100 km or 700 mile) indentation in Australia's southern coast; a part of the Indian Ocean. The bight and the arid coastline it touches were first explored by Pieter Nuyts in 1627 and were charted by Matthew Flinders in 1802. *See* AUSTRALIA; FLINDERS, MATTHEW.

Great Barrier Reef A coral reef that lies east of Australia, stretching from the Torres Strait south of New Guinea (Irian) to the Tropic of Capricorn. It is more than 1,750 km (1,100 miles) long and the world's largest reef. The Great Barrier Reef is really a reef system, not a single entity, since it encompasses channels, islands, and smaller reefs. The islands of the system's western boundary have navigable channels between them and the mainland. These islands are not all just sandy bars but vary in both size and elevation. The coral accretion to the reefs grows eastward and provides an incredible series of undersea landscapes, habitats, and exotic animals.

Geologically, the entire region was once a terrestrial landscape of undulating country with ranges of small hills. The Queensland plateau sank—most likely a geologically recent event, and the coral upbuilding on the drowned plain then began. The entire area is one of continuous change, with more upbuilding by coral than there is subsidence. *See* BARRIER REEF, CORAL, CORAL SEA, FIJI, ISLAND.

great circle The largest circle that can be drawn on a sphere. When the sphere is the Earth, the meridians of longitude and the equator are all great circles. The great circle is of navigational importance since the shortest distance between any two points on the Earth's surface is the great circle that both of those points lie on.

While the theory of great-circle sailing has been known and understood since classical times, it was not technologically possible until the 19th century. Earlier sailors had to be content with inadequate charts and chronometers, which could only give estimates of longitude, as well as the vagaries of wind and current. *See* NAVIGATION.

Great Lakes Five large lakes in North America that collectively form the largest body of freshwater on Earth; they cover an area of about 245,000 km², or 95,000 square miles. Because of their size, they are tidal and also subject to storm surges, as is the ocean. The lakes descend in elevation from Lake Superior, the westernmost, to Lake Michigan, Lake Huron, Lake Erie, and Lake Ontario. The last empties into the St. Lawrence River.

The lakes were formed in the Pleistocene epoch (about 4 million years ago); the exact mechanism of their formation is explained by a number of somewhat differing theories. The present lakes are remnants of larger and more extensive bodies of water. They have had their present conformations, however, since the last retreat of the glacial ice, about 10,000 years ago. *See* TIDE.

Greece *See* MARINE ARCHEOLOGY.

green algae (*Chlorophyceae*) Algae in which the chlorophyll is not masked by another pigment. They include Caulerpa, or sea grapes, found near corals in warm, temperate-to-tropical seas. These plants do not have discrete cells but a multinucleate protoplasm. *Ulva,* sea lettuce, are found in temperate waters; *Enteromorpha* are flattened tubules found in temperate waters; Cladophora are algae with branching filaments; and there are numerous planktonic species.

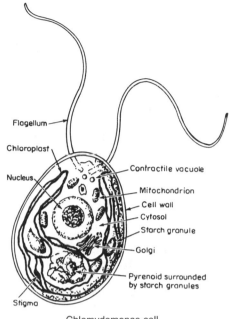

Chlamydomonas cell

Nowhere in open ocean waters are the green algae dominant. They are overwhelmed by the brown species, since algae with more pigments than chlorophyll alone are more efficient in utilizing the light they receive. Although green algae are not usually dominant in marine environments, they can create a problem. The *Caulerpa taxifolia,* an alga commonly used in aquaria, was accidentally introduced into the Mediterranean Sea, where it is devastating the normal marine organisms; it grows wildly and has no predator in the new environment. This mutated Caribbean species was noticed in 1990 near Monaco and has spread about 10 kilometers each year since then. Attempts to control it have not been successful, and it is now widespread. In smaller infestations, it is possible to remove this aquatic protist pest by use of chlorine or dry ice applied to small infested areas. *See* ALGAE, BLUE-GREEN ALGAE, BROWN ALGAE.

greenhouse effect A blanket term used for describing a worldwide heating trend linked with a general rise in the level of atmospheric carbon dioxide. During the last 200 years the increased use of fossil fuels such as coal and oil, and particularly the industrial use of these fuels, has increased the atmospheric carbon dioxide content by about 10%. Concurrently, the mean temperature has increased by 0.2°C. At the same time, the amount of airborne pollution generated in large part by aircraft and automotive exhausts, vapors, and aerosol propellants has increased. These gaseous, long-lasting additions to the atmosphere prevent insolation while trapping infrared radiation close to the surface of the Earth. The total effect is one of blanketing and raising the temperature, although preventing insolation by itself would cause the temperature to fall. *See* CARBON DIOXIDE, POLLUTION, TEMPERATURE.

Greenland The world's largest island has a land mass of over 2×10^6 sq. km. (840,325 sq. miles), of which only about 16% is not covered with ice. The surrounding ice mass is the source of most of the North Atlantic icebergs. Geologically, Greenland is considered part of the North American landmass. *See* ATLANTIC OCEAN, EVOLUTION OF OCEANS.

Greenland-Scotland Ridge An east-west ridge in the North Atlantic. It forms a sill that directs one branch of the Gulf Stream–warmed Atlantic Ocean waters between the Shetland and Faroe Islands. *See* ATLANTIC OCEAN, MID-OCEAN RIDGES, NORWEGIAN CURRENT, SILL.

Greenland Sea A body of water that lies east of Greenland, north of Ireland, west of Jan Mayen Land, and south of Spitsbergen. The sea is divided by the western Jan Mayen Rise (with minimum depths of about 1,000 m or 3,300 feet) into the Iceland and Greenland basins. The latter is the larger and deeper, with maximum depths of 2,800 m (9,200 feet) in the Iceland Basin and 4,800 m (16,000 feet) in the Greenland Basin. The sea has been referred to (only by its admirers) as an arctic Mediterranean.

Since Greenland and Scandinavia are being actively separated by the mid-ocean ridge, there are active volcanoes under the sea that produce lava. When arctic air, which is cold and dry, and polar maritime air, which is warmer and moist, meet over the sea, there is heavy precipitation. Most of this falls as snow.

The water system in this area is a complicated one. There are four distinct bodies of water: *Polar water,* which is of low salt content and cold (-1.85°C), forms the Greenland Current and moves south; *Intermediate water,* which is derived from the Norwegian Current, is saline (relative to polar water) and of higher temperature, averaging about 5°C; *Norwegian deep water,* which is colder and less saline than the intermediate water; and *North Atlantic water,* which comes from the Irminger Current, is 8°C or higher and has a salt content of at least 3.5%. It exists near Iceland and overlies the Polar water.

Spring and summer runoff further decreases the salinity of the Greenland Sea. The glaciers of Greenland produce icebergs which almost totally cover the sea's surface in April. *See* CURRENTS, GLACIERS, ICEBERG.

green turtle A large pelagic reptile, this tropical animal is in danger of extinction. The adults have been hunted extensively in the Caribbean Sea, and increased use of beach areas has demolished the green turtle's nesting sites. Since the female does not produce eggs every year and is faithful to her particular native beach, the population has diminished markedly in the 20th century. Attempts are being made by conservation societies to reintroduce turtle eggs to protected new areas that would become "home beaches." *See* REPTILE, SEA TURTLE.

Greenwich Mean Time (GMT, Coordinated Universal Time, or Zulu Time) GMT is determined at 12 noon when the sun is directly over the Greenwich meridian at 0° longitude. Time zones do not follow the lines of longitude exactly; therefore, all of Europe is in the same time zone. When it is 12 noon in GMT, it is 11 A.M. at 15° west longitude, 12 P.M. in London and Rome, and 7 A.M. in New York City.

The development of accurate clocks was very much a British endeavor. Consequently, the Royal Obeservatory at Greenwich became the fixed point from which all other locations could be determined. All locations on Earth are measured as some number of degrees east or west of this fixed point.

groin A man-made rocky structure built perpendicular to a beach to inhibit the movement of beach material and to prevent beach erosion. Groins extend from the landward side of the highest high water and continue into the water. They are usually placed in a series of parallel lines, which as a unit encourage the deposition of sand and other beach materials, thus expanding the beach. Care must be taken in the placement of the groins, or the growth of one beach may result in the erosion, undercutting, and ultimate destruction of another. *See* BEACH, JETTY.

grouper Any of a large number of species of fish belonging to the family Serranidae, order Perciformes. Two of the most numerous genera are *Epinephelus* and *Mycteroperca*. Groupers live in warm seas in all oceans. They are large, heavy-bodied animals that are usually dark brown or green. Some species are either flecked with red or are completely red. Some of the Caribbean varieties can vary their color.

Although the groupers are hunted as food fish, some species manufacture toxic substances and are dangerous.

Experimental stations in the Mediterranean have been marginally successful in attempting to raise grouper, but while the juvenile fish can be reared in a lagoon or pool, the adults need considerably more room. The average fully grown adult is about 2 m (7 feet) long and weighs over 200 kg (440 pounds). *See* FISH.

growth The increase in numbers of individuals in a population (members of one species), or the enlargement of an individual. The first depends on the existence of favorable conditions; the second also involves hormonal control without which there can be no differentiation of cells. Growth also encompasses the series of transitions whereby an egg divides in specific ways to produce different tissue and systems in the organism.

grunion A small (15-cm or 6-inch long) silvery fish (*Leuresthes tenuis*), found in Pacific Ocean waters off the California coast and known for its unusual spawning behavior. The eggs are deposited and fertilized in sandy beds at the highest high-tide mark in spring (April and May). The next high spring tide washes the hatchlings out to sea. When the fish spawn, they are present in such numbers that they can be scooped up by hand. *See* FISH, TIDE.

grunt One of about 175 species of edible fish related to the snapper. Grunts are usually tropical and small; the average individual is 20 to 22 cm (8 to 10 inches) long. Most grunts are elongated and slender, but a few species are deep-bodied. The color varies from white to black, though yellow, blue, striped, or mottled species are also known. The inside of the mouth is often red.

Grunts make noise, from which they derive their name. They grind their teeth and resonate the sound through their swim bladder, thus producing their characteristic sound. *See* SNAPPER, SWIM BLADDER.

guanine This nucleic acid is one of the four bases that forms DNA and RNA. When it is not part of larger structures, this compound crystallizes. Guanine plates made of crystallized material are found incorporated into the scales of many fish, where they act as mirrors. This renders the fish silhouette invisible to a predator. If the ambient light is dim or diffuse, the prey (the fish with the guanine plates) guano

melts into background. *See* COLOR, CAM-
OUFLAGE, PREDATOR-PREY.

guano The excrement of seabirds. In
areas of large bird population, notably
the islands on the western coast of South
America, the guano is commercially
mined for fertilizer because its nitrogen
and phosphorus content is very high. In
some places that have little rain, notably
coral reefs, the calcium carbonate of the
coral is altered by contact with quantities of bird excrement and becomes more
crystalline and sometimes chemically
changed. *See* EL NIÑO.

Guinea Current A current found on
the west coast of Africa that changes
its location with the season, being more
northerly and broader in summer than in
winter. The current moves east, bending
along the southern edge of the west African "bulge" as it moves into the Equatorial Countercurrent.

The path of the Guinea Current is
well documented. By 1850 it had been
noted that sailing ships moving west in
the current were held back about 60 km
(40 miles) a day. Although it does vary
somewhat with season, the current can be
generalized as a warm, highly saline, and
relatively oxygen-poor body of water. *See*
EQUATORIAL CURRENTS.

gulf A large coastal indentation caused
by the downwarping or subsidence of continental shelf areas or a rise in the sea
level. The latter occurred most often after
the melting of the Pleistocene ice sheet
about 10,000 years ago.

Gulf Stream A current in the North
Atlantic, first reported by Ponce de Leon
(1460–1521). Its point of origin was
assumed to be the Gulf of Mexico, based
on Ponce de Leon's statements and those
of other Spanish navigators. It is really the
northern and then western swing of the
North Atlantic Current. In the early 19th
century, British explorations, using Humboldt's observations, rightly attributed the

high seas and resulting current of the West
Indies to the trade winds.

In the 1840s, Alexander Dallas Bache
began a long study of the Gulf Stream,
thus following a tradition established by
his great-grandfather, Benjamin Franklin. In the 1880s Lt. J. E. Pillsbury of the
U.S. Coast Survey measured currents of
the Stream from an anchored ship, using
Benjamin Franklin's phrase "a river in the
ocean" to describe it.

The Gulf Stream separates the warm,
salty Sargasso Sea from the cold, less
saline inshore water. It moves north along
the edge of the Blake Plateau at about 2.0
to 2.5 m second (or 4 to 5 knots/hour).
This huge stream is 800 m (2,650 feet)
deep. As it rushes along, it brings with it
sudden storms, deep blue water, and phosphorescent displays at night.

About the latitude of Cape Hatteras,
North Carolina, the Gulf Stream leaves
the coast and changes character. It broadens and deepens to about 1,000 m (3,300
feet). The popular concept of the Gulf
Stream as a warm current is true only of a
part of this huge flow.

North of Cape Hatteras the Gulf
Stream flows over the continental rise and
is deflected eastward by the Labrador Current. It also spreads out over the abyssal
plain and increases greatly in width. There
are large meanders and eddies that form,
vanish, and reform. Countercurrents form
on the right (south) of the stream. By the
time the system reaches the Grand Banks
of Newfoundland, the Gulf Stream has
lost about 20% of its volume.

The path of the system and its direct
effect on Europe is not clearly understood. Although the North Atlantic gyre
brings warmer water into the waters of
northern Europe, it is not clear whether
this water is of "Gulf Stream" origin. *See*
BACHE, ALEXANDER DALLAS; CARIBBEAN
SEA; CURRENTS; EQUATORIAL CURRENTS;
FLORIDA CURRENT; FRANKLIN, BENJAMIN;
HATTERAS.

gull A long-winged, web-footed seabird
of the family Laridae with a relatively

Some commonly known gulls are as follows:			
Common Name	Scientific Name	Distinguishing Feature	Habitat
Black-headed	L. ridibundus	Red legs	Iceland, Europe
Bonaparte's	L. philadelphia	Black bill/red legs	North America
California	L. californicus		Western North America
Franklin's	L. pipixan		North & South America
Glaucas	L. hyperboreaus	White/pink legs	Hawaii & Mediterranean
Great black-headed	L. marinus	1.6 m (63-inch) wingspan (largest gull)	Polar
Herring	L. argentatus	Gray, fleshy legs (most common gull)	Northern hemisphere
Kelp	L. dominicanus	Black head	Southern hemisphere
Laughing	L. atricilla	Black head/red bill/ red feet	Maine to South America
Little	L. minutus	Smallest, 60 cm (24 inches)	Europe & North America
Pacific	L. pacificus		Australia & Tasmania
Ringbilled	L. delawarensis		North America inland
Sooty	L. hemprichi	Dark brown	Indian Ocean

The following look like gulls but are not in the Laridae.			
Common Name	Scientific Name	Distinguishing Feature	Habitat
Ross's	Rhodostethia	Pinkish white	Siberia, seldom in North America
Sabine's	Xema sabini		Arctic
Swallowtailed	Cregrus furiatus	Unusual tail	Galapagos

heavy body and strong, slightly hooked beak. Gulls are best known in the Northern Hemisphere. About 30 to 40 species live in temperate to polar regions. Gulls migrate but are less likely than terns to travel long distances to their winter grounds. Gulls are scavengers: they eat insects, crustaceans, mollusks, worms, fish, the eggs of other birds, and garbage.

gulper A highly specialized eel of the suborder Saccopharyngoidei and the order Anguilliformes. There are 11 known species in three families, and only one is large: 185 cm or 6 feet. The others are typically less than 75 cm long, some only 10–12 cm. They are deep-sea predators with the uniting characteristic of an enormous mouth in relation to their overall body size. They are well adapted to low-light environments.

gum A commercial product extracted from algae. The word is usually used for a single substance that might be agar, algin, or carrageenan. These are all extracted from one or another species of algae and used as stabilizers and/or gels in the production of foods, cosmetics, pharmaceuticals, paper, and textiles.

guyot Also called a tablemount. A flat-topped seamount rising from the seafloor. These structures were named in honor of Arnold Guyot, a 19th-century geologist, by their discoverer, Harry Hess. *See* HESS, HARRY; SEAMOUNT.

Gymnosomata Small pteropods, sometimes called sea angels, that are part of the zooplankton. They are gastropods like snails but have no shells. These carnivores may grow to about 1 cm (0.3 inches) long and prey for the most part on shelled pteropods. Although they are widespread, the life cycle of these organisms is largely unknown. Most genera are concentrated in warm waters. *Clione,* however, is one

gymnosome that is well established in Arctic regions. *See* PLANKTON, SNAIL.

gyre The roughly circular path of water circulation in the open ocean. The North Atlantic gyre comprises the Gulf Stream, the North Atlantic Current, and the Canaries Current. The return flow is the Equatorial Current. Although this is probably the best-defined gyre, gyres exist in all oceans. *See* CURRENTS.

Ocean gyres

habitat   The space occupied by an individual or group of individuals, including everything in that space both living and nonliving. Typical marine habitats are mangrove swamps, marshes, estuaries and mudflats, intertidal zones, and benthic (depth of the ocean) and pelagic (ocean surface) zones. Within each habitat are specific variations due to the temperature, light, the substratum, and other physical variables. *See* BENTHOS, ESTUARY, INTERTIDAL ZONE, POLAR BIOME.

hadal zone   The name used for the ecological zone at ocean depths greater than 6 km (20,000 feet). There is no light at this depth; the water pressure is over 600 atmospheres and can be as high as 1,200 atmospheres in ocean trenches. Both the temperature and salinity of hadal zones are fairly constant; the salt content is about 3.47%. The bottom itself is composed of red clay of terrigenous origin and radiolarian ooze. The dominant animals are omnivorous crustaceans, echinoderms, polychaetes, and mollusks. Present in very small numbers, these animals feed on one another or on detritus and the mud-and-detritus-eating organisms that exist even under these difficult conditions. *See* BENTHOS, PELAGIC ZONE.

haddock   A large North Atlantic fish of the cod family. The haddock resembles the cod but has a different coloring, usually dark brown or green, with a noticeably darker lateral line and white or creamy undersides. Haddock are about 84 to 95 cm (30 to 35 inches) long and are bottom-dwelling carnivores. Like the cod, they are an important component of the commercial fishing catch. *See* COD, FISH.

hagfish   Primitive fishlike animals that, like the lampreys, have a cartilaginous skeleton, no jaws, and belong to the class Agnatha. Also like the lamprey, which is a freshwater animal, hagfish have sucker mouths with many teeth. Hagfish are cold-water organisms that live in burrows and feed on invertebrates, detritus, and dying animals. Hagfish are also called slime eels; when disturbed, they produce massive quantities of mucus, which makes fishing for these animals a messy business. They are fished commercially and eaten in the Far East. The skins make good leather that is frequently sold as eel skin. *See* CYCLOSTOMATA, FISH, MUCUS.

hake   A fast-swimming carnivorous fish that is caught commercially off the Canadian coasts and in the eastern Pacific near New Zealand. Hake are sometimes classed as a family distinct from the cod, the *Merlucidae*. Another genus also commonly referred to as hake is *Urophycis*. *See* FISH.

Hakluyt, Richard (1552–1616)   An English collector of travel diaries. His famous work was *Principal Navigations, Voyages, Traffic and Discoveries of the English Nation,* written in 1589 when Hakluyt was the chaplain of the British ambassador to the French court.

Hakluyt continued to collect and publish accounts of the sea after the publication of his great work. He was convinced of the patriotic necessity of exploration and continued to write extensively to support this position. The Hakluyt Society, a British group founded in 1874, was named for Richard Hakluyt as a memorial. His original 12-volume work, by then much expanded, was included in the society's publications of original texts of travel logs. See EXPLORERS AND EXPLORATIONS.

Haldane, John Scott (1860–1936) An English physiologist who, in conjunction with J. G. Priestley, studied the effect of the partial pressure of carbon dioxide on the control of breathing, a body of research that was published in 1905. Haldane then carried out research on respiration under stress, in mines and caissons. He worked out the method of slow decompression still used in all ocean descents and particularly in ascents from ocean depths. His work was essential to the design of ventilation systems for diving equipment and submarines. See DIVING.

half-life The length of time required for half of some substance to disappear. If the substance disappearing is a radioactive isotope of some chemical element, then the half-life is that time necessary for half of the atoms of a given sample to decay and form another element by losing alpha and beta particles and gamma radiation. Depending on the isotope in question, the half-life can range from fractions of a second to billions of years. Since the half-life of radioactive isotopes is a constant, the rate of decay of these isotopes can be used to measure time. This is the basis of radioactive dating.

The concept of half-life is also used to measure the disappearance of pollutants, particularly those that are not easily attacked by naturally occurring enzyme systems. See GAMMA RAYS, RADIOACTIVE DATING.

halibut A large, right-eyed flatfish of the genus *Hippoglossus*. At maturity the fish is either brown or dark green, with a light underside. Halibut may be larger than 2 m (6 to 7 feet) and weigh over 300 kg (700 pounds). However, since the halibut currently caught commercially are considerably smaller than this, there is the possibility that halibut is being overfished in the North Atlantic.

Other genera are also known as halibut, but they are usually smaller than *Hippoglossus*. The California halibut (*Paralichthys californicus*) may have eyes and color on both sides but is nevertheless related to the left-eyed flatfish. Its range includes Japan and Indo-Pacific ocean water. See FLATFISH.

Halobacteria The halophiles, or salt-loving bacteria, are gram-negative rods or cocci that live in salt water. They are aerobic, with a metabolism based on amino acids rather than carbohydrates. Halophiles usually contain carotenoids that tint the colonies yellow, orange, pink, or red. They range in salt requirements from slight (the open seas) to moderate (saltier seas) to extreme (salt lakes). The Dead Sea, however, is too salty. See BRINE POOLS.

halocline A layer of water in which the salt content changes rapidly. It forms a boundary between regions of different salt content.

halogens The elements fluorine (F), chlorine (Cl), bromine (Br), and iodine (I). At room temperature and atmospheric pressure fluorine and chlorine are gases, bromine is a liquid, and iodine is a solid. These elements are important as ions (F^-, Cl^-, Br^-, I^-) in seawater; the chloride ion is the most abundant and noteworthy of the four. See BROMINE, CHLORINE, ELEMENT, IODINE, WATER.

hammerhead shark A type of shark named for the bizarre shape of its head: a grotesquely flattened T-shape with an eye and nostril at each end of the T. Hammerheads live in warm seas and have a worldwide range. They feed on fish, other sharks, rays, and skates. Since they range into coastal waters and will eat anything, they are potentially dangerous to humans but prefer a fish diet. Hammerheads are sometimes fished commercially for their skin and oil. The common hammerhead, *Sphyrna zygaena,* grows to lengths of 4 m (over 12 feet) or more. See ENDANGERED SPECIES, SHARK.

harbor seal The common or spotted seal (*Phoca vitulina*), frequently found near shores or in harbors, particularly fishing harbors, of the Northern Hemisphere. Male harbor seals are much larger than females; the average male is most often about 2 m (7 feet) long and weighing 120 to 130 kg (250 to 270 pounds). The young are dirty white or gray and darken to dark grey or black as they mature.

The harbor seal is a social creature, easily tamed, and a voracious feeder on fish, squid, and crustacea. *See* SEAL.

Harpacticoida A benthic free-living order of copepods. These tiny organisms range in size from 0.2 to 2.5 mm (0.0008 to 0.1 inch) and can be found in almost all aquatic environments. In marine habitats, they range from tidal pools to the abyssal zone. The approximately 2,500 known species are an essential part of the food web. They feed on diatoms, bacteria, and small protozoans. *See* COPEPODA.

harp seal A migratory seal (*Papophillus groenlandicus* or *Phoca groenlandica*) found in the Northern Hemisphere. It is a far-ranging animal that is a strong swimmer. The breeding grounds of the harp seal are Newfoundland, Greenland, and the ice shelves of the White Sea. The adult is a grayish yellow to brown-black color. Harp seals have strong, light markings, which are more distinct on the males than on the females. Males are about 1.75 m (6 feet) long and weigh about 180 kg (400 pounds). The harp seal is killed for its fur. The young are white for about two weeks after birth, and it is the fate of these pups, who are clubbed to death by hunters, that has stirred conservationists and humane societies to force sealers to agree to quotas and less ghastly killing methods. *See* SEAL.

hatchet fish Fish belonging to two groups—a freshwater group related to the carp, and a marine group akin to the salmon and trout—and belonging to the family *Sternoptychidae*. The range of the marine group is from warm to temperate waters and from depths of 200 to 1,000 m (650 to 3,300 feet). They are flattened, deep-bodied, small animals with slender, shapely tails. The overall length of the marine hatchet fish is most often less than 10 cm (4 inches). Many hatchet fish have light organs on their heads. All of the hatchet fish are carnivorous. *See* LIGHT ORGANS, SALMON.

Hatteras The Hatteras Abyssal Plain is part of the floor of the northwest Atlantic Basin east of the continental shelf of North America. The Bermuda Rise forms its western boundary.

Cape Hatteras is a sandbar forming a promontory on Hatteras Island, North Carolina. The shallows are a navigational hazard. It is at this point that the Gulf Stream, which has moved north along a line about 150 m (600 feet) off the coast, bends east and meets the cold coastal Labrador Current from the north. The cold air over warm water and the cold water meeting warm water at Cape Hatteras create both water and atmospheric disturbances that produce storms, fogs, high tides, and swift, short-lived eddies. *See* ATLANTIC OCEAN, GULF STREAM.

heat Energy that passes from one body to another because of the temperature difference between them. Heat may be generated by sunlight, chemical or nuclear reactions, or the conversion of electrical or mechanical energy into heat. Heat waves are part of the electromagnetic spectrum. *See* ENERGY.

Heezen, Bruce Charles (1924–1977) An American oceanographer associated with the Lamont-Doherty Laboratory from its inception in 1949. Heezen was involved in many of the 20th century's discoveries about the seafloor. He and Marie Tharp mapped the North Atlantic bottom and then all the world's oceans. He participated in the discovery of the role of turbidity currents, proposed the meteorite explanation of tektites, and was coauthor of *The Face of the Deep*, a book on undersea photography. Heezen's most significant work, done with William Maurice Ewing of the Lamont Doherty Laboratory, was the elucidation of the ridge-rift valley complex that is the most continuous feature on the Earth. In plotting the ridge profiles, Tharp noticed a dip at the centerline of the ridge. The plotting of undersea earthquakes coincided with these rifts. *See* EWING, WILLIAM MAURICE; LAMONT-DOHERTY LABORATORY; MID-OCEAN RIDGES; THARP, MARIE.

helium An inert chemical element that occurs as a gas in the Earth's atmosphere, of which it composes 0.0005%. Helium is used in high-pressure deep-diving equipment because its solubility in blood is much lower than that of nitrogen, and its use can therefore prevent nitrogen narcosis, or "the bends." *See* AIR, DIVING, ELEMENT.

Hellenic Plate A small section of the Earth's crust that bears Greece and the Ionian islands. It is being pushed against the Eurasian Plate by the African Plate. Seismic activity is frequent at the edges. *See* CRUST, PLATE TECTONICS.

Hemichordata A small phylum of several hundred identified living species and a considerable number of fossil species. They are very important in the study of vertebrate evolution. Acorn worms

are the most common living organisms in this group. The name Hemichordata was first proposed because the protruding mouthpart was assumed to be a notochord. They are characterized by a three-part body plan: a preoral lobe (the front end); a collar, a thick band that contains a notochord-like structure called a stomochord; and a trunk, or body, with branchial (breathing tube) openings, the analogs of gill slits. A dorsal nerve cord is present, as is a much smaller ventral nerve cord.

The best-known acorn worms are solitary burrowers and are closer to echinoderms than to chordates. Not only do the larval forms resemble echinoderm larvae, but DNA analysis upholds these observations. The Enteropneusta have worldwide distribution in intertidal zones with a few deepwater species. They usually range in length from 3 to 250 cm (1.5 to 8.3 feet). Some are spectacularly colored, with the oranges, reds, and yellows predominating. The Pterobranchia tend to live at greater depths and are usually sessile colonies. These were once grouped with the sea fans, which they resemble superficially. *See* ACORN WORM, ECHINODERMATA, ENTEROPNEUSTA, GRAPHOLITE, PTEROBRANCHIA.

Henry the Navigator (1394–1460) The patron of Portuguese navigation, Prince

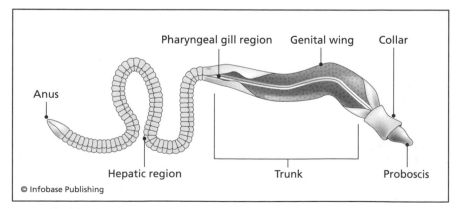

Pharyngeal gill region   Genital wing   Collar

Anus

Hepatic region   Trunk   Proboscis

© Infobase Publishing

Hemichordata were once classified with worms. They are now a separate phylum.

Henry, the Infante Dom Henrique, was the third son of John (João) I, king of Portugal.

As governor of the Algarve, Henry established an information center at Sagres. He had the first telescope-equipped observatory in Portugal and assembled an international court of geographers and mathematicians.

Henry's interest and support led to voyages of exploration to the Madeiras, the Azores, and the coast of Africa. The interest and support of his court and the concentration of knowledgeable people made these voyages of exploration possible. Henry continued to inspire the Portuguese in feats of navigation, and after his death voyages were dedicated to his memory. *See* EXPLORERS AND EXPLORATIONS.

herbivore The designation of an organism that derives its nourishment by consuming primary producers. Plankton that feed on other plankton that are photosynthesizers are herbivores, as are all the other grazing organisms. A better designation is *primary consumer. See* AUTOTROPHE, CARNIVORE, CHEMOSYNTHETIC BACTERIA, HETEROTROPHE.

hermaphrodism This denotes either a change of sex or sexual flexibility. Many invertebrates have both male and female gonads; this is commonly found in the mollusks, barnacles, copepods, and other arthropods, worms, and echinoderms. It is less usual in vertebrates, but it does occur. Some deep-sea fish produce both eggs and sperm, and a number of fish change sex during a lifetime. Coral reef dwellers such as parrot fish, groupers, and particularly wrasses and damselfish exhibit this pattern. Large males dominate the population. Fish tend to continue to grow throughout life, so if a female becomes old and large, it may become a functional male. It is thought that this change of sex conveys an evolutionary advantage by increasing the number of offspring of a survivor. *See* WRASSE.

hermit crab A type of crab that uses an empty snail or whelk shell as the protective shield for its body. As the crab grows, it changes the shell for a larger one several times during its life. The two families of hermit crabs, Paguridae and Coenobitidae, are most often found on sandy bottoms of shallow water in almost any warm to temperate region. Some hermit crabs have complex commensal relationships with sea anemones: When the crab moves from a smaller shell to a larger one, it takes its anemone along to its new home. A few species of hermit crabs are terrestrial, and some of these use plant stems or coconut shells as protection. *See* COMMENSAL RELATIONSHIP, CRAB, MOLT.

heron A fairly large, long-legged wading bird that lives in shallow water and feeds on fish, crustaceans, and amphibians. The range of herons is worldwide, but most species are tropical. The great blue heron of North America, *Ardea herodias,* has a wingspan that exceeds 1.5 m (5 feet). The slightly smaller gray heron, *Acinerea,* is typical of Europe. *See* BIRD, ESTUARY.

herring Members of the numerous Clupeidae family, herring are silvery-blue fish that move in huge schools. Their diet is plankton, and in turn they provide the food for other fish and mammalian predators.

The common or Atlantic herring, *Clupea harengus,* and *C. pallasi,* the Pacific species, account for the largest single fish catch taken commercially. Herring produce vast numbers of eggs that hatch at varying times ranging from December to July, depending on latitude and temperature.

The adult fish are 10 to 40 cm (4 to 16 inches) long and are mature at about three to four years. They have a long lifespan: Tagged individuals older than 20 years have been found. Several related species of fish are also called herring, including sprats, alewives, summer herring, lake herring, and shad. *See* ANCHOVY, SARDINE.

Hess, Harry H. (1906–1969) American geologist and president of the American Geological Society, Hess was part of the MOHOLE project and the person in charge of the choice of drilling sites. Hess discovered that the ocean bottom was composed of rock younger than that on the continents. The age of the bottom also varies, he found, being younger close to the mid-ocean ridges and increasingly older farther from them. To explain the origin of this younger rock, Hess assumed that the seafloor was spreading. This theory was eventually proven by F. J. Vine and D. H. Matthews, of Cambridge University, England, who, by means of a chronology in stone that constituted a permanent record of changes in the magnetic field of the Earth, verified that the seafloor was spreading out from the mid-ocean ridges. The deposition of new seafloor reflects the magnetic orientation of the Earth. It is a pattern of stripes, each corresponding to the position of the magnetic poles at the time of the deposition of that part of the seafloor. *See* MID-OCEAN RIDGES.

heterotrophs Organisms that cannot manufacture their own food from inorganic material; animals, bacteria, and fungi. Heterotrophs ultimately depend on autotrophs for their energy supply (usually in the form of food). In most cases, this means that they digest organic material made by autotrophs or other heterotrophs. *See* AUTOTROPHS, CARNIVORES, DETRITUS FEEDERS, ECOSYSTEM.

Hexactinellida Glass sponges, a class of the phylum Porifera. They have radial, six-way symmetry, and their spicules (support structures) are silaceous, or glasslike. There are over 450 species. *See* PORIFERA, SPONGE.

Heyerdahl, Thor (1914–2002) A Norwegian adventurer who in 1947 became famous for his raft voyage across the Pacific. He developed a number of theories describing Polynesian migrations and with five companions sailed from South America to the Tuamoto islands. The very successful book he wrote about his trip is *The Kon-tiki Expedition*. He went on to another raft voyage across the Arabian

Sea—the Tigris expedition. His last adventure was on Easter Island. Although he was a very popular figure, his anthropological work and theories of migration patterns were seriously questioned by professional anthropologists. For example, one of Heyerdahl's works gave South America as the source of Polynesians; therefore, Easter Island would have been the first island settled. Heyerdahl's entire migration theory was confused, and the Easter Islanders moved artifacts to suit his theory of the moment. According to anthropologists, he was a meddling amateur.

high water High point of the high tide at a particular place. *See* TIDE.

Hirudinea The leeches, a class of annelids with few body segments and no setae. They are parasites on vertebrate fish, both cartilaginous and teleost.

These hermaphrodite worms have sucking devices at either end. Most of the coelom is taken up with gonads. The single eggs are laid in cocoons and deposited in the water. The adults live on the hosts' blood.

Leeches have been used for centuries and are still used in some countries by physicians to bleed patients. A constant blood flow is ensured by the leech, which produces an anticoagulant that keeps the patient's wound open. The anticoagulant compounds produced by leeches are being explored for possible pharmaceutical applications.

While the leech is a parasite, it also has a commensal relationship with the bacteria that live in its gut pockets. These bacteria digest the blood that the leech ingests, feeding both themselves and the leech. *See* ANNELIDA, COELOM, COMMENSAL RELATIONSHIP, HERMAPHRODITE, PARASITE, SETAE.

Holocene epoch The younger portion of the Quaternary period and the most recent time in the history of Earth, it follows the Pleistocene. The significant development of the Holocene was the melting of the great sheets of glacial ice. This resulted in the springing back of large parts of continental landmass, such as Scandinavia, and a dramatic rise in the sea level. This rise in sea level obliterated the land connections between Alaska and Siberia, Europe and England, Sicily and Italy, Sri Lanka and India, and many other neighboring locations.

The beginning of the Holocene, about 10,000 years ago, was marked by increasingly warm weather worldwide. It was also wet: Areas of Africa, India, and the Arabian peninsula that are now desert were arable. The temperature declined, then rose once more from about 1000 to 1250 C.E. It then declined again about 1300 to 1800 C.E. and has been rising since. This latest rise may be normal or may be due to the world's increasing use of fuel or a combination of these factors. *See* GREENHOUSE EFFECT, *APPENDIX: GEOLOGIC TIMESCALE.*

Holocephali A subclass of bottom-dwelling carnivorous fish related to sharks. Many of these fish have light-emitting organs. Like sharks, they can concentrate urea in their body tissues, and it has been suggested that they use this as a flotation mechanism. *See* CHIMAERA, LIGHT ORGANS.

Holothuroidea This class of the phylum Echinodermata comprises the sea cucumbers. Although they have an ancient lineage, like other soft-bodied organisms, they have left a very incomplete fossil record. Some of the lineages are thought to have separated in the late Paleozoic era, about 350 to 250 million years ago.

The six orders represent hundreds of species. Many of these are of commercial importance and are fished to such an extent that the industry is collapsing in some areas. *See* SEA CUCUMBER.

homoiotherms (homeotherms) Warm-blooded animals. Birds and mammals are homoiotherms. Homoiotherms have much greater flexibility in selecting their environment but require much more insulation and much more food than do the poikilotherms (cold-blooded animals). The alternative to a constant and abun-

dant food supply is hibernation or migration. There is evidence that constant body temperature is exhibited by some reptiles; some crocodiles and alligators exhibit this property. Continuing study of dinosaurs has led some researchers to posit that this property existed in those extinct creatures as well. A case has been made that dinosaur lineages with constant temperatures are the ancestors of modern birds. *See* POIKILOTHERM, TEMPERATURE OF ANIMALS.

hook A spit or narrow peninsula, the end of which bends toward the land. The Hook of Holland and Cape Cod, Massachusetts, are examples.

Hoplocarida A superorder of Crustacea containing one extant order, the Stomatopoda, and one suborder, Unipeltata—the mantis shrimp. These animals have the typical crustacean body plan, including flagellated antennae, stalked eyes, and a long, narrow heart. The suborder is divided among 12 families, or about 350 species. These raptorial shrimp live in shallow, tropical waters. The larval stages are pelagic, but adults settle on the bottom, where they actively forage or lurk in burrows, catching whatever soft-bodied prey comes within striking distance. The raptorial claws are either smashing or spearing weapons. The velocity of the strike may be as high as 12 m/sec. Adults range in size from 1.5 to over 30 cm (0.4 to 12 inches), but most individuals are small. *See* CRUSTACEA.

horny coral *See* ANTHOZOA.

horse clam A large, edible mollusk, most often about 20 cm (8 inches) long and 3 to 5 cm (1.5 to 2 inches) wide, found on the west coast of North America from British Columbia to Baja California. The horse clam burrows into the sand to depths exceeding 1 m (40 inches).

horse latitudes Atmospheric belts of high pressure found in both the Northern and Southern hemispheres at about 30 to 40° latitude. These regions are known for good weather and light winds. In the Southern Hemisphere, where there is more water and less land, the high pressure belt is more complete than that in the Northern Hemisphere. *See* ATMOSPHERE, INTERTROPICAL CONVERGENCE, WIND.

horseshoe crab An arthropod belonging to the class *Merostomata,* but not a true crab. These animals have a three-part body consisting of the prosoma, which is the largest segment and includes the head; the middle segment, or opisthosoma, which is the heavy shell; and the spiny tail, or telson. Like lobsters, horseshoe crabs have compound eyes. They have numerous legs for movement, for capturing prey, and for feeding. Their circulatory and digestive systems are also similar to those of the lobster.

Horseshoe crabs spawn on beaches in the summer, at high spring tide; the next high spring tide then washes the larvae out to sea. Their growth is a complex series of body changes and moltings, and the horseshoe crab does not reach adulthood until it is 10 or more years old. The best-known genus is *Limulus.* All five existing species are ancient, having existed in their present forms for 200 million years or more. *See* ARTHROPODA.

horst A segment of crustal rock that lies between two parallel faults and is thrust upward. *See* CRUSTAL BLOCK, GRABEN, PLATE TECTONICS.

hot spots Centers of volcanic activity beneath crustal plates. The idea of hot spots was formulated in an attempt to explain the existence of seamounts in areas where there are no volcanoes. According to this idea, as a plate moves across a hot spot, volcanic eruptions occur, and a seamount forms if the plate is an oceanic (and therefore thinner) plate. The seamount can then become the basis for a coral atoll.

One explanation for chains that have different directions in the Pacific is that there are "lost plates." These different alignments of Pacific crust mean that a crust swept across a hot spot in different directions during past ages, pushed by plates no longer there but leaving island chains in place to mark the passage. Since the plates move over these spots slowly, liquid rock solidifies and accumulates. The plate may stay over a hot spot for 2 to 3 million years. Depending on how much material is deposited, the result is a seamount or an island. In Hawaii, the big island is exhibiting less volcanic activity because the Pacific Plate has moved. The hot spot is now building a new undersea volcanic peak southeast of Hawaii, and this soon-to-be island is still a seamount named Loihi.

The hot spot theory is also used to explain the existence of some Atlantic Ocean islands, notably the Atlantis Seamount and Ascension Island. *See* CONTINENTAL DRIFT, EVOLUTION OF OCEANS, PLATE TECTONICS.

hot springs Thermal springs with temperatures of approximately 100°F (40°C) or above. They occur under water and on land. They are frequently in areas of volcanic activity or where fossil water (water that has been trapped in rock for years) makes contact with hot magma (liquid rock). *See* ARTESIAN WATER, RED SEA, THERMAL SPRING.

Huang Ho (or Hwang Ho or Huang He) The Yellow River, China's second longest and northernmost large river, flowing over 4,800 km (3,000 miles) from Tibet to the Gulf of Chihli in the Yellow Sea. It is the world's leading river in sediment carried, hence its name. The approximately 2,000 km² (1,100 square miles) delta of the Yellow River grows rapidly as a result of the enormous sediment load deposited. The course of the river has changed repeatedly in recorded time (since the sixth century B.C.E.).

Since the river frequently shifts, the delta is a lobed one, as sediment buildup goes in one direction for a number of years, after which the direction of buildup then reverses. Disastrous flooding along the Yellow River has been well documented in the past, and not surprisingly, the River is also called China's Sorrow. Because of the torrent of water and the silt it carries, the Huang Ho has produced a vast rice-growing plain that supports China's huge population. Irrigation dams and hydroelectric installations have also been built along the river. *See* YELLOW SEA.

Hudson, Henry (1565?–1611) English navigator and explorer of North America. Hudson sailed for English interests on three of his voyages, in 1607, 1608, and 1610. He is best remembered—certainly in North America—for his voyage in 1609 for the Dutch East India Company.

On his first voyage Hudson's employer was the Muscovy Company. He sailed north to Spitsbergen (now Svalbard) in search of a Northeast Passage to China. The next year he tried to sail around the Arctic island group and found the way blocked by ice. On his third voyage, for the Dutch, he tried to find a sea lane through the North American landmass. He sailed his ship, the *Half Moon*, into the magnificent bay discovered by Verrazano and then sailed up the river now named for him. He realized that it was not a strait leading to the Pacific and returned to Europe.

This third voyage was the basis for the Dutch claim to territories in North America. Hudson's other expeditions were the English rationale for claims to Canada.

Hudson's last voyage was commissioned by the British East India Company and the Muscovy Company as a joint venture. He left London in April 1610, sailing on the *Discovery*, to explore a northern river in North America whose outflow was very voluminous. It was hoped that the volume of water from that river would indicate a water passage to the Pacific. The *Discovery* entered Hudson Bay, the sailors hoping that the sheer size of the waterway was a good sign. Unfortunately they moved south and entered the embayment known as James Bay, where the severe winter overwhelmed the crew. They

mutinied and set Hudson, his son, and several still-loyal crewmen adrift on June 22, 1611. When the *Discovery* returned to England, the leaders of the mutiny were not aboard; they had been killed in a fight with the Inuit (Eskimos). *See* NORTHEAST PASSAGE, NORTHWEST PASSAGE.

Hudson Canyon A canyon extending southeast of the mouth of the Hudson River to the edge of the continental shelf in the North Atlantic. The canyon is about 90 km long and terminates in a fan valley, formed by Hudson River sediment, at the bottom of the continental slope. The area has been explored for possible oil or gas deposits, but these have not so far been of commercial interest. *See* ATLANTIC OCEAN, CANYON.

Hudson Strait A strait linking Hudson Bay with the Atlantic Ocean via the Labrador Sea. The Strait is about 700 km (440 miles) long and ranges from 75 to 210 km (40 to 150 miles) wide. It is clogged with ice most of the year. The first exploration of the strait was attempted by Martin Frobisher in 1578, in an early attempt to find the Northwest Passage between Europe and Asia. *See* FROBISHER, MARTIN; HUDSON, HENRY; NORTHWEST PASSAGE.

Humboldt, Friedrich Heinrich Alexander von (1769–1859) A German explorer, geographer, and naturalist. Humboldt was influential in a number of fields and had sufficient money and energy to explore all of them. His training as an engineer strengthened his interest in natural science. His first book, published in 1790, was on geology. Humboldt also worked in chemistry, designed mine-rescue equipment, and in 1795, on a trip to the Alps, realized the relationship of altitude to climate and plant zonation, finding that the plants and animals found at higher altitudes were similar to those found at higher latitudes. Geomagnetism was also one of Humboldt's interests and gave rise to his work toward standardization and precision in geographic measurements.

In 1798 Humboldt went to Spain to "measure" the country with sextants, chronometers, thermometers, and barometers. He then obtained permission to "measure" the American colonies of Spain. Landing in Venezuela in July of 1799, he conducted extensive exploratory voyages into the tropical forests and sent home plants, made thousands of observations, and collected data on magnetism, meteorology, climate, geology, minerals, oceanography, zoology, and ethnography. Some of his travels were very hazardous. Traveling in native canoes, on packhorses, and on foot, he moved through the basins of the Orinoco and Magdalena rivers and proved the connection of the latter with the Amazon River.

Humboldt was well received in his own time. He visited the United States in 1804 and appeared at the American Philosophical Society. He later went on to do experiments with Gay-Lussac in Paris. In 1829 he again embarked on a grand voyage of exploration, this time through Siberia.

While not a great innovator, Humboldt was a "connector" who recognized the links between various fields of study. *See* CURRENTS.

Humboldt Current *See* PERU CURRENT.

humidity A measure of the quantity of water vapor in the air. It is reported as grams/kilogram (g/kg) of water. *See* ABSOLUTE HUMIDITY, AIR, ATMOSPHERE, FOG, RELATIVE HUMIDITY.

humpback whale A relatively short, 12 to 15 m (40 to 50 feet) long, but stubby baleen whale, *Megaptera modosa* or *novaeangliae*, with a hump on its back. It is black or sooty brown above and cream or white on the underside. This whale has long, thin flippers and a few hairs on its head; its throat and underside are deeply ridged. The humpback breeds in summer in warm waters, migrating there from the polar regions. It has been hunted almost to extinction. *See* BALEEN, WHALE.

hurricane A tropical cyclone in the western Atlantic Ocean. In the western Pacific the same kind of storm is called a typhoon. The storms develop around the Equatorial Convergence after the summer solstice in each hemisphere (June 21 or December 21). Therefore, the hurricane season is July to September in the Northern Hemisphere and January to March in the Southern.

One theory of hurricane genesis is that the storms are set off by the westerly winds invading the doldrums. The winds circulate around the low-pressure center (counterclockwise north of the equator, clockwise south of it) in a swirling column that rises to a height of 15 km (10 miles). The entire system moves west at a velocity of 5 to 14 m/second (18 to 54 km/hour or 30 to 90 mph). Not all hurricanes develop to their full potential, however.

The accuracy of predicting hurricanes has increased greatly, making protective measures more meaningful. Cloud seeding with silver iodide crystals has been attempted in an effort to promote rain and break the rush of the wind in these storms, and it has had sufficient success to warrant further experimentation. *See* DOLDRUMS, INTERTROPICAL CONVERGENCE, STORM.

Hutton, James (1726–1797) A Scottish geologist and chemist who originated the theory of Earth's development called uniformitarianism. This theory postulates that while millions of years were required to develop the Earth, the processes were identical with current ones. First a law apprentice, Hutton changed careers and became a medical student and finally a manufacturing chemist. He prospered and retired to be a gentleman-farmer, later moving to Edinburgh, an important center of scientific study.

Hutton presented his theory of the formation of Earth's layers to the Royal Society of Edinburgh in 1785. He published some of his work in two volumes in 1795; some remained unpublished at the time of his death. John Playfair condensed and edited Hutton's notes, and this version became a foundation stone in the study of geology and of the history of the Earth. *See* LYELL, CHARLES.

hydrocarbons Simple organic compounds (molecules) composed only of the two elements hydrogen and carbon. Hydrocarbons can, however, be much larger than these molecules and quite complex. As a general rule, the larger the molecule (and the more carbons and hydrogens it contains) the higher the boiling point. Molecules with chains of five to fifteen carbons are liquids at room temperature. Hydrocarbons whose carbon chains are 25 or more atoms long, are waxy solids. Very long chains of hundreds or thousands of carbon atoms (i.e., polymers) are plastic materials, such as polyethylene or polystyrene.

Hydrocarbons in the form of oil are of interest to marine scientists. Some algae store a food supply in the form of oil; other organisms use hydrocarbon molecules to build other organic molecules, notably sugars and amino acids. Oil, when spilled from tankers or other vessels by accident or as the result of a shipwreck, presents a serious threat to marine and

beach organisms. *See* AMINO ACID, POLYMER PROTEIN, STARCH, SUGAR.

hydrofoil A boat with wings held to the hull by struts. The hull is partially or totally above the water surface. This displacement of the hull causes hydrofoils to move faster than traditional boats. They move best on calm waters with little or no chop.

hydrologic cycle The path by which water moves in nature from one physical state to another and from one place to another. Water is present on land in large aggregates in lakes and rivers, as well as being dispersed in soil and trapped in rock. It exists in the atmosphere as water vapor and is the dominant feature of the Earth's oceans. About 97% of the world's water is in the oceans. An estimated 75% of the fresh water—only 3% of the world's entire water supply—is tied up in glaciers and polar ice.

Water vapor in the atmosphere condenses and appears as dew or precipitates as rain, snow, or hail. Liquid water and ice or snow evaporate (i.e., water changes from the liquid or solid state into a gas, or vapor). The evaporation rate is greater if

The smallest and simplest hydrocarbons are these gases.

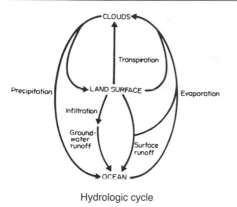

Hydrologic cycle

the water is fresh than if it is salt because the dissolved salt slows down the rate of evaporation. The movement of air, in the form of wind, increases it and humidity decreases it.

The movements of water from place to place or from one state (gas, liquid, solid) to another are physical changes. At no time in this process does the water molecule, H_2O, change. *See* WATER, WEATHER.

hydrothermal vents Vents of mineral-rich water, common at the ridge lines along the edges of crustal plates. Hydrothermal vents are present in the Pacific Ocean on the East Pacific Rise, in the Red Sea, and in the Caribbean Sea. Since the discovery of hydrothermal vents in 1977, researchers looked for evidence of former vent communities. One that was found is in the San Rafael Mountains in California. It is believed to date from the Jurassic period. The fossil finds include worm tubes, brachiopods, and gastropods. The primary producers are thought to have been chemosynthetic bacteria, as those are the primary producers in a modern vent ecosystem or in the community based around a cold seep. *See* COLD SEEPS, HOT SPRINGS, VENT COMMUNITIES.

hydrozoa Organisms that are taxonomically arranged in five orders and about 2,700 species. The usual hydroid exhibits two characteristic body plans. One is a tube-within-a-tube body with a tentacle-surrounded mouth at one end and a hold-fast structure that anchors the other end to a rock. The other body plan of these creatures is a medusal, or bell-shaped, body. The first of the five orders of hydrozoa is Trachylinida, those that have a very short medusal phase and a dominant polyp stage. They live within a chitinous exoskeleton and therefore are represented in the fossil record. Hydroida is an order of colonial organisms that exhibits alternating generations of hydroid and medusal forms. There are many genera and species, mostly marine, but the best-known hydrozoan, *Hydra,* is a freshwater organism. The hydrozoans are frequently mistaken for jellies when in the medusa phase of their life cycle. These creatures have left almost no fossil record. The Milleporina and Stylasterina orders will adhere to any hard surface—a rock or the shell of a snail or crab—and form delicate, branched colonies that are often mistaken for seaweed. Milleporina and Stylasterina species form large aggregates that are part of the coral reef

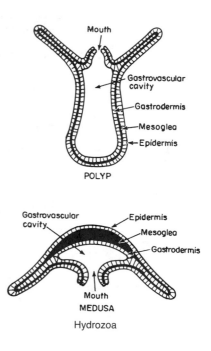

Hydrozoa

ecosystem. Their exoskeletons are aragonite, a crystalline form of calcium carbonate. Another name for these colonies is *fire coral,* since they do sting. Siphonophorida are the most complex hydroids; they form colonies wherein some organisms perform specific functions. The best-known member of this order is the Portuguese man o'war—a colony that has hundreds of individuals. Some of the individuals are tentacles bearing stinging cells, others are the digesters of captured prey, some are the reproductive organs, and others form the characteristic inflated air bag, the blue "float."

The hydrozoa have probably been present since the late Precambrian era, although definitive fossils date only from the Cenozoic and most of these are "impression" fossils. Since many hydrozoans have no hard structures, fossil finds are extremely unlikely. *See* CNIDARIA CORAL, GORGONIANS, NEMATOCYST, PORTUGUESE MAN O'WAR, SEA ANEMONE, SEA FAN.

hygrometer A device used in meteorology to measure the quantity of water vapor in the air.

ice   Frozen or crystalline water. It forms in regions of low temperature. Terrestrial ice is glacial, and if it enters the sea it forms icebergs. If the temperature is low enough, seawater will freeze despite its salt content. At a salt concentration of 3.5%, the freezing point of water is -1.91°C.

The formation of ice crystals segregates the salt in seawater. Salt does not fit well into the crystal lattice (the crystal structure) of ice, so it remains in the water, increasing the salinity of the water and further depressing the freezing point. The more saline, denser water sinks toward the bottom, and this brings up less salty water, which then freezes.

Sea ice forms as needle-shaped hexagonal crystals that first accumulate as slush and then in sheets. The thin sheets are dispersed by wind and waves as pancakes. The pancakes coalesce and form floes. The faster the ice forms, the more salt is trapped in it: The salt content of ice varies from 0.4 to 1.5%. As the ice ages, the salt trickles out, so that by the time a floe is a year old it is almost pure ice with no salt. Some floes are almost permanent, and ice stations have been put on them.

The composition of ice in glaciers is used as an indicator of large-scale weather change. The oxygen in water is largely O^{16} with some O^{18}. The ratio of the two isotopes is measurable. Both isotopes form H_2O (water). The water formed from the heavier O^{18} evaporates more slowly and is therefore concentrated in glaciers and in carbonates in marine structures. This is yet another means to determine the age of glaciers and of coral formations. Since the age of the ice can be "read" vertically, the size of a particular era is an indication of the precipitation falling at that time, whether it was rain or snow, and the inclusion of pollen and other air-blown material is an indication of what was happening on the Earth at that time. *See* CLIMATE, FAST ICE, FLOE, GLACIER, ICEBERG, PANCAKE ICE.

Ice Age   This was a term coined by Jean-Louis Agassiz, the 19th-century naturalist. He used it specifically to denote the most recent of the ice ages, a period of cooling that ended about 10,000 years ago. Glacial ice has been retreating since then. Agassiz had no knowledge of the earlier cold periods in Earth's history. These earlier cold episodes are also referred to as ice ages.

iceberg   A large mass of floating ice. Ordinarily a river brings water from landforms to the sea. A glacier is a frozen river bringing ice to the sea. At the shore, large blocks of ice fall into the water, forming icebergs. The process is called calving and is noisy. Just as a liquid river carries its greatest load in spring, so does a glacier. The greatest production of icebergs occurs in April in the Northern Hemisphere and in October in the Southern. *See* GLACIER.

ice shelf   A thick, floating platform that is to a glacier what a delta is to a river. Glacier ice and snow deposit on this shelf, which grows seaward until it is stopped by a combination of warm air or water or by ocean currents. In sheltered areas, an ice shelf can exist for millenia. The Ross Ice Shelf in Antarctica is the only example of such a shelf. It is firmly attached to the sea bottom and has been extensively drilled in studies of ice formations of previous geologic times. The

Ross Shelf is huge, having roughly the same area as all of France.

The structure of an ice shelf provides a theoretical basis for fiord formation as a result of glacial action. As the glacier continues to grow and depress the land at the edge of the sea, a wall of ice forms and grows from the sea bottom up. *See* GLACIER.

ichthyosaur One of several extinct marine reptiles. Ichthyosaurs flourished in the Mesozoic era, which ended about 65 million years ago. Good specimens have also been found in deposits of the Jurassic period. The average ichthyosaur was about 3 m (10 feet) long and looked something like a porpoise. It had a very sleek outline and a strong, propelling tail. The animal had large eyes and impressive jaws lined with numerous teeth. Some fed on mollusks, others on fish. *See* MESOZOIC ERA, REPTILE.

igneous rock A product of volcanic activity, it is formed from rapidly cooled magma and can be characterized by its crystalline appearance. Some of the crystals of igneous rock can be quite large.

IGY (International Geophysical Year) A period of worldwide geophysical research that began on July 1, 1957, and ended December 31, 1958. The 1957-8 IGY is the most recent of this series that has been extensively analyzed. The next such year of study and investigation began on March 1, 2007. That multinational interdisciplinary study will concentrate on the Arctic, the region of Earth that originates almost all the weather on the planet. The most important are to be examined is the accelerated rise in mean global temperature. More than 70 countries participated in the research, which included glaciology, seismology, geodesy, and many aspects of oceanography.

Several important projects were carried out simultaneously during the IGY. One was to explore the variability of solar radiation and sunspots, others collected information on cosmic radiation, the nature of the upper atmosphere, the nature of ice in glaciers and ice sheets, and the interpolation of this data with weather fluctuation. These projects used orbiting artificial satellites for the first time.

In terms of oceanography, the IGY yielded more new data on the oceans than had been produced since the HMS *Challenger* voyage of 1872. Charting was done using a variety of sounding devices. Undersea landslides and turbidity currents were studied, undersea volcanic eruptions were noted, and their output charted. Ocean currents were mapped, including subsurface currents, which have profound effects on all life on Earth and move vast quantities of water over great distances.

The exploration of the oceans during the IGY produced the evidence of continuous mid-ocean mountain ranges. Continuing research was one very vital legacy of this period of international research; the current year will focus on global warming. *See* HMS *CHALLENGER*, CURRENTS; EWING, WILLIAM MAURICE; HEEZEN, BRUCE CHARLES; MAGNETIC POLE; MID-OCEAN RIDGES.

Indian Ocean An ocean south of India, west of Australia, and east of Africa. Its origin traces back to the Mesozoic era, with the breakup of Gondwanaland into South America, Africa, Australia, Antarctica, India, and Madagascar. The large islands in the Indian Ocean—Madagascar, Sri Lanka, Sumatra [or Socotra]—are all continental.

The Indian Ocean has a surface area of about 76 million km^2 (33 million square miles). Unlike the Atlantic and Pacific oceans it has only a few seas: the Persian Gulf/Gulf of Oman and Red Sea/Gulf of Aden combinations in the northwest, the Arabian Sea to the west, the Bay of Bengal on the northeast, and the subdivision of the Bay of Bengal, the Andaman Sea. Since its division from the Pacific is indistinct, one can include in the Indian Ocean the Arafura and Timor Seas, the Gulf of Carpentaria, and the Great Australian Bight. From east to west the Indian Ocean lies roughly between 140° and 20° east latitude.

The Indian Ocean is the site of a three-plate juncture. The ridge lines—the marks of the juncture—are the outstanding ocean-bottom features. The Chagos-Laccodive Plateau runs along the mid-ocean ridge. It is a north-south feature, resembling an inverted Y, which breaks into two legs: the Southwest Indian and Mid-Indian ridges. The Southwest Ridge runs into the Atlantic-Indian Ridge at the Mozambique Fracture Zone. The Mid-Indian Ridge merges with the Southeast Indian Ridge at the Amsterdam Fracture Zone. The 90° east longitude Ridge runs north-south, east of the fan of the Ganges River down to the Southeast Indian Ridge, and is at right angles to the Broken Ridge and Diamantina Fracture Zone, which skirts southern Australia. There are several aseismic ridge systems in the ocean: the Madagascar, Seychelles, Mauritius, Agulhas, Kerguelen, and West Australian.

The 90° east longitude ridge is thought to be the evidence of India's move north before its collision with Asia. The ridge is the scour line marking the subcontinent's movement. The Mascarene or Madagascar plateau, which is northeast of Madagascar, includes the Seychelles Islands and Mauritius. These islands have "continental" bottoms, and their crust is thicker than the ocean crust. This leads back to the Atlantic myth, in that some authorities believe that this "microcontinental" region may be left-over, unconsolidated pieces of what was once Gondwanaland, the primitive southern continent. Deep core samples taken during the international Indian Ocean Exploration in 1965 have shown that the continental breakup of Gondwanaland was not instant; the separation of Madagascar from Africa dates to the Jurassic.

Winds in the Indian Ocean are a function of the monsoon pattern. In the Northern Hemisphere in winter, the Asiatic high pressure means a northeast wind blowing north of the equator. Between 35° south latitude and the Intertropical convergence, the winds are southeast. These winds take over in the northern summer, in the form of southwest monsoons bringing rain. Circulation to some extent follows the wind pattern and the West Wind Drift. The salinity of the Indian Ocean is highest near the Persian Gulf and lowest at the Bay of Bengal. The surface water in the ocean is of local origin, but the intermediate and deep water is of polar origin. Since there is no connection with the north polar regions. Antarctic water is found almost exclusively in the Indian Ocean. Some Greenland water has been sampled south of Africa. *See* ANTARCTIC WATER, EVOLUTION OF OCEANS, GANGES RIVER, ISLANDS, MID-OCEAN RIDGES, PACIFIC OCEAN.

indicator species One or more organisms used to assess the ecological health of a biome.

Indonesian Islands The world's largest island group with approximately 3,500 islands. The archipelago is about 5,500 km (3,400 miles) long and 170 km (1,100 miles) wide. Most of the islands are south of the equator. The Indonesian landmass is part Asian and part Australian. The Sunda Shelf in the west is separated from the Asian mainland by shallow seas; in the east, the Sahul Shelf is part of the Australian geologic platform. The center of the island chain has deep sea basins. The entire area, on the edges of crustal plates, is subject to frequent seismic activity. *See* KALIMANTAN OCEAN FLOOR, PLATE TECTONICS.

Inland Sea Also known as the Seto Naikai, a sea that separates the Japanese main islands, Honshu, Kyushu, and Shikoku. It has a very broken coastline and many small islands. *See* SEA OF JAPAN.

insolation The quantity of solar radiation that arrives at some given area on Earth. It is measured in calories per square centimeter, and varies with the several factors of latitude, type of surface (land, grass, snow, water, etc.), and wind. The type of surface is the most unpredictable variable in determining insolation, since reflecting surfaces such as ice and water bounce much of the Sun's radiant energy back into the atmosphere. This reflectance is further dependent on the incident angle of the Sun's rays.

interface A surface forming a boundary or demarcation between two bodies, substances, or phases. For example, there are interfaces between water layers of different saline concentration or different temperature. *See* HALOCLINE, INTERNAL WAVE, THERMOCLINE.

internal waves Waves caused by the flow of water over an irregular surface, or sharp atmospheric differences in a small geographic area or a rapid tide flow. Internal waves or tides can occur in bodies of water where different layers have different densities. They are noticeable on the surface as a moving slick or a glassy area that is not wind-driven. These waves are usually 7 to 15 m (20–50 feet) high, but in deep water they can be 10 times that high. They damp out to zero at the water bottom and at the surface.

Benjamin Franklin observed internal waves and created them by carefully moving a glass of water. The roiling of the liquid below the surface was obvious in the transparent container.

Frequent internal wave activity does occur in some oceanic areas. This phenomenon has been offered as a rational explanation for some otherwise inexplicable losses of ships at sea. *See* WAVE.

International Polar Year (IPY) A "year" of scientific study beginning March 1, 2007, and ending March 1, 2009. It includes climatology, chemistry, ecology, geodesy, geology, meteorology, microbiology, and zoology. The object of this multinational group of scientific societies is to study the air, water, land, and organisms that are found between the polar circles and the poles themselves. The impact of change in these variables since the last major study and the impact of such change on the Earth's populations will be assessed. The date for this multination effort coincides with the 125th anniversary of the First Polar Year, 1882–83. *See* IGY.

intertidal zone Also known as the littoral, it is a part of the seashore usually under water at high tide and exposed at low tide. *See* TIDE.

Intertropical Convergence Also called the Equatorial Convergence. This is the term used for an area in both air and water at or near the equator. It changes in latitude with the season, a change that is most obvious in the Indian Ocean. Because hot air rises, thunderstorms are produced as water vapor, which is carried aloft and then cools rapidly. These storms, which are typified by heavy rains, alternate with the calm surface winds that are a noticeable feature of the area called the doldrums. *See* ANTARCTIC CONVERGENCE, DOLDRUMS, WEATHER.

invertebrate An animal without a backbone. More than 95% of all animals are invertebrates. There is no definitive total of the known species, but scientists estimate the total number to be above 10 million. In general, individual invertebrates tend to be smaller than individual vertebrates, but some giants do exist both in the present and in fossil remains. *See* ANNELIDA, COELENTERATA, CRUSTACEA, ECHINODERMATA, MOLLUSCA.

iodine An element present in seawater as iodide ion (I^-) and iodate ion (IO_3^-), in concentrations of about 0.06 mg/liter. It is commercially extracted from seawater. Some organisms incorporate iodine in metabolically important compounds such as the thyroid hormone thyroxin and its various analogs. *See* ELEMENT, HALOGEN.

ion An atom or group of atoms that has an electrical charge resulting from the gain or loss of one or more electrons. *See* ATOM, ELECTRON.

Iran Plate A crustal region south of the Eurasian Plate and between the Turkish

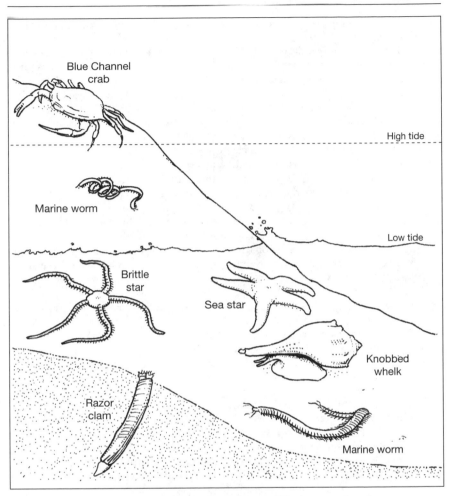

A rocky shore in warm-temperate waters will harbor a number of
unrelated species that are not in direct competition for the same food supply.

Plate to the west, Arabian Plate to the south, and Indian Plate to the east. The Iran Plate moves east-northeast and overrides the eastern edge of the Eurasian Plate. It is in turn pushed northward by the Arabian Plate approaching from the southwest. *See* CRUST, PLATE TECTONICS.

Irish moss A dark red alga, *Chondrus crispus,* also called carrageen. It is usually 5 to 25 cm (2.75 to 8 inches) long and grows around rocks of the North Atlantic on both the European and American coasts. Carrageenan is a gelatinous material commercially extracted from Irish moss and used as an emulsifying agent and thickener. *See* ALGAE.

Irish Sea A body of water that separates Great Britain from Ireland. Its southern end is arbitrarily divided from the English Channel and the Celtic Sea between England and Brittany by the St. George's Channel. The northern end of the Irish Sea is the channel between Bangor, Ireland, and the Mull of Galloway, Scotland. The two large islands in the sea are the Isle of Man and Angelsey.

The Irish Sea is not very deep. Its depth averages less than 100 m (330 feet), with a maximum of 175 m (576 feet) just off the coast of Ireland.

While the terrain on the surrounding shores is old (Paleozoic) the sea bottom is Triassic. The rising sea level at the end of the most recent glacial activity flooded large areas in the eastern parts of Ireland. A fanciful story has placed leprechauns and "drowned villages" in the area that was once flooded. *See* ENGLISH CHANNEL, PALEOZOIC ERA, TRIASSIC PERIOD.

Irminger Current A branch of the North Atlantic water that flows around Iceland. *See* ATLANTIC OCEAN, CURRENTS.

Irminger Sea A sea south of the Iceland-Greenland Ridge and north of 60° north latitude. The water is present in a series of layers, with distinct regions of warm and relatively saline Atlantic water, cold Arctic water, and a mixture of the two. The name *Irminger Sea* is not recognized as a distinct entity by all geographers. *See* ATLANTIC OCEAN, CURRENTS.

iron An element present in seawater as a hydrated oxide to the extent of about 0.01mg/liter (milligrams/liter). This is true for the ocean as a whole. However, at the edges of continental plates, regions where fresh magma is exposed to ocean water, there is considerably higher iron content in the water. These areas are the sites of hydrothermal vents or cold seeps. The vents are colonized by iron-oxidizing bacteria; the iron in the basalt is the energy source for these chemosynthesizers.

Iron is the most prevalent element in the Earth and is critical in several of Earth's processes; Earth is a giant magnet because of its iron content. Changes in the polarity of the Earth's magnetic field are evident in the alignment of crystalline iron in the Earth's crust. Only recently has the variation in the isotopic state of iron been recognized as a means of measuring a variety of biochemical changes that have occurred on Earth.

The change in iron that is evidenced by the change in color of iron-containing rocks from grey to red is striking. This change in the iron that started in the Precambrian period (3.5 to 3.0 million years ago) signaled the beginnings of oxygen production. The iron isotopes of the sulfur-rich waters of Precambrian seas were preserved in the rocks that date from that time, both iron and sulfur coming from volcanic events. The iron oxides formed from these deposits resemble those presently formed at hydrothermal vents. The inference is that the ancestors of chemosynthetic organisms that currently populate the vents might have changed the iron compounds millions of years ago and thus changed the very nature of the Earth we now see. The history of Earth may be written in the basement rock iron-to-sulfur ratios that are being studied.

Iron is also present on the sea floor in the form of consolidated deposits in the manganese nodules and in the remnants of metallic meteors. Some of the unconsolidated deposits have been commercially utilized; nearshore iron deposits have been dredged when the cost of dredging is less costly than the relatively inexpensive ore. *See* CHEMOSYNTHETIC BACTERIA; MANGANESE NODULES; MATTHEWS, DRUMMOND; VENT COMMUNITIES.

island A relatively small body of land surrounded by water. Islands may be either continental or oceanic in origin. A continental island is one that is part of a large, nearby continental mass of basement rock. Thus, Greenland is related to the rest of North America, and New Guinea is part of the Australian continent. The islands off the coast of Maine are hilltops of the submerged continental shelf.

Oceanic islands are those that rise from the ocean floor as volcanoes, or more properly as a series of volcanoes. The organisms living on any specific island depend on a number of factors, including the closeness of the island to other land, the temperature on and around the island, and the location of the island in terms of drift or wind patterns.

Island arcs are either strings of volcanoes at the edge of a crustal plate, the tops of submerged mountains, or the evidence of a hot spot. *See* ARCHIPELAGO, ATOLL, CORAL, HOT SPOT, OCEAN FLOOR, SEAMOUNT, VOLCANO.

island arc A number of islands, most often volcanic, arranged in a curved chain. The concave side of the arc is usually close to a continent, while the convex or ocean side is most often bounded by a deep trench. *See* ALEUTIAN ISLANDS, ARCHIPELAGO, HOT SPOT, TRENCH, VOLCANO.

isobar A line on a weather map that joins points of equal pressure. The closer together the isobar lines appear, the greater the velocity of the wind parallel to the isobars. A series of isobars that are very close together is an indication of a great change in barometric pressure in a relatively small area. The greater the change in pressure, the more noticeable the change in weather. *See* STORM.

Isopoda An order of arthropods in the class *Crustacea*. There are about 4,000 species, ranging from terrestrial to freshwater and marine organisms. The sow bug is a terrestrial isopod and a good example of these almost flat, ovoid organisms.

Members of the Isopoda are generally segmented, with many similar segments, the abdominal segments looking very much like the thoracic ones. The number of legs varies from 4 to 16. Most specimens are small, ranging from 0.7 to 3.5 cm (0.25 to 1.5 inches) long. One stunning exception is *Bathynomus giganteus*

of the Caribbean, which is 35 cm (14 inches) long. Many of the marine species are parasites that live on fish or shrimp. *See* CRUSTACEA.

isostasy The state of equilibrium of the constituents of the crust and mantle of the Earth. Isostasy explains the concept of continental flotation. The continents float in the lithosphere, the rocky layer of the Earth, as ice floats in water. *See* CONTINENTAL DRIFT, CRUST, LITHOSPHERE, MANTLE.

isotherm A line on a weather map connecting points of equal temperature. Isotherms are sometimes plotted against months of the year to show climatic variation over the entire year at a specific location. *See* TEMPERATURE, WEATHER.

isotope One of two or more variations of atoms of a chemical element that differ from each other in their atomic mass but not in their chemical properties. Atoms are composed of three principal particles: protons and neutrons in their nuclei and electrons orbiting the nucleus. Isotopes of the same element have identical numbers of protons and electrons but differ in the number of neutrons they contain.

The most common form, or isotope, of the element hydrogen has one proton, one electron, and no neutrons. The isotope deuterium contains one neutron in its nucleus, in addition to a proton, and a third isotope, tritium, has two neutrons. Of these three isotopes, only tritium is radioactive.

The decay of radioactive isotopes, or radionuclides, is used as a means of dating rocks and fossils. If radionuclides are present in water, the water mass can also be dated. *See* ELEMENT, RADIOACTIVITY.

jaeger A skua. A large, dark-colored northern seabird, of the family Stercorariidae, that forces weaker birds to give up their food by making them drop or disgorge it.

Jan Mayen Fracture Zone An island in the Greenland Sea, about 500 km (300 miles) east of Greenland, that belongs to Norway. It is the peak of the volcanic rise also called Jan Mayen. The last big blast of the volcano occurred in 1732.

The fracture zone runs northwest-to-southeast across the mid-ocean ridge north of Jan Mayen Island. It is a seismic site. *See* MID-ATLANTIC RIDGE, MID-OCEAN RIDGES.

Jan Mayen Island In the Greenland Sea about 500 km (300 miles) east of Greenland. It is considered part of Norway. The island is the top of the volcanic rise also called Jan Mayen. The last large eruption of this volcano occurred in 1732.

Japan, Sea of A body of water that separates the Japanese islands from the rest of the Pacific. The Korea Strait, the sea's southern boundary, has depths of more than 100 m (328 feet). The other connections to bodies of water are to the Sea of Okhotsk to the north and to the Pacific via shallow saddles. The average depth of the Sea of Japan is about 1,350 m (4,400 feet), although depths of over 3,000 m (9,800 feet) are found off the island of Honshu. The Pacific Polar Front runs through the sea at about 38° north latitude. The water south of this line is warm and blue, while north of the line it is cold and green. The water of the Sea of Japan is also divided into deep, intermediate, and surface layers. The surface water shows considerable seasonal fluctuation in temperature and salinity. The oxygen content of the sea is high, due in part to the extraordinarily high population of plankton.

Geologically, the Sea of Japan—particularly the northern half—seems to be an area of continental subsidence. The Korea Strait may also have been a land bridge in the Quarternary period. This would account for the remains of elephants and mammoths that have been found on some of the islands. *See* KUROSHIO CURRENT, OYASHIO CURRENT, PACIFIC OCEAN, TSUSHIMA CURRENT.

Japan Trench A trench east of the Japanese islands and part of the almost continental trench line that is found in the western Pacific. The Kurile Trench is north of the Japan Trench; the Bonin and Marianas trenches are south of it. The deepest point of the Japan Trench is the Tuscarora Deep, at about 8,500 m (28,000 feet). *See* MARIANAS TRENCH, TRENCH.

Java Head A headland at the western end of Indonesia at the entrance to the Sunda Strait. This headland is a famous landmark and was used as such by clipper ships on the way to China. *See* CLIPPER.

Java Sea A part of the western Pacific Ocean between Java and Borneo (Kalimantan). The entire bottom of the Java Sea is relatively flat and incised with well-defined river channels indicating the area's former state. This consisted of low-lying but dry land that was the route of the Asiatic faunal migration into the Indonesian islands.

The sea is even now fairly shallow, with average depths of about 45 m (150 feet).

The current in the sea flows west in winter—if one can call the colder season in a tropical region winter—and east from May to September due to the large inflow of river water. For a tropical sea, the Java Sea has low salt levels.

The oil deposits in the Java Sea are continuations of the oil fields on the island of Java and are presently being exploited. *See* FLORES SEA, SUNDA SHELF.

Jeanette A 19th-century American exploration vessel. On a polar exploration the *Jeanette* became frozen in the ice in September of 1879. The captain, George Washington deLong, and his crew eventually died of cold and starvation on the Lena River Delta (Siberia). They were attempting to sail close to the North Pole by plotting a course from the Bering Strait to Wrangell Island. Then they would go overland. The idea of a pole under water had not yet been attacked. The *Jeanette* was stuck in the ice in early September (1879) and drifted north of Wrangell. Bits of wreckage eventually wound up near Greenland. The *Jeanette* sank near the New Siberian Islands after drifting for two years. Frijdtof Nansen, the Norwegian explorer who established the fact that the North Pole was not on a continental mass, used the tragedy of the *Jeanette* in his arguments for this fact. *See* FRAM; NANSEN, FRIDTJOF.

jellyfish No longer an approved name. *See* SCYPHOZOA.

jet stream A long, narrow current of high-speed winds in the troposphere (the lowest segment of the atmosphere). The jet stream blows from a westerly direction, and can reach speeds of 400 km/hour (250 mph) or higher.

jetty A man-made rock structure usually at a harbor or river mouth or other area of ship passage. Jetties are constructed primarily to direct currents and to inhibit the shoaling or silting up of navigable channels. *See* COAST, GROIN.

JGOFS (Joint Global Ocean Flux Study) An international, multidisciplinary program that began in 1987. Its aim is to study the flow (movement) of carbon from water to atmosphere, the reverse flow, and the effect on climate of both of these flows. Each of the more than 20 nations involved in these studies has a working group that coordinates the activities in that country; the home base for the whole group is at the University of Bergen in Norway.

jib A triangular sail. It is set on a stay (supporting rod) extending from the first or foremast of a ship to the bow. As a verb, "to jib" means to swing such a sail or boom around to change the course of the ship. This is also called tacking. *See* BOOM, MAST, SAIL.

John Dory A food fish of worldwide range. The John Dory tends to live in deep water, to depths of about 200 m (650 feet). It is gray, with a characteristic black spot on the side behind the operculum, or gill cover, and has a large ugly head and spines on both its dorsal and ventral surfaces. Most specimens are about 10 cm (3 to 4 inches) long and are very thin laterally but deepbodied. *Zeus faber* is the Atlantic and Mediterranean species. In the Pacific the most common species is *Z. japonicus*. *See* FISH.

JOIDES (Joint Oceanographic Institution for Deep Earth Sampling) This 1970s project was a combined effort of the Scripps, Woods Hole, and Lamont-Doherty oceanographic institutions and several distinguished university oceanography departments, notably that of the University of Florida at Coral Gables. The object was to take core samples in various places and categorize the bottom. The ship designed to do this was the *Glomar Challenger*. A research ship named the *Joides Resolution* was launched in 1985 as an ongoing coring vessel. It has visited hundreds of areas, removing core samples and storing them for further studies (geological, meteorological, etc.). The deepest core taken to date was removed from a site in the western Pacific from a depth of 5,989 m (2.4 miles). *See Glomar Challenger,* LAMONT-DOHERTY GEOLOGICAL OBSERVATORY, SCRIPPS INSTITUTION, WOODS HOLE INSTITUTION.

Juan de Fuca Ridge   A site of volcanic activity in the Pacific Ocean near Vancouver, Canada. The Juan de Fuca Plate is being subsumed by the Pacific Plate. This creates a ridge that is seismically unstable. The area has been extensively explored in the 1980s. Sulfurous water of about 300°C has been found issuing from vents along the ridge at depths of almost 2 km (over 1 mile). The colonies of organisms found near these vents are unusual and were until very recently unknown. *See* MID-OCEAN RIDGES, PLATE TECTONICS, SULFUR BACTERIA, VENT COMMUNITIES.

Juan de Fuca Strait   A narrow international waterway between the Olympic Peninsula of Washington State and Vancouver Island, British Columbia. The average depth of the strait is less than 100 m (about 330 feet). It is named for a Greek seaman in the employ of Spain who was reputed to have explored it in 1592. For the next 200 years the strait appeared now and again on navigational charts as part of a possible route between the Atlantic and Pacific oceans. However, it is a complicated waterway to use, since strong rip currents run in it. *See* EXPLORERS AND EXPLORATIONS.

Juan Fernandez Islands   A group of islands that are outcroppings of the Juan Fernandez Ridge in the southeastern Pacific Ocean near Chile. Alexander Selkirk, a sailor who was left on the islands in 1704 after a fight, is the accepted model for Robinson Crusoe, the fictional character. The islands are typical of exposed rocky outcrops. *See* DAMPIER, WILLIAM; ISLAND.

junk   A flat-bottomed sailing vessel used by traders from the Malay Peninsula to China. The junk has square bows and a squared-off, built-up stern and is equipped with rudders. The sails are large and strutted with long horizontal rods of bamboo; they are carried on either two or three masts.

Because of their flat bottom and low length-to-width ratio (usually 3:1), junks are slow-moving.

The Chinese have used these vessels for centuries as merchantmen, warships, and vessels of exploration. Chinese mariners and traders of the 15th century brought the junk to Borneo, the Philippines, Sri Lanka, the Malabar Coast, and Zanzibar in east Africa. *See* EXPLORERS AND EXPLORATIONS, SAIL.

Jurassic period   The second oldest portion of the Mesozoic era. The name is based on the limestone of the Jura Mountains of France, which contains a splendid record of ammonite (fossilized cephalopod) evolution. Gibraltar is another Jurassic formation, as is much of Yosemite in California.

The Jurassic period lasted about 30 million years and ended about 135 million years ago. The location of continents and oceans in this period of geologic history is a matter of some controversy. The Precambrian shields in Canada, Scandinavia, Central France, and the Harz Mountains of Germany were dry land, as was some of Australia. Much of the rest of what is now the Earth's continental landmass was submerged, according to some geologists. Geologists do agree that mountain building was occurring in several locales, including North America, the Caucasus, and Southern Europe during the Jurassic.

The fauna of the Jurassic included some of the most spectacular dinosaurs: the Brontosaurs, Stegosaurs, Plesiosaurs, Ichthyosaurs, and the proto-bird Archaeopteryx. The most common fossil invertebrates were, in addition to the very characteristic ammonites, the nautiloids, belemnites, gastropods, pelecypods, and echinoids. Silaceous sponges have also left a fossil record of Jurassic sea life, as have numerous corals. *See* AMMONITES, MESOZOIC ERA, *Appendix: Geologic Timescale.*

K

kairomone A chemical associated with one organism that acts as a signal to another organism of a different species. Thus, a predator would receive a chemical signal announcing the presence of available prey. *See* PHEROMONE.

Kalimantan The southern three-quarters of the island of Borneo is politically part of Indonesia. Indonesians, however, use the word as a geographic term for the entire island. The origin of the name Kalimantan is obscure. In Sarawak, the term *Kelamantan* refers to the sago-eating peoples of northern Borneo. Indonesian Kalimantan is divided into four *propinsi-propinsi* (provinces): Kalimantan Barat, Kalimantan Selatan, Kalimantan Tengah, and Kalimantan Timur.

Kamchatka Peninsula A peninsula between the Sea of Okhotsk and the North Pacific Ocean-Bering Sea area. It is 1,200 km (750 miles) long in a roughly north-to-south line and about 500 km (300 miles) wide at its widest point. The Kamchatka Peninsula is volcanic; there are two mountain ranges lying along the long axis of the landmass. Like Iceland it has geysers and geothermal springs. Unlike Iceland, the climate is much colder and there is considerable snowfall. *See* BERING STRAIT, EAST CAPE, SEA OF OKHOTSK.

Kara Sea A body of water marginal to the Arctic Ocean east of Novaya Zemlya and south of Franz Josef Land. The Siberian plain lies to the south and southeast of the Sea.

The average depth of the Kara Sea is about 125 m (400 feet). It was formed after the last ice age, making it the world's youngest sea, which accounts for its very shallow depths. Almost half of it is less than 50 m (160 feet) deep, and its maximum depth is 620 m (2,000 feet). It is dotted with sandbars and islands. It is fed by the Ob, Yenisey, and Kara Rivers. The volume of freshwater brought by these rivers means that the surface waters of the sea are almost fresh, with salinity levels as low as 1.0% near the mouth of the Ob. The salt content is about 3.3% near Franz Josef Land.

Commercial fishing and hunting are major activities on and around the Kara Sea. Petroleum deposits have been found in the area. The commercial activity of the region is to a great extent dependent on the weather, which is arctic and known for frequent storms year-round. *See* ARCTIC OCEAN, BARENTS SEA.

kelp A group of large brown algae, or brown seaweeds, of the orders Laminariales and Fucales. Many are of commercial interest both as primary foodstuffs and as industrial raw materials.

The kelps resemble vascular plants upon superficial examination. They have a differentiated cell structure and distinct gamete forms. In the reproduction of *Laminaria,* called devil's apron, both sporophyte (having the full normal adult chromosome number) and gametophyte (haploid) generations exist. The large and visible kelp of the ocean is the sporophyte, which in late summer produces zoospores. These zoospores constitute the haploid gametophyte generation and may be either male or female (sperm or egg producing). The motile sperm are produced by

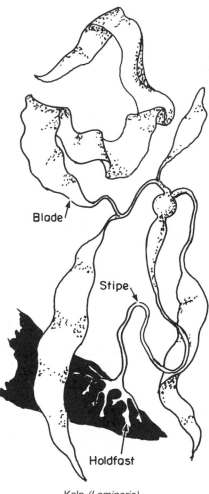

Kelp *(Laminaria)*

land plants. The function of the tubes is the same: They are part of the kelp's circulatory system and do what phloem does in vascular plants, transporting the products of photosynthesis to the various parts of the structure. *See* ALGAE, BROWN ALGAE.

Kelvin waves   The waves of warm water moving eastward that are associated with El Niño events. They form when the prevailing westward-moving trade wind either stops or slows dramatically for several days or weeks. This results in a buildup of warm water in the South Pacific—the warm pool. This increase in water temperature in Asia results in increased air temperature that triggers thunderstorms and more rainfall than is usual for that area. The warm water generates an eastward slide, a Kelvin wave. As it approaches South America, the warmed water splits into a northbound and southbound segment, warming the coasts as they slide past. These currents are sometimes known as Rossby waves. All are part of the El Niño phenomenon. *See* EL NIÑO.

Kepler, Johannes   (1571–1630)   A German astronomer and student of Tycho Brahe. Kepler's greatest contribution to astronomy was the mathematical treatment of the Copernican model of the solar system. His first important work was the *Rudolphine Tables of Planetary Motion* (the work Brahe began), named for Emperor Rudolph II, Kepler's patron. These tables were eventually published in 1627.

Kepler did careful observations of Mars for many years. These were finally published in 1609, along with observations of the nova of 1604. In his book *The New Astronomy,* he used the Copernican philosophy, his own excellent mathematics, and Brahe's careful measurements to plot the elliptical orbits of the planets. He also formulated three laws, known as Kepler's Laws, to describe the motion of the planets in the solar system. These laws later became the basis for Newton's theory of gravitation. All of this work was essential raw material for those who later devised

antheridia and caught and retained by the oogonia, or egg-producing cells. The fertilized egg or zygote undergoes immediate growth into the recognizable kelp. This complex life cycle where the zygote and the adult plant are diploid (a pair of each chromosome) and the zoospores (sperm and eggs) are haploid is analogous to the reproductive cycle of ferns on land.

The kelps grow in large structures called macrocysts that may be as long as 60 m (200 feet), or they grow as elaborately branched mats. These plants have structures that resemble the sieve tubes in

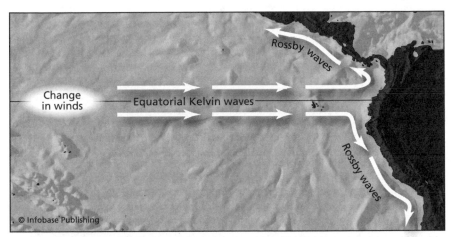

Kelvin waves are the water component of an El Niño phenomenon.

accurate tables with which mariners determined position. *See* COPERNICUS, NICOLAUS; NAVIGATION.

Kerguelen Islands A group of roughly 300 islands southwest of Australia that are volcanic outcroppings of the plateau that divides the South Indian Ocean Basin from the Atlantic-Indian Basin. The area is now a scientific base; in the 19th century the islands were known only to the occasional whaler and seal hunter. *See* INDIAN OCEAN, ISLANDS, PLATEAU.

Kermadec trench Also known as the Tonga-Kermadec trough, this ocean deep constitutes a border of the southwest Pacific basin. It is associated with seismic activity. Like its Atlantic Ocean counterpart, the Puerto Rico trench, the Kermadec trench is part of a large slip-strike region. It is much deeper than the Puerto Rico trench, with maximum depths of about 10,000 m (33,000 feet). *See* FIJI PLATE, TRENCH.

ketch A two-masted sailing vessel. Ketches were coastal fishing and trading ships that later were adapted to naval use as "bomb" vessels; the bombs were the missiles fired by mortars—cannons with a very short barrel that fires in a very high trajectory. With the end of this military

use, the square-sailed ketch was redesigned for recreational sailing. Ketches are now exclusively pleasure boats. *See* SAIL.

key A low or flat island or reef. Usually a coral-and-sand islet (e.g., the coral islets off the southern coast of Florida). *See* CORAL.

killer whale The largest of the dolphins; the males may exceed 9 m (28 feet) in length. It is an attractive black-and-white carnivorous animal that lives and hunts in packs. The large dorsal fin is a distinguishing characteristic. The range of *Orcinius orca* is worldwide. It has no natural enemies.

A great deal of writing about killer whales is highly exaggerated; in spite of their name, they do not attack everything that moves in the sea. Their usual prey are birds and small dolphins, although a large pack of killer whales may attack a weak baleen whale. There is little information available about the breeding behavior of *O. orca* other than that the young are black and orange. There is an unexplained decline in the population of *Orca* since the early 1990s.

king crab A large decapod crustacean of the order *Lithodidae*. This family of anomuran crabs inhabits cold waters. The commercially known (Alaskan) king crab of the North Pacific is *Pacalithodes camtschatica*. *See* CRAB.

King William Island One of the large islands of the Canadian Arctic. It lies east of Victoria Island and was the site of the death of John Franklin, the famous explorer of the Arctic whose disappearance prompted many rescue expeditions. *See* ARCTIC; FRANKLIN, JOHN; NORTHWEST PASSAGE.

Kinorhyncha A free-living phylum of tiny (less than 1 mm, 0.04 inches long) organisms. They have been previously classified as aschelminthes and nemataminthes. They are segmented but have a chitinous exterior. They are not annelids; unlike true segmented worms, they have an incomplete coelom (gut) but a complete ventral nerve cord. These animals have a worldwide distribution in subtidal buttoms, where they feed on diatoms. *See* ASCHELMINTHES.

kittiwake A type of North Atlantic gull. *See* BIRD, GULL.

knot A measure of nautical speed. One knot = 1 nautical mile (6,080 feet) per hour. Originally, a log attached to a rope with evenly spaced knots on it was heaved overboard to measure the speed of a ship. By counting the number of knots let out in a specific time period, standardized at 28 seconds, the sailor could calculate the speed, hence the term "knots." The space between knots was standardized at 47 feet, 3 inches.

Korea Current A warm, north-moving surface current in the Sea of Japan. It turns east and becomes the Tsugaru and Soya currents. The former moves through the Tsugaru Strait into the Pacific Ocean. The Soya Current travels along the Japanese islands of Honshu and Hokkaido into the Sea of Okhotsk. *See* CURRENTS.

Korea Strait A body of water that separates Korea and Japan. The Tsushima islands are in the center of the strait. The passage between the islands and Japan is also referred to as the Tsushima Strait. This waterway was the site of a major naval battle in 1905, when the Japanese destroyed the Russian fleet in one of the determining engagements of the Russo-Japanese War.

Krakatau, island and volcano Krakatau, or Krakatoa, is a small, uninhabited island in the Sunda Strait west of Indonesia that erupted in a series of blasts culminating on August 27, 1883. Most of the island disappeared, and the resulting tsunamis drowned over 30,000 people on other islands and as far away as India. This explosion, while not as great as that of the volcano Tambora in 1815, was very well documented, probed, and photographed. Krakatau is still active as a volcano today.

It has been postulated that the surrounding islands and Krakatau itself were the remnants of a 17th-century eruption that left only the fragments of a broken volcanic cone visible above sea level. By the 19th century a new volcanic cone had been created on one fragment, and this was the site of the famous eruption of 1883. A series of subsequent eruptions, with debris falling into the caldera (lava-filled crater), has produced a new island called Anak Krakatau, or Child of Krakatau. The new cone emerged in 1928 and is still growing. *See* TSUNAMI, VOLCANO.

krill Small shrimplike crustaceans of the genus *Euphausia* or *Thysanoessa*. They are distinguished from shrimp by their bristled tails and are largely confined to the southern oceans. Krill are pelagic organisms ranging in size from 1 to 6 cm (about 0.5 to 3 inches). Most are transparent, and many have light-producing organs, possibly as an aid in keeping the swarms of these animals together. *E. crystallorophias* has a wide vertical range; while most organisms live in the band 300–650 m (1,000–2,500 feet) deep, others have been found at depths of almost 4,000 m.

It is a swarming organism that feeds on phytoplankton—mostly diatoms—and in turn supports the Antarctic populations of whales, seals, penguins, and fish. *E. superba* is another significant organism of this genus; it is largely coastal in habitat.

The enormous numbers of krill make these organisms a significant food base that supports a large bird, fish, and whale population. The great krill swarms occur in the December to February summer season. The size of the swarm is enormous: It can be thousands of meters across and deep. The krill population is dependent on the available phytoplankton density. That population is the food supply, and it varies with the temperature of the water. Therefore, the presence of warm Kelvin-Rossby waves and accompanying El Niño events may alter the size of the krill population because warmer water in polar regions inhibits the cold upwelling of nutrients. A smaller photosynthesizing population adversely affects the krill population.

E. superba is the favorite food of baleen whales, who can ingest tons of krill in a day. Krill is harvested commercially for human food. Pink tofu is a krill product. *See* EL NIÑO, EUPHAUSIACEA, LUMINESCENCE, WHALE.

Kurile Trench A trench in the ocean floor that lies east of the Kamchatka Peninsula (Russia) and the Kurile Islands north of Japan. The maximum depth is 10,542 m (34,587 feet). The Kurile trench is part of the broken line of deep water surrounding the Pacific. *See* PACIFIC OCEAN, TRENCH.

Kuroshio Current Also known as the Japan Current, it starts east of the Philippines and branches from the North Equatorial Current—the Pacific gyre. The Kuroshio Current moves northeast from the Philippines to the east coast of Japan. The name, *Kuroshio*, is from the Japanese word for black; the water is very dark and lacks biota. It has a high temperature of about 20°C and a high salt content of 3.45%. The current moves east of Taiwan and the Ryukyu island chain, In the summer, a branch known as the Tsushima goes through the Korea Strait. This current is known for both high-volume water movement and wide meanders that are temperature dependent. As a result, current greatly influences the weather patterns along almost all the southern Japanese coast.

At about 35° north latitude, the Kuroshio Current turns east, deflected by the cold Oyashio current. The turned stream moves east as the North Pacific Current and disappears as a recognizable entity by the time it reaches Hawaii.

The waters of the Kuroshio are much darker blue than the water it moves through. Its rate of motion varies from 50 to 300 cm/second (1 to 6 mph). The Kuroshio has been called the Pacific Gulf Stream, but, it is not as well-defined as the true Gulf Stream; it does not transport as much warm water, and it does not travel as far as the Gulf Stream does. It is, however, the major west-to-east current in the North Pacific Ocean. *See* CURRENTS, GYRE, OYASHIO CURRENT.

Labidocera A genus of calanoid cope-pods. These pelagic organisms are noted for the long periods of dormancy their eggs can endure and still produce viable hatchlings. *See* CALANOIDA, COPEPODA, DIAPAUSE.

Labrador Current A cold, low-salt (3.0 to 3.4%), surface flow moving south from the Davis Strait between Greenland and Baffin Island and along the Canadian coast. The rapid introduction of very cold water into the warmer water coming up from the south creates the well-known fog banks off Labrador and Newfoundland. There is an area of upwelling off the Greenland coast that creates a source of nutrient mate-rials, and these are the base for the high population density of arctic animal life. *See* ATLANTIC OCEAN, CURRENTS.

Labrador Sea A body of water bounded by the coast of Labrador on the west, Greenland to the east, and Baffin Island to the northwest. At the Davis Strait, the sea is only about 800 m (2,500 feet) deep, whereas most of the Labrador Basin drops to deeps of 3,500 m (10,000 feet). The bottom sediment is composed largely of *Globigerina* ooze.

The climate of the Labrador Sea is polar (i.e., very cold, dry, and windy). The water on the shelf is 2° to 3.5°C, which is colder than the water in the depths of the sea. The oxygen content of Labrador Sea water is high and the salinity, particularly on the shelf, is low. In the Labrador Basin the salinity is a fairly constant 3.49%. Ice floes are a common feature of the shelf area, while the eastern part of the sea is generally clear of floating ice. *See* ARCTIC OCEAN, GLOBIGERINA.

lagoon A warm, shallow, quiet water-way separated from open sea either by sandbars, barrier islands, coral reefs, or some combination of the three. If the sep-arating land is a bar or barrier island, the lagoon is long and narrow; if the lagoon is surrounded by a coral atoll, it is circular. Coastal lagoons can become marshes if deposited sediment cuts off their connec-tion with the sea. *See* ATOLL, CORAL, REEF.

Lamarck, Jean-Baptiste Pierre Antoine de Monet, chevalier de (1744–1829) A scientist, taxonomist, and developer of an evolutionary theory, Lamarck was born to a family of the minor French nobility with a long military tradition.

Lamarck was destined for the Church as a career, but on his father's death he escaped into the army, where he served for six years. After trying banking as an occupation, he turned to medicine, then a common first step toward a scientific career, but found science more to his liking than medicine.

Lamarck worked on physical principles following some already outmoded chemical theories and finally, in 1778, published a major work, *Flore française*, which he had begun while in the army. This established his reputation, and he became a member of the French Royal Academy of Science.

In the Revolutionary decade, Lamarck worked in the Jardin du Roi (which became the Jardin des Plantes in 1793–1794) and the Royal Cabinet, which reemerged as the Musée d'Histoire Naturelle. While there, his major work was the reclassification of animals, which the Linnean classifica-tion scheme had put into broad categories. Lamarck worked on systemization and orderly grouping, and published this work in *Système des animaux sans vertèbres* (1801), *Philosophie zoologique* (1809), and the multivolume *Histoire naturelle des animaux sans vertèbres,* which appeared over the period from 1815 to 1822.

In these works, Lamarck was the first to successfully categorize most known invertebrates of his time. Linneaus lumped these many animals together as *Wormes* ("creepy crawlies"), undifferentiated boneless creatures. In his later works, Lamarck tried to develop a mechanism for the change in structure of organisms. He, like Erasmus Darwin (grandfather of Charles), was groping for a means to explain the transmission of characteristics from one generation to the next. He speculated about the possibility of inheritance of acquired characteristics but had no means of explaining how a distinct change would breed true in succeeding generations. He, like both Darwins, predated the Mendelian gene theory, although all three men were reaching toward something like it.

Lamarck's scientific reputation was nearly destroyed by attacks by his colleague, Cuvier, who rejected evolution as a concept and ridiculed attempts to use it as an explanation for the development of new species. Cuvier made special efforts to avoid any attempts to explain relationships between species and the relation of fossil remains to current organisms, as Larmarck was attempting to do. Since Cuvier used the official, clerical "age of Earth" as a foundation of his own work, there was no acceptable long time span for the development of variations. Since Cuvier assumed that all his own work was correct, he figured any other author who disagreed with him had to be wrong. There was no room in Cuvier's world for alternative solutions to problems.

As the leading French biologist of his day, Cuvier was invited to speak at Lamarck's funeral. He delivered a long funeral oration that was later printed. That speech was a denunciation of almost all of Lamarck's work and relegated that large body of material, both published and unpublished, to be a footnote in the history of biology for a century. *See* CUVIER, GEORGES; DARWIN, CHARLES; EVOLUTION; LINNAEUS; TAXONOMY.

Lamellibranchia The largest subclass of the class Bivalvia in the phylum Mollusca. Lamellibranchs are ciliary filter feeders with elongated gill filaments that are held together in parallel series to form folded lamellae. The 30,000 living species constitute the most successful of group of mollusks, in terms of both individual numbers and their importance in the marine food chain.

The tube worms that are characteristic of hydrothermal vents and cold seeps are *Lameliiobranchia.* Since the discovery of vent communities, the great question among scientists is, How does this very specialized creature find a suitable home where there is a fairly constant supply of hydrogen sulfide, the material on which their metabolism is dependent? The worms produce eggs or sperm that unite in the water to form embryos that drift until they find a suitable source of hydrogen sulfide at a new black smoker or cold seep, are eaten by a predator, or miss the hydrogen sulfide source and die. A microscopic larval tubeworm can drift in the ocean currents for about three weeks before exhausting its yolk (food supply). The adult tubeworm can live and continue to be reproductively active for about 200 years. *See* MOLLUSCA, SEEPS, TUBE WORMS, VENT COMMUNITIES.

Laminariales The brown algae known as kelps. There are more than 5,000 species of these complex organisms. They are algae, and some taxonomists classify them as protists. Others call them plants, looking at their complicated reproduction, which involves alternation of a microscopic haploid generation followed by the large and obvious diploid generation, and the presence of xylemlike sieve tubes.

Laminariales are the dominant photosynthesizers in the lower littoral and sublittoral areas of rocky shores in temperate to polar waters. *See* KELP.

Lamont-Doherty Geological Observatory An institution founded in 1949 by Columbia University, in New York City, to study earthquakes, submarine geology, marine biophysics, oceanography, chemistry, physics, and the body of work called polar studies. The Observatory and Laboratory maintains research vessels, houses a piston core collection of samples carefully removed from the Earth's crust, and maintains an index of organisms and fossils. The Lamont-Doherty Observatory pioneered the use of undersea photography, and its collection of photographs in this field is definitive. It also houses the reports of the JOIDES work, as well as the Atlantic, Antarctic, and Indian Ocean Surveys—total-ocean surveys done in the 1960s and 1970s. *See* EWING, WILLIAM MAURICE; EXPLORERS AND EXPLORATIONS; HEEZEN, BRUCE; JOIDES; THORP, MARIE.

lamprey A primitive, jawless, cartilaginous vertebrate of the order Cyclostomata in the class Agnatha. These jawless fish are generally found in coastal, temperate waters on every continent except Africa. Most are burrow dwellers 15 to 100 cm (6 to 40 inches) long. They start life in freshwater and, with the exception of a few species, migrate to the sea as adults. Most adult lampreys are parasites on fish. Lampreys have been partially responsible for a serious decline in the commercial fisheries of the Great Lakes. *See* CYCLOSTOMATA, HAGFISH, PARASITE.

Lampridiformes Order of bony fishes that dates from the Cretaceous period 135 million years ago. This group includes the ribbonfish, oarfish, and opah. There are about 21 species. The Lampridiformes are mesopelagic animals. They have distinctive trailing and retractible maxillae (ventral streamers attached behind the gills), but their outstanding characteristic is the absence of distinct fins. They have a dorsal ray that extends the entire length of the body and a spiny construction above the head that looks like a crown. Their scales are cycloid but small or absent entirely. There is a ductless swim bladder.

These fish are colorful and frequently very large. They may be 6 to 7 m (20 to 24 feet) long. They may have contributed to myths of sea monsters or sea serpents. *See* SEA MONSTERS.

lancet A long, thin, deepwater fish that has a long, spiny, dorsal fin. The adults are 1.5 m (60 inches) and more in length. Lancets also have sharp and prominent teeth, feeding on other fish and invertebrates. The most common Atlantic species is the longnose lancet, *Alepisaurus ferox*. The Pacific species are *Alepisaurus richardsori* and *Alepisaurus borealis*.

La Niña The obverse of El Niño. It is a colder-than-usual surface temperature in the eastern and central Pacific Ocean. The shift from El Niño to La Niña and back again occurs in an approximately four-year cycle. The consequences of this oscillation and its combination of water and air disturbances have significant weather effects worldwide. *See* EL NIÑO, METEOROLOGY, WEATHER.

lantern fish Mesopelagic fish that have light organs around their disproportionately large heads and on their undersides. The pattern of distribution of the light organs is species specific and may therefore be a mating recognition factor for the more than 150 species of lantern fish. The average size of the fish is about 2.5 to 15 cm (1 to 6 inches). They are usually found at depths of about 500 m (1,600 feet) during daylight hours; at night they migrate vertically and feed at the surface. *See* LIGHT ORGANS.

La Pérouse (Jean-François de Galoup), comte de (1741–1788?) A French explorer and naval captain. After his service with the French Navy in the American Revolution, he was given command of a voyage of exploration to the Far East and sailed with two ships, the *Boussole* and the *Astrolabe*. The explorers hoped to find the western approach to the Northwest Passage through the North American mainland.

La Pérouse touched the Asian mainland near what is now Vladivostok, and sailed through the strait between Sakhalin island and Japan. He landed on Kamchatka, in what is now Alaska, before turning south. The captain of the *Astrolabe* and some of its crew were killed by Samoans. La Pérouse continued on to Botany Bay, Australia, and disappeared some time after February 7, 1788, his last dated correspondence. Subsequent French expeditions were sent to find La Pérouse, and one of these, led by Dumont D'Urville, succeeded in finding the remains of the two ships but no trace of La Pérouse. *See* DUMONT D'URVILLE, JULES SEBASTIEN-CÉSAR; NORTHWEST PASSAGE.

Laptev Sea Formerly known as the Siberian Sea, this body of water lies east of the Kara Sea and is led by several rivers. The largest volume of water entering the sea comes from the Lena River. The Laptev is largely a shallow plain, and more than 50% of it is less than 50 m (160 feet) deep. The entire area is still recovering from glaciation; there is "relic" ice in its eastern region.

A thick layer of fresh water overlays the Laptev Sea. In the southeast the salinity ranges from 2 to 2.5%, falling below 1.0% in summer with the rapid inflow of river water. Very severe weather is the norm: The seashore has permafrost year-round, and the sea itself is covered with ice for most of the year. The water temperature is below 0°C for most of its depth. Deep pockets of water from the Atlantic have been traced to depths of about 250 m (800 feet). These waters are significantly warmer than the very low-salinity water of the Laptev. *See* ARCTIC OCEAN, KARA SEA, LENA RIVER.

larva The preadult stage of many organisms, which does not structurally resemble the adult. The larva usually exploit a different food source from that used by the adult.

Larvacea The subphylum of chordate hermaphrodites grouped with the sea squirts, tunicates, and salps as Urochordata. They do have notochords and feed by collecting debris on a layer of mucus. The larvaceans construct a "house," a structure in which the adult lives. This is a giant filter that is largely a protein and cellulose-like mucus-covered polymer. This house acts as the filter for debris-containing seawater, and when the organism discards it, this now-clogged film accounts for the rain of debris from the surface to the deep benthos. The houses are quite fragile and are discarded frequently, perhaps more than once daily.

The question of what the food source is for the abyss is answered at least in part by the availability of the protein and carbohydrate-rich houses shed by larvaceans. The deep-sea populations are large enough to need a constant rain of carbon compounds. It is now thought that the discarded houses of larvaceans supply much of this needed food supply. The houses are discarded when they become clogged with debris and therefore visible. This mucus mass is relatively heavy and sinks quickly.

The larvaceans are found in all the world's oceans. Most of them are small, though the giants are about 50 to 60 mm (2–2.25 inches). Their mucus houses are much larger and can be up to a meter in diameter.

Since these organisms are so fragile, their populations were underestimated until better plankton nets—the usual gathering devices for ocean biota—were designed. (Older plankton nets destroyed most larvaceans and their mucous structures.) The first giant was identified in 1898, and research on these interesting organisms is ongoing. *See* BIOFILMS, CHORDATA, FOOD WEBS, TUNICATES.

lateen A triangular sail. The word is a mangled version of "latin sail," because these sails were used on Mediterranean vessels. *See* SAIL, SQUARE-RIGGED.

lateral line A sensory organ along the head and sides of a fish. Some amphibians also have lateral lines. The "line," which shows on the side of a fish as a distinct streak of differing pigmentation, is made up of individual units called neuromes. These are groups of hairlike structures that move in response to a change in the surrounding water, caused by the movement of another animal or by a change in current flow. *See* FISH.

latitude The angular distance north or south from the Earth's equator, measured in degrees, minutes, and seconds of arc. Points on the Earth's surface are located as being at a particular latitude by measuring the angle of incident light from the Sun or a specific star above the horizon at a specific time. In the Northern Hemisphere the star used as a locator is the Pole Star; in the Southern Hemisphere it is the dominant star in the constellation of the Southern Cross. Lines on a globe or map connecting points of identical latitude are called parallels of latitude. One degree (1°) of latitude = 111.1 km or 60 nautical miles; 1 nautical mile = 6,076 feet or 1852 m. *See* ASTROLABE, CHART, LONGITUDE, MAP.

Laurasia An archaic northern continent that resulted from the breakup of the still-older supercontinent Pangaea. Laurasia eventually broke up and formed North America and Eurasia. *See* CONTINENTAL DRIFT; GONDWANALAND; PLATE TECTONICS; WEGENER, ALFRED.

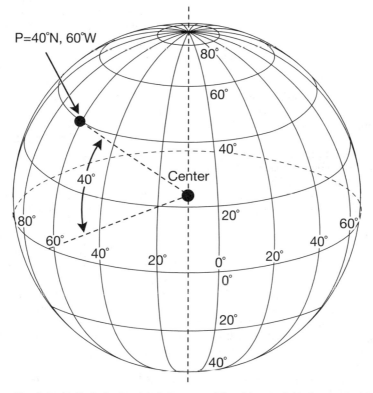

Parallels of latitude (horizontal circles around the globe parallel to the equator) and meridians of longitude (vertical circles passing through the poles)

lava Magma, or molten rock that pours out of a volcano or any other fissure in the Earth's crust. Lava is basaltic or rhyolitic, meaning that it has either a low or high content, respectively, of siliceous material. The former flows more readily, and when it cools it forms rock that is characterized as smooth (sometimes glassy, if it cools very quickly), rippled, or braided. The last type of rock is known by the Hawaiian word pahoehoe. Undersea lava flows are almost always of the pahoehoe type. This flows outward through tubes or natural vents or pipes in the crust. Clinker lava is a rough, pebbly type of lava and is fed by open lava streams. *See* BASALT, MAGMA, OCEAN FLOOR, VOLCANO.

laver A genus of red algae, *Porphyra,* that grow in sheets in water rich in nitrogen. Masses of laver are collected at the highwater marks of temperate seas. The laver is the "seaweed" most commonly used as food for humans both in Europe and Asia. It is called *nori* in Japan. *See* ALGAE.

Law of the Sea Conference A series of United Nations conferences, the first of which met in Geneva in 1958. The initial agenda was the standardization of national limits to territorial waters. Chile was the first nation to claim a 200-mile strip of territorial water along its shore, and other nations soon followed this lead. Historically, the limit to territoriality had been the three-mile strip, because this was the furthest reach of naval artillery in the 17th century.

Although Chile's claim was not resolved, the conference attempted to arbitrate claims to territorial waters and to look into the question of national rights to coastal mineral deposits. The 1960 meeting again tried and failed to achieve conformity.

The third meeting began in 1973 and continued for more than 10 years. The great issue of this conference was the possibility of settling the various nations' territorial claims by establishing a 12-nautical-mile limit that would be surrounded by another 12-nautical-mile–wide "protected fishing zone." Clearly, this left unresolved the problem of areas of overlap between adjoining nations. Furthermore, the proposed arrangement would have permitted six nations to claim over half the desirable coastal and continental shelf areas of the entire Earth. Despite adverse publicity, the United Nations Convention on the Law of the Sea was opened for signature in 1982 and entered into force in 1994. Agreement between developed and developing countries over outstanding seabed-mining issues was reached in July 1996.

lead A heavy metal. Significant quantities of this element are not normally found in oceans. It has been introduced as a pollutant resulting from human activities, mostly manufacturing. The addition of lead compounds emitted by automotive engines has also contributed to the oceanic load of this element. As a heavy metal, this element and its compounds tend to sink to the ocean floor and collect at nonvolcanic sites. Since that part of the ocean is so sparsely populated, the effect of this introduction of lead compounds is difficult to assess. *See* POLLUTION.

leathery turtle *See* SEA TURTLE.

leeward In the direction of the moving wind or a wave. *See* WINDWARD.

Lena River A river in Siberia, the second longest (2,734 mi; 4,400 km) in Russia. The Lena, flowing from mountains west of Lake Baikal, is one of the world's great rivers. It forms a huge delta in the Laptev Sea. Most of the water in the Lena is very cold because the water has its origins in melting snow. *See* ARCTIC OCEAN, LAPTEV SEA.

leopard shark A small shark (90 to 150 cm or 3 to 4.5 feet) long whose habitat is in shallow waters of the eastern Pacific Ocean. It is considered harmless and feeds on fish and invertebrates. The gray-and-black barring pattern of the shark gives it its name. *See* SHARK.

Lesser Antilles *See* ANTILLES, CARIB-BEAN SEA, WEST INDIES.

light *See* ELECTROMAGNETIC SPECTRUM.

light absorption and scattering Seawater absorbs some of the light passing through it. It is a medium that is transparent to short wavelengths (the blue through ultraviolet end of the spectrum), and which absorbs longer wavelengths in the red and infrared regions of the spectrum.

The quantity of light reflected from the surface of water depends on the angle of the incident light and smoothness of the surface. If the Sun is directly overhead (90°), the reflectance is less than 5% and the remaining light is absorbed. As the angle of the Sun drops, the reflectance increases. At an incident angle of 30° the reflectance is 6%, and at 20° it is 12%; at 5° it is greater than 40%. The rate of attenuation of the sunlight in a water column is determined by depth, absorption by pigments, and scattering by particles—plankton or debris. Blue light is absorbed least, and red light absorbed most. Since coastal waters contain the most suspended material, that is where light is attenuated most. The "back-scattering" of light is a complex phenomenon and is dependent on a large number of variables: living organisms in a variety of sizes, biological debris, bubbles, and inorganic dirt.

The color of the water is also an indicator of what is happening to light within it. The photosynthesizing pigments in phytoplankton absorb light, but the organisms themselves scatter it. The two processes change the color of the water, and this can be used to track the phytoplankton. The satellite Sea-viewing Wide Field-of-view Sensor (SeaWiFS) has been measuring and recording the movement of phytoplankton since its launch in 1997. *See* COLOR, INSO-LATION, PHYTOPLANKTON.

lighthouse A structure on land that is a mariner's aid, producing a light as a danger signal and navigational aid. The precursors of lighthouses were shoreline bonfires maintained to warn sailors of treacherous rocks or currents. The most famous ancient lighthouse was the Pharos of Alexandria, built in the reign of Ptolemy II (283–247 B.C.E.). The Romans built lighthouses at Ostia, Ravenna, Messina, and in the outposts of the Empire at Dover and Boulogne.

The light produced by ancient lighthouses was firelight; only in the 18th century did oil lamps and parabolic mirrors first become used. Modern lighthouses use electric light. The electric lamps that are used are either incandescent or sodium or mercury vapor lamps similar to the ones used on modern highways. The optics are very similar to those in use hundreds of years ago. A catoptric (reflected) light uses a light source and mirrors to produce a light beam. The mirrors may be plane, parabolic, or spherical. Some lights use a prism to concentrate the beam of light and increase its carrying power. This is a dioptric light. A combination prism-and-mirror light is called a catodioptric light.

The first structures that bore lights were simple wooden towers. The characteristic tower that most people associate with lighthouses—inhabited by the keeper of the light—was first designed for England's Eddystone Light by John Smeaton in 1759.

While the necessity of warnings for navigational hazards has not diminished with the passage of time, the use of lighthouses has. Lightships, offshore platforms, and buoys perform the same sort of function, and lighthouses are now automated structures that produce powerful beams of electric light and are designed to work in the worst of weather conditions with a minimum of human attention. The romance is gone. The increasing use and sophistication of onboard navigational equipment such as loran, radar, and sonar and GPS (global positioning systems) has also somewhat decreased the need for warning lights. *See* BUOY, LORAN, NAVIGATION, SONAR.

light organs Luminous or light-emitting structures on an animal. Deepwater fish are examples of animals that have these structures. The structures consist of a light-emitting body, a color filter, and a reflector; the organ is rather like a headlight. Light organs form from dermal tissue; the usual sources of the light are pigments known as luciferins, which are species-specific. Fireflies are examples of terrestrial producers of luciferins. When luciferins are oxidized by an enzyme known as luciferase, light is produced as part of the chemical reaction. This light is "cold light" (i.e., no heat is produced, such as when a lightbulb burns).

Marine organisms ranging from bacteria through dinoflagellates and from crustacea to fish—both teleosts with bony skeletons and cartilaginous fish—have been known to produce light; however, there is no other class of vertebrate animals that does this.

Some hadal organisms that emit light seem to do so by a different mechanism: They contract and expand the pigment cells, or melanophores, in their bodies. When the melanophores contract, light is emitted; when they expand, the emission of light is blocked. *See* BENTHOS, COLOR, HADAL ZONE, PHOTIC ZONE.

Ligurian Sea The northern end of the Tyrrhenian Sea, which is in turn an arm of the Mediterranean. The Ligurian is surrounded by the province of Liguria, Italy, to the north. Tuscany to the north and east, and Corsica to the south. The Tuscan archipelago separates the Ligurian from the Tyrrhenian seas. The maximum depth of the Ligurian Sea is found near Corsica and is greater than 2,800 m (9,300 feet). *See* MEDITERRANEAN.

Limacina Pteropods in the class Gastropoda. Some have shells, others do not. There are several organisms that have this name. *Clione limacina,* or the sea angel is a pteropod that is found in polar waters.

It is a very efficient storer of lipids, and this enables the creature to survive long periods of starvation. Its prey is *Linacina helicina,* a shelled pteropod. The *L. helicina* is a significant part of the zooplankton of both Arctic and Antarctic waters. Each polar region has its own variety of this tiny snail that is preyed on by both other *Limacina* and small fish, especially herring. *See* POLAR BIOME, PTEROPOD.

limestone One of the most plentiful rocks on Earth, made up principally of calcite ($CaCO_3$) and dolomite [$CaMg(CO_3)_2$]. About 20% of all sedimentary rock on Earth is either limestone or dolomite or both. The two types of rock form the typical dome structures that are associated with petroleum deposits.

Limestone is a sedimentary rock, and there exist three types which have different origins: clastic limestone, which is derived from other calcareous minerals; organic limestone, which consists of the accumulated calcareous remains of once-living organisms; and precipitated limestone, which is the end-product of chemical reactions occurring in seawater.

Since the deposition of limestone has gone on continuously through much of the Earth's history, organic limestone beds are excellent fossil records. Fossilized blue-green algae, called stromatolites, can be traced from the Precambrian to the Recent Period in such limestone beds. Most of the limestone was deposited in the Precambrian and Lower Paleozoic eras, but this process is ongoing. Stromatolites are still forming in the Bahamas, Florida, and western Australia. Reef structures are other examples of organic limestone.

Dolomite is most often deposited in areas of high salt content such as the Persian Gulf and the Caribbean Sea. Since temperature is a controlling factor in the deposition of all carbonate rock, the kinds of rock and their location are a means of examining the climate of various places during the Earth's history. *See* CAMBRIAN, DEVONIAN, GEOLOGIC TIME, PALEOZOIC.

limpet A small, flattened snail found on littoral rocks. These animals are found worldwide in temperate zones. They all share, to varying degrees, the ability to trap water beneath their shell and can therefore survive exposure on bare rock from one high tide to another. *See* LITTORAL, MOLLUSCA, SNAIL.

Limulus *See* HORSESHOE CRAB.

Lingulida Very ancient brachiopods. Lingulida and Acroteida are the two living orders of Inarticulata. These small (41–44 mm, 1.6–1.75 inch) burrowers are suspension feeders. They are found in shallow waters, largely in tropical and warm-temperate seas. They have survived essentially unchanged for more than 400 million years. *See* BRACHIOPODA.

Linnaeus, Carolus (Carl von Linné) (1707–1778) A Swedish naturalist and the founder of the presently used taxonomy—the naming of living things. His father was a serious botanist, and Linnaeus pursued his own botanical research while at the University in Lund and then in Uppsala, Sweden. His trips of exploration, undertaken while he was an improvished young professor, were to Lapland, in 1732, and to parts of Sweden unknown to him, where he collected plants and minerals. Linnaeus finished his medical degree in Leiden and stayed on in Holland until 1738, writing voluminously. His *Systema naturae (System of Nature)* (1735) and *Genera planterium* (1737) were his major works during this period. Linnaeus then returned to Sweden and gained a chair at Uppsala. He spent the rest of his life there, teaching, studying nature, writing, and encouraging his students, who loved him.

Linnaeus knew that his designations of living things were arbitrary. He used several criteria to categorize flora and fauna: Sexual organs and "cortex versus bark" in plants were but two. His use of Latin names came early, but in the last edition of the *Systema naturae* he introduced the binomial nomenclature for genus and species that made him and his classification scheme unique. He also used common names where possible. Linnaeus's treatment of animals is not nearly as good as that for plants, and his works on minerals are even less useful to a modern scientist. Not all scientists were convinced of the efficacy of the binominal nomenclature, particularly the influential comte de Buffon. *See* BUFFON, GEORGES-LOUIS LE CLERC; TAXONOMY.

Linschoten, Jan Huygen van (1563–1611) A Dutch explorer. Linschoten was a merchant in Spain, and in connection with his business he went to India in 1583. He did not return to Europe until 1590, having been shipwrecked on his return voyage. In 1594 he accompanied Willem Barents on a voyage to find the Northeast Passage to the Far East, but they reached only the Kara Sea. The next year saw Linschoten on another attempt at discovering the Northeast Passage. But the seven-ship fleet turned back again because they encountered impassable ice. This was Linschoten's last attempted long voyage.

Like Richard Hakluyt, Linschoten collected accounts of voyages. As a chronicler and early explorer he was an inspiration to other Dutch navigators. *See* BARENTS, WILLEM; HAKLUYT, RICHARD.

lipid Any of a large, heterogeneous group of molecules including fats and waxes that are involved in a number of cellular functions and that, together with proteins and carbohydrates, are the principal components of living cells. The simplest lipids are long-chain fatty acids. Fats and oils are all lipids but are only two types of substance within this group. Biological membranes such as cell membranes are constructed of complex lipid molecules. *See* HYDROCARBON, MARINE OIL.

lithium A soft, silvery white chemical element and the lightest metal known. Lithium is present in seawater (0.017%)

as the lithium ion, Li$^+$. The incorporation of lithium into rock is very slow: The turnover time of atomic lithium into the lithium ion is estimated at 20 million years. *See* ELEMENT.

lithosphere The Earth's crust and upper mantle. The lithosphere is about 200 km (125 miles) thick and is the layer thought to be broken up into the various crustal plates whose movement is referred to as plate tectonics. *See* CONTINENTAL DRIFT, PLATE TECTONICS.

littoral A term first used in the mid-19th century to denote the intertidal zone, or the part of the shore that is under water at high tide and exposed when the tide is low. This region has different biotas, depending on the type of bottom and temperature. Rocky bottoms, sand beaches, protected bays, salt marshes, mangrove swamps, and coral reefs all are littoral environments. Each has a different spectrum of plants and animals. Some littoral zones, such as the Galapagos islands, are so specialized that the animals and plants found on them exist nowhere else. *See* BEACH, COAST, CORAL, ESTUARY.

littoral currents *See* LONGSHORE CURRENTS.

lobster One of a number of families of the class Crustacea. True lobsters are *Homaridae,* spiny lobsters *Palinuridae,* slipper or Spanish lobsters *Schyllaridae,* and deep sea lobsters *Polychelidae.* All are decapods with stalked eyes and have chitinous, segmented exoskeletons that cover a three-part body. All members of these families also have five pairs of walking legs, of which one pair is the chelae, or claws. Usually, one of these claws is larger than the other.

Lobsters are most abundant in deep water of about 300 m (1,000 feet) or more, but the latest find of a species of lobster at the Mid-Atlantic Ridge was near a hydrothermal vent. Its taxonomy is still in doubt, and some researchers call it a crab. This find has greatly increased the range where these animals might be found. This latest possible lobster, tentatively classified as *Kiwa hirsute,* was found by the *DS Alvin* in waters 2,300 m (7,500 ft) deep. Its body is about 15 cm (6.5 inches) long with pincers (the pinching claws) about as long as the body. The shell is yellow-white and covered with hair, or setae, that is thought to be some part of the detoxifying structures that enable the animal to live on the minerals in the surrounding water. This animal is eyeless, as it lives in depths where there is no light. In general, lobsters live on carrion or live fish when they can get them. They are long-lived, maturing at about the age of five years, and 50-year-old specimens have been found. If they survive the predation of sea stars, cod, or rays, their only enemy is man.

Lobsters have been of interest to mariculturists for years. Attempts to "grow" them commercially have not been successful. While the animals have been studied by both pure and applied biologists, some aspects of their behavior are not well understood. The North Atlantic lobster, for example, has been observed moving in columns on the ocean floor. The reason for these marches is unexplained, as is the seemingly random cyclic rise and fall in this animal's population abundance. *See* CRUSTACEA, DECAPODA, VENT COMMUNITY.

locomotion Animal movement accomplished by a number of means. Vertebrates swim, walk, or crawl on sea bottoms or rocks using much the same muscle groupings that terrestrial animals employ.

Scallops use a rapid backward thrust to jet away from a predator by pumping water through their siphons; cephalopods also propel themselves by water jets. Octopi move by making and breaking hydraulic contacts as they adhere to surfaces. The medusae also propel themselves by pumping water in and out of the body cavity, although they have no obvious muscle. Animals like the sea star (echinoderms)

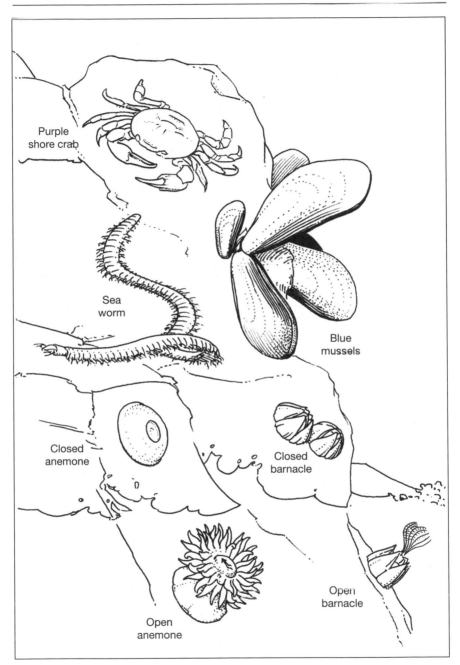

Temperate zone littoral. The rocky shore in temperate waters is characterized
by animals that arrange themselves in relation to the length of time
they can spend out of water. For example, the crab is mobile and
hard-shelled, but the anemone cannot be dried out at all—it will die.

move or hang onto prey using their tube feet as suction devices. Burrowers like worms expand their soft bodies with water inflow and then release it by exerting steady pulsed pressure on sediment, moving it slowly aside; the closest mechanical equivalent to worm locomotion is the operation of a hydraulic ram. *See* SCALLOP, OCTOPUS, WORM.

loggerhead turtle *See* SEA TURTLE.

Lomonsov Ridge An 1,800-km (1,100-mile)-long ridge line that divides the Arctic Ocean into the Canadian and Siberian basins. The crests of this mountain range rise to heights in excess of 2,000 m (6,600 feet) above the ocean floor and come to within about 1,000 m (3,300 feet) or less of the surface. Unlike the Mid-Atlantic Ridge, the Lomonsov Ridge is not associated with volcanic activity. Nor is it part of the mid-ocean ridge system as a site of new rock formation. *See* ARCTIC OCEAN, ASEISMIC RIDGE.

longitude The angular distance east or west from a reference circle, or meridian, drawn from pole to pole around the Earth and passing through Greenwich, England—the so-called Zero Meridian, or Prime Meridian. Longitude is used, along with latitude, as a measure of location on the Earth's surface. Locations on the Earth's surface are described as degrees, minutes, or seconds of arc east or west of the Zero Prime Meridian. The International Date Line is based on the 180° (E or W) line. It is distorted considerably, as are all other meridian-based time zones, to avoid bisecting part of the same country or island group.

The concept of longitude was understood long before it could be accurately measured. That development had to await the standardization of accurate clocks that would work reliably on ships. *See* CHART, MAP, NAVIGATION, PTOLEMY.

long-line fishing A method of fishing used by commercial fishermen. The drifting of long lines is a practice used in the Far East, mainly in the tuna industry. The lines are 150 to 400 m long in any one section and are dropped in several interconnected sections. The number of sections varies with the operator, the size of the boat, the number of crewmen, and the fish being taken. There may be as few as 10 or as many as hundreds of sections. Hooks are attached to the branch lines, and the number of hooks is also very variable.

The line, with its floats, is shot out of the boat at the start of the day's fishing. The crew then spends hours reeling in the line and the catch as it is marked by the floats. As the catch increases, the floats become submerged. Bottom long lines—set for cod, haddock, and flatfish, most often at depths of about 80 m—are attached to floats in the same manner as are the floating lines.

Long-line fishing probably has not contributed to the overfished state of some areas of the world's oceans. The tonnage of fish taken in this manner is decreasing. *See* FISHING.

longship A galley used by the Vikings. This was their all-purpose coastal raiding, trading, and emigrant ship. The heyday of the longships was about 600 to 1100 C.E. They ranged in size from about 40 to 80 oars, and also carried a single, square sail, but in fjords, on rivers, and in narrow channels depended solely on their oarsmen.

Longships were not used in winter, when they were pulled up on the coast. They were essentially open boats with no cover for their oarsmen, who were also assault troops. They were not very well designed, but the large crew made them workable. *See* GALLEY.

longshore currents Currents produced by waves landing on a shore at some angle with the shore other than zero. The direction and velocity of the rebound wave deflected by the shore is largely a function of the local topography. Longshore currents can move coastal sediment and are responsible for a good deal of the natural "drift" or erosion of some sandy beaches. *See* BEACH, CURRENTS.

Lophiformes The order of angler fish. Most of these predators live in the deep ocean. The largest species may reach 1.2 m (4.3 feet) in length. The distinguishing feature of these fish is a highly modified dorsal fin that resembles a bit of floating food; this lure attracts smaller organisms.

lophophore A ring of tentacles that forms a characteristic feeding mechanism. It is present in brachiopods, in bryozoans, and phoronids. The organism's mouth is in this ring, and the ring is covered with moving cilia that create a current, directing water toward the mouth. The distinctive and unusual feature of the tentacles is that they are hollow. The likelihood that brachiopods, bryozoans, and phronids have a common ancestor is based on the presence of this complex feeding structure in all of these bottom dwellers. *See* BRACHIOPODA, BRYOZOA, PHORONIDA.

loran Long Range Navigation, developed by the United States Coast Guard in the 1940s. Loran consists of pulsed transmissions from two pairs of radio stations. Measurements of the time interval between the arrival of each of these signals makes it possible to take a position line and plot it on a chart, which covers about 1140 km (700 miles) of sea. The technique is also used in charting, in which the ionosphere is a "bounce-back" or rebound medium. If one is aiming at the sky, the range of loran is 2,280 km (or 1,400 miles). *See* ECHOLOCATION, NAVIGATION, SONAR, TRIANGULATION.

Lord Howe Ridge A feature of the ocean floor in the South Pacific, also known as the New Zealand Rise. It rises to 1,000 to 2,000 m (0.6–1.2 miles) above the seafloor. This elevation runs from about 40° South latitude to Lord Howe Island to the northwest, and then north into the Coral Sea. *See* CORAL SEA, PACIFIC OCEAN.

Loricifera A group of tiny (400 µm) bag-shaped organisms first identified in Europe in 1983. They have since been found in other places in warm-temperate to tropical waters where they burrow in shallow-water sediments. They are probably related to Priapulida or Kinorhyncha. Some taxonomists have suggested that they represent a link between proto-anthropods and their Aschelminthes ancestor. *See* ASCHELMINTHES, KINORHYNCHA, PRIAPULIDA.

low water The lowest level of the tide before it turns and starts to rise again. *See* HIGH TIDE, TIDE.

lugworm A marine annelid, or polychaete worm of genus *Arenicola,* that lives in sandy, intertidal waters. It is a bottom dweller and, like earthworms, can grow quite large. Common specimens are over 20 cm (8 inches) long. *See* ANNELIDA.

luminescence *See* BIOLUMINESCENCE, LIGHT ORGANS.

lunar day The time from one lunar transit over the meridian to the next. This time interval is 24 hours and 50 minutes.

lunar tide The part of the tide that is due to the gravitational attraction of the Moon. The path of the Moon's motion is fairly complex, which explains the variation in the tide at any given location from one day to the next.

The Moon takes a lunar month—28.5 days—to revolve around the Earth. During this circuit it reaches a point where it is closest to the Earth—its perigee—and a point at which it is farthest away, its apogee. The change in the distance between the Earth and the Moon during the Moon's revolution is reflected in the tidal range at a given location. The closer the Moon is to the Earth, the greater its gravitational attraction.

The plane of the Moon's orbit intercepts the ecliptic (the plane of the Earth's orbit) at an angle of 5°. The plane of the Moon's orbit also rotates (precesses) in an

18.6-year cycle. This means that a complete tidal cycle is 18.6 years. *See* SOLAR TIDE, TIDE.

lungfish A bony fish of the subclass Dipnoi. Lungfish are freshwater species related to the coelacanth and can survive for long periods in deoxygenated water or in almost dry streams. In earlier times, there were nine families of these fish, and some were marine. The coelacanth, the sole surviving marine relative of the lungfish, is, however, thought to have originated as a freshwater animal. The bony mouthparts of some fossil lungfish suggest a diet of shelled invertebrates, which these fish crushed in their mouths. They are important in the evolutionary history of the Earth because they may be ancestral to terrestrial animals.

These ungainly, thick-bodied animals move sluggishly and are air breathers. Most have a single lung that is dorsal to the viscera. They breathe by gulping air. However, the juvenile forms of lungfishes have gills. The African and South American species estivate: In summer, when their rivers become very shallow, they bury themselves in mud capsules that trap both air and moisture. This enables them to survive the hot, dry weather.

Lyell, Charles (1797–1875) A British geologist, born in Scotland, who was originally a member of the bar and did his geological work as he and the courts traveled in England. By the time he published his very influential work, *Principles of Geology,* in 1830, he was established as a scientist. This basic three-volume work was very popular. The first volume was a part of the small library that Darwin brought with him on the HMS *Beagle.* Darwin arranged to have the second volume of this serialized book delivered to him when the *Beagle* arrived in Sydney harbor. Since geology was a rapidly growing field, Lyell added to his classic work and modified it extensively in 12 editions.

Lyell's great contribution to the field was the unification of the two dominant theories of Earth's processes, uniformitarianism and catastrophism. The first group held that the changes in the Earth were all gradual processes that took millennia; the second held that violent occurrences such as earthquakes, floods, and volcanoes were the great changers of the Earth. Lyell saw that both happened and demonstrated just how this had created the European landmass.

As part of his work, Lyell estimated the age of Earth using the time it took to create particular rock formations. His estimate of Earth's age was very close to the current one—about 4.5–4.85 billion years. Later he was one of the friends who finally persuaded Darwin to publish the work he had been engaged in for over 20 years and was afraid to bring into print. Lyell, like Darwin, knew that there would be a storm of controversy, but he was confident in both his own work and that of his friend. *See* DARWIN, CHARLES ROBERT; EVOLUTION; GEOLOGY; HUTTON, JAMES; MURCHISON, RODERICK.

Mackenzie River A river in northwest Canada. The Mackenzie and its tributary, the Finlay, form the second largest river system in North America. From its headwaters near Great Slave Lake, the Mackenzie moves northwest, keeping to the east of the Rocky Mountains along its 1,700 km (1,000 mile) run to Mackenzie Bay on the Beaufort Sea. The bay is east of the boundary between Alaska and the Yukon Territory of Canada. *See* BEAUFORT SEA.

mackerel Commercially important torpedo-shaped, carnivorous fish of the North Atlantic and also other waters. They are related to tuna and are distinctive because of the row of small fins posterior to their symmetric anal and dorsal fins. The common mackerel (*Scomber scombrus*) is about 30 cm (1 foot) long with wavy black or dark blue bars on a sea-green background. The underside is silvery white. Large schools of mackerel are exploited commercially, but the fish is also sought as a sport fish. Related species are the chub (*S. choos*), Pacific mackerel (*S. japonicus*), Spanish mackerel (*S. commerson*), and king mackerel (*S. cavalla*). The more tropical fish tend to be larger and heavier. *See* FISH, FISHING.

mackerel shark A large, fierce, pelagic shark of the genus *Lamna*. These shark are fished for food. They are colored various shades of gray, are about 3 m (10 feet) long, are ovoviviparous, and feed on fish of the temperate zone such as herring and mackerel. The name *mackerel shark* comes from the patterning on this shark, which resembles that of the mackerel.

Other sharks in the family *Isuridae* are the white and mako sharks. The mako is fished for food. The porbeagle (*L. nasus*), a small viviparous shark, is notorious for voracious eating and the destruction of commercial fishing nets. *See* OVOVIVIPARITY, SHARK.

macrobenthos Large organisms that live on or near the ocean bottom. Macrobenthic organisms are further classified by size. The organisms that pass through larger mesh screens in an array of screens of graduated size openings go into a larger size classification. The classification is as follows:

Megabenthos	large bottom dwellers: lobsters and crabs.
Macrobenthos	animals that are larger than a mesh opening of 0.5 mm: clams, oysters, annelid worms, barnacles, and amphipods.
Meiobenthos	organisms that are small enough to pass through a screen of mesh size 0.5 mm but large enough to be retained by a sieve of mesh size 0.05 mm. These include flatworms, roundworms, polychaetes, and tiny crustacea.
Microbenthos	benthic organisms that are small enough to pass through a sieve of 0.05-mm mesh opening. The bacteria, protists, and algae are in this group.

See BENTHOS.

macroclimate The study of meteorology and climatology on a worldwide scale, over long periods of time, or both. *See* CLIMATE, WEATHER.

Macruridae Deep-sea fishes related to the cods, the grenadier or rat-tail fish. This group of about 300 species are bottom dwellers in deep ocean waters. They are 30 to 60 cm (1 to 2 feet) long and have huge heads that taper into small, slender bodies. Some grenadiers have light organs, others grunt. *See* BENTHOS, FISH, LIGHT ORGANS.

Magellan, Ferdinand [Magalhães, Fernão] (1480?–1521) A Portuguese navigator who discovered the straits (named for him) linking the Atlantic and Pacific oceans. Magellan's earliest voyages were undertaken on behalf of his native Portugal. The first of his journeys was in 1505 to the Indies, and in 1511 he took part in the capture of Molucca. Magellan was in the employ of Spain during his subsequent voyages to the Far East.

A papal bull divided the world's unknown regions between Spain and Portugal, with all the territory east of the Cape Verde Islands reserved for Portugal and the rest for Spain. Magellan believed that the Moluccas were close to the Americas and approachable through Spanish lands. He therefore applied to Spain to finance an expedition that would look for a way through the American landmass to the Moluccas, or Spice Islands. Like Columbus, he hoped to find a passage in temperate waters. Magellan's ships left Spain in 1520 and followed the South American coast into cold, inhospitable waters through the autumn. In October Magellan found the straits between the South American mainland and the Tierra del Fuego archipelago, and sailed through them in 38 days. The flotilla almost starved on the long run across the Pacific to Guam and the Philippines. They landed there in March 1521 and were cordially received. Magellan, however, intervened in a local war and was one of the casualties; he was killed April 27, 1521. The surviving men returned to Spain later that year and were thus the first men to have circumnavigated the globe. *See* EXPLORERS AND EXPLORATIONS.

Magellan, Straits of [Estrecho de Magallānes] A sea channel, discovered by Ferdinand Magellan, through southernmost Chile and Argentina, connecting the Atlantic and Pacific oceans. The passage is a 560-km (350-mile)-long, rocky, 3.5 to 30 km (2- to 20-mile)-wide, circuitous course through islands, bars, and sometimes shallow channels at the southernmost reaches of South America. The average latitude is 54° south latitude. The weather of the region is dominated by almost constant cold and fog. *See* EXPLORERS AND EXPLORATIONS; MAGELLAN, FERDINAND.

magma Fluid rock beneath Earth's crust. Magma is the liquid material that is produced at the mid-ocean rift valleys. When it solidifies it forms igneous rocks. *See* MID-OCEAN RIDGES, VOLCANO.

magnesium (Mg) A light, malleable ductile metal used industrially. It is commercially extracted from seawater because magnesium is the second most plentiful metallic element in seawater. Its concentration in the sea is about 1,350 milligrams per liter of water. Some magnesium is also found in sediments, and some is present in the manganese nodules found on the seafloor.

The chlorophyll molecule incorporates magnesium as its central atom. The importance of this metal therefore cannot be overestimated, since with the exception of the vent or seep communities, all oceanic life depends on chlorophyll and the process of photosynthesis. *See* CHLOROPHYLL, ELEMENTS, MANGANESE NODULES, MINERALS, WATER.

magnetic anomaly The change of the Earth's magnetic field. This phenomenon was discovered by magnetic measurements of the ocean floor. Strips parallel to the mid-ocean ridges were found to have stronger or weaker magnetism, depending on the magnetic iron content of the ocean-floor rocks. This was explained in the 1960s (1963) by Drummond Matthews of Cambridge University and Fred Vine, as being the result of basalt formation. As new rock is extruded in ridge areas, it is affected by the Earth's magnetic field. This is now positive, or normal, although in some previous geologic times the Earth's magnetic field has been negative or weaker than normal. The alignment of iron-containing crystals in the rocks points one way in normal periods and is reversed in negative periods. This alternation of iron-crystal formation has been correlated with the formation of new rock at mid-ocean ridges. Since alignment occurs wherever new rock is formed, the same patterning also occurs with volcanic creation of new rock on land. *See* MID-OCEAN RIDGES.

magnetic field The region over which a magnetic body exerts its effects. The axis of rotation can be thought of as a bar magnet with the northern end arbitrarily called the positive end, the southern one is then the negative end. Compasses or magnetic (iron-containing) minerals point to the positive end of magnets. This includes the earth as magnet. The geographical position of the magnetic poles varies irregularly. Also the magnetism of Earth changes in intensity—the North Magnetic Pole attracts compasses less strongly and goes to almost no magnetic declination and gradually reverses polarity. A magnetic field produces detectable magnetic waves. *See* GEOMAGNETISM, MAGNETIC ANOMALY.

magnetic poles The two regions, in the Northern and Southern hemispheres, respectively, where the Earth's magnetic field is strongest. The magnetic poles are close to the geographic poles.

The North Magnetic Pole, the place toward which all compass needles point, was first found by Sir James Ross in 1831 on the Boothia Peninsula. While it is close to the geographic North Pole, the two are not coincident, and the position of the magnetic pole is not fixed, nor is its varia-

tion constant. The location of the pole and the magnetic field of the Earth are constantly changing. This variation is called secular variation. It is the end-product of the movement of the liquid ferrous core of the Earth. Since the variation is erratic, regular surveying is necessary (using a ship with no iron on it) to ascertain the location of the magnetic pole.

In 1947 the North Magnetic Pole was 320 km (200 miles) further north than it had been when Ross found it. Twelve years later, it was another 320 km north.

In addition to wandering, the magnetic field of the Earth has at times in the past reversed polarity. The North Magnetic Pole has, in the Earth's history, sometimes even been in the Southern Hemisphere. At present, the strength of the Earth's magnetic field is varying and also diminishing. If this continues, the magnetic field may weaken to the point at which compasses cease to work and the aurora disappears.

At present, we are in a period of weakening magnetic strength. This affects not only compasses (albeit this is not as much of a problem as in the past, when navigation depended on them) but also many migrating animals that move from one biome to another that is thousands of miles away (i.e., birds, sharks, whales, butterflies). These animals are dependent on the Earth's magnetism for their navigation. It is not known how these creatures adapt to the diminishing effect of the Earth as a magnet or to the subsequent change in its polarity. There is evidence that this has happened in the past; iron-containing abyssal rocks show the striations as iron crystals lined up when they cooled and pointed to what was the Earth's North Magnetic Pole at the time of their formation. *See* AURORA; FLINDERS, MATTHEW; MAGNETIC ANOMALY; MATTHEWS, DRUMMOND; MIGRATION; ROSS, JAMES CLARK.

mako shark *Isurus oxyrhinchus,* a mainly tropical mackerel shark. It often swims just below the surface of the water, with only its characteristic dorsal fin protruding. The mako does not approach shore, living only in deep waters. These sharks are fairly large, usually being longer than 3.5 m (11 feet). The mako is increasingly used as a food fish. Its numbers are declining. *See* FISH, SHARK.

Malacostraca Subclass of the Crustacea, in the phylum Arthropoda, comprising heavily armored crustaceans. There are more than 20,000 species of Malacostracans, making this group of organisms one of the most abundant among Earth's biota and certainly the largest group of crustaceans. In general, these organisms have about 20 body segments; five constitute the head, another eight the thoracic region, and the rest are abdomen. Some organisms have a heavier shell than others. The five classes of Malacostraca are the Stomatopoda (mantis shrimp), Amphipoda (sandhoppers), Isopoda (wood lice), Decapoda (lobsters and crabs), and Euphausiacea (krill).

Since this is such a huge grouping, there are many variations in size, shape, feeding mechanisms, and reproductive styles. In general, the gas exchange occurs in gill structures and digestion in stomachs. They are diecious; sexes are separate, and most malacostracans hatch into tiny versions of the adult. A few have juvenile forms, nauplia, that differ markedly from the adults. Most malacostracans are active. Some are burrowers, but most of the organisms move in search of food.

As a group, the malacostracans are very important to humans. Huge industries are centered on these organisms, which are either taken as wild food or farmed. Both therefore represent an environmental danger; overfishing on one hand and the pollution of marine environments by mariculture on the other. *See* CRUSTACEA, DECAPODA, ENDANGERED SPECIES, EUPHAUSIACEA, MARICULTURE, STOMATOPODA, *individual species.*

Maldives A string of coral atolls southwest of the southern tip of India. The Maldives are the overwater part of the Chagos-Laccadive Ridge. There are 15 principal islands and hundreds of uninhabited islets in the group. *See* INDIAN OCEAN, MID-OCEAN RIDGES.

mammal An air-breathing endothermic (warm-blooded) animal whose young are born alive and fed on milk. There are over 4,000 species of mammals. The aquatic mammals have lost all or the greater part of their hair but have otherwise retained the mammalian characteristics: a four-chambered heart, biconcave red blood cells, a diaphragm breathing muscle, and both the hard and soft palates of the oral cavity.

Sea-living mammals have special adaptations to their marine existence, such as powerful tails; lowered metabolic rates, permitting the use of less oxygen; a great fat layer for insulation; and milk that is very high in both fat and protein content.

There is no evidence of mammals before the Triassic period. They appeared first as land animals; modern marine mammals occurred later. Seals and walruses (the order *Carnivora*) date from the Paleocene epoch, toothed whales (*Odontoceti*) and baleen whales (*Mysticeti*) from the upper Eocene. The *Sirenia* (sea cows) are earlier, probably of mid-Eocene origin. *See* CETACEA, PINNIPEDS, PORPOISE, SEAL, SIRENIA, WHALE.

manatee A large, herbivorous, tropical mammal, of the genus *Trichechus,* of coastal waters, estuaries, and slow rivers. *T. manatis* is found in Florida and the Caribbean, *T. inunguis* in the Amazon and Orinoco systems, and *T. Senegalensis* in west Africa, where it lives in small herds. These slow-moving animals are from 2.5 to 4.5 m (8 to 15 feet) long, and weigh about 700 to 800 kg (1,500 to 1,700 pounds). Manatees have poor senses of sight and smell and therefore are likely to be hunted or killed in accidents. Attempts are being made, notably in Florida, to preserve the manatees. *See* MAMMAL.

manganese Manganese is an essential trace element. It is found in plant and animal tissue and is classified as a micronutrient. Its metabolic function is unknown. As a metal it is hard and brittle and resembles iron but is not magnetic. Manganese is also found in trace amounts dissolved in seawater.

manganese nodules Nodules found on the ocean floor, in which manganese is the dominant element. These nodules, which are greater than 3 cm (1.5 inches) in diameter and contain at least 26 other elements, were first discovered by the *Glomar Challenger* expedition. Manganese also occurs in the form of sand or rocks, as coatings on rocks, and as a filler in coral reefs.

The nodules seem to be built up in layers, although the mechanism of their formation is still only a matter of conjecture. It is possible that when metallic elements are added to seawater saturated with these elements, the elements, or salts of other compounds containing them, precipitate. As these precipitated substances descend to the ocean floor, they then attract other, similar materials that have reached the saturation point. At the ocean bottom, the precipitated material crystallizes and sometimes incorporates biogenic material nodules.

Manganese nodules are formed in areas of little or mild bottom current, and since nodules of specific composition form in specific areas, they represent signatures of the areas where they are formed. The rate of nodule formation has been estimated as about 6 million tons (metric) per year, and manganese nodules therefore represent a potentially significant source of minerals.

mangrove A tropical to semitropical plant (genus Rhizophora) that grows in salt marshes and tidal estuaries. Mangroves are found in Florida, in the Caribbean, along the eastern coast of South America, in West Africa, in Southeast Asia, and in Australia. They grow as trees, with their roots in the water, to heights of 10 to 12 m (30 to 40 feet) and have edible fruit. The roots produce offshoots that project several feet away from the parent and produce a new tree. The mangrove also grows "knees"—root structures that prop up the tree. Since the mangrove

tree stands in water-soaked ground, the tree needs such props to keep its heavy trunk upright. The "knees" are also fairly porous and provide an air supply for the submerged root system. The bark of the red mangrove contains tannins and is used in the manufacture of leather. *See* ESTUARY, SALT MARSH.

mantis shrimp These aggressive reef dwellers are stomatopods and distantly related to the edible shrimp, crabs, and lobsters. Most species live in shallow, warm seas—few range below about 40 meters (125 feet)—and are famous for their ability to kill prey, which is almost always hard-shelled mollusks such as clams and snails. The animal uses its second leg as a crushing or kicking device, in a manner similar to the hunting posture of the praying mantis, which it superficially resembles. This weapon is deployed both in feeding and defense. A mantis shrimp will defend its territory from other predators that are much larger than itself. They range in size from 1 to 2 cm (0.4–0.9 inches) to the giant, *Lysiosquilla*, which is about 30 cm (12 inches) long. The body is bent and resembles a saddle. This shape gives additional force to the kick, making it formidable and strong enough to break a snail shell. Part of the force delivered to the unlucky snail comes from the rapid strike. This motion creates bubbles of negative pressure at the point of impact, and the exploding bubbles break the shell—this is called cavitation. The bubbles literally explode. This happens with such force that sound is generated.

These animals are also notable for their ability to use fluorescence as a species-specific signal and for their remarkable eyes. They can see colors that are well beyond the human wavelength sensitivity range. Unlike humans, who can receive three pigments, the mantis shrimp can distinguish at least eight.

Despite its pugnacity, or perhaps because of it, stomatopods are collected by some enthusiasts. If annoyed, this pugnacious decapod can break the glass of aquarium walls. *See* CRUSTACEA, HOPLOCARIDA.

mantle *Geologically,* it refers to the layer of the lithosphere (the rocky part of the Earth) that lies between the two discontinuity layers: the Gutenberg discontinuity below and the Mohorovicic, or Moho, discontinuity above. The mantle consists of peridotite, which is largely olivine, a dark green, dense rock composed of magnesium and iron silicates.

A very recent theory concerns the beginnings of life on earth and how they shaped the crust. This theory examines the creation of free oxygen by living, photosynthesizing organisms and relates that to the evidence of difference in density of mantle rock and crustal rock. Crust is less dense than mantle; according to this theory, the reason for the difference was the creation of oxygen, which then was available to react with some of the constituents in the mantle and literally rise above it.

Biologically, the mantle or pallium is the membrane in a mollusk that surrounds the soft body parts. The entire body, other than the foot, is within the space called the mantle cavity. The mantle in the adult organism is lobed, with the lobes secreting the upper and lower halves of the shell (calcareous valves) in species that have a shell. The mantle isthmus connects the lobes of the mantle and secretes the ligaments that hold the valves in place.

The molluscan mantle is folded at its margin, which is the site of growth of the organism. The outermost fold secretes the shell (if there is one), while the middle fold is the sensory perception system. The middle fold also carries any tentacles that the animal may have and, in the scallop, bears the animal's eyes. The inner fold, which is closest to the gut, is muscular and controls the flow of water into and out of the animal.

The cephalopods, which lack shells, have a very thick, strong mantle covering. The pteropods and other unshelled molluscans such as nudibranchs have a highly modified mantle that functions as a propellant and food trap in one group and as a gill-like structure in the other. *See* CLAM, CEPHALOPODA, NUDIBRANCH, OYSTER, PELECYPODA, PTEROPODA, SCALLOP, SNAIL.

map A rendition of the Earth or part of it, usually on a flat surface. Mariners have almost always made and used maps. Maps have even been constructed by primitive peoples lacking a written language, such as the Polynesian islanders of the South Pacific, whose stick charts were accurate representations of island groups and showed island landfalls at the intersections of the sticks.

Ptolemy's map of the ancient world (second century C.E.) is a reasonably accurate depiction of the world he knew. The decline of Western Europe after the disintegration of the Roman Empire and the early Middle Ages brought the mapmakers' art to a standstill. Scientific cartography revived with the Age of Exploration. Maps made by Mercator in 1569 showed the congruence of the American and African coasts. The concept of longitude appeared in 1690 and was a function of the development of good chronometry. Sailors' maps, or portolans, gave accurate representations of the coastlines by the 13th century, and showed ocean depths by the 18th century.

Devices for measuring distance and speed, and for determining direction, were attempted in several societies. The Chinese had a compass by the first century C.E. Astrolabes of varying complexity and functionality were used by Arab sailors to calculate latitude. The oldest known European sea chart (1184) was shown to Louis IX as he prepared for his departure on the Crusade of 1270.

To measure the speed of a vessel at sea, a log attached to an evenly knotted line was heaved overboard. By counting the number of knots run out per hour—using an hourglass—the Renaissance sailor could calculate the distance his vessel traveled and the time required for the journey.

Maps are made in several styles. They depend on the emphasis the maker puts on some specific feature. Since it is not possible to accurately depict all of the features of a curved Earth on a flat surface, some sacrifice of one feature is necessary. Cylindrical projections, of which the Mercator projection is the best known, have rectilinear parallels of both latitude and longitude. This produces great distortions at the poles but is accurate in the depiction of coastal features. Conic projections begin from some focal position: if the position is in space, the projection is said to be orthographic; if the position is on the surface of the Earth, the projection is said to be stereographic. A projection from the center of the Earth is a gnomonic projection.

The international authority for nautical maps is the International Hydrographic Bureau, with headquarters in Monaco. *See* CHART; LATITUDE; LONGITUDE; MERCATOR, GERARDUS; PTOLEMY.

marginal seas Large areas of relatively shallow waters near continents on the boundaries of oceans that have effects on the entire ocean. The Caribbean, including the Gulf of Mexico, and the entire Mediterranean are marginal Atlantic seas. The South China Sea and the Sea of Japan are marginal seas of the Pacific, and the Arafura and Timor are marginal seas of the Indian Ocean. *See* CONTINENTAL SHELF, OCEAN.

Marianas Islands A group of islands in the Pacific Ocean, east of the Philippines. The Marianas are small, tropical, typical coral-reef and volcanic islands. They were found by the Magellan expedition (1521) and called Los Ladrones (the thieves) but were renamed in 1668 in honor of the widow of Philip IV of Spain and opened for colonization. Their existence was largely unnoticed by the rest of the world until World War II. Saipan, the site of a major battle, and Guam, are two of the principal islands. *See* CORAL, PACIFIC OCEAN.

Marianas Trench A deep ocean trench east of the Marianas Islands. It was first sounded in 1899. Ever-increasing depths have been recorded in the trench, the lat-

est figure being 11,708 m (38,635 feet). *See* PACIFIC OCEAN, TRENCH.

mariculture The cultivation, under appropriate environmental conditions, of marine plants and animals, in analogy to agriculture. For example, marine algae are raised by mariculture in Korea, Japan, and other locales where they are important dietary items. The Chinese have been raising carp in ponds for centuries.

The formation of lagoons for trapping fish is an ancient, marginally maricultural technique. It is, however, not very successful, since the creation of a lagoon changes the living conditions of the trapped organisms. The establishment of appropriate living conditions is a difficult problem, but the success or failure of any mariculture project depends on it.

The Roman raised oysters in baskets almost identical to the baskets used today, using an identical technique in which larval oysters were gathered, placed in the baskets to provide mechanical protection, and left in their natural environment to grow. This technique is successfully used today in Brittany, the Mediterranean, and in Japan.

In Japan, oysters are raised for both food and the pearl industry. The same methodology also works for mussel cultivation, and some clams and scallops are also cultivated under this type of system. Direct use of the natural environment is simpler than trying to duplicate this environment, since different larval stages of an animal require differing growing conditions and a different food supply. Another approach to mariculture involves the release of tank-raised juveniles into their natural habitat, followed by recapture of the adults, a technique that is more like ranching than cultivation. This method is used routinely to increase the yield of salmon, trout, and striped bass stocks.

While most of the fish catch is essentially the hunting of wild species, some care is entering the industry. Specially designed nets, which will not entangle air-breathing mammals, are now in use. Others are designed to allow juveniles to escape; this ensures their survival in the hope that they will live to maturity and reproduce. This simple technique has already increased the pollock stocks in the north Pacific.

Currently, the Japanese are experimenting with the raising of squid and octopus by mariculture. While the cultivation of lobsters is still eluding experimenters, shrimp can be raised in semi-wild surroundings. The larva are fed on plankton in indoor tanks and then released to protected waters.

Nevertheless, it is estimated that at least 20% of the fish and shellfish taken commercially can be raised in tanks or other enclosures. French efforts are concentrated on the turbot, sole, bass, shrimp, scallop, and lobster. Carnivorous species are more difficult to maintain in captivity because it is necessary to maintain a captive population as foodstuffs and to cultivate the necessary bacteria and plankton as foodstuffs for the desired crop or the intermediate feeder population.

Pollution is one major problem encountered in mariculture; there is excess food that the farmed population does not eat, and this attracts other species and bacterial decomposers. The other major problem is overcrowding. When an animal population is enclosed either on land or in a weir, there are more animals per unit of living space than is normal in a wild population. This overcrowding is a disease vector, and there are research stations attached to organizations promoting mariculture. These facilities study the bacterial, fungal, and viral pathogens that might infect the population being farmed. It is essential to maintain the health of the farmed population, since that is the money crop; its health also will ultimately affect that of the wild population that lives in the same waters.

The range of mariculture is expanding from farming finned fish and bivalves, particularly oysters and mussels, to include other invertebrates that form part of the human diet. A variety of shrimp and clam species and sea cucumbers are now in commercial production, as is seaweed. Asians have eaten seaweed for centuries; it contributes a considerable fraction of the protein and minerals needed for human metabolism. The increasing popularity worldwide of Asian cuisines has increased the demand for these products.

Algae, while important as food, have other uses in commerce—some of them also food-related. Agars, alginates, and carrageenans are extracted from a variety of red and brown algae for a number of uses. They are added to many products to increase spreadability (salad dressing, for example), to produce a product that is not oily but pours slowly, as thickeners and emulsifiers (as in other prepared foods or in cosmetics), as culture media for microbiology, or as finishing surfaces in the manufacture of paper products. *See* ENDANGERED SPECIES, FISHING, OYSTERS, SALMON.

marine archeology The systematic study of layers of material on the ocean bottom, with the careful mapping, siting, and sifting of artifacts. Some marine archaeological expeditions have historical and cultural objectives, others focus on sunken treasure. People have attempted to salvage materials from sunken ships for centuries. One of the many stories about Alexander the Great tells of his exploring a shipwreck using a very primitive diving bell. Certainly, Roman shipping had its salvage collectors, and when a vessel went down in waters shallow enough to allow free diving, people salvaged what they could retrieve.

The first large-scale, successful hunt for sunken treasure was made by William Phips, who in 1687 brought a cargo of more than 31,000 kg (68,000 pounds) of gold and silver to the Royal Dock near Greenwich, England. Phips had searched for and found, at a depth of about 12 m (40 feet), the wrecked *Nuestra Senora de la Concepcion,* a Spanish galleon that sank in 1641 north of the island of Hispaniola, in the Caribbean Sea. It had been a part of the annual treasure convoy from the Spanish colonies back to Spain. Phips's salvage was the result of a 40-day diving session using native divers, with no equipment other than the "Bermuda bell," a primitive diving bell made of a weighted wine barrel.

The first modern marine archeological exploration was initiated in 1958, when Peter Throckmorton, an archeologist who learned to use scuba equipment, used the reports of sponge divers to locate a Bronze Age wreck that had gone down around 1200 B.C.E. near the island of Rhodes in the Mediterranean. An expedition led by George Bass of the University of Pennsylvania extensively explored the site 280 m (90 to 95 feet) down. An archeological grid was established and photographed. The calcareous accretion on the wreck was removed and new maps and photographs taken. This process was repeated as each artifact emerged. The remaining wooden parts of the ship were raised to the surface, cleaned, and preserved.

In 1961–62 the same two archeologists and a team of 15 others worked on a seventh-century ship, a Byzantine wreck located 36.6 m (120 feet) below the surface near Bodrum (Turkey). A three-dimensional grid was laid over the cargo ship, and each section was carefully examined. The cargo of amphorae and sarcophagi was cleaned in situ by carefully breaking away calcareous accretions and then vacuuming them away, examining each artifact repeatedly, and sifting the removed material. Cameras were used extensively. The structure of the ship was preserved down to its makeshift repairs. It was then raised and immediately immersed in a solution of polyethylene glycol to fill in the cellular structure of the ship's wood and keep it from crumbling once it was out of the water and dried.

The techniques developed by Bass and Throckmorton were used again on other ships. A Roman cargo vessel found on the bottom in the Gulf of Taranto (between the "heel" and "sole" of the Italian boot) was carefully examined in this way. The

Swedes have spent years working on the warship *Vasa*, the pride of Gustavus Adolphus's navy, which sank upon launching on August 10, 1628. It was rediscovered in Stockholm harbor in 1956 and brought to the surface in 1961. The intact ship has been completely preserved and studied and is now in a museum that has been built around it near its original launch site. The hull of Henry VIII's flagship, the *Mary Rose*—which sank in Portsmouth harbor in 1545 with its full complement of 400 men—was found in 1968. The almost-intact hull was injected with preservatives to keep the wood from disintegrating as it dried out. The salvaged ship is now on display in the museum built around it in Portsmouth Harbor.

In 1985, after years of searching, Mel Fisher, who had been surveying the Florida waters for downed galleons, finally found the silver store of the *Neustra Señora de Atocha*, which had sunk near Florida in 1622. Fisher was certainly more careful not to damage his site than William Phips was, even though his objective, like Phips's, was to recover treasure rather than practice archeology.

The efforts of the last several years have contributed enormously to the understanding of ancient shipbuilding techniques. Before the careful examination of ancient wrecks, the only information available was based on contemporary drawings and coins, from which it was not certain just how much artistic creation and how much adherence to the model had entered into each rendition. The raising of wrecks has also opened up an entirely new area of social history, the extent and variety of ancient maritime commerce, which formerly could only have been speculative. The range of Chinese ceramics that have been collected from the waters of the China Sea and around Indonesia is astounding. This export ware was a regular item of commerce, and the largely intact remains are so plentiful that they are available for purchase—the museums have as much as they can use. Elegant Chinese ceramics dating from the ninth through the early 20th centuries can be obtained by private collectors.

Marine Biological Laboratory Located in Woods Hole, Massachusetts, it is the oldest private marine laboratory in the United States. Founded in 1888, it is an independent institution for teaching and research. Primarily a summer laboratory in its early years, the MBL today retains its renowned summer program while simultaneously serving as the year-round home for a cadre of world-class biologists and ecologists. Web site: http://hermes.mbl.edu. *See* AGASSIZ, ALEXANDER; GIANT SQUID.

marine oils Fish oils, characterized by a high fatty acid content. The fatty acids that constitute these oils are carbon compounds whose molecules consist of chains of carbon atoms to which hydrogen and oxygen atoms are attached by chemical bonds.

At one or more points, these chains are linked through atoms of oxygen to a molecule of some alcohol, usually glycerin (glycerol). Whether a fat is a liquid or a solid depends on its length and structure and on the temperature at which it exists. Animal fat of either terrestrial or marine origin is composed of fatty acids that are 12, 14, 16 (the predominant number), or 18 carbon atoms long, with a small quantity of acids of shorter or longer carbon length. The significant difference in marine oils is the degree of unsaturation, or number of carbon atoms per molecule that are doubly rather than singly bonded to neighboring carbon atoms. This is much higher (there are more double bonds) in marine oils than in most other fats.

The natural marine oils have a variety of uses. They are of commercial interest as edible oils, as drying agents for the paint and coating industry, and in the tanning of leather. Marine oils must, however, be actively sought, as opposed to the fat of terrestrial animals, which is largely a by-product of the meatpacking industry. The fish most widely sought as oil-producers are the California sardine (pilchard), menhaden, and herring.

Whale oil is becoming increasingly rare due to overhunting.

Fish liver oil, which is rich in vitamin A, is stored in the livers of cod and some sharks. Shark oil has become increasingly sought after since there is an as-yet-unproven claim that it is has cancer-preventing properties. Sharks were endangered before this claim for their oils was made, and the claim has increased shark-fishing. Generally speaking, fish rich in liver oil do not have very oily flesh, whereas oily fish do not store great quantities of oil in their livers. *See* ENDANGERED SPECIES, FISH OILS.

marine reserves or sanctuaries Areas set aside where no fishing or mining is allowed. This is a technique used in areas that have been seriously overfished or where the original biome has been disrupted because it gives the depleted commercially significant population a chance to build up its numbers. Since marine species need large areas in which to disperse eggs, spores, or larvae, a marine reserve can be more than 50 km (20 miles) on a side. Some are even larger, but in terms of total ocean areas the marine reserves constitute about 0.01% of total ocean area.

One bit of evidence that these marine parks make a difference is that the protected population doubles and individuals grow larger within 20 to 40 years. The growth in size of individual fish is a good indication of the improving health of the population as a whole. Many of the fish now taken in the North Atlantic are much smaller than those previously caught; the juveniles are being caught before they reproduce and thereby the numbers of that population are driven down. For many wild species, the minimum numbers necessary for a viable and sustainable population are not known. *See* ENDANGERED SPECIES, EXTINCTION.

marine snow The name given to the finely divided solid matter that is carried along by the ocean's water. It was called that because divers noticed that when they looked up from the sea bottom, the water appeared turbid rather than clear; it seemed to have tiny particles suspended in it. This phenomenon is the rain of debris from the surface layer. It was named by Japanese researchers who observed the phenomenon from a submersible several years after World War II. Their lights revealed particles—like snowflakes—in the water, well below the euphotic zone. At first it was thought that these particles—fecal pellets, formaniferan tests, radiolarians, pteropod shells, other shell fragments, dust blown from continents, coccoliths, and the like—would be too light to sink and would drift around in ocean currents for decades. By the 1980s, after a Woods Hole Oceanographic Institution experiment in the Sargasso Sea, it became obvious that the tiny particles repackaged themselves into large aggregates, and these sank fairly rapidly.

Marine snow particles are held together by sticky and/or fibrous strands of protein that may be the remains of pteropods or phytoplankton. This sticky material, mucus or biofilm, is the glue that holds the aggregate together while it sinks. During the descent through the water column, an aggregate may fall apart and its contents may become reassembled into another somewhat differently composed aggregate. Eventually the clump will fall to the bottom, providing a source of nourishment to bottom dwellers.

This movement of carbon from the ocean's surface, where photosynthesis occurs, to the deep, dark bottom is the essential part of the ocean's carbon cycle. The formation of aggregates means that surface material will land on the bottom in a matter of days or, at most, a few weeks. *See* BIOFILM, CARBON CYCLE, EUPHOTIC ZONE, PHYTOPLANKTON, PTEROPOD.

marlin   A group of large, tropical food and sport fish of the order Perciformes of the family Istiophoridae. The several species of marlin are characterized by a long body with a long nose that terminates in a spear. Marlin are carnivorous, deep-sea animals that are sought by both sport and commercial fishermen. The most common species are listed below. *See* FISH.

marsh   A brackish-water area adjoining salt water. The ecosystem of a marsh is highly specific for climate, tidal height, and the amount of salt in the water. Typical tropical marshes are dominated by mangroves. More temperate marshes have grasses as the preeminent plant species. The most usual halophyte flora (salt-tolerant plants) are cordgrass (*Spartina*) and its relatives. The regions near tidal creeks have the least iron, sulfur, or both in the soil and the greatest plant growth. In turn, they support large animal populations.

The ecological relationship of tidal marshes and open sea are only now beginning to be understood. The marsh supports a large aquatic population of plankton, which is eaten by crustaceans and mollusks, which in turn either supply food for or feed on the insect population. The insects, in turn, are prey for egrets, gulls, rails, and ducks. The amphibians, snakes, and turtles of the marsh exist on this base, as do the rabbits, mice, and deer and the carnivores that prey on them. The

MOST COMMON MARLIN SPECIES		
Blue marlin (*Makaira nigricans*) (Atlantic Ocean)	400+ kg (900 pounds)	Blue with a silvery white underside
Black marlin (*M. indica*) (Indian Ocean)	700+ kg (1,500 pounds)	Blue-gray to black, with stiff pectoral fins
Indian marlin (*M. audax*)	120 kg (260 pounds)	Barred upper surface
White marlin (*M. albida*) (Atlantic Ocean)	40–45 kg (90–100 pounds)	Greenish-white

use of marshes as a breeding ground for many species that are oceanic as adults is slowly making the study of marshes and their life-forms important to commercial interests. Until very recently, the average approach of the marine engineer to a marsh was to drain it. *See* DUCK, ESTUARY, HERON, LITTORAL.

Marshall Islands An island group east of the Caroline and Marianas islands; the most easterly part of Micronesia. The Marshall Islands form two parallel chains lying in a northeasterly to southwesterly direction. They are typical volcanic and coral oceanic islands. The islands were first noted by European sailors in 1529 when Alvaro Saavedra of Spain claimed them for his king, Charles I (Emperor Charles V). Kwajalein is one of the largest islands; others in the group include Bikini and Eniwetok—both of which were used as sites for early above-ground atomic tests. *See* ATOLL, CORAL, ISLAND, PACIFIC OCEAN.

Mastigophora The class of flagellated protists, consisting of unicellular, tailed organisms. There are over 5,000 species. *See* DINOFLAGELLATES, FLAGELLA.

mating The union of males and females for reproduction. Mating is different in a watery world without boundaries. It is also a topic that has received little study beyond observation of the mating behaviors of large mammals, birds, and an occasional pair of turtles. But this area of study is important in a number of species, particularly those that are endangered. Sharks, for example, are now being studied because little is known about them and their population is dropping precipitously. Sharks and other elasmobranches produce few offspring and mature slowly. In that regard they are more like mammals than like other fish. The results of shark watching have produced some preliminary results. There is great diversity among shark mating behavior—some species exhibit polyandrous matings (several males and one female)—and it seems that overfishing has taken a disproportionate toll on male sharks.

Among the fish, there is species-specific mating behavior. Many fish species announce their availability by making noise; they grunt using swim bladders as voice boxes, or they grind their teeth. Since sound travels better in water than in air, this is a useful technique. It is postulated that invertebrates manufacture pheromones, chemical lures that bring males and females of the same species together in an area. Thus, the eggs and sperm have a chance to be in the same place at the same time. Several related species can then spawn in the same locale; eggs have chemical recognition proteins on their surfaces that will allow bonding only with sperm from the appropriate species. The chemical signals that bring mating pairs together also trigger responses from predators, who then sense where the new eggs or larvae will be. *See* ENDANGERED SPECIES, KAIROMONE, PHEROMONE, SHARK.

Maury, Matthew Fontaine (1806–1873) An American oceanographer known for establishing the first "sea lanes." As a midshipman he circumnavigated the globe in the expedition of 1826–30. He was promoted to lieutenant, but an accident prevented his going to sea, and he was given shore duty at the Depot of Charts and Instruments, an office that later became the U.S. Naval Observatory and Hydrographic Office.

Maury issued form-logbooks to captains and used their notes to produce wind charts, which were published in 1848. These charts made possible shorter passages using areas of the ocean that had the most favorable winds—the first establishment of "sea lanes." Maury was the American representative at the first meeting of the International Hydrographic Bureau. He attempted soundings of the Atlantic in preparation for the work involved in laying the transatlantic cable. His classic work, *Physical Geography of the Sea*, first appeared in 1855. It was to be a standard in the field and went through many editions. Although it contained errors, it was the best compendium of information available, and was written in a very readable, informative style by someone who obviously liked what he did. Maury's recommendation for different travel lanes for eastbound and westbound steamers created internationally honored sea lanes designed to avoid collisions. The vastly increasing use of the ocean for travel and freight shipment made this kind of regulation necessary.

During the Civil War, Maury worked on the electric torpedo, which the Confederate Navy was trying to develop, and went to England as a special envoy. He served with Emperor Maximilian in the attempt to establish a European empire in Mexico. He returned to the United States in 1868 and taught meteorology at the Virginia Military Institute in Lexington, Virginia until his death.

Mediterranean Sea An extension of the Atlantic Ocean between Europe and Africa. The term *Mediterranean* is also used to describe a type of landlocked sea: an arm of an ocean surrounded by land. The Caribbean is an example of a Mediterranean-type sea.

The area of the Mediterranean is about 2.88 million km², or 1.15 million square miles. The tides are semidiurnal and usually very shallow; the average tide is no more than 1 m (3.3 feet) high. The predominant surface current is an inflow of cold, oxygenated water from the Atlantic Ocean through the Strait of Gibraltar. The water then circulates along the North African coast to the eastern end of the Mediterranean, where some breaks off to make the circuit of the Ligurian Sea. The return flow is along the European coast back to Gibraltar, where the warmed-up, oxygen-poor water leaves the Mediterranean below the inflow of fresh cold Atlantic water. The Mediterranean loses vast quantities of water through evaporation and is therefore considerably more salty than the Atlantic. The salinity at the eastern end is about 3.6%.

The present Mediterranean is a much smaller version of the primordial ocean that encompassed almost half the Earth in the Jurassic period. The Aral, Caspian, and Black seas were once integral parts of the Mediterranean rather than totally landlocked lakes. The present sea may be thought of in terms of two distinct basins, an eastern and a western one, which may be subdivided into distinct regions. The western Mediterranean may be subdivided into the Alboran Basin, between Gibraltar and the Gulf of Lions; the Balearic Basin to the east of the Alboran and ending at Corsica and Sardinia; and the Tyrrhenian Basin east of Corsica and Sardinia and west of Italy. The eastern end of the western Mediterranean is the Strait of Sicily. The eastern Mediterranean may be subdivided into the Adriatic Basin, east of Italy; the Ionian Basin, from the Strait of Sicily to Greece and Crete; the Aegean Basin, north of Crete; the Sea of Marmara to the east of the Dardanelles; and the Levantine Basin east of Greece.

The average depth of the Mediterranean is about 1,500 m (4,600 feet), although there are considerable deeps. Those of the French coast near Marseille (the Calanques) have been well explored. The greatest depths, of more than 5,000 m (16,000 feet), are in the Hellenic Trough, which makes a wide arc around the Greek peninsula and Crete. The Balearic Abyssal Plain covers a triangular area encompassing most of the bottom of the Balearic Basin. The base of the triangle lies along the Algerian coast, and depths of more than 2,700 m (8,000 feet) are reported. Most of the rest of the Balearic bottom is the delta cone of the Rhône River. This is one of the few western rivers that delivers any appreciable volume of water into the Mediterranean.

The Mediterranean is relatively low in nitrates, phosphates, and other nutrients, since most of its water source is the Atlantic, not land runoff. The eastern basin is more impoverished in terms of dissolved minerals than the western and is also saltier.

The dominant winds in the Mediterranean are the mistral, a northwesterly cool wind that blows into the sea from the Alps and brings very clear skies; the bora, a cold, dry winter wind; and the khamsin (humid) and sirocco (dry) hot winds that move from Africa into the sea and into Europe and, depending on their velocity, may transport large quantities of sand.

The dominant bottom feature of the eastern Mediterranean basin is the delta cone of the Nile. The Nile delta is not as imposing as that of the Rhône. Nile silt has not built up as much area as would otherwise be expected, because of the active subsidence in the region, which is near the edge of a crustal plate. The quantity of Nile sediment has not significantly increased since the construction of the Aswan Dam in 1970.

There are many islands in the Mediterranean, which range in size and historical importance from tiny ones in the Greek archipelago to Mallorca, Crete, Corsica, Cyprus, Sardinia, and Sicily—the largest islands. The general climate of the whole region is warm and relatively arid. *See* CURRENTS, DELTA, ISLANDS, SEAS, TETHYS SEA.

medusa One of the two forms of all adult cnidarians. Jellies (jellyfish) have the umbrella or bell shape of the medusa with their mouth at the bottom center of this structure. The medusa moves by expelling water. *See* CNIDARIA, JELLIES.

Meinesz, Vening (1887–1966) A Dutch geodesist known for undersea gravity measurements. His full name was Felix Andries Vening Meinesz. Some references to his work list his name as Vening-Meinesz.

Geodetic measurements are concerned with the shape and size of the Earth, with the exact position of points on the Earth's surface, and with variations of the Earth's gravity field. Meinesz began his career by doing a gravimetric survey of the Netherlands in 1911. He designed a dual pendulum apparatus for measuring the effect of gravity, but surface waves created excessive perturbations. He then used submarines, since waves are damped exponentially by depth. This technique of submerging his pendulums made very exact mapping of the ocean bottom possible.

Between 1923 and 1939 Meinesz used eleven different ships and made over 800 observations of the seafloor. His apparatus was used by all explorers until the 1950s. After that, spring gravimeters were used on ships equipped with shock-absorbing platforms.

The Earth's gravity field was extensively explored by Meinesz. He plotted gravity anomaly belts of island arcs that were thousands of kilometers long. They lay along trenches and volcanic islands and clustered around earthquake foci. Meinesz thought that gravity anomalies were a result of compression of the Earth's rigid crust. He did not think in terms of plate tectonics. Although the theory of plate tectonics had been proposed by Wegener before World War I, it was generally ignored by most geophysicists and geologists. This idea of a dynamic Earth did not gain acceptance until the late 1940s. *See* OCEAN BOTTOM, PLATE TECTONICS, SOUNDING.

meiobenthos *See* MACROBENTHOS.

Mekong River A major southeast Asian river. The Mekong rises in the Tibetan highlands; flows south through Tibet and the province of Yunan, China; then forms the boundaries between Burma and Thailand and between Thailand and Laos; and eventually flows through Cambodia and the southern end of Vietnam to the South China Sea.

For much of its roughly 4,200-km (2,600-mile) course, the Mekong is a turbulent stream with falls and white-water regions. The delta is vast and is certainly Vietnam's greatest asset: It is the best-producing rice-growing area in the world.

Melanesia The collective name of the group of Pacific islands northeast of Australia. Fiji, the Bismarck archipelago, the New Hebrides, and the Solomon Islands make up this group. *See* ATOLL, ISLAND, PACIFIC OCEAN.

Mendocino Escarpment A sharp drop in the seafloor of the North Pacific Ocean. This "break" extends from the California coast toward a point north of Midway Island. *See* CRUST, FRACTURE ZONE, SUBSIDENCE.

menhaden *See* HERRING, MARINE OIL.

Mercator, Gerardus (Gerhard Kremer) (1512–1594) A Flemish cartographer who produced a global map in 1538. Very much influenced by the global map of Ptolemy, Mercator made a determined effort to produce a map that was as good as his 2nd-century inspiration. He followed the production of the map by a globe of the Earth in 1541 and a celestial globe in 1551.

After his appointment as professor of cosmography at the University of Duisberg, Mercator changed his approach. He redesigned his map on the basis of parallels of latitude perpendicular to the meridians of longitude. His new maps were published in 1569, and he planned an atlas, the first installment of which appeared in 1585; it was never finished.

The increasing use of Mercator's rectilinear maps by French explorers led to their general adoption. These maps show coastlines very distinctly in spite of their distortions of shape and size. The error or distortion on Mercator maps is the result of the shape of the map. The Earth is spherical, not rectilinear. Thus, on a map on which latitude and longitude lines form right angles, the north-south distances become more exaggerated than do the east-west distances as latitude increases. The best example in the appearance of Greenland. On a Mercator map Greenland appears to be larger than South America, which is certainly not the case. *See* MAPS, NAVIGATION.

mercury A chemical element which in its usual state is a reactive, silvery-white, liquid metal. Mercury has been used extensively in the chemical industry since ancient times in spite of its known toxic effects.

While the toxicity of the compounds of mercury has been long established, it was not until the mid-1960s and 1970s that the potential of mercury for environmental contamination was fully understood. The manufacture of chlorine, sodium carbonate, and certain other chemicals are among the industrial processes that use mercury. The effluent of plants manufacturing these products eventually contaminated aquatic as well as terrestrial environments. Minimata Bay in Japan became famous because of the tragic effects of mercury poisoning observed in many residents of this heavily industrialized area. This was the first documented example of the entrance of mercury metal into the ecosystem of a region. Mercury effluent entered the bay from industrial runoff. Compounds of mercury—which are much more soluble than is the metal itself—were formed in the water by anaerobic or sulfur metabolizing bacteria. When they were incorporated and concentrated in the fish population, they eventually poisoned all of the consumers of the fish, including humans. *See* ELEMENT, POLLUTION.

meridian A circle of longitude passing through both the North and South Poles of the Earth.

Meristomata A class of the subphylum Chelicerata (antenna-less), phylum Arthropoda. These animals are large and helmet-shaped. The surviving members of this once widespread group are the well-known horseshoe crabs that appear on the temperate to cool Atlantic beaches in the spring. *See* HORSESHOE CRAB.

mermaid's comb This beautiful shell (*Murex pecten*) is sometimes called a Venus comb. It is found in tropical Pacific waters. *See* MUREX.

mermaids' purses (sailors' purses) Dried brown, almost flat, leathery ribbons found on Atlantic beaches. They are the egg cases of skates. The specimens found are usually empty—the embryos either hatched or died. If the egg case is ripped from the underwater rocks to which it is attached, it becomes dried out. The small hooks at the edges of each egg case hold it in place on rocks beyond the intertidal zone. The egg case is shaped to maximize the flow of clean, aerated, food-containing water through it. *See* SKATE.

mermaid's wineglass This is a tropical alga of the genus *Acetabularia*. It is in the order and family Danycladaceae and the phylum Chlorophyceae. These plants have a considerable fossil presence going back to the Ordovician. Population blooms occurred in the Carboniferous, Jurassic-Cretaceous, and Eocene periods. The long stipe (stemlike structure) is surrounded by whorls of calcified gamete-bearing branches. These structures survived long enough to become fossilized. Since this organism was found in the Mediterranean, it was widely available to the early naturalists who studied it at length. *See* CHLOROPHYCEAE.

mesoderm The central tissue layer of an embryo. In vertebrates it gives rise to the tissues that form the skeleton, circulatory system, and muscles.

Mesogastropoda Large snails. This order encompasses marine, terrestrial, and freshwater snails in 17 superfamilies of at least 10,000 species. A superb example is the queen conch, a Caribbean specimen.

mesopelagic A term used to delineate the part of the water column from depths of 200 to 1,000 m (70–3,300 feet) below the surface. It is the ocean below the euphotic, or sunlit, zone and the dark depths. Until recently, it was assumed that not much happened in this part of the world's ocean, but a number of expeditions and better collecting methods have revealed a considerable number of organisms; many of them are transparent.

Bioluminescence is much more a factor in this zone than it is closer to the surface. Some fish can produce light, and it is

often used as a lure. However, the hatchet fish has photophores on its ventral surface that provide counterillumination. The underside of the fish looks like the dim light filtering down from the surface and becomes invisible to anything below it.

There is considerable daily migration from this part of the water column to the surface, where the phytoplankton photosynthesize and the carnivorous plankton of varying size eat them. The salps and siphonophores are among the most conspicuous organisms of this region of the ocean, although they have only recently gained in research interest. Since they are almost totally transparent and are destroyed by the dredges and other large collecting equipment, they were not noticed until submersibles scanned the region without attempting collection. These filter feeders trail long nets of biofilm with which they encircle their prey.

Another considerable fraction of the biomass of this part of the ocean is small crustaceans—amphipods, copepods, and euphausiids. While individually they are small, there are millions of them and therefore their biomass is staggering.

mesozoa Very simple animals that usually contain from 10 to 20 cells. These cells are usually undifferentiated. The mesozoan group is an intermediate one between unicellular and colonial organisms. There are about 50 species. The phylum was created to encompass those few organisms that did not seem to fit into the taxonomic classification of either Monera or Protista. *See* COLONIAL ANIMALS, MONERA, PROTISTA.

Mesozoic era The time in the Earth's history following the Paleozoic and followed by the Cenozoic era. The Mesozoic lasted about 130 million years (from 230 million to 100 million years ago). The name was suggested by British geologist John Phillips in 1840, to commemorate intermediate life forms. The Mesozoic is divided into the Triassic, Jurassic, and Cretaceous periods.

Landforms of the Triassic were eroded in the Jurassic, and then overlain by warm seas in the Cretaceous. The greatest botanic variations were in the development of seed-bearing plants. The gymnosperms (coniferous plants) developed first and were then followed by the angiosperms (flowering plants). The Mesozoic is the age of the vertebrates. On land, the spectacular diversification of reptilian forms led to the development of the dinosaurs. Most of these highly specialized reptiles lived in the Jurassic and Cretaceous periods.

In the sea, Paleozoic corals were replaced by modern types during the Jurassic period. *See* Cretaceous, Jurassic period, Triassic period.

Messina, Strait of A waterway separating Italy and Sicily; Homer's much feared Scylla and Charybdis. The main current runs south for six hours and then, after a short period of slack water, reverses. The north-to-south current is less strong. When, however, the current is running northward and is augmented by a sirocco, it is possible for it to be a navigational risk even today.

During the Pleistocene epoch the strait did not exist. The water level in the entire area around Sicily was lower, and Sicily was not an island but an extension of the Italian peninsula. *See* Mediterranean, sirocco, strait.

metals Elements that chemically have atomic orbitals with fewer than half the electrons required for completion. They are almost always found as compounds, and the compounds have varying solubility in water. There are traces of almost all metals in the seawater. Concentrations are higher near shore than they are in open water, where concentrations of about 0.5 micrograms per liter are average. These are sequestered by different organisms, and often in startling ways. The most prominent metals in seawater are iron, cobalt, nickel, copper, and zinc. Others are present, such as manganese, molybdenum, and vanadium in tunicates. The Antarctic copepod *Metridia* is an indicator organism. It seems to sequester heavy metals, and therefore analyzing it is a measure of the metal content of its surrounding waters. Since lead is a pollutant and a heavy metal, this organism is a useful monitoring system for monitoring waters that receive industrial residues. *See individual elements.*

metamorphic rock Igneous or sedimentary rock which on subjection to heat, pressure, or both recrystallizes or metamorphoses.

Meteor Research Expedition An oceanic exploration voyage funded by Germany, which took place from 1925 to 1927. This was one of the most significant ocean voyages of exploration, and its many traversals of the Atlantic resulted in carefully mapped bottoms of both the North and South Atlantic basins. The expedition's chemical analysis of seawater established the circulation patterns of all the oceans and the resultant effect on nutrient dispersal and in turn on plankton growth. *See* Antarctic Water; Atlantic Ocean; Ekman, Vagn Walfrid; mid-Atlantic Ridge; Wüst, George.

meteorology The study of the atmosphere and its phenomena: weather and climate. Weather and climate depend on the interactions between the atmosphere and the oceans. These interactions are extremely complicated, and meteorologists tend to consider the atmosphere and the oceans as a single system.

Such ocean currents are wind-driven. In summer the ocean absorbs most of the radiant heat from the Sun, and this heat flows back into the air in winter, particularly where warm water meets cold air and cold water. The interface between cold North Atlantic water and the warm Gulf Stream is an example. Another is the Kuroshio Current off the east coast of Asia. The addition of warm, moist air to a region of dry, cold air produces major weather instabilities, such as the Polar Fronts, which are more or less permanent features of the high latitudes.

Atmospheric pressure, or the pressure of a column of air, is about 1 kg/cm^2

(14.7 pounds/square inch) at sea level. If the air is cold and dry, that figure is correct, but if the air is warm or wet it is less dense and the pressure is lower. Therefore, on a stationary Earth, air flow would be horizontal from high-pressure areas to those of low pressure. However, the Earth's rotation produces the dominant effect on wind, and with increasing altitude, the Westerlies increase in velocity and reach a maximum at about 32° north latitude, where they are known as the jet stream.

Clouds are formed as warm air and water vapor rise from the ocean and the mixture expands because the atmospheric pressure decreases. If large quantities of water rise rapidly, they condense as ice crystals in the high troposphere forming small wispy cirrus clouds. Cumulus clouds are the result of very local, rapidly (6,000–12,000 m; 20,000 to 40,000 feet) rising warm, moist air and these large, elongated clouds can grow into thunderclouds.

Wind movement in the atmosphere is of two types: either continuous (i.e., of the jet stream type) or seasonal, as are the monsoons. Local manifestations, in addition to cloud formation, are cyclonic winds.

Horizontal movements of air are called frontal advances. If cold air displaces warm, the front is a cold front; if warm air moves a cold body of air aside, it is a warm front. In some areas the heat-flow pattern is such that fronts are stationary. The polar convergence is such a zone.

Meteorological instruments are used for the study of weather patterns and the prediction of weather phenomena. These instruments include the two oldest ones: the thermometer and barometer, used for the measurement of temperature and air pressure, respectively. The 20th century has added radar for tracking storms, infrared heat sensors, a variety of recording and photographic devices, and sophisticated methods, such as balloons, kites, rockets and satellites, for putting these devices in the air and maintaining them in place. *See* CORIOLIS EFFECT, CYCLONE, JET STREAM, MONSOON, WEATHER, WIND.

methane Possibly the basis of the community that gathers at a cold seep; the break in the Earth's crust that is similar to a hydrothermal vent but is not hot. The gas is the simplest hydrocarbon; its formula is CH_4. The compound that is normally a gas and a component of crude petroleum is present as methane hydrate, where the hydrocarbon is encased in a cage of frozen water molecules; as such it is an energy source. Whole communities of bacteria, bivalves, and worms live in and on the frozen methane hydrates that emerge from cracks in the Earth's surface along with droplets of oil. These in turn are preyed on or live in commensal relationships with other bivalves, worms, anemones, or crabs.

The quantity of methane hydrate present under the sediment layer of the ocean bottom is unknown. Methane is a greenhouse gas; it has the potential to create great change in the world's ecosystem if the water warms sufficiently and if this ice cage around the methane melts. At present, the methane is under great pressure in the ocean deeps and is considered by some to be a potential source of fuel. While the methane hydrate is usually below the sediment, in some places, notably in the Caribbean, it is exposed on the ocean bottom. There it is found as mounds, 6 to 8 feet in diameter. These mounds are covered with the biota, notably, the worms that live on and in the mounds, feeding on the methane. *See* CHEMOSYNTHETIC BACTERIA, EXTREMOPHILES, ICE WORM, VENT COMMUNITIES.

methanogenesis The generation of methane. Some is the result of human industry, but most is a product of microbial metabolism in swamps, marshes, and tundra. Methane is also generated by bacteria in caves with low oxygen levels and at crustal plate edges in areas called cold seeps. Methane is a greenhouse gas; its concentration in the atmosphere is increasing at a rate faster than that of carbon dioxide. *See* GREENHOUSE EFFECT.

Mexico, Gulf of A body of water encompassed by the eastern coast of

Mexico and the southwestern United States, with outlets to the Carribean at the Yucatan Channel and to the Atlantic Ocean through the Florida Straits. The gulf is roughly 1.6 million km^2 (620,000 square miles), and relatively shallow; the Sigsbee Abyssal Plain has an average depth of about 3,600 m (11,500 feet).

There are wide continental shelves south of Texas and Louisiana, north of the Yucatan (the Campeche Shelf), and east of Florida. The delta of the Mississippi River is built of silty sand and extends almost to the edge of the continental shelf. The Sigsbee Knolls, in the center of the Sigsbee Plain, are thought to be the tops of salt domes similar to those on the Texas-Louisiana shelf. The domes, like those on the continent, are an indication of oil deposits. The knolls are composed of foraminiferal deposits. The silty deposits of the Mississippi and other rivers emptying into the gulf are spread by bottom currents and surround the knolls but do not cover them.

The major water inlet to the gulf is through the Yucatan Channel, where the sill depth is between 1,500 and 1,900 m (4,500 to 5,700 feet). Cold South Atlantic water comes into the gulf through the passages between the Windward and Leeward Island chains. The average salinity of the gulf is high, usually above 3.5%, although areas near shore may be brackish due to heavy river outflow. The temperature on the surface varies for the most part with latitude; the water temperature is fairly constant below 1,000 m (3,300 feet).

The tides in the gulf are fairly gentle. Storm surges due to hurricanes are the exception. Tides in many places are diurnal, with only one high and one low tide each day. *See* CARIBBEAN SEA, FLORIDA CURRENT, GULF STREAM, YUCATAN CURRENT.

Mexico Trench   A cleft in the seafloor that marks the eastern edge of the Pacific crustal plate where it abuts the small Cocos Plate. This depression runs northsouth, into the Gulf of California. *See* CRUST, PACIFIC OCEAN, PLATE TECTONICS, TRENCH.

Michaelis-Menton equation   Used to estimate relative nutrient uptake rates:

$$V = N / ([N] + K_N),$$

where V is the relative rate of nutrient uptake, [N] is the concentration of the nutrient, and K_N is the half-saturation constant of the nutrient.

microbial loop   Food webs of the bacterial populations that are the detritus

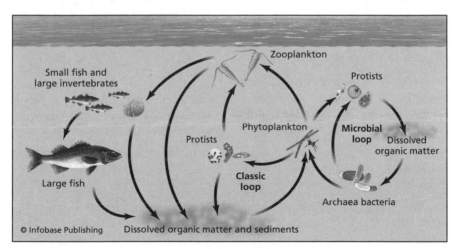

The microbial loop links the inanimate parts of any biome to the visible organisms in that biome.

feeders and other decomposers. These are the organisms that reduce all living matter to its basic inorganic components. Most of these tiny organisms were unknown and unavailable for study before the development of better microscopes with which to see them and better collection methods and instruments that did not destroy them in the collection process. The microbial loop is an integral part of the food web in nutrient-poor parts of the ocean. In regions that cannot support photosynthesis or in mid-ocean, where the organisms tend to be small, the archaea and bacteria are essential links between the dissolved organic matter and protists. *See* BACTERIA, CARBON CYCLE, FOOD WEB.

microfossil A fossil that is too small to be seen without magnification. Most, but not all, ancient dinoflagellates are microfossils.

Micronesia A group of western Pacific islands east of the Philippines; it includes the Marianas, Carolines, Marshalls, Gilberts, and some smaller islands. *See* ATOLL, PACIFIC OCEAN.

Mid-Atlantic Ridge The undersea mountains of the Atlantic Ocean first found and charted by Matthew Maury, U.S.A. *See* MAURY, MATTHEW FONTAINE.

mid-ocean ridges An interlocking series of continuous undersea mountain ranges. Together they constitute the longest series of mountains on Earth, extending over 56,000 km (35,000 miles). Each mountain range has its own name. The north-to-south ranges are the Mid-Atlantic Ridge in the Atlantic Ocean, the Reykjanes in the North Atlantic and Arctic, and the East Pacific and Mid-Indian Ocean rises in the Pacific and Indian Oceans, respectively. The Carlsberg Ridge runs southeast into the Mid-Indian Rise from the Arabian Plate. The Southwest Indian Ocean Ridge and the Southeast Indian Ocean Ridge separate the Antarctic Plate from the South American, African, and Nazca

Plates. The Southeast Indian Ridge rises between Antarctica and the Indian and Australian Plates.

Seismic sounding has shown that the thickness of the Earth's crust under a ridge is not greater than that elsewhere on the ocean floor. Thus, the current theory holds that molten rock rises under the ridge crest, is ejected as lava into the water, and then flows onto the valley floors on either side of the ridge, solidifying as it cools and gradually widening the oceans as it pushes continental blocks apart. The slopes of the ridges eventually form all of the suboceanic crust. There is constant creation of new rock at the crest of a ridge and disappearance of old rock at the edge of the slope under a continental plate.

The ridge itself is a mountain complex rising from the abyssal floor in a sharp ascent. If the Mid-Atlantic Ridge is taken as the model ridge system, a ridge bisects the ocean between continents, with the ridge crests at the midpoints. The crests are the high rift mountains; the rift valleys between them, and the plateaus from one crest to another, are scarred by deep fracture zones called slip faults. The crests drop down to a series of lower mountains and plateaus that form the flanks of the ridge.

Rift valleys, which are depressions that lie parallel to the sharp peaks of the ridge center, are the sites of shallow-focus seismic activity. The line of epicenters of such seismic activity follows the rift valleys, including the East African Rift Valley, which is terrestrial. The existence of a rift valley and its accompanying ridge was predicted near Easter Island because of the earthquake activity in the region. Other characteristics of rift valleys are high heat flow and magnetic and gravitational anomalies.

The Pacific Ridge system has a more rounded crestline than does its Atlantic counterpart. The ridge system of the Indian Ocean is extremely complex since there are intersections in it. The junctions of ridge systems are areas of great interest: for example Macquarie Island, south of the southern island of New Zealand, has literally been pushed to the surface by an abutment of crustal plates.

The *Challenger* voyage "discovered" the Earth's undersea ridges, which had been unknown in all the millenia that men had sailed on the ocean's surface. The first evidence that the sea bottom was not featureless did not come until the 1850s, when Matthew Maury recorded what he called the Dolphin Rise. (A ridge and rise system are roughly the same. The difference is one of intensity: A rise is gentler and does not have the axial rift valleys. It is the result of a greater volume of lava that spreads more rapidly.) This was the first indication of the existence of the Mid-Atlantic mountain system.

The rift valley is the dividing line between crustal plates and the scene of much geologic activity. The rock as it cools and is moved away from the very edge of crust is considered flank, and the flattened area still further from the crustal edge is ocean basin crust. This area is still relatively hot at temperatures less than 100°C in comparison to the surrounding water at 4°C and under great pressure. This relatively young, 3.5 million-year-old crust supports microbial life in the forms of bacteria and archaea. Nitrate- and sulfate-reducing organisms are present as well. *See* AFRICAN PLATE, AUSTRALIAN PLATE, EURASIAN PLATE, NORTH AMERICAN PLATE, PACIFIC PLATE, SOUTH AMERICAN PLATE; MAURY, MATTHEW FONTAINE; MID-ATLANTIC RIDGE; OCEANIC CRUST; PLATE TECTONICS.

migration A term used to describe regular animal journeys along well-defined routes, particularly those involving a return to breeding grounds.

There are three types of migrant fish: (1) *Oceanodromous fish* (always in the sea), such as the herring. Herring all belong to one species, but there are non-mixing local populations that live in one area and spawn in another and at different times. Thus there are several groups of herring living in the North Sea. The cod is also an oceanodromous migrant. It

travels, as does the tuna, greater distances than does the herring. (2) *Anadromous* fish (which spend most of their lifespan in the sea). These include the salmon that return to their native rivers to spawn and, in some cases, to die. (3) *Catadromous* fish are essentially freshwater species that spawn in the sea, such as the eels.

Other animals also migrate. Sea turtles spend their entire lives at sea, usually in deep water. Some have been tagged and were found to travel over 2,000 km (1,400 miles) to their native beaches to lay eggs. The most familiar migrants are birds. Coastal birds stay on the continental shelves, pelagic ones have small nesting areas and great flying ranges. Some petrels fly almost from pole to pole. The prime example is the Arctic tern. Its breeding areas are northeastern Europe, Asia, or North America but it lives in Antarctica.

Humpback whales are good examples of migrating mammals. They winter in warmer water and move toward the poles for the greater food supply that follows the planktonic bloom. Not all individuals move in such a migration, and not all have specific breeding areas. Seals and walruses do, however. The Pribilof Islands are vast seal nurseries.

The greatest migratory movement in oceans is that of the plankton. It has long been known that euphausiids, plankton that depend on visual cues, move to lower depths in daylight and that plankton with less-developed eyes, such as copepods and chaetognaths, do not. This was called diurnal vertical migration, but that name has been changed; it is now diel vertical migration, since it occurs over the day-night cycle, not just during the day. Euphausiids and fish move up at night and down before sunrise. Some of the distances are large in terms of the size of the traveling organism: *Metridia*, a tiny swimming copepod, will travel about 350 m (1,000 feet) up and down every day.

Some polychaetes move from inshore rocks to the open sea to reproduce. The palolo worm rises vertically from coral reefs to the surface to cast off its tail segments, which contain the eggs or sperm. Crabs mate close to shore and lay eggs in brackish water.

A great deal of research has been devoted to unraveling the question of how animals know when to migrate and what clues lead them back to some communal or ancestral spot. There is evidence that the development of gonadal tissue is stimulated by changes in light intensity, and thus that migratory animals respond to a change in season before they respond to a weather change. They do use the Sun's position to navigate. There are theories of the use of "radar" and the magnetic field of the Earth by flying flocks of birds.

Many organisms that depend on the Earth's magnetic field in their migrations will have difficulties in the future. The magnetism of the Earth is weakening at present, and this may indicate polar reversal. This has happened in the distant past, and its effect on migration is unclear. *See* ANADROMOUS, CATADROMOUS, EEL, HERRING, PLANKTON POLYCHAETES, SALMON.

Milne Seamount A submarine mountain of volcanic origin in the North Atlantic, to the west of the Mid-Atlantic Ridge. It is east of the mid-ocean canyon that separates the ridge from the Grand Banks off the coast of Newfoundland. *See* CANYON, GRAND BANKS, SEAMOUNT.

minerals Naturally occurring chemical elements or compounds, usually solid and crystalline, obtained from the ground or from the sea. The sea has been used as a source of minerals for centuries: Ancient coastal civilizations made salt. The winnowing of sodium chloride from the sea has largely been abandoned, but some other alkali salts are now obtained by evaporation of seawater. The commercial production of different minerals from the sea is recent. The sea is divided into regions, each of which yields different products that must be extracted using different methods.

Seawater contains almost every element; the criterion in deciding which to extract is cost. For sodium, magnesium, and bromine, the removal from seawater is cost-effective.

Beaches yield building materials, such as the sand used for concrete and limestone (calcium carbonate). A variety of useful minerals is found either on shore or just off it. Diamonds are mined near the coast of Southwest Africa, for example; iron ore off the Japanese coast; and gold near Nome, Alaska. The drowned river valleys of Indonesia and Malaysia are a source of tin. The continental shelves are mined for sulfur and explored for petroleum. Some of these probes have yielded good results, notably in the Gulf of Mexico and the North Sea.

The sea bottom is harder to get to commercially and will therefore be the last region to be exploited for minerals. However, the research vessel *Glomar Challenger* is part of a project to develop deep-sea petroleum on a commercial scale. Also on the bottom are manganese nodules, which contain copper, nickel, and cobalt in commercially attractive quantities. Since there are large reserves of the nodules and new ones accumulate, they are a renewable resource.

The red clay of the ocean bottom contains some rare metals in amounts of less than 1%. This is enough to make such a find exciting, but recovery would be expensive and awaits a cost-effective technique for bringing the clay from the bottom to the surface.

The methods of retrieving minerals from the ocean bottom vary with the mineral in question and the depth at which it is found. Oil is pumped up onto offshore platforms by the same mechanism used to pump it out of the ground. The platforms are large and heavy enough to supply support for the pumping machinery. They are most often fixed in place, but they need not be. Sulfur is also obtained by pumping. For removing tin ores, oceangoing dredges and draglines are used.

The effort put into a sea-mining installation is a function of the value of the material retrieved. For rare metals, more costly methods, such as the construction of a tunnel from nearby land or the use of submersibles or work crews that spend long periods (a week or more) underwater might be considered. In Antarctica there exist mineral deposits of major international importance; however, the difficulties of a sustained effort in that very hostile environment require a multinational project. Such a project is currently under discussion. *See* DIVING, MANGANESE NODULES, SEAWATER.

Miocene epoch

The fourth epoch of the Tertiary period of the Cenozoic era. The other epochs of this era have been the Paleocene, Eocene, Oligocene, and Pliocene—the last being the last epoch of the Tertiary period. The name Miocene was suggested by the geologist Charles Lyell in 1833. He meant to imply a smaller number and diversity of fossils than had been dated from the Pliocene. The epoch lasted more than 10 million years, during which the climate was generally mild, with warm to temperate conditions experienced almost as far into the polar regions as the poles themselves. There was considerable volcanic activity at this time, and much mountain building. The Caucasus, Himalayas, Apennines, and Pacific Coast Range all grew during this time, and during the Miocene the European Alps became folded and deformed. The ocean overlay considerable parts of the globe, and when it receded, left great banks of sediment thousands of meters thick.

On land the Miocene was the time of the dawn redwoods. The great forests receded and left large grassland areas, the savannas. The horse and the camel, typical grassland animals, evolved in this time, while the oreodonts—carnivorous doglike mammals—disappeared.

To a great extent, the marine fossils of the Miocene resemble the present marine biota. There was a luxuriant expansion of gastropod families in this epoch. Oysters and sea urchins were also abundant and distributed worldwide. The seagoing mammals, sea cows, dugongs, and whales adapted further to the marine environment. *See* OLIGOCENE EPOCH, PLIOCENE EPOCH, TERTIARY PERIOD.

Mississippi River

A 4,120 km (2,470-mile-)-long river flowing south through the United States from northern Minnesota to the Gulf of Mexico. The Mississippi and its largest tributary, the Missouri, form the largest river system in North America, in terms of both length and volume. While the annual output of the Mississippi-Missouri river system varies enormously from one year to the next, it is the great river system of the continent, draining about 12% of North America.

The source of the Mississippi is Lake Itasca in northern Minnesota. The lake is at an elevation of over 440 m (1,463 feet) in a region glaciated during the Pleistocene. The river moves over a series of falls in the mountain ranges between its source and the Minneapolis-St. Paul region. It is navigable from that point to the Gulf of Mexico.

The upper Mississippi is fed by Minnesotan snowfall, but south of its confluence with the Ohio, which drains the well-watered northeast, over 50% of the volume of water in the combined river is of eastern origin. The Missouri River component is less than 20% of the total, since the sources of the Missouri's water are the Rocky Mountains and the more arid prairie.

South of the junction of the Ohio with the Mississippi, the river is about 1,360 m (4,500 feet) wide. It meanders considerably, changing its channel as it moves over the floodplain. The floodplain is the part of the river valley which can flood in periods of high river volume, notably in spring. A floodplain can be miles wide. It has changed its course many times in the past, and much of the center of North America bears scars and marks of the ancient river's lateral movement.

The junction of the Mississippi with the Missouri near St. Louis is roughly 1,000 km (600 mi) from the sea, but with its meandering the river's true path is almost three times that length.

The Mississippi Delta begins in Louisiana at the junction with the Red River. The lowest reaches of the river break up into a network of small, shifting channels called bayous. *See* DELTA.

mixotrophe An organism that can be either a consumer or a producer. For example, a lithotrophic bacterium is an organism that derives energy from a particular mineral or from organic compounds depending on which is available. *See* AUTOTROPHE, CHEMOSYNTHETIC BACTERIA, HETEROTROPHE.

mizzen The mast behind the main mast on a sailing vessel. The term also refers to the sails carried on the mizzenmast. *See* SAIL.

MODE Mid-Ocean Dynamics Experiment, a study of short-term, small-scale weather phenomena in the North Atlantic. The United States and Great Britain cooperated on this venture in 1973. The objective was to identify oceanic-atmospheric interactions so as to establish what were called "weather" patterns in both air and water. This makes the analogy between the two and equates oceanic eddies with cyclonic storms, such as hurricanes. *See* OCEAN-ATMOSPHERE INTERACTIONS, STORMS, WEATHER.

Moho layer An abbreviation for Mohorovičić discontinuity, a region within the Earth in which the density changes suddenly because there is a change in the composition of rocks from those of the Earth's crust (granites and basalts) to those of the mantle below (olivines and pyroxenes). The crust of the planet is from 4 to 60 km (2.5 to 40 miles) thick, varying from one location to another but thinnest beneath the sea. The discontinuity was named for its discoverer, Andrija Mohorovičić. *See* DISCONTINUITY LAYERS, MANTLE, PLATE TECTONICS.

Mohorovičić, Andrija (1857–1936) A Croatian geologist and seismologist who studied the 1909 earthquake in the Balkans. He was trying to establish a method for detecting the epicenters of earthquakes by measuring the seismic waves they generated. He correctly surmised that these waves were generated in the uppermost layers of the Earth. There are two sets of these waves, P and S, and they obviously bounced back toward the surface; from this he discovered that there was a boundary layer deep in the Earth that bounced these waves back to the surface. This return wave is then used as a locator for the quake's epicenter. Much of this early work was later incorporated into seismic studies—a field that includes the study and prediction of tsunamis. *See* EARTHQUAKE, TSUNAMI.

molecule The smallest unit of a chemical compound that retains the characteristic properties of that compound. A molecule is a group of atoms held together by strong chemical bonds. Molecules vary in size and composition from the very small hydrogen (H_2) molecule, composed of two atoms of hydrogen, to proteins and starches, which have thousands of atoms in a single molecule. *See* ATOM, PROTEIN, SALT.

Mollusca A very large phylum of invertebrates. Next to the insects, this group has the greatest number of living species and individuals. They inhabit both terrestrial and aquatic ecosystems. The individual members of the phylum are usually encased in calcareous shells produced by their body covering, the mantle. The mantle is the distinguishing feature of the mollusks. The marine mollusks are found on rocky or sandy bottoms, from dry land to the littoral region to deep sea.

The taxonomic classification of the mollusks is being revised. There is agreement that these animals have a long geologic history. Their fossil record dates from the early Cambrian. The mollusks include the well-known groups of single-shelled animals (snails), the two-shelled (oysters, clams, scallops), the plated animals (chitons), and unshelled animals (the squid and octopus).

Most mollusks have bilateral symmetry: The right side is a mirror image of the left. In the gastropods (snails), however, one side is highly developed as the foot, the large muscle that the animal uses for locomotion. Many mollusks move slowly. Some, like the snails, slide along on a mucus layer; cephalopods, the swiftest group, move by jet propulsion; scallops flap their shells in a half jet propelling, half hopping movement; oysters are sessile (stationary) as adults, as are mussels; clams burrow.

The internal structure of the mollusks is as varied as their exterior. Basically, all have a combination head-foot organ that consists mainly of the foot, and mantle-covered viscera (internal organs). They have a mouth, salivary gland, esophagus, and a stomach that digests their food, which among the shelled members of the group usually consists of detritus. Snails and their relatives have a specialized mouthpart, the rasping radula. The mollusks respire through paired, ciliated gills, which are also in the mantle cavity (the space within the mantle). The rate of respiration is controlled by sensory organs, also part of the mantle, which regulate the incurrent and excurrent siphon, the means by which the animal brings in water and expels the products of digestion, carbon dioxide, and nitrogenous waste in the form of ammonia or urea. The molluscan heart lies dorsal to the viscera. One of its prime functions is to maintain turgor (muscular rigidity) in the foot. The blood, which has some red pigments, like the blood of the annelids, also controls the tentacles, siphons, and other appendages. There is a nerve net in all the members of the phylum, with that of the cephalopods being quite complex and specialized.

Reproduction among the mollusks is quite varied. Chitons, some snails, scaphopods, most bivalves, and the cephalopods have separate sexes (dimorphism). Most gastropods are hermaphrodites, but fertilization usually requires two individuals, although each animal has male and female gonads. In many organisms, particularly the bivalves, fertilization is external. Most molluscan larvae are free-swimming.

The bivalve mollusks are commercially fished, and have been part of the human diet for thousands of years. They have also been raised commercially by mariculture. Some gastropods (e.g., the abalone) are eaten, as are the cephalopods. Oyster drills and teredos are pests. Some members of the phylum Mollusca are highly toxic, such as the cone shells; others prey on each other. Many are hosts to parasites, such as the foraminiferans that cause red tides, and most support large predator populations. See ABALONE, CHITON, CEPHALOPODA, CLAM, OYSTER, PEARL, SCALLOP, SEA HARE, SEPIA, SHELL, SNAIL.

mollusk An animal belonging to the phylum Mollusca. All members of the phylum have soft bodies that are covered by a mantle, a thick membrane. In many mollusks the mantle secretes a calcareous shell. Clams, snails, and scallops are all examples of shelled mollusks. Octopi and squid are mollusks that have mantles but no shells. See CEPHALOPODA, CLAM, GASTROPOD, MANTLE, MOLLUSCA, OCTOPUS.

molt Also called ecdysis; the shedding of an exterior. On land the familiar animals that molt are birds and reptiles. In a marine context, the significant animals that molt are the arthropods. In order to grow larger, an arthropod splits its chitinous (hard exterior) shell at its fuse lines and walks away from it. A new shell is secreted to replace the older, smaller one. The soft-shelled crab is merely the ordinary blue crab caught soon after molting. The hermit crab does not molt: It moves out of a smaller shell and looks for a larger one. See ARTHROPODA, CHITIN, CRAB.

molybdenum Present in phosphorite rocks along with manganese. The ratio of the two metals is in the range of 0.004 to 4.5. The content of the rocks ranges from 2 to 128 (Mo) to 12 to 195 (Mn).

Phosphorites formed in reducing atmospheres (low or no oxygen) contain more molybdenum than manganese. Higher molybdenum content in ferromanganese crusts is a result of both metals precipitating out of the seawater together. Phosphorites that did not weather either on land or in oceanic shelves show a higher Mo to Mn ratio. *See* MANGANESE.

Monera A kingdom of very simple organisms that lack distinct cell nuclei. Bacteria and cyanobacteria are examples. The genetic material of these organisms is interspersed with the rest of the protoplasm of their cells, rather than contained in a nucleus.

Monoplacophora Very ancient animals belonging to the phylum Mollusca. The monoplacophorans are "living fossils," along with the horseshoe crabs and coelacanths. They have been found only in the Pacific, in warm temperate-to-tropical areas of Central and South America. They were differentiated from Paleozoic limpets in the 1940s but were known only as fossils until recently. A live specimen of the species *Neopilnia* was found in mud off the Pacific coast of Central America at a depth of about 2,900 m (11,700 feet).

These single-shelled organisms look like limpets. Their relatively flat shell is about 4 cm (1.6 inches) in diameter and covers a very un-molluscan body. The digestive, excretory, and muscular systems of the animal do not resemble those of the limpet.

The diet of the monoplacophorans may be detritus. *See* FOSSIL, LIMPET, MOLLUSK.

monsoon A wind that changes direction with the season, most often associated

Typical monsoon weather conditions

with Africa and South Asia. Winter or dry monsoons tend to blow from northeast to southwest. In summer the wind direction is reversed, and the monsoon brings rain.

Monsoons have their source in the shapes of land and water masses, the general circulation of winds about the Earth, and the different heat-retaining properties of land and water.

The Indian monsoon produces warmth and rain as it moves north until July, after which the entire subcontinent is hot and wet. By September or October, colder air moving from Tibet reverses this pattern, and the prevailing weather is hot and dry. Since larger land areas produce greater monsoonal effects, the monsoon in Indonesia is not as obvious as the Indian monsoon in terms of temperature change and rainfall accumulation, but the seasonal change in wind direction is marked.

While the basic monsoon pattern has not changed, in the last several years it has become more dramatic. Monsoons can and have in the past brought incredible winds and precipitation, some areas receiving more than 400 inches of rain in a season. Recently, the combination of El Niño events and global warming has resulted in hotter winds blowing toward India from Australia, resulting in several higher-than-normal precipitation years. These higher levels are becoming the norm. Unfortunately, Australia is becoming hotter and drier.

In Africa the Hamattan is a hot, dry wind from the Sahara. It moves from northeast to southwest to the Intertropical Convergence, where it meets hot, moist air from the Bight of Benin. This line of convergence moves from the equator in winter to about 15° North Latitude in July.

In the Americas, monsoonal developments in the Gulf of Mexico are weak and barely distinguishable from normal seasonal variations. *See* WIND.

Moon The Earth's sole natural satellite. It has a diameter of 3,476 km (2,160 miles), and its mean distance from the Earth is 384,000 km (238,857 miles).

The Moon rotates on its axis in 27 1/3 days, and revolves about the Earth in 29 1/2 days. It shines by reflection of the Sun's light, and during a complete revolution about the Earth, four phases are observed: the new Moon, first quarter (half Moon), full Moon, and last quarter (half Moon). The Moon exerts an attractive force on the Earth. This gravitational force is most obvious in its effect on that part of the Earth's surface that moves easily, the water. Since the Earth is not perfectly spherical, the gravitational force exerted on different parts of Earth varies and gives rise to different tide heights in different places.

There have been several theories of the genesis of the Moon; one that was reasonably popular around the turn of the century was that the Moon was "ripped out" of the Earth's crust and that the Pacific Basin is the result of the Moon's departure. Currently it is believed that the Earth and the Moon formed simultaneously. *See* LUNAR DAY, LUNAR TIDE, TIDE.

mosasaur A Cretaceous aquatic lizard of worldwide distribution. Mosasaurs were contemporary with the ichthyosaurs and plesiosaurs. While most were about the size of the modern porpoise, some were much larger, at 9 m (30 feet) long. Mosasaurs had paddlelike limbs and elongated heads and bodies. Their jaw, like that of their modern relatives the monitor lizards, had a hinged mandible (lower jaw). This enables the modern lizard to consume large prey quickly and presumably functioned in the same manner for the ancient lizard. The mosasaur preyed on fish and cuttlefish. *See* CRETACEOUS PERIOD, ICHTHYOSAUR, PLESIOSAUR.

Mozambique Current The part of the Indian Ocean gyre that flows between Mozambique on the African mainland and the Malagasy Republic (Madagascar). It is a south-flowing stream along the southern coast of Africa, and most of it eventually doubles back and flows north along Madagascar. It feeds into the Agulhas Current.

See AGULHAS CURRENT, CURRENTS, GYRE, INDIAN OCEAN.

mucus *See* BIOFILM.

mud Sediment, consisting either of rock fragments or organic detritus, in which the particles are smaller than 0.06 millimeters. *See* MUD FLAT.

mud flat A stretch of level ground, behind the backshore, that on its sea side merges gradually into the slope of the beach. The characteristic plants of mud flats are eel grass (*Zostera*) and turtle grass (*Thalassia*). The best-known animals of the region are the fiddler crabs (*Uca*). *See* BACKSHORE, BEACH, ESTUARY, FIDDLER CRAB, ZOSTERA.

mullet A worldwide, abundant food fish of the family Mugilidae; about 100 species are known. Their usual habitat is tropical to temperate waters. Mullet can tolerate low salt conditions well and are frequently found in brackish water. The usual adult fish is a silvery, large-scaled individual about 30 to 90 cm (1 to 3 feet) long.

Mullets are omnivores and do well on a largely plant diet. The common species, *Mugil cephalus,* has been successfully cultivated. *See* FISH.

Murchison, Roderick Impey (1792–1871) A Scottish geologist who defined the earlier periods of the Paleozoic era (395 million to 570 million years ago). He joined the Geological Society of London in 1825 and then explored extensively in France and in the Alps; first with Adam Sedgwick and then with Charles Lyell.

Murchison's major work centered on the Silurian strata. He published this in 1835 and followed it with a long investigation of Old Red Sandstone of England.

Profound disagreement with Sedgwick over their joint work in Devon led Murchison to a long controversy with his former partner in the assignment of dates to the Early Silurian and Late Cambrian systems. The disputed layer was named Ordovician by Charles Lapworth well after the deaths of the disputants.

Murchison was commissioned to form a party with the prominent geologists, Philippe de Verneuil and Alexander Keyserling to survey European Russia. This group explored and named the Permian deposits. Murchison returned to Britain, where he served as director of the Geological Survey of the British Isles from 1855 until his death. *See* CAMBRIAN PERIOD; DEVONIAN PERIOD; LYELL, CHARLES; SEDGWICK, ADAM; SILURIAN PERIOD.

murein Murein is a peptidoglycan—the lattice formed by repeating disaccharides (two sugar molecules linked together) that are interconnected by polypeptides (more than 10 amino acid units linked together). This is the structure that makes up the cell wall of bacteria.

murex Any of a group of widely distributed, tropical-to-temperate marine snails of the family Muricidae that live in rocky, shallow water. They have a characteristic attractive, heavy, spined, or frilled shell. The snails feed on other mollusks by boring a hole in the shell of these prey and sucking out the contents. *Urosalpinx cinerea,* known as the oyster drill, is a murex species of commercial importance since it destroys a valuable cash crop.

Many murex species produce a yellow fluid that turns to red or purple in sunlight. *Murex branoris* is a Mediterranean species that is the source of Tyrian purple—the royal purple of the ancient European and Near-Eastern world. *M. pecten* is the Venus comb, an Indian Ocean shell much prized by collectors. *See* GASTROPODA.

Murray, John (1841–1914) A Scottish oceanographer born in Canada. He went to Scotland at the age of 17 to study medicine at the University of Edinburgh but left without a degree in 1872. By that time he had already been on one major oceanographic collecting voyage and was therefore a reasonable choice as an assistant to Charles Wyville Thomson, professor of natural history at the University of Edinburgh and organizer of the HMS *Challenger* expedition sponsored by the Royal Society. Murray spent three and one-half years on board this ocean-going research vessel.

He carefully established that the already-known single-celled protozoans of the genus *Globigerina* were ocean surface dwellers whose skeletons eventually sank. To do this, Murray had extremely fine nets constructed that were rigged to lie on and just below the surface and that could be towed without fouling. He collected and named radiolarians, diatoms, and pteropods. His work established the terrestrial origin of the red clay of the ocean bottom, and the amazing slowness of its sedimentation rate.

By the time Thomson died in 1882, Murray had spent years sorting the hundreds of HMS *Challenger* specimens and was named editor of the forthcoming report of the voyage. He also wrote the summary of this report and the section on deep-sea deposits with Alphonse Renard, Belgian geologist. Murray's deduction that the antarctic ice had a continent beneath it reinvigorated exploration of the South Pole.

In addition to the HMS *Challenger* report—a massive, multivolume work—Murray and Hjort produced in 1912 a famous textbook on oceanography, *The Depths of the Ocean,* which became a standard in the field. *See* HMS CHAL-LENGER; THOMSON, CHARLES WYVILLE.

mussel A bivalve mollusk that is well represented in temperate to tropical waters worldwide. *Mytilus edulis* is the edible mussel; it is a blue-black, slender bivalve, 5 to 15 cm (2 to 6 inches) long, that attaches itself to pilings or rocks by byssal threads. This is the species that is widely cultivated. The California species (*M. californianus*) is brown. An Atlantic species, the horse mussel, is *Modiolus capex.* Zebra mussels were brought to American waters accidentally. They are a pest species that is invasive.

Strands of mussels can be seen in the water of intertidal zones. Unlike the oyster, the mussel is not permanently fixed in place. Mussels are raised commercially. Besides being consumed by humans, they are also eaten by birds, such as gulls and ducks, and by starfish and flounder. *See* BYSSAL THREADS, MANTLE, MARICULTURE, MOLLUSCA.

Mydocopa A subclass of Ostracods. These crustaceans are usually small (0.9 to 32 mm or 0.04 to 1.5 in.). They are worldwide in distribution and occur at all ocean depths. Many species make vertical daily migrations to feed, and almost all can swim. Depending on their feeding habits, they have differentiated mouthparts. Some are filter feeders, others are scavengers and detritus feeders, and the rest are predators. *See* CRUSTACEA, MIGRATION.

Mysiidacea An order of tiny, shrimp-like, free-swimming crustaceans. The average adult is about 18 to 20 mm (0.25 to 0.80 inches). These animals are found in temperate waters and range from the intertidal zones to hadal regions. The deep-dwellers are usually larger. There are more than 600 species of *Mysiidacea.* Some form commensal relationships with spongers and/or sea anemones. *See* BENTHOS, CRUSTACEA.

Myzostomaria Flattened, almost disk-shaped annelids. They are tiny and are parasites on echinoderms, particularly crinoids. Fossil sea lilies bear the scars of these exterior parasites. These organisms date from the Ordovician. They had been classified by some as annelids but are now assigned to a separate phylum of their own. This phylum resembles that of the flatworms. It is possible that at some time in the distant past they had developed and then lost some features that are characteristic of particular phyla such as coeloms and body segments. This theory complicates the possible evolutionary links between organisms and makes relationships difficult to determine. *See* EVOLUTION.

Nansen, Fridtjof (1861–1930) A Norwegian oceanographer, polar explorer, and humanitarian. He was trained as a zoologist and was a professor of zoology at the University of Christiania (now Oslo), Norway. Nansen's interest in oceanography as a separate science led to his directorship in 1901 of ICES—the International Council for the Exploration of the Seas. In 1908 he became professor of oceanography at the renamed University of Oslo.

Prior to his directorship of ICES, Nansen had undertaken extensive explorations in the arctic. He tried to reach the North Pole, and in 1895 achieved 86° 14' North Latitude, the farthest north anyone had previously reached. His work in the arctic included charting parts of the Kara Sea and doing careful sampling work while aboard the *Fram,* a vessel built to withstand ice pressure—an ever-present danger in Arctic exploration. After observations of the movement of pack ice from the unanchored ship, Nansen realized the relationship of wind to current but could not adequately explain this relationship without mathematical expression. He suggested the problem to his colleague Ekman, who developed equations to explain what is known as the Ekman spiral—current moving at a 45° angle to the wind.

Nansen was also known for his humanitarian efforts on behalf of World War I refugees, and in 1922 he was awarded the Nobel Peace Prize. *See* EKMAN, VAGN WALFIRD; *FRAM*; KARA SEA; NANSEN BOTTLE.

Nansen bottle A device designed by Fridtjof Nansen and Otto Pettersson for collecting deepwater samples. It contains a series of concentric insulated rings with a thermometer mounted in the assembly. The bottle sinks, mouth down and closed, and is then tripped open at a particular depth. It thus collects water at a given depth, then closes to isolate the sample, and maintains the sample at the temperature it had when collected. The thermometer gives a reading of that temperature. This device has undergone much redesign in the years since its introduction. *See* NANSEN, FRIDTJOF; OCEANOGRAPHY.

Nares, George Strong (1829–1915) A British admiral and surveyor, and the ship's captain for most of the voyage of the HMS *Challenger* from 1872 to 1876. Nares had gone to sea as a boy, and served on many scientific and technological voyages. He was in the Arctic from 1852 to 1854, then in the waters north of Australia, surveying that continent's coast. Nares was then assigned to the Mediterranean, where he did surveying that was the first step to laying telegraph cables.

Prior to his service on the *Challenger,* Nares was involved in the project of charting the currents off Gibraltar. He left the *Challenger* in 1875 for an Arctic expedition (1875–76). His son, John Dodd Nares (1878–1957), was also an admiral and a hydrographer. *See* CHALLENGER, HMS

narwhal An arctic porpoise related to the beluga whale. The narwhal is a large mammal, about 5 m (16.5 feet) long at maturity. The young narwhal is usually quite dark—almost black—but turns gray with a light underside as it grows. The adults are almost white.

The narwhal's incredible characteristic is its visible tusk: Males exhibit a long, spiral-grooved left tusk that may be 2.5 m (8 feet) long, while their other tusk remains buried in the gum; females have both tusks in the gum. The tusks are probably sexual attributes and are sometimes broken or scarred in fights between rivals. The tusks are thought by some to have originated the unicorn legends, or at least to have contributed to them. Narwhals live in herds and feed on fish, usually cod and halibut, and crustacea. *See* MAMMAL.

nauplius The free-swimming or drifting larval stage of an organism that as an adult is sessile or lives within another animal, either as a parasite or commensal. The nauplid juvenile may look very different from the adult. *See* PLANKTON.

Nautical Almanac This is the name of several publications, all of which are compilations of data such as tide tables for particular areas, shoals, movement of seacoasts, regular weather phenomena, ocean currents, and the like. See the official British version at http://www.nao.rl.ac.uk or the American one at http://aa.usno.navy.mil/publications/docs/almanacs.html.

nautilus A cephalopod mollusk most frequently found in the South Pacific and the Indian Ocean. The true nautilus, also called the pearly or chambered nautilus, is differentiated from the paper nautilus (Argonauta), which is more octopoid. The pearly nautilus is the only genus of the Nautiloid order present in today's oceans. Paleontologically, the true nautiluses are an important group that once had many genera and species, now largely extinct; however, the animals alive today are not living fossils but rather are different from the individuals that lived in ancient seas.

Present-day adult nautiluses have coiled shells that are about 25 cm (10 inches) in diameter. The shell is usually brown with white bands, and is composed of 35 to 40 compartments. The animal always lives in the largest, last-constructed compartment and uses the empty ones as a flotation device by controlling gas flow

within the shell. The nautilus is striking in appearance: it has a lensless eye, can move fairly quickly, and has upward of 90 tentacles. It lives in relatively deep waters, 100 to 400 m (330–1,300 feet) below sea level, on slopes of fringing or on barrier reefs. It migrates up at night and feeds on carrion, to which it is probably attracted by smell. Until fairly recently there were few living nautilus specimens known, and therefore much of the lifestyle of the nautilus was conjectural. *See* CEPHALOPODA, MOLLUSCA, PAPER NAUTILUS.

Nautilus The name of several famous ships. The first was a reasonably functional submarine designed by Robert Fulton, who is remembered as the designer of the first steam-powered ferry. The fictional Captain Nemo's submarine was named *Nautilus,* as was the first U.S. nuclear-powered submarine. *See* FULTON, ROBERT.

navigation The application of observation to the problem of plotting a course for a ship or airplane moving from one place to another. Bearings are taken with respect to the position of the true North Pole (from the point to the Pole), or to the magnetic North Pole, which is the place on Earth that a compass points to.

While the properties of magnetic iron were known in antiquity, the relationship between iron and the Earth as a magnetic object was unknown. Compasses were introduced possibly as early as the 12th century. Without them, ocean voyages were very adventuresome and much more dangerous than they are today.

Early sailors did their navigation by simply moving from one known point on the shore to another. The difference between such navigating and piloting is almost nonexistent. The only aids available to the early coastal sailors were sounding rods or lead lines to determine depths, and windvanes.

Travel across the open ocean was much more difficult: A sailor would fix on an established point to determine his position and then take aim on his destination by dead reckoning (i.e., taking the straight line or course from the starting point to the destination). Since the surface of the Earth is curved, the straight line may be a great circle, and without any fixed object to sight on, holding a course was very difficult.

Long voyages that took ships out of the sight of all land were nevertheless undertaken. The Phoenicians were very experienced in navigating the Mediterranean, and, late in their history, they also moved out of sight of land to Cornwall, England, and its offshore islands. The Norsemen and the Polynesians—at roughly the same time—also moved out of sight of land. All three seafaring cultures knew and used celestial navigation.

Celestial navigation, which was a great advance for scientific navigation in general, involves the position of a ship relative to the position of the Sun, or the Moon, or a star on the celestial sphere—an imaginary, hollow construction surrounding the Earth and, like the Earth, making one rotation about its axis every day.

The rapid development of navigational aids and instruments in the 15th, 16th, and 17th centuries displaced the astrolabe with a succession of other instruments such as the backstaff, the quadrant and its variant the octant, and the sextant. With these instruments a good estimation of latitude was possible. The user of a sextant measures the altitude of a celestial body from the horizon as the reference point. The Sun (or Moon, or some particular star such as the North Star, Polaris) appears at a point some degrees above the horizon at a particular time of a particular day. Tables have been constructed which, when compared with the data read from the sextant, translate these observations into the number of degrees that some position is above the horizon and north (or south) of the equator. At the equator, each of the 360 degrees of longitude of the circumference of the Earth corresponds to about 110 km (1° = 110 km) or 70 statute miles. As the meridians of longitude converge with increasing latitude, the degrees of longitude (and the time zones) shrink. Good clocks and exten-

sive tables for the correction of instrument readings have resulted in the accurate location of ship positions with respect to the vessels' longitude.

The use of clocks to calculate longitude from a prime or first meridian was recommended in 1530, but it took another 250 years before the Harrison chronometer finally made it possible to measure longitude in this simple manner. Trigonometric tables (epherimids) are needed to solve these spherical trigonometric problems, and the first were published by the decree of King Alfonso X of Castile in 1252. The standard text on spherical trigonometry, a reworking of Greek mathematics by John Holywood (Johannes Sacrobosco), a famous 13th-century cleric and mathematician, appeared in Europe at roughly the same time as the tables.

The idea of a prime or 0° meridian of longitude predated Ptolemy, but became internationalized only in the 17th century. An observatory was created at Greenwich, England by Charles II in 1674, and John Flamsteed, the royal astronomer, was assigned the task of updating and revising the lunar tables, which were finally published as the *Nautical Almanac* in 1765.

The 20th century was another period of rapid technological development in the field of navigational instruments. The 20th century also brought radio and several other techniques that are based on radio waves. Radio has made dead reckoning much simpler on the sea, and it is a vital tool for the flight navigator. Radio waves, long wavelength radiation bands, are used in hyperbolic navigation, a technique based on the observation that radio waves move around the Earth's surface in great circles. When they intersect, they indicate position. Radar can be used as a distance measure in addition to its locating capability. Loran is a radio-based technique that operates by triangulation (i.e., establishing a position relative to that of two known signal broadcasting points), and it is used for both surface navigation and flight.

Once a position on the surface of the Earth is determined in degrees of latitude populations in that region. *See* COPEPOD, FOOD WEB.

Neogastropoda These snails are one order of Prosobranchia. Like the Mesogastropoda, they are asymmetric (they only have one gill, one auricle, and one kidney). Part of their mantle is photosensitive. Some members of this order produce a toxic substance that allows them to paralyze and engulf prey. Many of these animals are sought after by shell collectors. *See* MESOGASTROPODA, MOLLUSCA, WHELK.

Neridae A family of worms of the class Polychaeta, of the phylum Annelida. The most common genus is *Nereis*. These worms are usually bottom dwellers that live on other worms and zooplankton. They range in size from about 2 to 90 cm (0.75 to 33 inches) and are found in the temperate waters of the North Atlantic or the Pacific Ocean. They are usually brightly colored; may be red, brown, or green; and are known for their strong jaws. Their respiratory systems are developed; they have gills. Like many other polychaetes, the majority of species of Neridae are dioecious, with distinct male or female individuals; some other species are hermaphroditic. *See* HERMAPHRODISM, POLYCHAETA.

neritic distribution A function of the available food supply. The neritic region of the ocean is the part of it that is near shore but not coastal. The waters over the continental shelves are a perfect example of neritic regions. These areas are close enough to the coast to have ample food supply and support much of the biomass of the ocean. The biomass of the neritic region is about 40 times that of the oceanic region, which is about 10 times greater in area. The base of this giant food web is phytoplankton. *See* BENTHOS, CONTINENTAL SHELF, FOOD WEB, PHYTOPLANKTON.

nervous system This is an essential part of almost every animal (and there are opinions about the capability of plants as well). A living organism is considered to have a nervous system if it has specialized cells, called neurons (nerve cells), that produce chemical triggers that electrically affect synapses. Synapses are the connections to other cells where some action occurs. Given this very broad definition of a nervous system, all animal phyla have one. In general, invertebrates have a few large nerve cells, and vertebrates have smaller, more numerous neurons.

Sponges, the epitome of sessile organisms, can move parts of the body and can regulate the size of their opercula (pores) to change the rate of water flow. Coelenterates have real nerve cells dispersed about the body in a "nerve net." This net connects neurons which may be bipolar or tripolar to each other and to muscles. The multiple polarity means that the neuron may carry impulses in more than one direction. Some medusal forms have photoreceptors and/or statocysts for detecting change in water pressure at the periphery of the bell. The nerve net communicating with musculature may regulate the duration of the signal; some signals are fast, others are slow.

Flatworms are most often parasites. Generally, they have almost no systems other than reproductive systems. However, this phylum exhibits the beginnings of cephalization—the development of a "head" end that concentrates brain development.

Nematodes have very specific, fixed neural responses. This makes them interesting as research animals. The first whole organism whose genetic code was determined is *C. elegans*.

Annelids have large ganglia, or nerve bundles fused at one end into a protobrain and nerve segments in each section that join into a ventral nerve cord. Sense organs are present, sensitive to light, heat, salt and mineral concentration, and pressure. Responses to these involve impulses to muscles that create motion.

Since they are segmented and have real legs and a hard exterior, arthropods require more neural connection, although their brains are still relatively small and they have a simple ventral nerve cord. The average arthropod has several types of specialized sense-receiving organs.

The Mollusca range enormously in size, shape, and body plan. That results in many specialized structures; some are analogous to those in annelids or arthropods. This leads to the thought that they have a common ancestor. That may or may not be true since the evolution of similar structures in different organisms that have similar need of such a body part is known. These structures in common are a bilateral, albeit small, brain that sends and receives information to and from the eyes, tentacles, foot (in snails) or whatever structure is present for locomotion, mantle, gut, and mouthparts. Like those found in coelenterates and annelids, some muscles are fast-reacting, others are slow. The nerves that serve them are also either fast or slow. Some have very highly modified nervous systems that correspond to a highly altered body part. The asymmetric snails are an example. Some mollusks have fairly complex nervous systems, and therefore their brains are larger. The octopus has the largest invertebrate brain in relation to body size. It can learn and adapt behavior based on learned information. Squid are famous (neurologically) for producing very large neurons and axons—the long processes leading from the neurons and axons to either another nerve cell or a muscle. These giant neurons and axons have been used in the basic research on nerve transmission.

By contrast, the echinoderms have received far less research attention. They apparently have a dual system of nerve nets, both sensory and motor, but just how information moves from one to another is unknown. Another unknown is how information from the sensory nets is communicated to the unique hydraulic system of movement. What travels between the sensory nerves and the tube feet? The neural involvement in other life processes is also unknown.

Vertebrates, those organisms most like humans, have received considerable study. The vertebrate nervous system's basic structure consists of sensory neurons leading to motor neurons via a spinal neuron connector. This is a reflex arc, the basic component of a fast response to stimulus. There is also another spinal connection to a large, well-organized brain. This is all well-documented vertebrate neural activity. The vertebrate brain is much larger than that of an invertebrate of similar size and weight. It may also be modified in size and shape depending on organism. Thus, some sharks—for example—have large olfactory lobes because smell is the dominant information-gathering sense. Eyes are complex structures, and the cranial development of optic lobes reflects that. The importance of cerebellar function is evidenced by the large, folded structure of that organ which maintains balance and muscle tonus. Both cerebellum and cerebrum are folded, which this increases surface area and therefore size.

The unique structure in a fish's nervous system is the lateral line. These organs are found on the heads and bodies of fish and juvenile amphibians. They are connected to cranial ganglia and, through the ganglia, to the brain. A lateral line organ is composed of a group of sensory cells that have a hair-like projection, like a cilium; these projections are in an epithelial pore or canal. Some fish also have taste buds. In mammals, taste is an oral function, but in fish, taste buds form along the head and body, as well as in the mouth. *See* SPONGE, ANNELID, ARTHROPOD, MOLLUSK, NEMATODE, FISH.

nets The use of nets to catch fish is older than written records. Nets range in size and type depending on locale, type of fish sought, and the level of industrialization of a culture. Nets of the surrounding type—seines and purse seines—are used to trap pelagic fish such as herring and tuna, which are found near the surface and in open water. The demersal fish, bottom dwellers such as cod and haddock, are trapped in Danish seines, which also surround their catch. These are bag-shaped trawls that are pulled along the bottom. Gill and drift nets are fish traps. Lift nets are used beneath a school of fish that is attracted upward by a light. Commercial fishermen, faced with diminishing fish populations, are increasing their usage of nets that allow marine mammals and undersized juveniles to escape. Better, ecologically sounder net design is yet another factor in responsible ocean resource management. Web site: http://www.swfsc.ucsd.edu/swfcpg.html. *See* FISHING.

neurotoxin *See* TOXIN.

neuston Organisms that use the surface film of water for support. Neuston includes snails, protozoa, and cnidaria. These animals attach themselves to the water-air boundary by adhering to the more cohesive layer of water molecules at the surface. *See* SURFACE TENSION.

New Hebrides basin A part of the Coral Sea located east of Australia, northeast of the Bellona undersea plateau, and west of the New Hebrides island chain. The islands in the basin are volcanic. While some parts of the islands have been dated to origins in the Tertiary, the still-active volcanoes are younger. *See* PACIFIC OCEAN, TERTIARY PERIOD.

New Hebrides trench The point of contact between the Australia Plate and the Pacific Plate. The trench runs down the western edge of the New Hebrides islands chain. The deepest point, 7,661 m (25,281 feet), is the maximum depth in the Coral Sea. The existence of the trench was deduced because of the volcanic activity in the area. Its location was finally established in 1910. *See* NEW HEBRIDES, TRENCH, VOLCANO.

nickel A relatively rare, silvery-white metallic element with commercial importance in alloys and as a catalyst. It is found dissolved in seawater in concentrations of about 0.002 mg/liter. It is also a constituent in manganese nodules, in which the average content of nickel is about 2% (ranging from 0.01 to 2.35%). Manganese nodules are nevertheless a good source of nickel, since their recovery may eventually become cost-effective. *See* ELEMENT, MANGANESE NODULE, MINERALS.

Niger River A West African river that empties into the Gulf of Guinea in the South Atlantic Ocean. The Niger is about 4,200 km (2,600 miles) long and is the great river of West Africa. The amount of water carried by it is enormous. The resulting delta has an area of about 24,000 km² (about 14,000 square miles) of built-up mudflats. The river is channelled into a network of waterways that are obstructed by shifting sandbars, making most of the waterways unnavigable. *See* DELTA.

Nile River An African river flowing from Lake Victoria in Uganda north into the Mediterranean in Egypt. The Nile is the most significant North African river; it is the longest river in the world with a length of about 6,650 km (4,130 miles). It runs almost directly north into an area of the Mediterranean that was once a great gulf. This area has since been filled by the Nile with silt from the Ethiopian highlands. The delta of the Nile is 16 to 25 m (50 to 75 feet) above sea level and about 160 km by 250 km (100 by 155 miles) in area, and consists of extremely fertile soil. The northern periphery is broken by extensive salt marshes. The Nile cone on the floor of the Mediterranean is not as well-defined as is that of the Rhône.

The river is of enormous importance to the countries that it flows through. For both Egypt and the Sudan it is almost the only source of water. In antiquity the incredible grain harvest of the Nile valley fed most of the populations living in the eastern Mediterranean, and for a considerable portion of its very long history, Egypt was a grain exporter. The construction of the Aswan High Dam in 1970 brought much change to the valley and the configuration of the river: Huge areas behind the dam were flooded permanently (Lake Nasser), while sites downriver of the dam are no longer inundated as they were for thousands of years. Seawater irrigation has brought significantly more salts into areas that did not have this problem before 1970. Since most of the mountain silt that comes down the upper Nile is now trapped behind the dam, the delta has ceased to enlarge. Moreover, because the silt also acted as a fertilizing material, the agricultural production of the Nile valley and delta now depends on chemical fertilization, which increases the problem of runoff. See DELTA, MEDITERRANEAN.

nitrogen A colorless, odorless gaseous element that constitutes about 78% of the total volume of the atmosphere. It is present in air as the inert diatomic N_2 molecule, and as a variety of nitrogen oxides, NO_x, where x is some number (0, 1, 2, 3) of oxygen atoms. Other nitrogen oxides are N_2O_3 and N_2O_5, both of which are products of combustion. Nitrogen is present in water largely as an inert constituent of dissolved air, but more importantly in the form of the dissolved nitrite (NO_2^-), nitrate (NO_3^-), or ammonium (NH_4^+) ions. The concentration of ammonia (NH_3), and to an extent also of NO_3^-, in water is a function of the decomposition of plankton. Thus, it is greatly increased in spring in shallow water, where most of the plankton population is found. Some oceanic water is richer in nitrogen than other waters; the water of the North Atlantic, for example, is relatively nitrate-poor.

Nitrogen is present in surface ocean water in low concentration as the molecule N_2. It is more prevalent, in concentrations of about 220 to 240 ppm (parts per million parts of water), in deep waters as amides and other higher–molecular weight compounds. These are the result of biochemical reactions involving living organisms. This nitrogen concentration is fairly stable; these compounds degrade very slowly.

Some free-living cyanobacteria are nitrogen-fixing organisms. They inhabit a variety of places, including tidal flats, marshes, coral reefs, coastal areas, and the open sea. While nitrogen fixing is more efficiently done on land, the volume of water in the ocean is so great that a considerable proportion of all nitrification occurs in the sea. The nitrogen-to-phosphorus ratio may be a factor in controlling plankton blooms, in addition to the temperature, trace metals, herbivores, and available light. See ELEMENT, NITROGEN CYCLE, PLANKTON.

nitrogen cycle The cycle of chemical processes that exchange nitrogen between the atmosphere and the Earth. Nitrogen is present in the atmosphere as a diatomic molecule, N_2. On land, nitrogen-fixing bacteria convert this relatively water-insoluble gas to a soluble ion, either NO_2^- or NO_3^-, which is taken up by autotrophic organisms and incorporated into protein. Bacteria do this in the sea. Upon the death or degradation of a nitrogen-containing organism or material, free nitrogen is produced and may again become part of the hydrosphere. This then provides the nitrogen necessary to form the proteins in marine plants and in the animals that feed on them. Denitrifying bacteria return nitrogen to its elemental gaseous state.

The movement of nitrogen in and out of the atmosphere is a complex series of reactions, some of them only recently discovered. A NASA study revealed that the amount of snow in the Himalayas determines the weather and climate conditions in much of the Indian subcontinent, the adjacent Arabian Sea, and the eastern coast of Africa. The causative agent is phytoplankton.

If there is less-than-usual snowfall in the mountains, less sunlight is reflected back into space; there is therefore more solar radiation warming the mountains, heating the landmass, and evaporating even more snow. This increases the temperature difference between the landmass and the adjacent Arabian Sea. The end effect is a persistent low-pressure area over the water and a high-pressure area on land. The usual product of this air pressure condition is wind, and the wind is normally essential in bringing rain to the

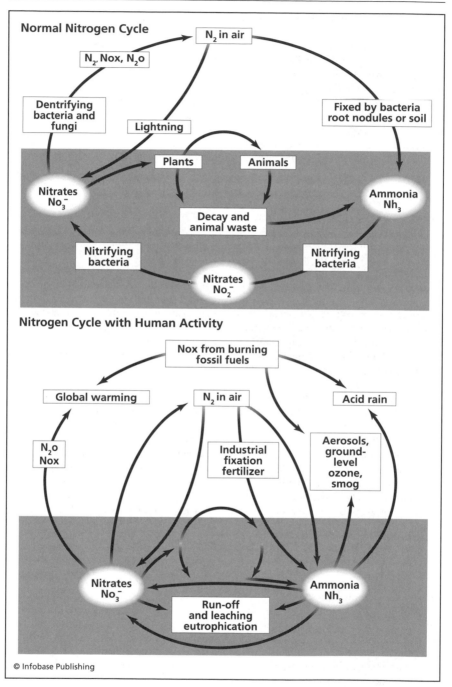

Normal Nitrogen Cycle

N₂ in air

N₂, Nox, N₂o

Dentrifying bacteria and fungi

Lightning

Fixed by bacteria root nodules or soil

Plants

Animals

Nitrates No₃⁻

Ammonia Nh₃

Decay and animal waste

Nitrifying bacteria

Nitrifying bacteria

Nitrates No₂⁻

Nitrogen Cycle with Human Activity

Nox from burning fossil fuels

Global warming

N₂ in air

Acid rain

N₂o Nox

Industrial fixation fertilizer

Aerosols, ground-level ozone, smog

Nitrates No₃⁻

Ammonia Nh₃

Run-off and leaching eutrophication

© Infobase Publishing

Nitrogen moves through the air-water continuum; human activity changes that and introduces compounds not found in nature.

landmass in summer. As the wind moves seawater, cold bottom water surfaces, bringing with it a high nitrate concentration that results in algal bloom.

This increase in the phytoplankton population also increases the fish population, and it is therefore desirable from the fishing industry's view; in the long term, however, it is detrimental, since the increase in phytoplankton in an oxygen-poor region such as the Arabian Sea uses up more oxygen than it produces. This leads to eventual die-off of the populations there, including the fish population.

The denitrifying bacteria also use nitrate ion. These bacteria consume nitrate ion and release nitrous oxide—a compound whose physiological importance is the object of a good deal of recent research. In seawater however, it ties up nitrate ion, which photosynthesizing organisms can use, and makes nitrate unavailable. Nitrous oxide escapes into the atmosphere as a greenhouse gas. In the atmosphere, it is much more of a problem than is carbon dioxide, the principal greenhouse gas. This denitrifying process increases global warming. See AUTO-TROPHS, CARBON CYCLE, CLIMATE, GLOBAL WARMING, GREENHOUSE EFFECT, HYDRO-SPHERE, PEPTIDE FORMATION.

nitrogen fixation Involving the chemical reaction of atmospheric nitrogen with oxygen to produce nitrite ion (NO_2^-) or nitrate ion (NO_3^+). The agents of change are cyanobacteria, *Trichodesmium*. This process occurs on land, and it was assumed incorrectly until fairly recently that it was not a significant reaction in the ocean. Nitrogen fixation contributes a large proportion of the nitrate in the euphotic zone, where is it essential for the phytoplankton. Quantifying the amounts of nitrate added to ocean water by bacteria is difficult and is an area of ongoing research. The mass of nitrate in the upper layers of the ocean not only is important as a part of the nitrogen cycle but also is directly related to the carbon cycle since the algal bloom supported by the increase in nitrate is the base of the food web for all other life in the upper layers of the world's oceans. See CARBON CYCLE, CYANOBACTERIA, ELEMENT, EUPHOTIC ZONE, PHYTOPLANKTON.

nitrogen narcosis A loss of critical thinking capacity, accompanied by giddiness and euphoria, which the French marine explorer Jacques Cousteau called "rapture of the deep." At depths of about 100 m the concentration of nitrogen in the blood acts as a narcotic, and a diver loses control of the capacity for critical judgment. The depth at which this phenomenon occurs varies with the diver and with individual dives. See COUSTEAU, JACQUES-YVES; DIVING.

NOAA The National Oceanic and Atmospheric Administration (NOAA), an agency established by the U.S. Department of Commerce to encompass a number of previously existing services and to give them all a common direction. These services include:

National Ocean Survey (formerly the U.S. Coast and Geodetic Survey), which creates nautical and air charts, conducts oceanographic surveys, makes tide and current predictions, and prepares navigational charts for American waters.

National Weather Service (formerly the U.S. Weather Bureau), which forecasts the weather and conducts storm watches.

National Marine Fisheries Service, which does research on economically important species of fish and other marine organisms. This agency also oversees Marine Mammals Protection and the Endangered Species Act.

Environmental Data Service, which studies the effect of environment on the nation's food and energy supply. It operates the National Climatic Center, the National Oceanographic Data Center, and the National Geophysical and Solar Terrestrial Data Center.

National Environmental Satellite Service (NESS), which maintains and collects data from the NOAA satellites.

Environmental Research Laboratories, which study ocean-atmosphere processes and the effect of pollution on coastal ecosystems.

Office of Sea Grants, which gives funds to research institutions.

Office of Coastal Management, which exists to help coastal states in the development of management systems for their coasts. It also operates marine sanctuaries.

NOAA is directly involved in the study of climate change and global warming. The Committee on Global Change Research is a part of this overall organization.

Noctiluca A worldwide genus of dinoflagellates; one of the most common is *Noctiluca scintillans,* whose common name is sea sparkle. The name is derived from the sparkling effect of seawater that is the result of the diatom's presence. *Noctiluca* produces a gas float that enables it to drift just below the ocean's surface, capturing its planktonic prey. They are the largest diatoms and range in size from 200 to 2,000 microns in diameter. The organisms are usually green or blue, but when they are red or orange, they are part of the toxic red tides and are implicated in shellfish, bird, and mammal mortality. *See* DIATOM, TOXIN.

North American Basin An Atlantic Ocean region of rather poor definition, west of the Mid-Atlantic Ridge and east of the continental rise to the Blake Plateau. The Bermuda Rise is almost central in the North American Basin. The Nares and Hatteras Abyssal Plains are encompassed by the basin, as is the Puerto Rico trench. The deepest point of the Atlantic is in this true, "island arc" trench. *See* ATLANTIC OCEAN.

North American Plate A continental plate that extends from the Mid-Atlantic Ridge to the west coast, where it abuts the

Pacific Plate. The southern boundary areas of the North American Plate include the South American Plate east of the Antilles chain and the Caribbean Plate. The southwestern boundary is the small Cocos Plate. There is seismic activity at the boundary edges of the Caribbean, Cocos, and Pacific plates. The famous San Andreas Fault is the result of the slip of the Pacific Plate northwest relative to the North American Plate. *See* ATLANTIC-TYPE MARGIN, CRUST, PACIFIC PLATE, PLATE TECTONICS.

North Atlantic Current Part of the surface current of the North Atlantic. When the Gulf Stream moves into the Grand Banks region, it becomes a weak, divided flow; it is also known as the North Atlantic Drift at this point and drifts eastward. The water is warmer than the surrounding water, which is of polar origin. The mixture results in the polar front, an area of permanent fog and low pressure that is relatively stationary north of the current. *See* CURRENT, GULF STREAM, GYRE.

North Atlantic Deep Water Cold water of low salinity sinking either east or west of Greenland and then spreading southward. Not until the influx of Mediterranean water does the salinity reach a more normal level for oceanic water. The North Atlantic water is characterized by a high oxygen and relatively low nutrient content. The latter is in striking contrast to the waters of the Antarctic, which support very large populations of living organisms. *See* ANTARCTIC WATER.

North Atlantic Oscillation The phrase used to designate the periodic, cyclic alternation of temperatures on the coasts of the Atlantic Ocean. In the years 1770 to 1778, when Hans Saaabye kept a diary in Jakobshavn, Greenland, he observed that the weather followed a fairly regular oscillation. A very mild winter in Greenland occurred in the same year that Denmark was having a very cold winter. The phenomenon was illustrated again in 1890 using the average mean temperatures of

Jakobshavn (Greenland) and Vienna (Austria) for a longer span of years. Monthly mean temperatures showed a seesaw effect over a 42-year span. This was repeated using temperatures from Oslo and Vienna. This oscillation of temperature affects the climate of both Europe and North America. Wind velocities, temperature, and precipitation all vary with the oscillation, affecting energy use, agriculture, commercial shipping, and human health. *See* CLIMATE.

North Cape Current A North Atlantic current that flows around the northern tip of the Scandinavian peninsula, bringing the warmer, saltier water of the Atlantic into the Barents Sea. It was discovered in the mid 19th century, when warm water was noted in the Barents Sea and the existence of this branch of the North Atlantic gyre was charted. *See* ATLANTIC OCEAN, BARENTS SEA, CURRENTS.

North Pacific Current The clockwise circulation in the North Pacific, also known as the North Pacific gyre. The circle is composed of the Alaskan Current and the California Current, both of which move south from the coldest parts of the Pacific to the coast of Mexico, where the water turns west and is eventually warmed by the waters south of the Philippines. There it turns north and east as the Kuroshio Current, branches of which move eastward past the Hawaiian islands and eventually dissipate within 1,000 km (600 miles) of the North American coast. *See* CURRENTS, GYRE, KUROSHIO.

North Pacific gyre The clockwise circulation that prevails in the Pacific Ocean north of the Equator. Its component parts are the Alaskan and California Currents that move cold water south to Mexico, where the gyre turns west as the North Equatorial Current and crosses to the warm regions of South Asia. There the North Equatorial Current moves in a northeasterly direction, passes the Philippines, becomes the warm Kuroshio Current, and continues east. It passes north of

Hawaii and disappears at a point about 1,000 km (600 miles) west of the coast of North America. *See individual currents.*

North Sea A fairly shallow arm of the North Atlantic between Great Britain and northern Europe. The southern end of the North Sea is the shallower part, averaging less than 40 m (150 feet) deep. At the Dogger Bank one finds average depths of 12 to 13 m (40 feet), with some spots as shallow as 5 m (about 15 feet). The area is an excellent fishing ground. The catch is herring, flatfish, mollusks, and crustaceans.

The bottom of the North Sea slopes downward north of the Dogger Bank to the Scandinavian shield and the shallow Viking Bank. A narrow, deep channel to the north encircles the southern end of Norway and runs into the Skaggerak—part of the North Sea between Norway and Denmark.

North Atlantic water enters the North Sea mainly from the north, not via the Straits of Dover, where a high sill acts as an effective barrier. The water is generally cold, from 2 to 7°C in winter and about 13 to 18°C in summer at the surface. The salinity, which is greatest in winter, is less than 3.5% in the north and west, and because of the relatively warm and salty Atlantic inclusions, the North Sea very rarely has a frozen surface.

The sea is an epicontinental one; it covers an area that had been exposed in the Wurm glacial period, the most recent "ice age." The bottom is continental and made up of the local material. Helgoland (or Heligoland), the rocky island off the coast of Schleswig-Holstein (Germany), is also a continental fragment. Peat bogs are submerged under the North Sea and are continuous with the peat deposits in England and Ireland and on the European continent.

The early history of the North Sea area goes back to the late Permian Sea (280 to 225 million years ago), which was in a hot region. Intense evaporation produced the salt domes in North Germany and Denmark, and those in the North Sea that are currently being tapped for oil. The closing of the Tethys Sea in the Mesozoic (225 to 65 million years ago) produced the beginnings of the current shoreline. The land bridge between Dover and Calais was broken about 10,000 years ago by the rise in sea level. The northern Netherlands drowned at roughly the same time, and so did other parts of the North Sea coast such as the Godwin Sands off the Kentish coast of England—submerged sand dunes that caused many medieval shipwrecks.

Charming tales exist in many North Sea areas that abut these submerged coasts: the Germelshausen legends in Germany and the "drowned village" stories in Ireland, Jutland, and the Netherlands. They are all related to the same legend of ghostly church bells sounding under the water. See ATLANTIC OCEAN, DOGGER BANK, NORWEGIAN SEA, PERMIAN PERIOD.

Northeast Passage A sea route to India and the Spice Islands around the north coast of Eurasia. Norse sailors had found their way to the Barents and White seas by the 10th century, and trade between Russia and Norway using a northern sea route was established by the 14th century.

By 1525 Russians were theorizing the existence of a sea route from the White Sea to China, an idea that had also occurred to the English, who tried to reach the fabulous lands to the east while avoiding Spanish and Portuguese ships. Several English mariners were involved in the attempt to find a Northeast Passage, but none went farther east than the Kara Sea.

Local Siberian sailors used the rivers to the Kara and the Laptev seas. In 1648 Dezhnev rounded the eastern tip of Asia and found the strait named for its rediscoverer, Vitus Bering. Russian attempts to map the far north were made in the 18th century and established the existence of a Northeast Passage. However, it was blocked by ice and too difficult to pass. The whole of the Passage was finally traversed by N. A. E. Nordenskjold on the *Vega* (1878–79). Others who used the passage were deLong in the *Jeannette,* Nansen using the *Fram,* and Toll in 1900.

Although modern ice-breaking equipment has made the passage more accessible, it is a shallow one, and the fact that there is little of commercial value in the area has left the passage largely unused. See BERING, VITUS; DEZHNEV, SEMYON IVANOV; NANSEN, FRIDTJOF; WILBOUGHBY, HUGH.

Northwest Passage A northern sea route from Europe to Asia sought by European sailors trying to reach China and the Spice Islands—Indonesia and Malaysia. Cabot's attempt in 1498, on his second voyage, cost him his life. Cartier, in a similar attempt, was first hopeful when he found the St. Lawrence River, only to be disappointed when the great gulf at the entrance to the river led to a progressively narrowing waterway.

The search for a sea route to the Orient intensified as the Turks gained total control of the eastern Mediterranean. After Pope Alexander VI (1493) divided the unknown world into future Spanish and Portuguese territories, as formalized by the Treaty of Tordesillas (Spain), the only known routes to the lucrative spice and silk trade were in the hands of the Spanish and the Portuguese. These were the route through the Straits of Magellan at the tip of South America and the sea route around the tip of Africa, the Cape of Good Hope. Both the English and the Dutch sought another way.

The attempts at a western passage were primarily but not solely made by English sailors in the 16th and 17th centuries. Gilbert's original petition for a charter called for the search for a Northwest Passage. He was followed by Frobisher, Davis, and Hudson. Luke Fox and Thomas James in 1631 proved that Hudson Bay was not a route to the Pacific. De Ulloa tried, as did Juan de Fuca, to find the "Straits of Anian" from the western edge of North America. After these unsuccessful attempts it was not until the 18th century that interest revived in finding the Northwest Passage. The North West Company, which had been formed to commercially exploit the northern regions of North America, revived the search as a public relations measure. One result was Mackenzie's discovery of the river named for him. James Cook on his third voyage charted the Alaskan coast, thus giving reliable information about it to the mapmakers of the day.

The British Admiralty established a prize, periodically amended, to reward explorations that approached the pole or particular meridians of longitude. It was the stimulus for much subsequent exploration. John Ross in 1818 sailed into the Davis Strait to Lancaster Sound. He turned back at a line of hills that he called the Croker Mountains. The secretary of the Admiralty, John Barrow, was skeptical of the existence of hills, and sent Ross's second in command, William Parry, to find the hills in 1819. They were gone. The hills were either icebergs or folded upthrust ice on floes. Parry wintered in the arctic, the first European to do so successfully. He later made two more Arctic journeys. George Lyon and Frederick Beechey also sought the passage by going through the Bering Strait. John Ross returned to the Arctic in 1829 and spent several winters icebound. During the winter of 1831 his nephew, James Clark Ross, discovered the magnetic pole. This series of voyages ended with the death of John Franklin, and James Clark Ross subsequently led the first of many expeditions that went in search of the missing Franklin. In 1850 Robert McClure discovered the existence of the sea lanes that made up the passage. Leopold McClintock in 1857 found the remains of the Franklin party.

The much sought-for passage between the Atlantic and Pacific Oceans was finally sailed through by Amundsen. He began in June 1903 at the Atlantic end, and arrived in Alaska in August of 1906. The passage is ice-free about three to four months a year. It is actually a series of interconnected channels that skirt the islands and permanent ice of the Canadian Northwest Territories. The submarine U.S.S. *Nautilus* traversed the passage while submerged in 1958. The southernmost route from west-to-east is used to supply American radar stations in the area. Some Alaskan oil is moved by tanker out of the Point Barrow area and through the Northwest Passage when it is relatively ice-free. This was first done in 1969 by the tanker *Manhattan*. That journey was the first realization of the centuries-old dream of a free New World passage for commerce. The rapid melting of the polar ice has made the possibility of open water in winter something that may come to pass before the end of the 21st century. *See* BAFFIN BAY, BERING SEA, CLIMATE, EXPLORERS AND EXPLORATIONS, HUDSON BAY, *individual explorers.*

Norwegian Current A continuation of the North Atlantic Current, which carries warm, salty Atlantic water north along the coast of Norway to Svalbard (Spitsbergen), where it becomes the North Cape Current. *See* CURRENT, GYRE, NORTH CAPE CURRENT.

Norwegian Sea A body of water that extends north-to-south from Svalbard (Spitsbergen) to the Faroe Islands. On an east-to-west axis it extends from Norway's coast to Jan Mayen Island and to Iceland. The deepest point in the Norwegian Sea is about 4,000 m (13,000 feet). The sea is divided into basins and separated from the rest of the North Atlantic by a series of undersea ridges. The Mohns Ridge is the mid-ocean ridge and the sea's western boundary. The bottom is volcanic, with foraminiferan and diatomaceous residues. An earthquake belt runs from Jan Mayen Island to Spitsbergen as a continuation of the Mid-Atlantic Ridge.

According to one version of the formation of the Norwegian Sea, the Scottish mountains, Norway, Bear Island, Spitsbergen, Greenland, and Newfoundland all constituted one continuous landmass that was then separated by faulting and oceanic spreading, a process that has gone on since the Permian period 280 to 225 million years ago. *See* ATLANTIC OCEAN, EVOLUTION OF OCEANS, FORAMINIFERA.

Norwegian Trench The undersea perimeter of Southern Norway from the Skagerrak to the Norwegian Basin of the North Sea; it is about 250 to 600 m deep. The origin of the trench is disputed. Some researchers believe that it is a glacial scour; others think it is a tectonic rift. *See* GLACIER, TRENCH.

notochord The single, rodlike dorsal support structure of chordate animals. The notochord develops after the infolding of the blastula—the hollow ball of embryonic cells. This phase of early embryonic development is called gastrulation. The notochord is almost always replaced. In vertebrates it is replaced by a spinal column. Other animals, such as the tunicates, absorb the notochord and develop into radially symmetric adults rather than bilaterally symmetric ones. Some primitive fish such as the hagfish and lamprey retain the notochord in adult life. *See* HAGFISH, LAMPREY, VERTEBRATE, TUNICATA.

nudibranch An unshelled marine snail, also known as a sea slug. These are extremely successful animals; over 3,000 species are known. It is further characterized by having no gills or mantle cavity. These shallow-water animals have striking branched body outgrowths, called cerata, which are involved in respiration and function like gills. The cerata may be complex in structure; they are frequently brilliantly colored, making the nudibranchs the most ornately colored mollusks.

The nudibranchs feed largely on sea anemones and sponges. Some families swim instead of crawling about. While most nudibranchs are small, there are some larger species, especially those of Australia's Great Barrier Reef, which may attain the giant size of about 30 cm (12 inches). While many nudibranchs are associated with tropical reef communities, they do populate colder waters. The Sula Shelf off the Norwegian coast is home to a considerable number of nudibranch species. Nudibranchs appeal to underwater photographers because of their spectacular shape and color. A large number of Web sites and photographs are available. *See* GASTROPODA.

Nummulites A genus of foraminiferans. Thousands of species of these animals flourished in seas of the Tertiary period. They were particularly successful in the Eocene epoch, and their fossils form a significant proportion of Eocene limestones. These tiny formaniferans are less than 6 cm (2.4 inches) long but formed huge limestone formations. The Giza pyramids were constructed of nummulite limestone. *See* EOCENE EPOCH, TERTIARY PERIOD.

nurse shark A shark, *Ginglymostoma cirrata*, common in the Gulf of Mexico;

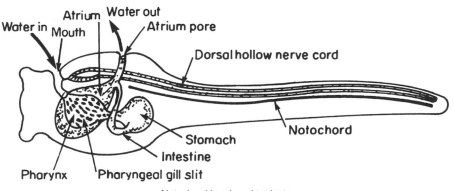
Notochord in a larval tunicate

also an arctic shark, *Laemargus microcephalus*. Nurse sharks are also found in the Pacific Ocean. These animals range from a yellow-gray to gray-brown, and they can grow to a length of 4 m (14 feet), although most specimens are smaller, in the 35–45 cm (1.2–1.5 feet) range. They are ovoviviparous and can cruise in relatively shallow water. Some attacks on humans have been reported. *See* OVOVIVIPAROUS, SHARK.

nutrients Chemical substances necessary for the maintenance of life in the sea. They are generally the same materials needed for the growth of marine plants, consisting of nitrites, nitrates, phosphates, ammonia, and silicates.

The surface layer of the water is relatively poor in all nutrients. An intermediate layer with a concentration of all dissolved material that increases rapidly with depth overlies the layer of maximal concentration at about 500 to 1,500 m (1,550 to 5,000 feet).

The varying concentration of nutrient material in this layer is dependent on many factors, including the inflow of river water, light, temperature, insolation, currents, and upwellings. All nutrients are recycled. The death of organisms or excretion of waste by organisms releases material into the water, where it is exposed to oxygen and the nutrients are regenerated. The turnover rate depends on all of the factors listed above and differs from one ocean to another. For example, phosphate is cycled about seven or eight times a year. A complete turnover of nitrate ion takes twice as long. Different materials in the water promote different types of growth: An increase of nitrogen and phosphorus compounds stimulates algal growth to overload; trace minerals are essential for growth of animals; and upwellings from black smokers support large vent communities. *See* ALGAE, CYANOBACTERIA, DEAD ZONE, METHANOGENESIS, SULFUR BACTERIA, TUBE WORMS, VENT COMMUNITIES.

Nyctiphanes A genus of euphausiids that are the major foodstuff of the Ant-

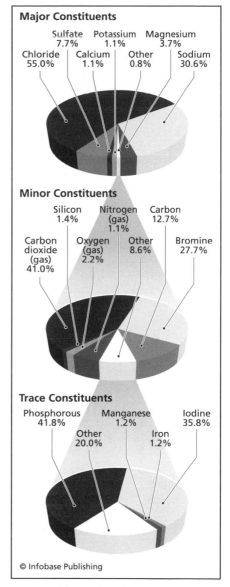

Major Constituents

| Chloride 55.0% | Sulfate 7.7% | Calcium 1.1% | Potassium 1.1% | Other 0.8% | Magnesium 3.7% | Sodium 30.6% |

Minor Constituents

| Carbon dioxide (gas) 41.0% | Silicon 1.4% | Oxygen (gas) 2.2% | Nitrogen (gas) 1.1% | Other 8.6% | Carbon 12.7% | Bromine 27.7% |

Trace Constituents

| Phosphorous 41.8% | Other 20.0% | Manganese 1.2% | Iron 1.2% | Iodine 35.8% |

© Infobase Publishing

The nutrients in seawater

arctic waters. (It is one of the organisms collectively known as krill.) *Nyctiphanes australis* is an essential factor in the Antarctic food web. It is the principal food of the jack mackerel, one of the species that is important in commercial fishing. The population of the krill is dependent in turn on a number of factors, particularly the phytoplankton, which are nourished by nitrates in the water. There is an expected upturn in the krill population in autumn. In years when La Niña events are observed, the population of the krill diminishes greatly, and the dependent larger vertebrates die off as well. Other species of *Nyctiphanes* are part of the food web in the North Atlantic and the North Sea, where they also support large fish populations. *See* FOOD WEB, KRILL.

ocean-atmosphere interchange The interchange of water and air between the oceans and the atmosphere, and the complex of causal relationships that affects the combined air-water system. On a large scale, the air currents or winds generated by temperature differentials move in specific patterns. Air is also moved into water and water into air. Heat, salt, and gases other than air are also exchanged from one fluid to the other.

Gross movements of large quantities of fluid (water or air) are influenced by the Earth's rotation and the shape of its ocean basins. They in turn produce ocean waves and currents. The ocean is the recipient and storage unit for a surplus of radiant solar energy. This is then released to the atmosphere at differing rates through the year. The consequence of this heat exchange is worldwide weather.

A direct and troubling connection has been investigated by NASA. If there is less snow than usual in the Himalayas, there is algal "bloom" in the Arabian Sea. Less snow means that solar radiation is reflected less, and less energy from the sun is used in melting snow. This means that the land becomes hotter in summer, and the effect is a greater temperature differential between the hot land and the cooler sea. This temperature difference triggers an air pressure difference with a low-pressure region over the Indian landmass and a high-pressure region over the Arabian Sea. The result is that winds push the upwelling of nutrient-containing bottom water, which feeds algal populations and encourages population increases. These algal blooms are apparent along the coasts of Somalia, Yemen, and Oman.

While this is good for the fishing industry, it is ecologically troublesome. The Arabian Sea is relatively oxygen-poor. As the excessive phytoplankton population produces organic matter, more decomposition of this material occurs. The decomposing bacteria use oxygen, and this in turn deprives the fish population of oxygen, resulting eventually in significant fish die-off. The water's lack of oxygen also creates an environment favorable to the growth of denitrifying bacteria. Denitrifying bacteria live on the nitrate ions in the water that the phytoplankton use and convert those ions to nitrous oxide. This compound is a greenhouse gas. It has the potential to worsen the already poor climatic conditions in the region. *See* CLIMATE, CURRENTS, GREENHOUSE GAS, METEOROLOGY, NITRATE, ODOR, PHYTOPLANKTON, WEATHER.

ocean basins Those parts of the Earth's oceans that are neither continental nor part of the structures of the mid-ocean ridges.

One theory of basin development, which depends on continental plate tectonics, is that the continents are granitic (siliceous or sialic) or acid igneous and float on the denser mantle rock or sima (silica/magnesium) by displacing their own weight. As the mid-ocean rift separates, the heavier basic mantle rises, becomes seafloor, and eventually dives under and into a trench at a specific region known as a subduction zone. The mantle is then remelted, releasing granitic (or sialic) lighter rock, which is deposited as lava or granitic intrusions. These accrete to form island arcs, which eventually coalesce to form continents. The regions between the continents are ocean basins. They are, according to some researchers, "crustless" (i.e., they are on the naked plates as compared to crust-bearing or continental plates). *See* MID-OCEAN RIDGES, PLATE TECTONICS, TRENCH.

ocean floor The substrate underlying the water of the world's oceans. The ocean floor is a three-layered structure overlaying the denser, more viscous mantle. The topmost layer extends down to a depth of about 2 km (1.2 miles) and is composed of unconsolidated sediment that is almost entirely of terrestrial origin. The middle layer is composed of lava of two types. The upper level of this layer is 0.5 km (0.3 miles) thick and consists of what is known as pillow lava. It consists of a series of rapidly cooled bubbles derived from the volcanic activity of the deep sea. The deeper portion of the middle layer is about 2 km (1.2 miles) thick and marked by feeder dikes or lava sheets intruded into other rock. The lowest layer of the ocean floor is approximately 5 km (3 miles) thick and consists of gabbro—coarse-grained igneous rock.

The major features of the ocean floor are the ridges or rises, trenches, and fracture zones. Fracture zones are troughs traversing the oceanic ridges. They are seismically active regions, sites of plate shifts. Frequently they are named; the Mendocino Escarpment, west of the California coast, is a sheer drop with clearly defined magnetic anomalies on either side.

A new computer-imaging technique for mapping the ocean floor is based on the same technique used in creating maps from satellite-transmitted information. It is so sensitive that it has revealed tiny bumps and valleys, which are clues to seamounts and trenches that had previously been undiscovered. *See* CRUST, FRACTURE ZONE, MANTLE, MID-OCEAN RIDGES, VOLCANO.

oceanic circulation The movement of water in the oceans. This includes regular large movements such as the Gulf Stream, seasonal flows such as the rush of water through the Drake Passage in the Antarctic spring, air-water interactions, coastal flows, surface currents, eddies, and deep-water currents. Recent oceanographic research has effectively mapped many of these water movements.

oceanic crust The floor of the ocean. It may, but usually does not, extend under the continental areas. The ocean bottom is only about 5 km (3 miles) thick. It is

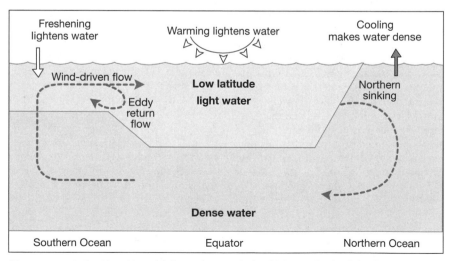

The movement of cold, nutrient-rich Antarctic water is the driving engine in all the Earth's oceans.

largely basalt and is the youngest rock on Earth. *See* BASALT.

oceanic islands Islands arising from the seafloor rather than from continental shelves. These islands are likely to be volcanic in origin. An undersea volcano will build a mound toward the surface as material moves up from the mantle to the plate surface, where it is subjected to erosion just as a terrestrial rock formation is subjected to erosion. The volcanic mount may be worn away as the volcano becomes extinct. If it is reinforced and augmented by coral accretion, or if there are several volcanoes in the same area, the aggregate structure may have a sufficient pile of material to emerge as an island. Hawaii is an example of this.

Lines of volcanoes of the above type exist in the Pacific. The best examples are the Marquesas and the Samoan and Hawaiian islands.

Oceanic rises may have several peaks that emerge as islands. These are near but not directly on the rise crest. Reunion, Ascension, and Easter Islands are such islands. Other islands, such as the Aleutians, are associated with the edges of crustal plates. They are in unstable areas at the edges of trenches and experience frequent earthquakes. *See* ATOLL, CORAL, ISLAND, MID-OCEAN RIDGES, SEAMOUNT.

oceanic rises *See* MID-OCEAN RIDGES, SEAMOUNTS.

oceanographic equipment Equipment and instruments used in the study of the oceans and their flora and fauna. Since oceanography is itself a composite field of study, the equipment for any investigation depends on what is being investigated.

For the analysis of seawater, if it is the physical properties of the water that are of interest, the equipment might consist of sampling bottles of the Nansen type or others; thermometers, both recording and ordinary; and photometers for measuring water transparency. Nansen bottles collect water at a specific depth and record the ambient temperature at the time and place of its collection. To determine the chemical properties of the water—such as the salinity, nutrient content, pH (acidity), and dissolved gases—chromatographs, spectrophotometers, and pH meters are used. Physical oceanographers use depth-sounders, satellite maps, and devices to measure the electrical fields and pressure of the Earth under its water envelope.

The velocity of ocean currents can be measured electrically by using a predetermined reference point. Since seawater is an electrolyte (i.e., conducts an electric current), the amount of current carried will change. Subsurface currents can be tracked by radio waves, a technique that is accurate down to depths of abut 8 km (5 miles).

Sediment samples can be scooped up fairly easily where the ocean is not very deep. Small samples can also be vacuumed up. If a large sample is needed, however, this presents a problem, since dredging is the only available technique for this, and the weight of the dredge requires very heavy duty equipment, which is difficult to control from a moving ship. Similar difficulties are encountered in core sampling, another useful technique, which produces samples arranged in chronological order, with the oldest on the bottom.

Biological sampling is done by means of large nets of the type used by the fishing industry. These nets are intended to capture rapidly moving fish traveling below the surface. For plankton, nets are designed to have a specific mesh-size, so that a given volume of water can be examined for its marine biota.

Oceanographic laboratories range from permanent ones, such as those at Woods Hole, Massachusetts, and the Scripps Institution at La Jolla, California, to those created for a specific expedition. They include floating laboratories on ships such as the *Glomar Challenger* or those on fixed platforms and on unmanned floating panels. The Floating Instrument Platform, or FLIP, is a University of California creation that is towed horizontally and then righted. It is 108 m long and when upright, 17 m is underwater. It contains laboratories, cranes for manipulating sea-sampling equipment, and living areas for the crew.

oceanographic vessels Ships used for scientific exploration of the sea. In the past, such vessels have been fairly nonspecific in design. They were either converted warships, yachts, or merchant ships. HMS *Challenger* was a corvette; the *Atlantis*, built in 1930 by the United States, was a ketch. Since World War II, the increased interest in oceanography has been reflected in a number of new ships designed to support a working team of scientists on board, and to act as support stations for submersibles. Since heavy equipment—including submersibles—is lifted by the mothership, the latter must have a sufficiently sturdy hoisting capability and stability. Coring equipment is also necessary, both for bottom sampling and probing for important minerals and petroleum.

Oceanographic platforms, which perform many of the same functions as research vessels, may be as simple as an unmanned buoy that houses recording devices, or as complicated as an offshore oil rig anchored to the ocean bottom. Like a ship, the platform functions as laboratory, workplace, and home for those based on it. NOAA now uses the *Ferrel*, which launched in 1998, has wet and dry oceanographic labs, and concentrates on environmental research designed to maintain sustainable seas. The *Ka'imimoana*, launched in 1996, monitors data collecting buoys at 125° through 140° west longitude and performs atmospheric research in the Pacific Ocean. The *Ronald H. Brown*, launched in 1996, works in the Atlantic Ocean on atmospheric and oceanographic research.

A number of organizations that are designated as research institutions are funded by U.S. governmental agencies for specific investigations. The organizations include Woods Hole Oceanographic Institution (WHOI), Scripps Institution of Oceanography, and universities such as the University of Washington—the last is actively engaged in the exploration of the Juan de Fuca crustal plate. Some of the government agencies involved are the National Science Foundation, the U.S. Navy, and NOAA (The National Oceanic and Atmospheric Administration).

The R/V *Thomas G. Thompson*, built in 1991, is a research vessel funded by the U.S. Navy and operated by the University of Washington. The R/V *Atlantis* is the second of that name and mother ship of the *Alvin*, the famous submersible that first explored hydrothermal vents. It is operated by WHOI, as are other vessels such as R/V *Knorr* and R/V *Oceanus* that are outfitted for specific areas of research. *See* ALBATROSS; ALBERT I; *ATLANTIS*; CHALLENGER, HMS; COUSTEAU, JACQUES-YVES; *FRAM*; ROVs; *individual vessels.*

oceanography An umbrella term for the science of the sea. This encompasses the study of currents, winds, and tides; sea-air interactions; seawater; the geology of seafloor features; and the biota of the sea. Since man has been exploring this environment for thousands of years, considerable information, particularly geographic and commercial (fishing), has been collected. However, the systemization of this body of information is less than 200 years old. The scientific societies devoted to the study of the sea are younger still: The first, the International Council for the Exploration of the Sea, began in Stockholm in 1899.

The voyage of HMS *Challenger* (1873–76) has been taken by oceanographers as the beginning of their science, because it was the first systematic government-sponsored expedition entirely dedicated to scientific exploration.

Earlier explorations frequently included naturalists: Captain James Cook travelled with Joseph Banks (among others), and Robert Fitzroy had Darwin with him. There were also captains who were amateur scientists, such as Alexander Scorsby, who contributed to information about the sea by bringing home materials and by recording observations for scientists to work on. An example of the latter activity was the careful recording of specific gravities of water samples by Chappe d'Auteroche, which resulted in Lavoisier's work on the salinities of diverse waters.

Preceding the voyage of HMS *Challenger*, the establishment of the U.S. Coast Survey (today the National Ocean Survey) in 1807 was an early realization of the importance of the ocean to the young American nation. The original charge of the Coast Survey was to chart coastal waters, since commerce was vital. Later, the Survey's chief, Alexander Bache, a descendant of Benjamin Franklin, directed studies of the Gulf Stream. He believed that it was a study that was essential to the American nation as a whole, not just to the coastal states.

In 1830 the U.S. Navy's Depot of Charts and Instruments, now the Naval Oceanographic Office, was founded as a library for information about the sea. Its director from 1842 to 1861 was Commander Matthew F. Maury of the navy, who made it a center for information on marine weather conditions, sea lanes, and currents.

S. P. Baird, an assistant secretary of the Smithsonian Institution, initiated the U.S. Fish Commission (today the National Marine Fisheries Service) to help maximize fish catches. The research vessel *Albatross* and its expeditions came from this, as did the laboratory at Woods Hole, Massachusetts. These early beginnings in the United States were all made before 1900.

The expedition of the HMS *Challenger*, led by Charles Wyville Thomson, himself a botanist, included a number of other professional scientists. Among them was John Murray, who finished the monumental task of producing the report of the voyage after Thomson's death in 1882. The 50 volumes of the report were finally completed in 1895.

It had by then become a matter of national pride to have an ongoing oceanographic research effort. The end of the 19th and the early 20th centuries saw the Germans mapping the oceans in the *Gazelle*, *Valdivia*, and *Deutschland*. The Norwegians explored the Arctic in the *Fram* and

the *Michael Sars*, the French were in the Pacific in the *Travailleur* and *Talisman*, and the Russians explored using the *Vitiaz*. Prince Albert I of Monaco used several yachts named *Alice* or *Hirondelle* on a number of expeditions. These were not the only expeditions; many more were financed for short periods or limited studies.

The first combined effort at oceanography was funded by ICES—the International Council for the Exploration of the Seas. The members were Norway, Sweden, Denmark, Finland, Germany, the Netherlands, Russia, and Great Britain. The United States returned to the scene with the foundation of the Scripps Laboratory in California in 1925. The Woods Hole Oceanographic Institution on Cape Cod was created in 1930.

There were significant expeditions by several nations between the world wars and directly after World War II. They concentrated on a number of subjects, such as the biological aspects of oceanography, the problems of mapping, currents, and weather observations. The combined efforts of many nations have been more the norm in the post–World War II era. The 18-month-long IGY (International Geophysical Year) in 1957–59 approached many problems. The ICITA (International Cooperative Investigations of the Tropical Atlantic) and IIOE (International Indian Ocean Expedition) were both done in the early 1960s. The International Oceanographic Data and Information Exchange (IODE) is a branch of UNESCO. It was established in 1961 for the express purpose of enhancing marine research, exploitation, and development by making it simpler for member states of the UN to exchange data about the oceans. *See* ALBERT I; AGASSIZ, ALEXANDER; CHALLENGER, HMS; IGY; *individual research institutions' Web sites for specific projects;* MAURY, MATTHEW FONTAINE; NOAA; EXPLORERS AND EXPLORATIONS; *specific organisms or phenomena.*

oceans The bodies of salt water that cover more than 70% of the Earth's surface. This vast area is divided into a number of oceans and seas. Their names depend on both the reference used and the convention followed. Attempts at standardizing the definitions and boundaries of the oceans is the charge of the International Hydrographic Bureau, based in Monaco. The criterion for calling a body of water a specific ocean or sea is that it have some definable and independent land contour, bottom, topography, and wind and water circulation in both the horizontal and vertical planes.

The evolution of oceans has been a topic of continuing interest for geologists. The many, often conflicting theories about this were clarified once the idea of continental drift was accepted. Many bits of evidence concerning the present size and shape of the Earth's ocean basins could then be unified. Thus, despite continual continental sedimentation, the ocean floor is about ten times younger than are the continents—200 million years old versus about 2 to 3 billion years old. The ocean floor is also somewhat denser than the continental rock.

One theory explaining the bottom topography of the ocean is that hot lava rising at the crests of mid-ocean ridges pushes older rock aside and drives the continental plates further apart. Another idea is that the cold, dense slabs of seafloor sink in zones of seismic activity (e.g., the Aleutian Islands and Peru-Chile Trench), and that this pulls the rest of a plate toward the sunken zone. According to the latter theory, seafloor spreading is not inexorably unidirectional but can be changed by major events. One such event was the collision of India with the Asian landmass, which altered the movement of all Indian and Pacific plates. The rate at which plates move is 2 to 20 cm/year, which is enough to account for the present size of the ocean basins.

The hydrosphere (the water component of the Earth's surface) may have come from the degassing of the primordial Earth, which occurred about 500 million years ago. The early, acidic ocean of the young Earth must have dissolved some silicates and carbonates due to its high hydrochloric acid level; it also dissolved sodium in rocks to produce a salinity level close to that of the present oceans.

The mineral content of the oceans has been in a steady state for about 100 million years. Thus, as materials are added by runoff from continental areas, new sedimentary material is deposited, which is subsequently recycled as new land areas or reintroduced as gases to the atmosphere. The concept of the ocean as a complex solution in equilibrium has replaced the earlier one of the ocean as a large collecting body. *See* CONTINENTAL DRIFT, EVOLUTION OF OCEANS, OCEAN FLOOR.

octopus Any eight-armed cephalopod mollusk of the order Octapoda, genus *Octopus*. These shell-less mollusks are bottom dwellers in warm-to-temperate shallow water, and some also occur at abyssal depths. Their range in size is from about 5 cm to 5.5 m (2 inches to 22 feet). The bag-like body conceals a sharp beak, and their muscular arms are equipped with rows of suckers. The largest ones can have an armtip to armtip length of up to 9 m (30 ft). The octopus is a clever predator. *O. vulgaris*, the common octopus, feeds on crustaceans. Some species are plankton feeders.

The octopus is a solitary animal that lives among rocks or in holes. Its major sense is keen eyesight; like other cephalopods, the octopus is deaf. It can alter its color to match the surroundings, and does so within minutes. If startled, it produces ink as a camouflage, and moves rapidly backward by water-jet propulsion. *See* CEPHALOPODA, MANTLE, PAPER NAUTILUS.

Odontoceti The family of toothed whales. *See* BALEEN, WHALE.

odor Produced and detectable by many aquatic organisms. Odor plumes are the streams of chemicals that are produced to act as signals; pheromones (odors that communicate sexual information from one individual to another of the same species) are an example. Some creatures use the chemical signals produced by other animals as leads to the food supply. Crabs, lobsters, and other arthropods apparently have excellent odor-sensitive receptors at the base of their antennae and employ this sensory equipment in their foraging. Fish also use odors as guides to food supply; the olfactory lobes of fish brains are large in relation to the size of the entire brain. This is the central processing unit for information in the form of odors.

Anyone who has smelled the sea knows that its odor is distinctive even if it is present in air in parts per billion. It is a combination of gases, several of which are produced by microalgae. The dimethylsulfide produced by organisms that live in the sea plays an important role in climate control, since it affects the backscattering of the sun's radiation and the formation of clouds.

The sulfur compounds in those algae and other protists that manufacture them deter the grazing plankton and other sea animals. Petrels and other birds are sensitive to the odor of the water, which is in turn an indicator of what organisms are in abundance just below the surface. This generation of sulfur-containing materials and their subsequent degradation is part of the sulfur cycle of the biome.

offshore drilling The exploitation on continental shelves of the geological features that are likely to yield oil and gas. Offshore deposits now account for about 13–15% of all oil and gas production. In addition to fixed platforms, floating platforms are being used. These are limited to use in waters that are no more than about 90 m (300 feet) deep because a floating platform is really not floating but is braced on the bottom on retractable legs. Floating drill ships can operate in shallow water of about 18–20 m (60–70 feet) depths. Mobile drill rigs in deeper waters are frequently submersibles. They are towed to a site and sunk. Their advantage is that they are very stable and unaffected by weather. Some use of semisubmersibles has been made. These rigs are designed to float below the water surface, and they too are weather-independent.

offshore zone A relatively flat zone extending from the breaker line on the beach to the edge of the continental shelf. *See* COAST, CONTINENTAL SHELF, PETROLEUM.

Oikopleura A family of Appendicularians; they are primitive tiny chordates and retain their notocords throughout life. The body of a typical oikopleuran is less than 0.25 cm (0.1 inch) long. These filter-feeding organisms are found in all pelagic environments.

The organism has a small ovoid body, rather like that of a tadpole, and a long tail. The distinguishing feature of oikopleurans is the fragile mesh net that it creates around itself. The long tail beats and creates a current that moves water and the particles suspended in it into the net, where they either pass through or adhere. The fragile net is frequently damaged; the organism is therefore constantly either repairing its net or constructing a totally new one. The fragments of this structure become part of the "marine snow" that feeds the abyssal organisms. Both the oikopleurans and their characteristic nets are phosphorescent.

These animals are interesting in terms of DNA studies; they have the smallest genome of all animals studied to date. *See* CHORDATA, FILTER FEEDERS, MARINE SNOW.

oil A term often used interchangeably with fat. Generally oils are liquids at room temperature, while fats are solids. Whether they are called oils or fats, these substances are composed chemically of triglycerides—compounds consisting of glycerol combined chemically with long-chain fatty acids. If, in the fatty acid chain, one or more of the methylene ($-CH_2-$) groups that—linked to one another—make up the "backbone" of the chain are replaced by a doubly-bonded carbon ($-CH-$) group, the oil is called unsaturated. When there are several $-CH-$ groups in the molecule, the oil is said to be polyunsaturated.

Marine oils are commercially important products. Whales are hunted for their oil, which is rendered from the blubber or body fat on factory ships—huge, mechanized versions of the 19th century whalers. Sperm whale oil is special because it contains alcohol esters called spermaceti, a material used in cosmetics and pharmaceuticals. Fish oils are generally extracted from the livers of sharks, cod, menhaden, sardines, tuna, and mackerel by steam distillation, a process that extracts the oil without heating it to temperatures at which decomposition might occur.

Fish oils are used extensively in the manufacture of lubricants, soaps, and paints. The medicinal use of cod liver oil is based on its high vitamin A and D concentration. Certain fish oils are also sought for their possibly beneficial effect in preventing or retarding atherosclerosis. *See* MARINE OIL, SQUALENE.

Okhotsk, Sea of An arm of the northern Pacific Ocean, separated from it by the Kurile Islands and the Kamchatka Peninsula. The Kurile Straits are the outlet to the Pacific. The Nevrlesky (Tatar) and LaPérouse (Soya) straits lead to the Sea of Japan. The Okhotsk Sea is surrounded by a mountainous shoreline. The average depth of the water is about 775 m (2,350 feet), with a maximum depth of over 3,370 m (10,000 feet). The bottom is in part continental shelf and the remainder is of island origin. The southern end is the most ancient and also the deepest part.

There is considerable precipitation in the Sea of Okhotsk, since it is an area in which cold continental air meets warmer maritime air. There is an ice cover for about six months of the year. The waves are noticeably higher than in the Pacific in general. The water temperature is relatively warm. The salinity, particularly at the surface, is low.

There is a warm deep current in the sea, that is also highly oxygenated. This anomalous water property supports a huge phytoplankton population. This, along with the resident algae, supports a large mollusk and echinoderm population that is in turn a food supply for the commercially important salmon, seal, and crabs. *See* DEZHNEV, SEMYON IVANOV; PACIFIC OCEAN.

Oligocene epoch The part of the Tertiary period that followed the Eocene epoch and preceded the Miocene. Very few terrestrial Oligocene fossils are closely related to modern biota; however, most marine Oligocene fossils are very similar if not identical to modern forms. The Foraminifera were extremely numerous in both species and individuals in this epoch.

Geologically, a major event of this span of time, which began about 38–40 million years ago and lasted about 12 million years, was the upthrust of the Tethys Sea bottom, resulting eventually in the formation of the European Alps.

Warm climates prevailed as the sea encroached on land areas and retreated again in several cycles of advance and retreat. These cycles are each named, usually for a locale. At some point in the Oligocene, South America was isolated from North and Central America, allowing divergent faunas to develop in these different areas. *See* TERTIARY PERIOD.

Oligochaeta A class of worms of the phylum Annelida. The common earthworm is an oligochaete. The class is widely distributed in water (both in fresh and marine environments) and in soil. The individual members of the class range in size from those that are only millimeters long to the great giants of Australia that are 3 m (10 feet) long and more.

Setae, or bristlelike projections, and a clitellum (the thickened band around the body, which is closer to the "head" than to the tail), are characteristic of the Oligochaeta. In marine specimens, the clitellum is much less conspicuous than in terrestrial ones. On the earthworm this organ produces the gelatinous material in which the eggs mature and hatch; however, this semiliquid nest is not needed in a water environment. Oligochaetes are hermaphrodites. *See* ANNELIDA.

omnivore An animal that ingests both plant and animal tissue. *See* HETEROTROPH.

ooze Soft sediment on the ocean floor, composed largely of the remains of minute organisms. Oozes are characterized by the organism that contributes the major part of their composition: They may be foraminiferan, diatom, or radiolarian. *See* DIATOM, FORAMINIFERA, OCEAN FLOOR, RADIOLARIAN.

operculum The hard, scalelike cover over the gills in bony fish.

Ophiuroidea The brittle stars, members of a class of echinoderms. These graceful relatives of the common starfish can live in almost all relatively shallow waters and have a long fossil record. Web site: http://www.ucmp.berkeley.edu/echinodermata/ophiuroidea.html.

opholite A basaltic fragment of old marine origin. It is the result of the horizontal spread of what had been a vertical slice of ridge crest. The Trovdos opholite on Cyprus is typical. Other opholite sites are in the Shetland Islands, the Bay of Islands in Newfoundland, the Ural Mountains, and numerous sites along the India-Asia suture line. *See* BASALT, CRUST, EVOLUTION OF OCEANS, PANGAEA.

Opisthobranchia A subclass of Gastropoda that includes the sea hares, cone shells, sea butterflies, and nudibranchs. Many species have no gills and respire directly through breathing organs usually situated behind the heart. In general, opisthobranchs lack a mantle cavity and an operculum. The lack of a shell in the nudibranchs is significant; they are the "slugs of the sea." *See* GASTROPODA.

Orca *See* CETACEA, KILLER WHALE.

Ordovician period Part of the Paleozoic Era, preceded by the Cambrian and followed by the Silurian period. The name became established in 1873, and is derived from a tribe of pre-Roman Britons. The epochs encompassed by the Ordovician are variously named in different locations, and these local names are used to describe the rocks found there. The entire period lasted about 70 million years, and ended about 430 million years ago. During the Ordovician period sedimentation on the continental shelves of Gondwanaland built up landmass and closed the Iapetus Ocean, which has been called the proto-Atlantic. Cincinnatian seas covered most of mid-North America, while in South America Ordovician formations occurred in Venezuela and Colombia, and south to Argentina. Well-defined fossils dating from this time have also been found in Australia.

The fossil record of the Ordovician is extensive: Brachiopods are the most characteristic fossils, graptolites are generally found, and cephalopods, trilobites, and corals are known, as are some crinoids, gastropods, pelecypods, and ostracods. Primitive fish also proliferated in the Ordovician. *See* CAMBRIAN PERIOD, PALEOZOIC ERA, SILURIAN PERIOD. *Appendix: Geologic Timescale.*

organic matter Compounds derived from organisms. The flow of carbon compounds through the ocean is being studied by an ongoing research effort that is part of the U.S. Joint Global Ocean Flux Study (JGOFS). It is following organic materials produced and found in surface waters or produced on land and washed into ocean waters. These materials and their chemical reactions with other compounds are then followed as carbon moves through the water column and into the sediments below it.

Water samples are collected from several fixed locations—in the Black Sea, the tropical Pacific Ocean, the eastern African coast, and Antarctica—and from varying depths in those locations. Sediments are sampled and cores taken. All of these specimens are examined to produce a complete picture of the biogeochemistry of the entire ocean.

The factors that create variation are the productivity in specific locales, what is living in particular waters, how minerals and trace elements are moved through the biome (or are not incorporated), and how fast sediment falls. The goal of this study is to determine the role of ocean in the climate and environment of Earth in the past and to use that information to make predictions about the continuing changes that will inevitably occur on Earth. *See* CARBON, CLIMATE CHANGE, MINERALS, WATER.

Orinoco River A 2,900-km (1,800-mile)-long river flowing through Venezuela from the Brazilian border to the Colombian border and east into the Atlantic Ocean. It is often compared to the Niger River in content and configuration. Its delta is quite large, built by its own sedimentation in a relatively tranquil area and augmented by deposits from the Amazon River brought to the northeast by offshore currents. *See* AMAZON; ATLANTIC OCEAN; HUMBOLDT, FRIEDRICH HEINRICH ALEXANDER VON.

Ortelius, Abraham (1527–1598) A Belgian (Flemish) geographer who was a friend and contemporary of Mercator, the famous cartographer; he was almost as influential as his friend. Ortelius had extensive, firsthand knowledge of geography, having traversed a considerable part of the known world. At Mercator's urging, he compiled the *Theatrum Orbis Terrarum,* an atlas of 53 maps, published in Antwerp in 1570.

Because the atlas collected information useful for sea voyages, and appeared at a most propitious moment in European history, it was a huge success in spite of its errors, which were the result of using locally obtained, and often inaccurate, recordings for depths and distances.

In 1575 Ortelius was invited by Philip IV of Spain to be the king's personal geographer. He accepted and continued his work by adding more maps to the atlas; eventually 17 more were included. Ortelius spent the remainder of his life collecting maps and attempting to codify nautical notation.

osmosis The movement of water across the barrier of a semipermeable membrane, such as a cell membrane. The water flows from a region with a greater concentration of water to one of lesser concentration. This means that water moves in the direction of a greater salt content. If water outside a cell has a greater salt content than water in the cell, water will move out of the cell, dehydrating it.

Animals in the sea have chloride levels in their bodies that mirror the chloride content of the seawater (i.e., they are isotonic with the seawater). Some animals can adjust to changes in the salt content of the water, not becoming bloated when it drops or becoming dehydrated if it increases. The hagfish is such an animal. In most bony fishes the kidneys and gills are the barriers to the entry of salt.

The mechanism by which migrating animals adjust to changes in the salt concentration of water is not completely understood. Fish such as the salmon may go only one way—such as from fresh- to salt water—without need for an adjustment in the mechanism to deal with an increase in the salinity of the ambient environment. Other animals live to return again to an environment of different salt (especially chloride) content. They can therefore change the permeability of the membranes of their gills, digestive systems, and kidneys.

Aquatic mammals have the lowest blood salt levels of all aquatic animals. It is, however, higher than that found in terrestrial mammals. *See* CHLORINE, MINERALS IN SEAWATER.

osprey Also known as the fish hawk. A large, fish-eating hawk known on all continents. It has a wingspan that ranges from about 60 to 65 cm (2 to 2.25 feet), is a dark brown bird with a white underside and some white head feathers. The osprey fishes by hovering over and then seizing its meal. It breeds on rocky islands, cliffs, and in some trees. The osprey population, which had been declining since about 1900 because of inroads on its habitat, was severely crippled by the introduction of chlorinated insecticides. As with other predatory birds, the effect of DDT accumulations in the adult birds disturbed the mechanism by which the eggshells are formed, resulting in thinner than normal shells. This meant that adult ospreys, in attempting to cover their eggs in order to hatch them, crushed them. The osprey has been making a comeback since the ban on the use of DDT. *See* FISH, POLLUTION, PREDATOR-PREY RELATIONSHIP.

Osteichthyes The class of bony fish. Their skeletons contain true bone; they have sutural (composites of several bones that fuse together) skulls and fins. There are more than 18,000 species of bony fish—they are the dominant modern form. *See* COD, HERRING, GROUPER, FLATFISH, EELS, SALMON, MACKEREL, POLLACK, WRASSE.

Ostracoda Tiny arthropods known as mussel or seed shrimp. The name is derived from their appearance. They have seemingly lost their segments, and have a large carapace. They are small (1 mm or 0.04 inches) bottom dwellers that feed on plankton or detritus. Some are parasitic. The 2,000-odd species that exist now include some freshwater and terrestrial forms. The 10,000 extinct species date from the Upper Cambrian to the Pleistocene. Some genera can be used as index fossils (characteristic reference points) in the exploration for mineral deposits, especially for petroleum. *See* CAMBRIAN PERIOD, FOSSIL, PETROLEUM, SHRIMP.

otolith These are the "earbones" of bony fish. Since they grow in layers that are deposited continuously, they are a measure of the age of the fish. The otoliths vary in size and color; they are white when new and darken with age. Since they continue to accrete as the fish lives, they are also a measure of the availability of food, pollution levels, and other environmental conditions.

otter A semiaquatic mammal of the weasel family; usually a riverine animal. The sea otter (*Enhydra lutris*) of the North American West Coast was hunted to near extinction in the early 20th century. Protection in 1910 led to a gradual increase in their numbers, and now they are claimed to be a threat to abalone beds.

The otter is a strong swimmer. It feeds on fish and also on mollusks (usually abalone), which it opens by smashing them on a stone. Otters are playful, large (75 to 120 cm or 30 to 50 inches long), thickset (15 to 40 kg or 32 to 70 pounds) animals. They have thick red-brown to dark brown lustrous fur. Their heads and faces are usually gray. They float on their backs in groups (between gorgings) and generally seem to enjoy themselves. *See* ABALONE, PREDATOR-PREY RELATIONSHIP.

overfishing *See* ENDANGERED SPECIES, FISHING INDUSTRY, WHALES.

ovoviviparity A property of some invertebrates, fish, and reptiles involving the incubation of growing young within the body of the female. The eggs are fertilized internally, and the developing embryos are nourished by large eggs rather than placental tissue. The young are hatched within or right after extrusion from the parent. Several species of shark exhibit this characteristic. *See* FISH, SHARK.

oxygen A chemical element that as a gas (O_2) forms about 21% of the Earth's atmosphere, and in combined form, about 50% by weight of the Earth's crust. It is also a key constituent of organic compounds and all living matter. Dissolved in seawater (H_2O) as a gaseous molecule (O_2), the concentration of oxygen varies considerably from 4 to about 6.5 ml/liter (milliliter of oxygen per liter of water). The concentrations of oxygen in the water covering the continental shelves are higher—in the 5 to 8 ml/liter range. The concentrations of oxygen in tropical surface waters are generally low.

The oxidation of organic material and the metabolism of animals depletes the Earth's oxygen supply, while the activity of algae in the photic zone of the oceans replenishes it. The zone of minimum oxygen concentration is that at intermediate depths (800 to 1,500 m).

Tiny amounts of the allotropic form of oxygen, O_3, known as ozone, are present in seawater. This form of the element is much more reactive than is O_2. The greatest concentration of O_3 on Earth—although the quantities are still very small—is in the stratosphere, where it is produced by a photochemical reaction.

The oceanic dumping of waste materials of all sorts—industrial, sewage, agricultural, and radioactive—removes oxygen from the water. A crude measure of the quantity of oxygen needed to react with waste material and incorporate it harmlessly into the environment is the biological oxygen demand, or BOD. Since the supply of oxygen is relatively constant, areas with a high BOD are those with a diminished flora and fauna.

The ratio of two isotopes of oxygen, O^{16} (the more common one) and O^{18}, is used as a dating mechanism. The water containing the lighter and more prevalent isotope is more likely to be in the vapor state. When that water vapor moves toward the poles and falls as snow, it is compacted into the glaciers. Thus, glacier ice is richer in O^{16}. The water in the oceans is richer in the heavier isotope, and this differential can be used to date ice cores since each annual layer begins with an O^{18} line; the snow that contains it fell first. Since oxygen is also incorporated into the tests of formanifera, the presence of the heavier isotope in ancient tests can be used as a measure of the temperature and salinity of ancient seas. This is done either indirectly—the oxygen-18 measure is compared with the concentration ratios of magnesium to calcium in shells, since the concentration of magnesium rises with increase in temperature—or directly, by examining the salinity of ancient seas. The saltier the water is, the lower the temperature at which it freezes, and this ancient frozen water is present in polar ice.

Part of the program to look for life on other moons or planets is to look at analogs on Earth and studying the early ocean. The primordial oceans had no free oxygen dissolved in the water. This condition persisted for about the first 2.5 billion years of Earth's history. At present, the oceans are relatively oxygen-rich and have been so for about the last 0.5 billion years. The intervening 1.5 to 2 billion years of relatively anoxic water are the subject of study. It is known that life began during this anoxic period, and some eukaryotes appeared as early as 2.7 billion years ago, but the advent of relatively oxygen-rich water spurred the development of eukaryotes—the precursors of multicellular organisms. *See* EUKARYOTE, MINERALS, NITROGEN, NITROGEN CYCLE, OCEAN, POLLUTION, PROKARYOTE, SALINITY.

Oyashio Current A northern Pacific Ocean flow, also known as the Kurile Current. It is a cold, polar current whose counterpart in eastern North America is the Labrador Current. The Oyashio flows southwest along the Kamchatka peninsula after having moved west across the Bering Sea. It meets the Kuroshio at about 35 to 37° north latitude, and some of the cold water sinks below the warm current. All of the current is traceable because of its low temperature (4 to 5°C) and low salinity, (3.37 to 3.4%) which is in contrast to the warm (about 12 to 18°C, depending on the season), and saltier (3.45 to 3.50%) Kuroshio.

The Oyashio carries along a much greater plankton and nutrient load than does the Kuroshio. Hence, the Japanese have given it the name it has, which means parent. *See* CURRENTS, GYRE, KUROSHIO CURRENT.

oyster A gray, bivalve mollusk with one convex and one flat shell, of either the Ostreidae (true oyster) or Aviculidae (pearl oyster) family. Edible oysters are native to almost all waters of the Northern Hemisphere. They have been a foodstuff sought after by humans for millennia and a cultivated foodstuff certainly as long ago as Roman times. Neolithic middens (garbage dumps) contain oyster shells.

Oysters range in size from 7 cm (2.5 inches) to the 30-cm (12-inch) diameter of *Crassostrea gigas* of Japan. The young spat is a ciliated, planktonic swimmer (the larva is called a veliger) for a few days after hatching. It then attaches itself to a rock and filters water for food. Oysters are harvested commercially when they are between three and five years old.

These animals breed in summer and produce incredible numbers of eggs per female: frequently over one million eggs per breeding season. The eggs are fertilized in the water. Some species are hermaphroditic (the *O. edulis* of western Europe) and some, *C. virginia* (of the American East Coast) are either male or female, with sex changes occurring several times in the lifetime of an individual.

All oysters form pearls, which are layers of calcium carbonate ($CaCO_3$) deposited around an irritant, such as a grain of sand. Only some pearls are lustrous; the best are from the species *Meleagrina vulgaris* of the Persian Gulf. Cultivated pearls result from an artificially introduced irritant. The oyster is then grown in a protected environment for several years before the pearl is taken out. This is a major industry in Japan.

Animals that are predators of oysters include the oyster drill (Murex), sea stars, fish, birds, and sea mammals. *See* MARICULTURE, MOLLUSK, MUREX, VELIGER.

ozone This is an allotrope of oxygen. The normal oxygen molecule is O_2 (dioxygen), and there are two atoms bonded together. In ozone, the formula is O_3. This trioxygen molecule is present in the normal atmosphere almost exclusively in the stratosphere. The concentration varies with the seasons. At maximum, it is about 10 ppm (parts per million). It is crucial in creating a shield that blocks some of the ultraviolet solar radiation. Ozone is formed in manufacturing and by the sparking of electric motors. In the troposphere, it is a pollutant. It has an odor even in minute concentrations; it is the "fresh" smell noticed in the air after a lightning storm. Ozone is extremely toxic.

The ozone "hole" that was first noticed in the late 1970s over the Antarctic was soon perceived as a danger since the absence of ozone in the stratosphere affected Earth's surface, as increased ultraviolet and cosmic radiation penetrated through the troposphere. Eventually all nations agreed to stop using chlorofluorocarbons as aerosol propellants. These very low molecular weight gases diffuse through the atmosphere and rise into the stratosphere, where they eventually react with the ozone, thereby reducing its concentration. The immediate effect of lessened ozone in the atmosphere is an increase in skin cancer incidences. The ozone concentration is expected to return to its former levels by about 2070. *See* STRATOSPHERE.

P

Pacific Bottom Water Cold water at the bottom of the Pacific Ocean. It originates in Antarctica and spreads north to become the major constituent of the Pacific bottom. It is detected as far north as Japan. Surface water sinks at the Antarctic and Tropical convergences. At the latter, the equatorial countercurrent spreads water laterally, and as it sinks, this water moves poleward. Upwellings of cold water are well known on both the North and South American coasts. *See* ANTARCTIC CONVERGENCE, EQUATORIAL CURRENTS.

Pacific Decadal Oscillation A phenomenon that is a long-term temperature oscillation. Its period is 20 to 30 years. This temperature change was first detected by the *TOPEX/Poseidon* satellite. The present ocean temperatures indicate that Earth is at the beginning of a relatively cool period that began in about 2004. It manifests itself as a lower-than-normal sea temperature and wave height in the eastern Pacific and higher-than-normal temperature and wave heights in the northern, western, and southern parts of the Pacific. The years 1977 to 1999 were a warm phase near Indonesia and Micronesia while the western Pacific was cold. This phenomenon shifts the jet stream and the attendant storms that it brings. This fluctuation of climate occurs in addition to the changes created by the El Niño and La Niña events. *See* EL NIÑO, LA NIÑA.

Pacific Ocean An ocean extending from the Arctic in the north to the Antarctic in the south and from the Americas in the west to Asia and Australia in the east—it is the largest body of water on Earth. It is

separated from the Atlantic Ocean by the Drake Passage and from the Arctic Ocean by the Bering Strait. The boundary of the Pacific with the Indian Ocean runs across the Timor Sea to Cape Londonderry, Australia, and across the Bass Strait between Australia and Tasmania. The Pacific Basin may be considered as three distinct areas: east, west, and central. The eastern Pacific is the area stretching from the American coast to the Hawaiian Ridge; the central Pacific is a plain at average depths of about 5,000 m (16,500 feet). The western Pacific extends from the line of trenches (the Aleutian, Kurile, and Tonga) to New Zealand. Most of the islands in the Pacific are in the western Pacific.

The Bering, Okhotsk, Yellow, and Timor Seas and the Bay of California are marginal to the Pacific Ocean.

The water currents of the Pacific are wind-driven. In terms of the quantity of water moved, the major current is the Antarctic Circumpolar Current. Others are the Kuroshio, North Equatorial, Equatorial Undercurrent, North Pacific, North Equatorial Countercurrent, California, and Peru currents.

The salinity of the Pacific varies with geography and temperature; it is highest in warm regions, where evaporation exceeds runoff, and varies between 3.55 and 3.65%. It falls below 3.3% at the poles. The oxygen content is highest at the surface, drops with increasing depth, and increases again at depths greater than 4,000 m (13,000 feet). The nutrient content, particularly of phosphates, is highest in the intermediate layer, where the oxygen concentration is lower. Upwellings along the American coasts bring nutrients up into the photic zone, and this results in the characteristic spectacular growth of plankton.

In addition to its great size, the basin of the Pacific is varied and geologically interesting. It has more oceanic bottom features (ridges, trenches, seamounts, mineral deposits, coral, hot spots, and islands) than do the Atlantic and Indian Ocean basins.

Voyages across large expanses of water in the Pacific began about 2,000 years ago with Polynesians in canoes using stick maps. A stick map is an accurate depiction of oceanic territory, with islands or reefs shown as focal points (i.e., conjunctions of several lines of rods). Information about winds, currents, and shoals was built into these maps. The adventurous, striking out over vast areas, were preceded by earlier coastal voyagers. Thus, stick maps came after a long period of sailing experience.

Others also sailed the Pacific. There were Chinese-Arab trade relations by the first century C.E. and the possible exploration of the Americas by the Chinese. The lure of exotic treasures and fortunes to be made led many others on expeditions. While trade was the accepted reason for most Pacific travels, there were later voyages of colonization, missionary activity, and piracy. *See* EVOLUTION OF OCEANS, MID-OCEAN RIDGES, NAMES OF INDIVIDUAL MARGINAL SEAS AND CURRENTS.

Pacific Plate The plate essentially coincident with the main or central Pacific Basin. This is a plain from which ridges and seamounts rise. It is cut by a series of fracture zones perpendicular to the ridge lines. The average depth of the plain is more than 5,000 m (17,000 feet).

The East Pacific Rise and its extensions form the eastern boundary of the plate. The North American, Cocos, and Nazca plates are east of the Pacific Plate. The plate edges are delineated by trenches and islands, frequently in double arcs. The inner one is seismically active, the outer one is an aseismic ridge line.

The southwestern edge of the Pacific Plate is the "Andesite line," or the dogleg described by the New Hebrides and Tonga island arcs. There is some dispute over the composition of the crust under the marginal seas of the plate, especially those of the southwestern Pacific. They have been called epicontinental by some; other researchers dispute this nomenclature. The Pacific is neither a "new" ocean nor the site of a "drowned" continent. *See* ANDESITE, CRUST, INDIVIDUAL PLATES, PACIFIC OCEAN.

Pacific-type margin The leading edge of a continent. At the convergence of two plates this leading edge is thrust upward. *See* ATLANTIC-TYPE MARGIN, CONTINENTAL DRIFT.

pack ice Any mass of floating pieces of ice, 3 m (10 feet) or less in diameter, driven together to form a solid layer. If there is a cover of pack ice in a waterway, it will stop all shipping except for icebreaker vessels. *See* ICE.

Paleocene epoch The oldest epoch of the Tertiary period of the Cenozoic era. The predominant terrestrial rock of the Paleocene is sedimentary. The name for this epoch, signifying "ancient recent time," was suggested by the botanist W. P. Schimper in 1874.

The Paleocene epoch lasted about 10 million years, commencing about 70 million years ago. It saw two cycles of mountain uplift and the retreat of the sea. The Laramide orogeny—the upthrust of the Rocky Mountains—occurred in this period. The fauna of the Paleocene included some holdovers from the Cretaceous: cephalopods, pelecypods, gastropods, echinoids, and foraminiferans found favorable growing conditions. They repopulated many ecosystems after the very noticeable disappearance of the belemnites, ammonites, and fishlike dinosaurs. Reptiles moved onto land and became snakes, lizards, and crocodilians. The outstanding animal development during the Paleocene was terrestrial. *See* CENOZOIC ERA, TERTIARY PERIOD.

paleogeography The study of the shapes of the earth's crustal plates and the changes that have occurred in them. Mapping projects have studied several areas, including the Caribbean. NOAA is studying paleoclimate. The JANOS databases are compilations of data taken from ocean drilling projects, and its areas of study are the paleontology, geology, and geophysics of ocean rocks and sediments. *See* NOAA; *specific plates, areas, or organisms.*

paleontology The study of fossil plants and animals, and of the rock formations in which these fossils are found. The science was named by Ducrotoy de Blainville and Fischer von Waldheim in 1834. Before that, it was an area of natural science that had absorbed the interest of many professional scientists. Linnaeus, Buffon, Cuvier, Agassiz, Hutton (qq.v.), and others used paleontological specimens in their work before the official name was given to their area of study. Large finds are always newsworthy events. Finding a fairly complete *T. rex* fossil was important, as was the find of an even larger sauropod in Argentina. Fossil finds in new places are also exciting. The plains of Kansas were the oceans of Kansas up until the late Cretaceous period. The predominant fossils found there are mosasaurs, plesiosaurs, pterodons, and giant sharks. The most prevalent fossils are the tiny ones—the nannofossils that were calcite or aragonite plates on the phytoplankton. When these plates fall, they become incorporated in the bottom sediment. When that hardens into rock, the amount of calcite determines what kind of rock it is. Rocks that are almost exclusively nanoplankton shells are chalks; those with fewer plates are limestone or shale. *See* COCCOLITHOPORES, CONODONTS, DIATOMS, DINOFLAGELLATES, FORAMINIFERA, FOSSIL.

Paleozoic era The period of earliest multicellular life on Earth. The name signifies "old life." Followed by the Mesozoic, the Paleozoic ended about 245 million years ago. It is divided into a Lower and an Upper segment. The Lower (or older)

Paleozoic is further subdivided into the Cambrian, Ordovician, and Silurian periods; the Upper Paleozoic is subdivided into the Devonian, Carboniferous, and Permian periods.

The climate during this very long era seems to have remained fairly constant and warm. While warm seas overlaid many land areas, folding and downwarping raised mountainous regions above the water.

The flora of the period was dominated by algae in the water; on land, mosses and ferns had emerged. The fauna, however, was entirely aquatic. Trilobite fossils dominate the Cambrian remains. By the Upper Cambrian, cephalopods, brachiopods, and graptolites were present, and bryozoans, gastropods, and ostracods developed. By Silurian times there were fish.

The Upper Paleozoic saw several cycles of advance and retreat of the seas. The climate was much more variable, and deserts appeared in the Permian period. The biota changed markedly as the climate changed. There were several ice ages in the late Paleozoic and a major extinction at its end. *See* CAMBRIAN PERIOD, CARBONIFEROUS PERIOD, DEVONIAN PERIOD, PERMIAN PERIOD, *Appendix: Geologic Timescale.*

Paleozoic marine revolution Referring to the gradual increase in the population of predators that eventually changed population dynamics in the oceans. Brachiopods were the dominant organism in the ocean from about 550 million to 250 million years ago. This is the span of the Paleozoic era. Other shelled creatures that drilled into the brachiopod shells and ate them began to appear early in the Paleozoic. Over time, the drills became better equipped, as did the prey organisms. Ever-heavier shells became standard in brachiopods, but over time the brachiopod population shrank; about 95% of it was extinct by the end of the Paleozoic. The drills have survived and prospered. About 25% of all mollusks in the oceans now are eaten by other mollusks that drill through their shells. *See* BRACHIOPODA, PREDATOR-PREY, WHELK.

palolo A polychaete worm of the genus *Eunice* that swarms in a pattern strictly geared to a lunar timetable. This worm is about 40 cm (16 inches) long and ordinarily is a bottom dweller in warm tropical seas. It produces an enormous sac of eggs or sperm in its tail segment. This is synchronized with the third lunar quarter in June or July in the Atlantic Ocean near Florida, and in October or November near Samoa in the Pacific. The entire tail segment of the animal rises to the surface and ruptures, releasing the eggs or sperm. The tail portion resembles a worm and even has eyes. It "lives" for a very short period, with the whole swarming period lasting only a day or two. The unruptured tail is considered a delicacy and is caught and eaten on the spot. The female tails, which are blue, are considered the best. Male tails are greenish blue. *See* POLYCHAETA.

pancake ice New sea ice that is formed in autumn. The round pieces of ice are between 30 cm and 3 m (1 to 10 feet) in diameter. These are in turn coalesced into larger units. When they are pushed together, the relatively flat surfaces may become buckled, and the resulting floes are not as flat as the pancakes that formed them. *See* FLOE, ICE.

Pangaea Alfred Wegener's name for the primordial supercontinent. He believed that the breakup of Pangaea in the Cambrian was the result of tidal action within the earth, or a centrifugal force. While he could see the present continents fitting together, he could not know of the crustal plates that the continents perch on or the upwellings of magma that constantly change ocean bottoms. When continental drift began, Pangaea broke up into Laurasia and Gondwanaland. *See* PLATE TECTONICS; WEGENER, ALFRED LOTHAR.

Panthalassa The primordial ocean that surrounded the supercontinent Pangaea before the latter's breakup in the Jurassic period. *See* CONTINENTAL DRIFT, EVOLUTION OF OCEANS, PANGAEA.

paper nautilus An octopus, *Argonauta*. The females manufacture a thin mucus envelope in which to brood their eggs, which hardens to an opaque, seamless cover, from which the name "paper nautilus" was derived. The males of the species are much smaller than the females; they were once thought to be a separate species. The sperm-transferring organ of the paper nautilus is a very specialized arm that in the 19th century was described as a species of worm. *See* OCTOPUS.

parapodia Projections on the segmented, marine annelid worms. Parapodia with cirri (curls) or setae (bristles) on their surface are characteristic of the polychaete worms. *See* ANNELIDA, POLYCHAETA.

parasite An organism living in or on another organism (the host) and depending on the host for nourishment without giving anything in return. Many if not most marine animals may at some time in their lives harbor parasites. A great number of the known parasites are worms. Because parasites constitute an enormous group of often unrelated organisms, their mode of attack and their hosts are also unrelated.

parasitism A relationship between two organisms in which one benefits at the expense and to the detriment of the other. *See* COMMENSAL RELATIONSHIPS, SYMBIOSIS.

parrot fish A group of tropical marine fish related to the wrasses. They are brilliantly colored and have oblong bodies with relatively small fins and tails. Their coloration and large beaky heads explain their name. Most parrot fish are herbivores, but some prey on coral fauna.

The largest parrot fish, *Pseudoscarus guacamia,* may grow to about 1 m in size and weigh about 14 kg (30 to 31 pounds). This species and several others are eaten, but on the whole the parrot fish is not considered an object for commercial fishing.

COMMON MARINE PARASITES		
Parasite		**Host**
Phylum	*Class*	
Platyhelminthes	Turbellaria	Internal parasites of crabs, mollusks, echinoderms
	Monogenera (fluke)	External on fish gills
	Cestoda (tapeworms)	Internal parasites of all crustaceans, and all vertebrates. Tapeworms are very specific about their vertebrate hosts.
Aschelminthes	Nematoda	All marine species
	Rotifera	External; crustacean gills
Acanthocephala		Internal; isopods, fish, birds, mammals
Annelida	Myzostomaria	External; crinoids
	Hirudinea (leech)	Internal; fish
Arthropoda	Crustacea (Branchiura)	External; bony fish
Chordata	Cyclostomata (lamprey)	External; bony fish

Passeriformes Nicknamed "perching birds," this is the largest order of land-based birds, comprising about 10,000 known species of finches, robins, rooks, sparrows, tits, and warblers. Most birds connected with the sea are not passerines; they are members of the following families: Laridae (auks, gulls, skuas, and puffins), Sulidae (boobies and gannets), Pelecanidae (pelicans), Fregatidae (frigatebirds), Spheniscidae (penguins), Gaviidae (divers and loons), and Procellariidae (petrels, albatrosses, and shearwaters).

pea crab The genus is *Pinnotheres*. These are hard shell crabs that are about as large as a green pea. The males are usually commensal, living within the shells of bivalves or tube worm shells, and are much smaller than the females. The females are most often parasites, living within the mantle cavity of bivalves.

peanut worm A small phylum of marine worms, the Siphonculida, that, with the annelids, are probably descended from the same ancestor. These, however, are not segmented. The exterior is an elongated body with a thickened trunk and a narrower front end where the mouth and tentacles are located. It cannot really be called a head because it seems to be able to turn itself inside out for part of its length. The worms are mostly gut. They are distributed in almost every benthic environment. They are dioecious (male and female reproductive organs borne on separate individuals of the same species) detritus feeders.

pearl A hard, smooth, lustrous, protective structure formed around an irritant by the mantle of a shelled mollusk whose shell interior is nacreous (mother-of-pearl). The coating applied to the irritant consists mainly of calcium carbonate and some proteins. A round or ovoid pearl forms in a part of the mollusk not subject to muscle pressure. If the pearl forms where it is affected by muscle action it is deformed (baroque) or flattened.

Gem-quality pearls are found in both marine and freshwater mollusks. They vary in color from white to cream, rose, lavender, green, gray, and black, depending on the locale and the producing organism. A cultured pearl is a pearl formed in a captive oyster by inserting an irritant into the oyster and then maintaining it in a protected environment for several years. This has become a major industry in Japan. The best natural pearls in terms of luster and color are supposedly those of the Persian Gulf. *See* MANTLE, MOLLUSCA, OYSTER.

Peary, Robert Edwin (1856–1920)

An American naval officer and explorer who is credited with being the first man to deliberately arrive at the North Pole. In 1891 Peary organized his first attempt at attaining 90° north latitude; it was unsuccessful. He explored the northern coast of Greenland in subsequent summers, and established the existence of water under the ice north of the island. He designed an ice-withstanding ship, called the *Roosevelt,* in which in 1905 he sailed north to 87°6' before turning back. He tried again in 1908, and finally arrived at the Pole on April 6, 1909. This expedition established the Pole's existence on an ice platform: the North Pole is not on land, as the South Pole is.

Frederick Cook, a doctor on one of Peary's expeditions, claimed that he, not Peary, had reached the Pole first, almost a year earlier than Peary did, but eventually his claim was discredited. *See* EXPLORERS AND EXPLORATIONS.

pelagic environment

The open ocean, excluding the ocean bottom and the shore. This is by far the largest area of "open space" on Earth. The organisms in this space are grouped into arbitrary categories by size and type. The categories frequently differ from one researcher to another. The main categories are plankton, both plant and animal, and nekton. Some biologists include fungi and nonphotosynthesizing bacteria, along with the diatoms, dinoflagellates, green, brown, and red algae, coccoliths, and autotrophic bacteria as members of the phytoplankton. The familiar zooplankton include copepods, foraminiferans, tunicates, cnidarians, amphipods, pteropods, and chaetognaths. Other, less familiar groups of animals are also in this category of pelagic organisms.

A wide range of organisms previously unknown to researchers has been identified since the 1990s based on improved collecting techniques. Plankton nets have been used down to depths of 500 m (1,600 feet), and ROVs (remotely operated vehicles) have been used to photograph life down to depths of more than 2,800 m (9,000 feet). The range of ctenophores, siphonophores, and hydromedusae in what was once called empty ocean is startling. Gelatinous creatures inhabit all oceans, though there are fewer species in polar water. The organisms in every biome are arranged in a column with the phytoplankton closest to the surface. These are fed on by copepods that are preyed on by ctenophores, and below them are the scyphomedusae.

The largest individuals in the open ocean are those capable of swimming against the current for some time. These constitute the nekton, which includes all open-ocean fish, sea mammals, large cephalopods, and some large crustaceans.

Pelagic populations in any one place depend on the availability of sunlight, current flow, nutrient concentration, salinity, and temperature. *See* INDIVIDUAL ANIMALS, NEKTON, PLANKTON.

pelagic sediment

An ocean-bottom deposit of material from seawater. The greatest contributions to this sediment are the remains of coccoliths and foraminiferans. Such deposits are calcareous, (calcium-containing). Diatomaceous and radiolarian remains also eventually drift to the bottom, producing siliceous, (silicon-containing) deposits. Phosphate deposits, which generally consist of apatite, originate from the skeletons of the bony fish.

Mineral deposits in the pelagic sediment are present as encrustations, usually consisting of oxides of iron and manganese. Minute grains of quartz and feldspar fall on the sea as dust and eventually form part of the pelagic sediment. Some extraterrestrial iron and nickel, which come from meteorites, are also parts of this layer on the sea bottom. See CARBON-ATE, COCCOLITH, OCEAN FLOOR, MANGA-NESE, MINERALS, RADIOLARIAN.

Pelecypoda Bivalves, one of the largest orders in Mollusca that includes clams, oysters, mussels, and scallops. They are found in all aquatic environments, fresh- and salt water, and at all depths. Their ancestors appeared in the Cambrian period and continued to produce new forms on through to the early Ordovician.

The most obvious characteristic of bivalves is their shell, which is formed from secretions of the mantle. Some bivalves have nacreous (mother-of-pearl) shells that are of economic importance. Many bivalves are edible and as such are sought after and economically important. Others are destructive: The shipworms, for example, are not worms at all but mollusks. See CLAM, MOLLUSCA, OYSTER, PEARL, SCALLOP, TEREDO.

pelican One of a small group of water birds of the family Pelicanidae. Pelicans live on rivers, lakes, and seacoasts in tropical to warm-temperate areas. They are related to the gannets and cormorants. Most pelicans nest in colonies.

Their nests are most often in scrub trees that are found in wetlands. The young are fed on regurgitated food—mainly fish and mollusks—held in the parents' pouched throat. Ancient depictions of the pelican have it feeding its young on its own heart's blood—a fable based on inaccurate observation.

The American white pelican is found in the Everglades of Florida. It is the largest example of the group and has an average wingspan in excess of 3 m (10 feet). The survival of the genus was in doubt in the early 1970s, but it has recovered from the effects of DDT poisoning, although its territory is still being encroached upon by human activities such as the channeling of the Everglades. This reduces the range of this bird and of other species. See BIRD.

penguin A flightless, erect bird of the family Spheniscidae, order Sphenisci-formes, that is adapted to swimming in cold seas and found mainly in the Southern Hemisphere. There are about 15 species of penguin.

Penguins breed in colonies on cold, rocky coasts from California and the Galápagos to Africa, Australia, New Zealand, and South America. The emperor and Adelie penguin are Antarctic species, and the most "typical" penguins.

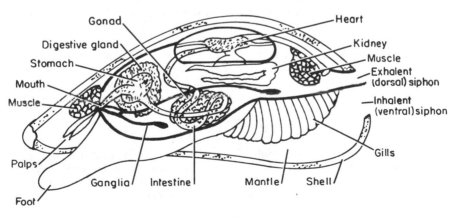

Typical pelecypod

The group ranges in size from the small blue penguin, which is about 40 cm (16 inches) high, to the largest, the emperor penguin, which stands 120 cm (4 feet) high. All of the penguins are dark on their heads and backs. Their dark backs range from light brown to black. Penguins are extremely skillful swimmers and divers. Tagged emperor penguins have been found at depths exceeding 200 m (660 feet). They have white undersides, webbed yellow or orange feet, and flipperlike wings.

Both sexes of adult penguins look alike. In the breeding season, the parents take turns caring for the one or two chicks. Depending on the species, the young are fed regurgitated meals of fish, squid, or crustaceans. The diet of the emperor penguin, which is the deepest diver, consists almost exclusively of squid.

Penguin chicks are herded together in nurseries by the attending adults. This is part of the attempt to protect the young, which are preyed upon by skua (seabirds related to gulls), which steal the eggs. The predators that attack adult penguins are seals and killer whales. *See* BIRD.

peninsula A landform that juts out into a body of water; an extension of a larger landmass. The Arabian, Iberian, and Florida Peninsulas are examples. *See individual peninsulas.*

Pennatulacea An order of octocorals or, as they are commonly known, sea pens or sea pansies. These colonial organisms are noted for their polyp dimorphism. The form in which they are seen is the polyp. Within the colony there is some differentiation. The organism that begins the colony grows quite large, sheds its tentacles, and is the anchor that attaches the rest of the structure to a rock. Other polyps, called autozooids, are the filter feeders for all the organisms. Other specialized colony members are the siphonozoiids, which are the water conduits for the colony. All the organisms create the central rod of calcium carbonate. This is the only hard structure in the colony. As such it has been identified in the fossil record. Pennatulaceans were well established by the Cenozoic and have been dated back to the Precambrian. *See* CORAL, SEA PEN.

Pennsylvanian epoch The part of the Paleozoic era that produced huge coal deposits. It was characterized in North America by a huge oceanic area in what is now Kansas and Missouri. The marine deposits there date from this period. *See* PALEONTOLOGY.

peptide formation The linking together of amino acids to form polymers. The shorter polymers are known as peptides; the longer ones as proteins and polypeptides. The amino acids themselves are produced by the same cells that manufacture nucleic acids and carbohydrates.

The peptide linkage is represented chemically by the $\overset{\displaystyle O}{\overset{\displaystyle \|}{\text{H—N—C—}}}$ bond in the structure:

This polymer chain has a repeating structure. The side chain, X or X^1, can be any atom or group. This means that considerable variation in the side chain and in the length or complexity of the main chain is possible, although only about 20 different side chains occur in nature. While the amino acids are common to all life, the particular peptide, polypeptide, and protein polymers they create are unique. Each such molecule is determined by the DNA and specific enzyme systems of a particular species of organism.

The enzyme systems that regulate protein formation are extremely conservative and do not differ greatly from one species to another. The proteins in related species do have differences, but they are slight. The more closely related two species are,

Peptide formation

the more similar are their proteins. Evolutionary studies based on protein structure are now quite common. *See* AMINO ACID, COMPOUND, DNA, PROTEIN.

Perciformes The order of spiny-finned fish. This is the dominant form of fish. It is the largest vertebrate order and the most diverse, and as a result, it is frequently redefined.

perigee That part of an orbit of a satellite (natural or artificial) at which the satellite is closest to the planet it is orbiting. *See* APOGEE, LUNAR TIDE, PERIHELION.

perihelion That part of the orbit of a planet or comet at which the orbiting body is closest to its sun. *See* APHELION, LUNAR TIDE, SOLSTICE.

period (Geologic) That part of a geologic era characterized and dominated by a particular life form. *See Appendix I.*

periwinkle A small, widespread, marine snail of the genus *Littorina,* found on both sides of the Atlantic Ocean on intertidal rocks and flats, and feeding on algae. The periwinkle is usually gray and has a thick whorl. It is about 2 to 3 cm (0.8 to 1.2 inches) across and is known to anyone who has walked on a beach. Periwinkles are edible and are used in Western European and Far Eastern cookery. *See* GASTROPODA, LITTORAL, SNAIL.

Permian period The last period of the Paleozoic era. This period is named for the Perm region of the western Ural Mountains in the Soviet Union. The best-known North American formations that date from the Permian are in Texas. There were long periods of drought in the Permian period that resulted in desert conditions on land. There were also several ice ages and glacial deposits, and scars of these exist on all of the Earth's southern continents. The presence of glacial scars, all from the same time period but on different landmasses, cannot be coincidental, and supports the continental drift theory. The continental landmasses must have been closer together than they are at present, and it is now postulated that the single supercontinent, Pangaea, was centered near the South Pole and broke up during the Permian.

The Permian period lasted about 20 to 30 million years and was known for mountain building: The Appalachians, Urals, and highlands of central France were pushed up at this time, as were the northern Himalayas.

Marine life during the Permian was dominated by brachiopods, bryozoans, and goniatites, all of which were either extinct or greatly reduced in number by the end of the Permian, which was also a time of "great extinction." It is estimated that 90 to 95% of all marine life on Earth became extinct at the boundary of the Permian period and the Triassic period. There is considerable ongoing research to determine the causes of this major change. One possible cause may be the change in the chemistry of the ocean. Pyrites are relatively rare in the layers of earth. Looking at the sulfide content of rock samples may provide a clue to the sulfur content of ancient oceans, and examining the C^{13} content of the carbonates may account for the possibility of a drop in CO_2 content in the water. It is proposed by some that sulfides from deep bottom water were introduced into the upper layers of ocean. This upwelling may have occurred several times. This change in the water chemistry would cause great damage to the ecosystems of most of the ocean and would particularly affect life on continental shelves.

Another research effort is directed toward finding the remains of a meteor crater. An alternative view of what precipitated the end of the Permian is that a large extraterrestrial body crashed into the Earth, causing a series of earthquakes, firestorms, and climate changes. A hollow in the Earth's crust near the northwestern Australian coast is being examined as a possible site of such a meteor strike. *See* CONTINENTAL DRIFT, EXTINCTION, PALEOZOIC ERA.

Peru-Chile Trench Also known as Atacama Trench, this ocean deep is in the Eastern Pacific. It parallels the coast of South America for part of its length, then turns south. The deepest point is the Richards Deep at 7,635 m (about 25,000 feet). *See* PACIFIC OCEAN, TRENCH.

Peru Current An eastern Pacific Ocean current also known as the Humboldt Current in honor of Alexander von Humboldt, who charted it and measured its flow in 1802. It is a cold, broad (900 km or 580 miles wide) surface current that flows north along the western South American coast, originating in the West Wind Drift in antarctic waters. The Peru Current separates from the major stream of the West Wind Drift as the latter heads toward the Drake Passage. The Peru branch moves north to about 4° south latitude, where it turns westward as part of the South Equatorial Current. The cold Antarctic water is augmented by upwellings that bring cold water to the surface, along with nitrates and phosphates. The upwellings of nutri-

ents provide the basis for an extremely rich and varied fauna.

The phenomenon known as El Niño is a disruption of the Peru Current. Its consequences are economically and meteorologically important even to regions far removed from the South American coast. *See* ANTARCTIC WATER; CURRENTS; DRAKE PASSAGE; EL NIÑO; HUMBOLDT, FRIEDRICH HEINRICH ALEXANDER VON.

petrel A widely distributed seabird of the order Procellariformes. The average petrel is a small black bird with a considerable wingspan. They are most often 25 to 90 cm (10 to 30 inches) long and weigh 100 to 400 g (0.2 to 0.9 pounds). Petrels have tubular nostrils and—uncharacteristically for birds—a good olfactory sense. Their beak is long and straight, with a hooked tip. They have webbed feet and swim well, and use the land only as a breeding ground. The average nest is a hole in a cliff, a spot on a cliff ledge, or a messy collection of seaweed, grass, and feathers on a rocky beach. Most petrel species produce only one or two eggs each year, and the young are slow-growing. The birds' diet is largely crustaceans, small fish, large plankton, and squid. The bulk of the species spend some part of their lives in Antarctic waters. While some petrels and fulmars are often associated with European, Mediterranean, or African shores, the most likely breeding ground for the average petrel is South America.

There are four families of petrels, comprising 92 species. They are the albatrosses, storm petrels (*Hydrobatidae*), diving petrels (*Pelecanoididae*), and petrels and shearwaters (*Procellariidae*). The storm petrel is also known as Mother Carey's chick or St. Peter's bird. It is about the size of a starling and skims above the waves, taking whatever small creature or bit of debris it can find. The diving petrel is also a small member of this order. These birds use their wings as flippers, diving headlong into the waves like penguins. Their diet is more likely to include crustaceans and fish. *See* ALBATROSS, BIRD, FULMAR.

petroleum An oily, viscous flammable liquid, consisting mainly of hydrocarbons, which is refined to produce gasoline, naphtha, and other chemical products. It is found in dome formations on continental shelves and on continents. Offshore petroleum comes mainly from the Gulf of Mexico, the Persian Gulf, and the North Sea. Underwater domes are found by echo sounding and magnetometric techniques. One theory of petroleum formation holds that petroleum is formed in undersea locations by a compacting of organic materials that are brought together and nourished by upwellings.

There have been petroleum findings in depths of 3,000 and 4,000 m (10,000 to 30,000 feet). This would indicate a yet unexplained source of the organic material that produces this petroleum. Such deep deposits would be extremely difficult to tap commercially because the lack of a stable platform would render the recovery of such oil extremely expensive. The cost of petroleum crude oil would have to be quite high before such an operation would be a commercial success. *See* MINERALS, OIL.

Pfisteria This dinoflagellate was first noticed in Chesapeake Bay in the late 1990s; unlike many other toxin-producing dinoflagellates, it is not associated with a dramatic color change in the water. Fish kills and still-living fish with large sores brought this unknown organism to prominence. Its life cycle includes an ameboid state and the more active flagellated stage in which it propels itself through the water. The extent of the effect of this algal pest on fish and shellfish populations is under study. The impact of this on water safety and quality and the results for human health are to be determined. *See* ALGAL BLOOM, TOXIN.

pH A measure of the "hydrogen ion" content of water, and therefore of its acidity. Neutral water, which is neither acidic nor basic, has a pH of 7. A pH lower than 7 indicates acidity, and a pH higher than 7 indicates basicity. The pH of the primordial ocean was probably lower than that of today's seawater. This is theorized on the basis of a greater concentration of carbon dioxide in the atmosphere and therefore in the ocean.

The pH of the ocean is today about 7.2—slightly basic—and is remarkably constant despite the production of ammonia by animals, the dissolution of metallic oxides that would increase the water's basicity, and the production of acidic sulfur compounds by bacteria. This constancy is the result of the action of phosphate and carbonate ions, which buffer the ocean system, or "cushion" it against a sudden, large pH change by accepting large quantities of the hydrogen (H+) ion that causes acidity and giving up hydrogen ion in response to an increase in basicity, thus stabilizing the pH. *See* AMMONIA, CARBON DIOXIDE, SULFUR, WATER.

Phaeophyceae The class of brown algae that are almost all marine and multicellular. These marine organisms contain xanthophylls—pigments that give them their characteristic brown color. Examples of phaeophytes are the thallose "seaweeds" that inhabit the intertidal zones, the Fucales or wracks (e.g., *Fucus*), and the laminariales, or kelps (e.g., *Laminaria*). *See* BROWN ALGAE, *FUCUS*.

Phanerozoic era The division of Earth's life span wherein macroscopic life evolved, also known as the Phanerozoic eon. It began about 540 million years ago and is characterized by finds of organisms that have hard exteriors or bony internal support systems—body parts that can be fossilized. The divisions of the Phanerozoic era are in order of their appearance: Paleozoic (the age of fish), Mesozoic (the age of dinosaurs), and Cenozoic (the age of mammals). The animals named are the dominant ones in each particular age but certainly not the only ones present. *See* CENOZOIC ERA, GEOLOGY, MESOZOIC ERA, PALEOZOIC ERA.

pharmaceutical research The pharmaceutical industry is seriously interested in the use of natural products as lead compounds for new drugs to be used as analgesics or anticancer compounds. Extensive exploration of terrestrial habitats is ongoing; the exploration of marine environments started later, but is now part of basic research in the attempts to find new classes of compounds extracted from marine plants and animals to test for their effectiveness in the treatment of bacterial, fungal, viral and

PHARMACEUTICAL COMPOUNDS OF MARINE ORIGIN		
Compound	*Source*	*Possible Use*
Aurantoside	*Siliquariaspongia* (sponge)	Antifungal
Phorboxazole	*Phorbas* (sponge)	Antifungal
Halishigamide	*Halichondria* (sponge)	Antifungal
Fascaplysin	*Fascaplysinopsis* (sponge)	Antifungal
Meridine	*Corticium* (sponge)	Antifungal
Bengazole	*Jaspis* (sponge)	Antifungal
Ptilomycalin	*Ptilocaulis* (sponge)	Antifungal
Haliclonadiamine	*Haliclona* (sponge)	Antifungal
8-Hydroxymanzamine	*Pachypellina* (sponge)	Antibacterial
Axispmotrole-3	*Acanthella* (sponge)	Antibacterial
Ergorgiaene	*Pseudopterogorgia* (gorgonian)	Antibacterial
Litosterol	*Litophyton* (soft coral)	Antibacterial
Puupehenone	*Verongida* (sponge)	Antibacterial
Jasplakinolide	*Notheia*	Antiparasitic
Geodin	*Geodia* (sponge)	Antiparasitic
Amphilactams	*Amphimedon* (sponge)	Antiparasitic
Sigmosceptrellin-B	*Diacarnus erythraenus*	Antiparasitic (nematodes)
Cyclic peroxides	*Plakortis* (sponge)	Antiprotozoan (Leishmaniasis)
Ascosalipyrrolidinone-A	*Ulva* (green alga)	Antiprotozoan (Chagas disease)
Kalihinol A	*Acanthella* (sponge)	Antimalarial
Squalamine	*Squalus acanthias* (shark)	Antitumor
Brp, psphaerone	*Sphaerococcus* (red alga)	Antibacterial
Jorumycin	*Jorunna* (nudibranch)	Antibacterial
Pestalone	*Pestalotia* (marine fungus)	Antibacterial
Ara-A	*Cryptotethia* (sponge)	Antiviral
Didemnins	*Trididemnum* (tunicate)	Antiviral
Eudistomins	*Eudistoma* (turnicate)	Antiviral
Solenolide-A	*Solenopodium* (gorgonian)	Antiviral

parasitic diseases. Some of the entries in the table are now in clinical trials and are on their way to becoming marketable drugs. Since much of this work is highly competitive, many more compounds are being studied and will appear in the near future.

pheromone A hormone that acts as chemical signal between animals of the same species. The scent trails that include the pheromones may be territorial markings, food locators, or sex attractants. Widely diverse organisms manufacture pheromones.

The typical pheromone is a fairly simple molecule that is produced in very small quantities. It is, however, picked up by the appropriate receptors on a given species of animal even when its concentration can be measured only in parts per trillion. Since artificial pheromones have been successfully used as pesticides by attracting insects into traps, there has been research into the use of pheromones to try to decipher the complex life cycle of the lobster, to act as shark repellent, and to direct the movement of schools of fish. *See* KAIROMONES.

Philippine Sea A marginal area of the Pacific Ocean south of Japan and surrounded by the islands of Taiwan, the Philippines, and the Ryukyus (Okinawa) on the west, and the Marianas on the east. The Kyushu-Palau ridge separates the Philippine Sea into two areas. Other delineating features of the sea are the Kuroshio Current north of the sea and the Equatorial Countercurrent, which forms its southern limit.

The Subtropical Convergence is at about 23° north latitude. It runs through the Philippine Sea and alters the prevailing weather. The weather pattern for the area is Siberian in the northern portion, and the winter temperature near Japan is below freezing. Near Mindanao in the Philippines, the weather is tropical. Typhoons and other cyclonic storms occur, typically in September, and move into the sea from the direction of the Caroline or the Marshall islands.

The Philippine Islands are a mixture of volcanic and coral material. Many are too small to have formal names. The chain of islands starts south of Taiwan. Of the larger islands, Luzon is the northernmost. As the chain arcs to the south, it breaks into two. The islands of Mindoro and Palawan are in the western portion, while Leyte, Cebu, and Panay are in the east. Mindanao is the southernmost large island. The major fault line in the region runs between the two arcs and through the island of Mindanao.

The edge of the Philippine plate is the Philippine or Mindanao Trench, which lies east of the islands at this point. Depths greater than 10,400 m (34,500 feet) have been recorded in this trench.

When Magellan first sailed into the Philippine Sea in 1521, he found a population that consisted of people of Malay, Melanesian, Chinese, and Arab ancestry. The islands were named by Magellan for St. Lazarus. After they were formally occupied by Spain in 1564, the name of the islands was changed to the Philippines after King Philip II of Spain. The Spanish occupation continued until the war with the United States in 1898. During World War II the Philippines were occupied by Japan. The United States recovered the islands after several major battles in 1944. The Philippines became an independent nation in 1946. *See* CURRENTS, ISLAND, PACIFIC OCEAN, PACIFIC PLATE.

Phoenicians The inhabitants of ancient Phoenicia (today's Syria and Lebanon); the biblical Canaanites. They first appeared in the 11th century B.C.E. along the length of the eastern Mediterranean coast. While not the first to sail into open water—Cycladic Greeks may have reached Spain by 2500 B.C.E. and Minoans had done extensive sailing and trading—the Phoenicians took up where these other Bronze Age cultures faltered. They founded colonies along the North African coast and were exploiting Spanish mines by 1000 B.C.E.

The Phoenicians' major scientific advance was celestial navigation. They were particularly likely to sight on the Pole Star, which the Greeks called the Phoenician Star for that reason. The ship these mariners used was the galley, usually a bireme (with two tiers of oars). Their longship, used as a warship, was long, slender, and most often equipped with a ram and with sails. The traders' vessels were shorter and wider, with rounded bows and sterns and no ram.

Long after their hegemony in the field had ended, Phoenicians were still being employed by others as experts in long-distance sailing and navigation.

phonoreception The physiological perception of sound. Vertebrates are the only animals that can hear underwater. Fish, and particularly mammals, have sound-receiving organs and cranial areas to process this information. The cetaceans, in particular, make and receive sounds from each other. This is called information processing. There is continuing research on dolphin language, too.

Sound travels faster in water than in air. In the water, the speed of sound is a function of temperature. It travels faster in warm water than in cold and will be deflected (refracted) by cold pockets within the body of the ocean. Therefore, listening devices must be calibrated accordingly. *See* LORAN, REFRACTION, SONAR.

Phoronida A phylum of small worms, less than 1 cm (0.4 inches) long. They live in membranous tubes of their own making, on the bottom of shallow waters. Some species are borers. They can penetrate calcareous rock, coral, or bivalve shells by literally "dissolving" their way in, using an acid secretion that breaks down calcium carbonate.

Phoronids protrude a complex, retractible lophophore, their food gathering organ, which resembles that of the bryozoans and the entoprocts. The life cycle of the phoronids is complex. Their larvae are planktonic and metamorphose into the sessile stationary adults. *See* BRYOZOA, ENTROPROCTA, LOPHOPHORE, SESSILE.

phosphorescence An emission of light that has nothing to do with the element phosphorus and should be called bioluminescence. Some organisms manufacture luciferin. This compound, when acted upon by the enzyme luciferase, produces light. Fireflies do this, as do some fungi. It is also a marine phenomenon since many planktonic species glow; those species including coelenterates (jellyfish) and mollusks (squid), that feed on this glowing plankton will then glow, too.

phosphorus A chemical element in the nitrogen family. It is present on Earth in rocks as phosphorites. Phosphorous in rock varies; therefore, its abundance in ancient rock strata can be used as an indication of climate change. Phosphorous is present in living tissue, and if large quantities of it enter water this results in the eutrophication of oceanic areas. These materials encourage explosive algal growth that ultimately depletes the oxygen in a body of water rendering it uninhabitable by the normal biota. In earlier, pre-human times, the presence of phosphorous compounds in rock indicated an ancient algal bloom. Some terrestrial phosphorus is contributed by glacial melts and coastal erosion. The phosphorus salts in some areas of ocean are higher than others because some of the terrestrial sources are richer in phosphorites than others. Undersea volcanoes and hydrothermal vent activity contribute a significant portion of the phosphate content of the ocean. Living organisms contain phosphorus compounds and the quantity of the element varies with the organism; plankton, the organisms that represent the greatest fraction of the biomass of the ocean, account for a disproportionate share of the phosphorus in the ocean. Although the greatest concentration of phosphorus compounds in the ocean is at its uppermost layer where the rate of photosynthesis is highest, the material does eventually fall and is eventually incorporated in the bottom sediment. The phosphorus content of bottom sediments varies from 0.01% to 10%; it is lowest in calcareous or diatomaceous muds and higher on continental shelves. Contamination of bottom water with sulfides or iron oxides will precipitate phosphorus out of the water. *See* CLIMATE, DNA, EUTROPHICATION, MINERALS, PHOSPHORUS CYCLE.

phosphorus cycle The circulation of the element phosphorus in the biosphere. Through the erosion of phosphate rocks, or the use of phosphate fertilizer, phosphorus enters the soil and is taken up by plants, which are eaten by animals. The dead animals or plants are then decomposed by bacteria, and the released phosphorus, in the form of dissolved phosphate salts, reenters the soil or is carried into the sea by runoff water. New phosphate rock is eventually formed from marine sediments.

The movement of phosphorus through the Earth and its relationship to the carbon and nitrogen cycles is being studied. High carbon dioxide levels and high ocean levels promote increased weathering that results in an increase of phosphorus in runoff water. This in turn encourages plankton bloom, and the living organisms remove the carbon dioxide from the atmosphere to incorporate it into living tissue. The decrease in the atmospheric carbon diminishes the greenhouse effect and leads to its opposite—the icehouse effect, a geological period in which the ambient temperature drops and so does the frequency of rainfall. The latter leads to desertification and the loss of arable land. Research now in progress attempts to relate the deposition of phosphorites on the ocean floor, climate, and atmospheric carbon. The phosphorite buildup of the Cretaceous has been correlated with intense chemical weathering of continents. *See* PHOSPHORUS.

photoadaptation Defines the relationship of an organism to its location in the ocean. It depends on the specific organism and the conditions of its biome. Almost all organisms in the euphotic zone respond to light. Plankton move up and down in the water column, and their adaptation to light occurs in a matter of seconds—or days. Since light penetration drops off with increasing depth, the planktonic response to less light is an increased production of chlorophyll and other light-absorbing pigments. Some photosynthesizing organisms are better adapted to this variation in light accessibility, and therefore the plankton population is a layered one: The organisms that have only chlorophyll are closest to the surface, and the diatoms and dinoflagellates are found at lower

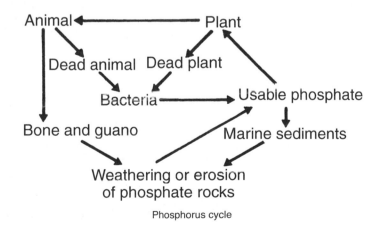

Phosphorus cycle

depths. Conversely, if there is too much light, plankton shield themselves and the chlorophyll molecules by also manufacturing carotenes and related protective pigments. These absorb the light and manifest it as heat that is dissipated in the water. See CAMOUFLAGE, COLOR, PIGMENTS, PLANKTON.

photography, underwater The operation of cameras below the surface of the water. The use of cameras underwater requires modifications in equipment and lighting: The cameras used must be watertight or in housings that are, and the cameras or their containers must be designed to withstand water pressure. The lighting for underwater photography is supplied by tungsten or electronic flash equipment, which must also be protected.

The refractive index of water (the angle through which it bends incident light) is greater than that of air, and as a result the camera lens manipulates light differently when under water. The angle viewed is actually smaller than that in air, and distances appear shorter than they really are. Direct contact lenses can compensate for this difference, but they do not work as well under water as they would in air.

Water clarity is also a factor in gauging the availability of light for underwater photography. Since water absorbs the red, orange, and yellow end of the visible spectrum, a flash must be used that overcompensates for these elements of the spectrum; or the photographs that are taken will lose the reds and yellows and appear as washed-out greens and blues.

Water is a denser medium than air; therefore, both the resolution of the image and transmission of light fall off faster in water. Thus, a large object is in focus only at short distances and wide-angle lenses are most often used. Unfortunately, the wide-angle lens is distorting. Cameras, too, must be protected from water, and since changing lenses is impossible under water, several cameras with lenses of varying focal length are necessary. An alternative is one 35-mm single-lens-reflex

camera with a 50–55 or 60 mm macro-focusing lens. Special telephoto adapters exist, as do auxiliary tools, for focusing on small objects. While camera bodies may be made of lightweight, metal, or plastic, the best optical elements are glass. For deep-sea photography, both cameras and light sources must be strong enough to withstand the increased water pressure. See MARINE ARCHAEOLOGY.

photophore A light-emitting organ. The animals that carry photophores inhabit the deep sea or abyssal depths. They use light as a lure or to camouflage themselves; for example, a glowing underside will render a fish invisible to predators below it.

photoreceptor See EYE.

photosynthesis The process by which plants and some single-celled flagellated organisms convert inorganic carbon dioxide (CO_2), water (H_2O), nitrite ion (NO_3^-), and phosphate ion (PO_4^{3-}) into sugars and amino acids, using the energy in sunlight. This occurs wherever chlorophyll-bearing organisms can exist. Light is a key ingredient in the chemical reaction that can be represented by the following greatly simplified equation:

$$CO_2 + H_2O + Light\ (Energy) \rightarrow (Sugar)\ (C_nH_{2n}O\ n)_x + O_2$$

where n is some small number. Photosynthesis can occur underwater to depths of about 100 m (330 feet), the photic zone, depending on the transparency of the water.

The extent of photosynthetic activity in any area determines the extent to which that particular area supports populations of organisms. The type of biome depends on the basic food supply manufactured by plants and plantlike protists. Thus, there can be a complex and well-populated food web in a seemingly inhospitable area. For example, phytoplankton blooms occur at the edge of ice shelves in both the north and south polar regions of the Earth. Upwellings of nutrients are an important

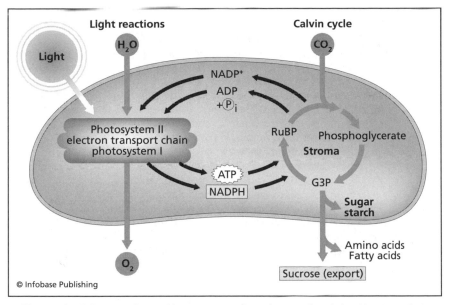

Light reactions

Calvin cycle

Photosynthesis is a complex, multi-step process. Sugar in some form is its ultimate product.

factor in Arctic water, where the surface water is nutrient-poor, but less so in the Antarctic, where it is not. Nevertheless, the phytoplankton bloom in Antarctic waters extends 400 km (250 miles) into the Ross Sea. This bloom constitutes a major food source for krill, the tiny organisms which, in turn, are the food of baleen whales. Algae in melting ice may act as a starter population for this food cycle.

Between 1994 and 1996 the Joint Global Ocean Flux Experiment (JGOFE) studied the Arabian Sea. This body of ocean water is typified by the monsoons that sweep through it, affecting light levels and therefore photosynthesis. The phytoplankton population is directly dependent on the amount of incoming sunlight and available nitrates and phosphates. Since the monsoon season in June and July produces increased aerosols and cloud cover, the experiment measured the effect on the algal bloom. This increase in the photosynthesizing population is an early summer event in the Northern Hemisphere. However, in the Arabian Sea the plankton population reaches a maximum in late August,

since July is almost always cloud-covered and minimal insolation is possible.

The JOGFE has attempted to construct a predictive model for the growth of plankton and the nekton that feeds on it as a function of available light as it varies from region to region. This information and model will be applied to other geographic areas such as the Caribbean and other regions where humans and endangered species interact.

Another group of oceanic organisms that photosynthesize has recently begun to receive research attention. These are the photoheterotrophic bacteria: They are closely related to the purple photosynthesizing bacteria, but their defining pigments are the orange carotenoids. Their habitat is the euphotic zone of all oceanic environments. These organisms can metabolize organic carbon if it is available, but in a carbon-poor region they will photosynthesize. Therefore, they are both primary producers and consumers. As photosynthesizers, they contribute to CO_2 fixation in the ocean. These organisms coexist with the phytoplankton, and both function in the movement of carbon through its cycle. How they interact is still unknown.

There have been many estimates and claims for the "fertility of the sea" and its capacity to be farmed as intensively as the land. Several factors argue against this theory. Phytoplankton do not normally feed terrestrial animals, including humans. They are eaten by zooplankton or die and fall through the sea to feed benthic organisms. The energy-storing compounds in marine organisms are oils rather than starches. This is exhibited throughout the marine food web. Only in shallow water and in warm weather is the photosynthetic capacity of the sea comparable to that on land on a square-meter to square-meter basis. The reason for this is that land plants are simply more efficient photosynthesizers. Ironically, the maximal photosynthetic activity in water occurs in those shallow, warm waters that are most likely to be contaminated by human economic activity. *See* ALGAE, CHLOROPHYLL, FOOD CHAINS, KRILL, MARICULTURE, PHYTOPLANKTON.

phototaxis Movement in response to light, either toward or away from it. *See* MIGRATION.

phycoerythrin The characteristic pigment of red algae. This pigment can photosynthesize using blue light, the segment of the visible spectrum that can penetrate to the furthest depths of the euphotic zone. *See* PHOTOSYNTHESIS, PIGMENT, RHODOPHYTA.

phycospheres Phytoplankton when dead, alive, or encysted can become colonized by bacteria; this combined structure of alga and bacteria is a phycosphere. There is now research investigating these structures, since some have more than one bacterial population while others have only one. That would indicate that the bacteria have some chemical defense against invaders. The singularity of a bacterial population means that it can kill off all others and therefore protect its food supply by chemical means. The compound (or compounds) that repel other organisms are then antibiotics. These

are being investigated. *See* PHARMACEUTICAL RESEARCH.

phytoplankton The major photosynthesizing organisms in the open ocean. Many of the species are so small that they were uncollectible until the advent of truly fine nets. The nutrient requirements of these organisms are nitrates or ammonia, phosphorus, and other specific trace elements such as silicon for some, iron for others. The presence and quantity of the required materials and certainly temperature are the growth-limiting factors for phytoplankton. The plankton blooms occur in spring, and a smaller bloom is observed in the autumn. They are most often the result of the nutrient-rich cold bottom water moving toward the surface euphotic zone. Cold core rings such as eddies spun off major ocean currents like the Gulf Stream will also provide a nutrient-rich region in which there will be a planktonic bloom; in a warm current the bloom can occur at any time of the year. The ultimate controlling influence on the growth of the phytoplankton population is the microzooplankton population that feeds on it. *See* PLANKTON, PREDATOR-PREY RELATIONSHIP.

Piccard A family of Swiss scientists known for their underwater and stratospheric research and exploration. Auguste Piccard (1884–1962), a physicist, designed a low-pressure balloon for stratospheric investigation. Jean Félix (1884–1963), his twin brother, a chemist who later became a naturalized American citizen, varied the construction of the balloon, and both brothers made high altitude balloon ascents in 1931, 1932, and 1936 in order to study cosmic rays in the absence of atmospheric interference. Auguste also designed the bathyscaphe, a submersible vessel for deep-sea exploration. The vessel was built after World War II and made its first dive in 1958 off the coast of Senegal, West Africa. The support ship's captain was Jacques Cousteau. Other vessels based on the design of the bathyscaphe were built by Jacques Piccard, Auguste's son, who also designed a submarine, called the *Ben Franklin,* that utilized subsurface currents as its motive power. *See* BATHYSCAPHE; COUSTEAU, JACQUES-YVES; OCEANIC SUBMERSIBLES.

pigment A substance that imparts color to an object. The most prevalent pigments in phytoplankton, the most numerous organisms in open water, are the varieties of chlorophyll. The quantity of the chlorophylls in seawater changes the color of the water. This change is indicative of the quantity of phytoplankton, and in periods of greatest population the water turns green, signaling a bloom. If other pigments are present, they too are detectable. These might be the carotenoids (yellows and oranges), fucoxanthin (brown), or the reds of toxic red tides. *See* CAROTENOIDS, CHLOROPHYLL, FUCOXANTHIN, PHYTOPLANKTON, SeaWiFS.

pilot fish *Naucrates ductor,* a pelagic, spiny-finned fish found in tropical to warm temperate waters. Pilot fish are best known for accompanying sharks but will also follow other large fish or ships to feed on the leftovers they discard. Pilot fish are distinctively banded, with vertical dark stripes on a blue-white body. The usual pilot fish is about 60 cm (2 feet) long and has a slender body. *See* COMMENSAL RELATIONSHIPS, PELAGIC ENVIRONMENT, SHARK.

pilot whale A small whale of the genus *Globicephala,* also known as the blackfish, that ranges in size from 4 to 6 m (13 to 20 feet). It lives in large groups in all but polar waters and feeds mainly on squid. It is the most dolphinlike whale and belongs to the family Delphinidae. These whales are trainable when captured, as are dolphins. *See* WHALE.

pinniped The name used to describe any of a group of fin-footed carnivorous aquatic mammals including the seals, sea lions, and walruses. The name means fin-footed.

PINNIPEDS		
Common Name	*Scientific Name*	*Habitat*
Sea Lions		
Stellar	*Eumetropias jubatus*	Alaska and the Pribilof Islands
California	*Zalophus californianus*	Western North America and the Galapagos Islands
South American	*Otaria flavescens*	Western South America
Australian	*Neophocus cinerea*	
Hooker's	*Phocartos hookeri*	New Zealand
Fur Seals		
Alaska	*Callorhinus ursinus*	Northern Pacific
Juan Fernandez	*Arctocephalus philippi*	
Guadaloupe	*Arctocephalus townsendi*	
Galapagos	*Arctocephalus galapagoensis*	
Australian	*Arctocephalus pusillus dorifer*	
Antarctic	*Arctocephalus gazella*	
Subantarctic	*Arctocephalus tropicalis*	
		(Continued on next page)

PINNIPEDS		
(Continued from previous page)		
Common Name	Scientific Name	Habitat
Phocids (True Seals)		
Baikal	*Phoca sibirica*	Lake Baikal
Ribbon	*Phoca fascinata* (white ribbon on black)	Siberian coast
Caspian	*Phoca caspica*	
Ringed	*Phoca hispida*	Arctic
Larga	*Phoca largha*	Northern Pacific
Hooded (also called bladder nose)	*Cystophora crustata*	
Bearded (large moustache)	*Erignathus barbatus*	
Harbor	*Phoca vitullina*	
Harp	*Phoca groenlandica*	
Gray or horsehead	*Nalichoerus grypus*	
Northern Elephant	*Mirounga angustirostris*	Western North America
Hawaiian Monk	*Monachus schauinslandi*	Protected
Mediterranean Monk	*Monachus monachus*	
Caribbean Monk	*Monachus tropicalis*	Probably extinct
Southern Elephant (largest phocid)	*Mirounga leonina*	South America
Weddell	*Leptonychotes weddelli*	Antarctic
Ross	*Ommatophoca rossi*	Antarctic
Leopard	*Hydrunga leptonyx*	Antarctic
Crabeater	*Lobodon carcinophagus*	Antarctic
Walrus		
Walrus	*Odobenus rosmarus*	Arctic

All pinnipeds are ungraceful on land and remarkably sleek and agile in the water. The differences between seals and sea lions are relatively minor ones. The former (Phocids) have no external ears and use only their forelimbs, while sea lions (the otarids) hobble around on all four limbs. The fur seals are an intermediate group between the seals and sea lions: They are eared and move on four limbs but have thick underfur. The walrus (an odobenid) has no external ear and moves ponderously on all fours. *See* CRABEATER SEAL, HARBOR SEAL, HARP SEAL, SEAL, WALRUS.

Placodermi An extinct class of armored, carnivorous fish; these are the earliest fish with jaws. These animals appeared when sharks did, about 400 million years ago, and flourished in the Devonian period. They rapidly became extinct after only about 50 million years of existence. Many individuals were not large, about 10 to 15 cm (4–5 inches) long, but some did grow to enormous size. Some fossils are over 6 m (20 feet) long. The large, armored head had a movable lower jaw and employed bony mouth structures instead of teeth. Like sharks, these animals had an asymmetric tail and were probably efficient swimmers.

plaice *Pleuronectes platessa,* the European flounder. An important, commercially taken flatfish. It lives in North Atlantic waters, blending well with the bottom sands it lives on. The upper, exposed side of the plaice carries red, yellow, and orange speckles on a brown background. The plaice is a large fish, its maximum size being about 85 to 90 cm (about 3 feet).

The relatives of the European plaice are the North American right-eyed flatfish, the American plaice or sand dabs (*Hippoglos-*

soides platessoides). They are also Pleuronectidae, but differ slightly in appearance from their European counterparts. This is another fish whose population is declining. *See* FISH, FLATFISH, FLOUNDER.

plankton Tiny protists of the sea. The name comes from the same Greek root as "planet," and both terms denote wanderers. The term was first applied to organisms in the sea by Viktor Hensen, the director of the German expedition of 1889, known as the Plankton Expedition, whose specific charge was to find and systematize the organisms of the sea.

Plankton comprises a vast group of diverse aquatic organisms; some are freshwater, others marine. They are further separated into photosynthesizing organisms or

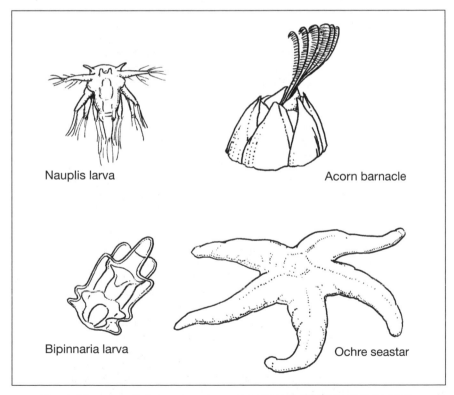

Nauplis larva

Acorn barnacle

Bipinnaria larva

Ochre seastar

The planktonic juvenile forms of some organisms bear no resemblance to the adults. The examples are a barnacle and a sea star.

phytoplankton, and nonphotosynthesizing organisms, or zooplankton. This division is not made on a plant versus animal basis, since some chlorophyll-containing organisms are also flagellated and ingest other organisms, both of which are animal-like traits. The euglenoids are prime examples of protists that contain chlorophyll, and are basically phytoplankton. If, however, the light intensity in the water drops, photosynthesis declines, and the euglenoids must then hunt for their food rather than manufacturing it.

Name	Size in micrometers (10^{-6} m = one millionth of a meter)
Picoplankton	0.2–2 μm
Nanoplankton	2–20 μm
Microplankton	2–200 μm
Mesoplankton	200 cm

The plankton are also graded by size, ranging from the smallest, which are invisible without magnification, to those visible to the naked eye.

Some plankton are unicellular, others are colonial, and still others are multicellular organisms. Plankton are further categorized as those species that spend their entire lives floating with the motion of the water and those that are active rather than passive. Both the photosynthesizing plankton and organic fragments tend to float with the movement of the water and accumulate either just below the surface or at frontal edges. Zooplankton tend to form dense aggregates well below the surface at sites of either upwellings or downward currents. These organisms propel themselves; therefore, this clumping behavior illustrates active effort to maintain their position at a particular depth.

Zooplankton migrate in response to the availability of food supply. Their ability to maintain position despite currents is an indication of their ability to locate food, operate almost as a colonial organism, and expend considerable energy in swimming.

The permanent floaters (i.e., those organisms that remain planktonic for their entire lives) are called holoplankton. Those that are planktonic only as larvae are called meroplankton. The meroplankton, or temporary plankton, may turn into adults that are benthic (fixed, or bottom-dwelling organisms) or nektonic (swimming organisms). *See* ECOSYSTEM, EUGLENA, FOOD CHAINS, MIGRATION, NEKTON, PHOTOSYNTHESIS, PHYTOPLANKTON, PROTISTA.

plant Although there is some disagreement about the placement of large photosynthesizing organisms—notably kelp—in the Protista, there are no real plants in the ocean. The organisms that photosynthesize are not at all like terrestrial plants; they are either single-celled organisms or colonial ones. Even the large kelp are undifferentiated cells; they have a holdfast cell structure that anchors the entire blade. That is the only set of cells that look different from the rest. All the other parts of the blade are alike, and all the cells, including the holdfast, can regenerate from one cell. *See* ALGAE, EELGRASS, FUCUS, KELP, SARGASSUM, ZOSTERA.

plate tectonics A theory of modern geology based on the movement of huge blocks, or plates, of the Earth's crust. Although the motion of continents was postulated many times in this century, this concept was put forth most clearly by Alfred Wegener in 1915. He "fit together" the known landmasses using the best geographic information of the time and called the primordial continent that had given rise to them Pangaea (all-Earth). However, Wegener could not explain the subsequent breakup of the continental mass of Pangaea and simply concluded that the continents had drifted apart, without proposing a mechanism to explain their drift. For many years his theory remained an idea that awaited corroborating evidence. By the 1960s there had been gathered a large body of supporting evidence for it. Fossils found in Antarctica proved to be related to Australian ones, and fossils found in both Antarctica and Australia were related to South American specimens. Magnetic anomalies in rocks of the ocean floor confirmed that there was motion of the Earth's crust, and indicated that Pangaea had indeed once existed as an equatorial landmass. The magnetic anomalies discovered by Vine and Matthews (of Cambridge University in England) and the nature of seafloor spreading explained by Robert Dietz established Wegener's continental drift theory, now called plate tectonics.

The outer shell of the Earth is the crust and upper mantle. This is the sialic layer consisting of light rock rich in silica and alumina, that floats on the denser sima, which consists of silicon and magnesium-containing rocks. The continents are sialic and therefore less dense than the lower mantle and asthenosphere (a semiliquid layer). When the crustal plates move, they slide past each other (transform), move apart (diverge), or move toward each other (converge). When the plates converge and meet, one plate overrides the other, which then sinks into the asthenosphere. This action is called subduction.

The crustal movements have been slow but dramatic. Examples of visible evidence of these movements are oopholites and terranes. An oopholite is a piece of what was once ocean floor. Finding in the Himalaya Mountains a part of what was once an island chain in the Indian Ocean is, for example, evidence of incredible change in the shape of the Earth's continents and oceans. Terranes are "exotics," or pieces that do not belong to the continents they are surrounded by. A major piece of the Carolina Slate Belt, for instance, was accreted onto the North American continent after the Cambrian period, when the primeval Atlantic Ocean opened and separated England from North America. Similar foreign inclusions on the west coast of North America are also known.

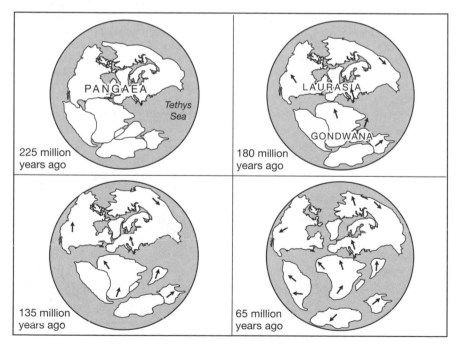

The original locations of landmasses. Each continent is on its own plate, and all plates continue to move.

Major plates of the Earth's crust and the directions in which they are moving

The single landmass of Pangaea, originally postulated by Wegener, existed during the Permian and into the Triassic periods from 280 million to 225 million years ago. By the end of the Triassic and in the late Jurassic, about 180 million years ago, it began to break up along what is now the Mid-Atlantic Ridge, a diverging zone where new material is introduced into the Earth's crust at the edges of the separating plates. According to one current hypothesis, this intruding material pushed the plates further apart. By the Cretaceous period, 135 million years ago, the land masses were well separated. Since then the continents framing the Atlantic have continued to move away from each other fairly steadily.

The Australian, Indian, and Antarctic plates have not had unidirectional motion. The paths and movements of parts of the Indian Ocean, which includes Madagascar, and of the Caribbean and Mediterranean

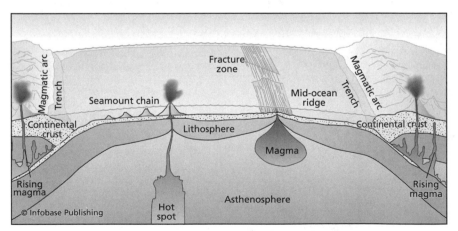

The movement of plates is started by hot liquid rock rising.
The moving plates collide and the resulting friction causes earthquake,
release of carbon dioxide and a redistribution of minerals.

seas are not as well defined as those of the Atlantic. *See* BASALT; CONTINENTAL DRIFT; CRUST; EVOLUTION OF OCEANS; EWING, WILLIAM MAURICE; HEEZEN, BRUCE CHARLES; HESS, HARRY H., SUBDUCTION; WEGENER, ALFRED.

plateau   A relatively level structure of the ocean floor, higher than the adjacent bottom, along a continent but beyond the continental shelf, or an extended flat portion of a mid-ocean rise. These marginal structures are extensions of the shelves they adjoin.

The Blake Plateau in the North Atlantic is probably a section of continent that has been subsiding since the Cretaceous Period. It is heavily overlain with limestone deposits and coral. The Bahama Banks are considered to be a marginal plateau, but exhibit the structure of a barrier reef and consist of coral. The underlying rock of Florida and that of the Bahamas islands is different, and not part of the plateau. The Campbell Plateau east of New Zealand is a large, shallow area with a continental bottom. It and New Zealand are both of continental origin. The Falkland Ridge, which geographically belongs to the same Devonian pattern found on the Falkland Islands, is a fairly sharp drop. The plateau to the east of Brazil may be a sinking portion of continental shelf, and therefore quite different from the structure of the Blake Plateau.

Some mid-ocean plateaus, such as the Coral Sea Plateau and the Bellona Plateau (about 1,000 m long) are aseismic and coralline. The Melanesian Plateau has some volcanoes. The Flores Sea Plateau is coral in its base. The Mascarene Plateau in the Indian Ocean reaches from the Seychelles islands to Nazareth Bank and Mauritius. It is partly volcanic, and some of its oldest parts are Precambrian. It is postulated that this plateau once formed a part of the isthmus between India, Madagascar, and Africa. *See* BLAKE PLATEAU, CORAL, CRETACEOUS PERIOD, DEVONIAN PERIOD, SUBSIDENCE.

Platyhelminthes   A phylum consisting mainly of parasitic flatworms. The freeliv-

ing members (Turbellaria) are of worldwide distribution and include fresh- and salt-water types as well as terrestrial organisms. Those that are parasitic, flukes (*Trematoda*) and tapeworms (*cestoda*), may have marine hosts. The tapeworms in particular may have fish hosts but are also known to infest humans. *See* TURBELLARIA.

Pleistocene epoch   The first epoch of the Quaternary period of the Cenozoic era, usually referred to as the age of ice. It was when much of the globe was covered by ice sheets extending from the poles toward the equator. Modern land animals such as the horse, the elephant, and the cattle family made their appearance during the Pleistocene. The beginning of the Pleistocene is variously given as 1 to 4 million years ago, and it is agreed that it ended about 10,000 years ago. *See* ICE AGE, QUATERNARY PERIOD.

plesiosaurs   Extinct Mesozoic marine reptiles that lived about 190 to 65 million years ago. The plesiosaurs evolved into two groups: One had small heads and very long necks, the other was short-necked and had an enormous head. Both were worldwide in range. The former, the elasmosaurs, half of whose body length was neck and head, preyed on schooling fish. The latter, the pliosaurs, fed on large fish and cephalopods.

Many plesiosaurs—and particularly the elasmosaurs—were gigantic; some specimens are known to have been 13 to 15 m (43 to 50 feet) long. They were excellent swimmers whose limbs were paddle-shaped and allowed precise maneuverability. These animals literally "flew" through the water as penguins do now. *See* REPTILE.

Pliny the Elder (23–79 C.E.)   A Roman natural historian whose full name was Gaius Plinius Secundus. He came from an aristocratic Roman family, traveled extensively in the service of the Empire, and developed a flair for the collection of information. He specialized in art and natural history, and his two works, which are still entertaining reading, were a compendium of fine art and a book of natural history.

Unlike Aristotle, who flagged those entries in his compendium of natural history he considered suspect, Pliny recorded everything.

Pliny's interest in natural history was the cause of his death. While in the service of the Emperor Titus, defending the Bay of Naples from pirates, he went ashore to inspect an interesting cloud. It turned out to be the eruption of Vesuvius, and Pliny died from the noxious fumes produced by the eruption.

Pliocene epoch The last epoch of the Tertiary period immediately preceding the Quaternary. The Pliocene lasted from about 10 million to about 4 million years ago. The British geologist Charles Lyell divided the epoch into the Older and Newer Pliocene. In the former, less than half of the remains of species from the former half resemble the species now on Earth, while more than half of the remains from the latter are like present forms of animal life. The newer Pliocene is now called the Pleistocene.

The major geologic change in the Pliocene was the retreat of the Tethys Sea, which occurred as the North Sea continued to creep north and the temperature dropped throughout the world.

The Pliocene fauna, particularly marine organisms, resemble the species of today. Echinoderms are common in fossil remains, as are other bivalves and gastropods. The colder temperature displaced corals and foraminiferans. On land, large mammals were dominant, and the land bridges between America and Asia and between North and South America allowed the migration of these animals. *See* CENOZOIC ERA, PLEISTOCENE EPOCH.

Pogonophora A small wormlike phylum of about 120 species. These animals were first discovered in 1900 off what was then the Dutch East Indies (Indonesia), and new ones are found periodically. The

huge tube worms of hydrothermal vents, the vestimeniferans, are now categorized in this phylum. The typical pogonophore is surrounded by a hard tube made of chitin and protein. It may be as small as a human hair—about 0.3 cm (0.1 inch) in diameter and about 100 cm long (30 inches)—or 4 cm (1.6 inches) in diameter and about 2 m (80 inches) long. The hydrothermal tube worms are 2 m long.

Since the pogonophores were first sighted, these animals have been recovered from all the world's oceans and from depths ranging from 30 m (90 feet) to depths of more than 9,000 m (30,000 feet). A considerable length of the worm's body is buried in the bottom sediment. That part of the worm most closely resembles annelid structure. The part of the pogonophore that extends above the bottom surface more closely resembles bryozoan structures. Some taxonomists combine the pogonophorans with annelids in a larger grouping called Trochozoa.

These are the only multicellular animals known that have no digestive system at all. Their nutrients are supplied totally by their symbiont bacterial population. The worms have tentacles that apparently absorb materials from the water, and the symbiotic bacteria are clustered in a trophosome—a structure in the tentacles. The products of the bacteria feed the worm. The digestive path is unknown. These animals have closed circulatory systems and blood that contains hemoglobins. These complex compounds move oxygen and sulfur, materials necessary for the metabolism of the symbiont bacteria. The worm's sexes are separate and discharge gametes directly into the surrounding water.

Another group of large pogonophores lives near cold seeps. That environment is much more stable and long-lasting than is the hydrothermal vent populated by the tube worms. The cold seep dwellers are not as large as their relatives in the vent communities but live longer and grow more slowly. These tube worms are found at depths of 5,000 m (1,800 feet) off the Louisiana coast, where hydrocarbons seep slowly upward through the sea bottom sediment. The hydrogen and sulfur is incorporated into the worm's body, and it uses this—again the mechanism is unknown—to fuel the bacteria that produce the energy-storing compounds needed for the worm's metabolism.

The presence of sulfur is important to the worm. It both provides an oxygen analog in the food material and is part of the chemical defense of the worm. Sulfur compounds are avoidants: Other animals reject as food those that taste of sulfur. Since the tube worms can live for hundreds of years, chemical defenses are necessary for their survival and protection from predation. *See* CHEMICAL DEFENSES, COLD SEEPS, TUBE WORMS, VENT COMMUNITIES.

poikilotherm An animal whose temperature varies with that of the environment. The blood of poikilotherms is often slightly above the ambient temperature, and these animals are therefore sometimes incorrectly called cold-blooded. Most animals other than birds and mammals are poikilothermic. *See* HOMOIOTHERMS, TEMPERATURE OF ANIMALS.

polar biome The community of plants and animals found in the north and south polar regions of the Earth. The Arctic supports more varieties of organisms, all ultimately dependent on a phytoplankton and zooplankton food chain, than does the Antarctic. Numerous predatory transients also use the territory for breeding colonies.

The rapidly diminishing ice cover in the Arctic is of concern to meteorologists and biologists. The International Polar Year 2007 will be one of intense study and projection of the effect of the warming on the ecosystems of the region and, ultimately, all the Earth. *See* ARCTIC, BIRD, GULL, PENGUIN, PINNIPED, PLANKTON, SEAL, TERN, WHALE.

polar front An atmospheric phenomenon that occurs at the latitudes near both poles, where polar cold water sinks below the surface water to form the intermedi-

ate layer of the world's oceans. Since there is a difference in the temperature of the air and the water in these areas, weather disturbances are frequently generated. The region, particularly the Arctic Convergence, is characterized by an almost constant cloudbank. *See* ANTARCTIC CONVERGENCE.

Polaris The North Star, Pole Star, or Phoenician Star (it has many names). It is part of the constellation known as the Little Dipper, or the Little Bear (these are only two of its names). It is at true north in the skies of the Northern Hemisphere. Because of its position and the fact that it is visible year-round, it has been used as a bearing in navigation. In the Southern Hemisphere the Southern Cross serves as the basis for celestial navigation. *See* NAVIGATION.

pollock A North Atlantic fish of the cod family (*Gadidae*). Pollock are carnivorous fish that are green or green-black on their upper surfaces and creamy white on the underside. The average adult individual is about 1 m (40 inches) long, and weighs up to 15 kg (33 pounds). Pollock are taken by both sport and commercial fishermen. Their meat is frequently sold as cod. Pollock are an important item in the prepared food industry; it is the fish most often used in frozen fish dinners. The flesh is also colored and shaped to be sold as surimi (imitation crab meat).

The stock of pollock has been seriously depleted by overfishing. The use of better designed nets is slowly increasing their numbers in Pacific waters. *See* COD, NETS.

pollution Contamination of the environment (air, water, or earth) by waste material or any other product as a result of human activity. Water-polluting wastes can be industrial, resulting from the construction or destruction of buildings, equipment or materials, refinery or paper mill waste, or sewage; agricultural, from runoff containing fertilizers or pesticides; or urban, from sewage and litter. Airborne pollution is a problem in the sea as it is on land. The airborne pollutants may be small particles of metal, wood pulp, manufacturing waste, or any other material fine enough to be windblown, or they are gaseous pollutants, such as sulfur and nitrogen oxides, that result from fossil fuel combustion. These affect seawater by altering its pH or by blocking out sunlight. By creating "acid rain," airborne pollutants also affect vegetation and human health.

Large cities are frequently major pollutors. They are almost always manufacturing centers as well as high-density population centers, and the local manufacturing processes use local waters to remove a variety of chemicals that are either end-products or byproducts of the manufacturing process. Cities also produce large quantities of garbage and sewage, both of which may enter the sea directly if the city is coastal, or indirectly if it is on a large river. Areas of the ocean that receive the effluent of cities are all too often "dead seas."

Besides rendering the shoreline aesthetically unappealing, floating garbage and chemical residues, including oil spills, prevent sunlight from reaching the phytoplankton in the water, and interfere with gas exchange between water and the atmosphere. Both factors decrease the available oxygen required by living organisms in the water. The chemical residues draw on the oxygen supply, since oxidation and eventual degradation of these chemical pollutants involves their reaction with oxygen. The decreased availability of atmospheric oxygen that is caused by floating pollutants means that natural cleansing of the water takes much longer.

Eutrophication is the result of an influx of phosphate or nitrate ions from either an agricultural area or an area of high population density. Phosphate and nitrate ions stimulate algal growth. For a while the algae increase the available oxygen supply, but as they grow they also cover the surface of the water, which means that other plants, down in the photic zone, begin to die. The death of some plants and the increase in the population of decompos-

ing bacteria consumes more of the available oxygen, which means that there is less available for animals (e.g., fish), which then begin to die. Without intervention and aeration, the eutrophied area will become a foul-smelling and lifeless swamp.

Where coastal areas can be flushed by the action of waves, the effects of pollution are lessened but not removed. Better disposal methods and more effective wastewater treatment are essential, but require commitment by governments and the investment of considerable funds. It is only recently that concerted efforts have been made to limit and clean up pollution.

In 1972 the first United Nations Conference on the Human Environment studied the effect of pollution, particularly industrial pollution. It was the Conference's conclusion that the effect of industrial wastes on water quality, and the resulting environmental impact, was most pronounced in enclosed marine areas, even very large enclosed areas such as the Mediterranean, Baltic, North Sea, and Caribbean. The mid-ocean effects of pollution were felt to be the least damaging.

A 1982 follow-up report by the U.N. Conference on the Human Environment stated that the open ocean was cleaner than it had been 10 years earlier. The 1972 Conference in Stockholm had been very concerned with the concentration of chlorinated insecticides in all of the Earth's ocean waters. By 1982, as a result of the ban of the use of DDT and the decreased use of chlorinated insecticides in general, there were less of these compounds in the water. However, other pollutants remain a problem.

With increased awareness and investment in environmental clean-up, the water quality of the Northern Hemisphere is improving, even in those areas that are known problem spots, such as the Mediterranean. But as industrial development moves to the Southern Hemisphere, so will pollution. Heavy-metal dumping, acid rain, and nuclear wastes all pose problems of oceanic pollution. Since large concentrations of people live near the

Earth's oceans, with an estimated 70% of the world's population living less than 80 km (50 miles) from a seacoast, and humans also depend on food that is taken from the sea, international cooperation and study are required for continuing pollution control and environmental cleanup. *See* ESTUARY, MEDITERRANEAN.

Polychaeta A class of the phylum Annelida, comprising the segmented worms. These animals range in size from less than 1 cm (0.3 inches) to longer than 3 m (10 feet). There are more than 15,000 species of polychaetes; most of them are marine. Among the polychaetes are animals with many varied life styles. There are free-swimming larvae that grow up to be free-swimming adults, other animals that can move but usually don't, and still others that don't move at all. Polychaetes can be divided into those that are soft-bodied and move to trap prey, and those that form tubes and wait for food to come to them. The outstanding characteristic of the polychaetes is the presence of parapodia, or projections on the body, that

bear setae (bristles). There is also distinct cephalization—the development of a head equipped with sensory organs. Most of the motile polychaetes are light-sensitive; they are attracted to light.

Like other annelids, the polychaetes have pigmented blood. It is often red, although some have green blood. These animals have closed circulatory systems and live lives that may exceed 10 years in length. The different species are sometimes brilliantly colored. Although polychaetes are usually dioecious, with separate sexes, some species are hermaphrodites. *See* ANNELIDA, NEREIDAE, PALOLO.

polymer A large molecule composed of repeating units of smaller molecules called monomers. Some polymers are built of repeating units of the same monomer, (e.g., repeating units of glucose form the polymers cellulose, amylose, and starch).

Proteins, polypeptides, and nucleic acids are also polymers but are mixed polymers composed of more than one monomer. Amino acids are the constituent monomers of proteins. There are about 20

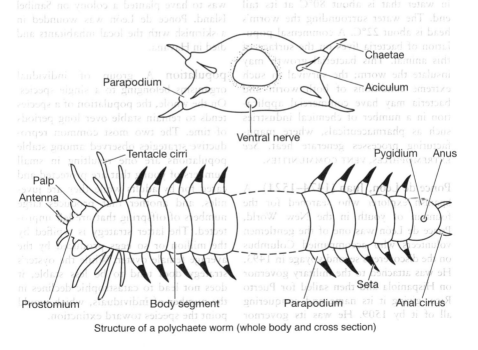

Structure of a polychaete worm (whole body and cross section)

amino acids commonly found in naturally occurring proteins. The specific proteins are manufactured by cells according to the code contained in the DNA of their nuclei. *See* DNA, RNA, SUGAR.

Polynesia A collective name for the islands of the central and southern Pacific Ocean. The largest islands in this group are New Zealand and the Hawaiian Islands. Others are the Society Islands (Tahiti), Fiji Islands, Samoa, Tonga, Easter Island (Rapa Nui), and the Tuamotus, Marquesas, and Line Islands (Christmas Island, Palmyra). *See* ATOLL, ISLAND, PACIFIC OCEAN.

polyp A column-shaped cnidarian or part of a colony of cnidarians. The corals are typical examples. *See* CNIDARIAN.

Pompeii worm The common name of *Alvinella pompejana*, a common inhabitant of hydrothermal vents. It is about 13 cm (5 inches) long and currently the multicellular organism that can withstand the highest temperature. It lives in a burrow in water that is about 80°C at its tail end. The water surrounding the worm's head is about 22°C. A commensal population of bacteria lives on the surface of this animal. This bacterial growth may insulate the worm; the survival in such extreme conditions of both worm and bacteria may have commercial application in a number of chemical industries such as pharmaceuticals, where manufacturing processes generate heat. *See* EXTREMOPHILES, VENT COMMUNITIES.

Ponce de León, Juan (1474–1521) A Spanish explorer who searched for the fountain of youth in the New World, Ponce de León was one of the gentlemen volunteers who accompanied Columbus on the discoverer's second voyage in 1493. He was attached to the military governor on Hispaniola and then sailed for Puerto Rico, giving it its name and conquering all of it by 1509. He was its governor until he was replaced by Diego Colón (the son of Columbus) in 1512. Ponce de León was then given an expedition to discover new lands and sailed to the northwest in March 1513. Ponce de León was looking both for treasure and the fountain of youth. This Eurasian myth had been transplanted to the New World; however, he seems to have been the only explorer to make serious attempts to find it.

The expedition moved north of Cape Canaveral before turning back south along the Florida coast and round it, finding the Florida Keys and naming them the Tortugas. Tortuga is Spanish for turtle. He thought the islands looked like turtles coming up out of the water. Rounding the tip of the Keys he sailed north into the Gulf of Mexico and north on the gulf coast to roughly the latitude of Fort Myers, Florida. Possibly still looking for the fountain of youth he went south again and sighted the Yucatán Peninsula of Mexico. Ponce de León returned to Puerto Rico in October, and sailed for Spain in the next year (1514).

A second voyage to Florida in 1521 was to have planted a colony on Sanibel Island. Ponce de León was wounded in a skirmish with the local inhabitants and died in Havana.

population A group of individual organisms belonging to a single species. On the whole, the population of a species tends to remain stable over long periods of time. The two most common reproductive strategies observed among stable populations are one resulting in small numbers of young that are protected and cared for by adults while they are juveniles, and another that produces large numbers of offspring that are left unprotected. The latter strategy is typified by the million or so eggs produced by the average female oyster. While the oyster's strategy does tend to be less stable, it does not lead to catastrophic declines in the number of individuals, which would point the species toward extinction.

Occasionally everything goes right for a particular population and one cohort, or age group, is very successful. The classic example of this phenomenon involved the 1904 herring spawn. The herring spawned in 1904 dominated the population of all herring catches for over 15 years.

It is not only the availability of food or the lack of predation that will affect the makeup of a particular population. Temperature, weather conditions, storms, and currents will alter the biome and make one year better or worse than another. Some species respond in ways that are upon first appearance may seem rather peculiar. One of these responses is a lack of breeding behavior. In others, sex changes with variation in the environment are a possible and viable option. The driving evolutionary explanation for a sex change is that the two sexes have different metabolic demands. It is metabolically "cheaper" to maintain a population with a large percentage of male individuals.

In many animal populations, notably birds, seals, and other predators, not all adults breed. Furthermore, since reproduction is an energy-expensive function, adolescence may extend for years.

In fish populations, the high mortality among the young is due to both disease and predation. In any stable fish population (i.e., one not perturbed by human fishing), adults tend to die as they age. The age of fish is determined by the annual rings on their scales. Some fish species can live to a great age, with some individuals having more than thirty rings not being rare.

In years of population "bloom," the rapidly increased population is one that has not had an accompanying increase in numbers of predators. While there is a large number of new individuals in the population, there is also greater competition for the same food supply and reproductive space. The initial effect is an increase in the die-off of young. This is followed by an increase in the predator population, and the end result is a much lower population in the next breeding cycle. Thus, a stable population is not one with no change but one where there is both fluctuation and migration.

Populations vary in their behavior as well as in numbers. Some, such as those of whales and porpoises, are cooperative societies. In others there is active competition, as in the communities of elephant seals. Still others have parallel lives, such as the thousands of frigate birds that nest in the same place, all doing the same thing at the same time while seemingly oblivious of one another and without helping one another much. Penguins, on the other hand, share their babysitting chores with their neighbors.

A current multinational project aims to take a census of all the ocean's species. This involves collecting and organizing data from more than 70 participating countries. The project's official title is Census of Marine Life; its compilation is the database Ocean Biogeographic Information System. That branch has accumulated information about more than 40,000 species; 15,500 of them are fish. Their estimate is that this total represents only about 20% of the organisms in the oceans. The project is expected to go on indefinitely. See BIRD, FISH, FOOD CHAIN, FOOD WEB, MIGRATION, PREDATOR-PREY RELATIONSHIP.

porgy A shallow-water food fish that is well distributed in warm and temperate waters in America and Europe. There are about 100 species of these relatives of the snappers. Their average size is about 30 cm (1 foot), but some tropical specimens may be more than twice that size. The European porgy is called a bream. See FISH.

Porifera A large phylum of multicellular, sessile (nonmoving), primitive marine animals, commonly known as the sponges.

Sponges are found in all of the world's seas. There are about 5,000 species, but Antarctic species are particularly abundant. While most sponges are found in fairly shallow water, some are present at all depths, and a few individuals are hadal. A few species live in brackish or even freshwater.

The Porifera are distinguished by their very simple body plan, in which there is only one large opening, the osculum, and by the many small pores in their body wall. They feed by filtering water and its dissolved and suspended material through the small wall pores, then into and through the cells of their bodies, and expelling it through the osculum. The various genera of sponges are characterized by the type of spicule that is found in the body wall. These are the hard support structures and may consist of silicon compounds, calcium compounds, or a proteinaceous material, spongin.

The Porifera are usually tan or brown, although some living in shallow tropical waters are exotic greens, reds, or oranges as a result of their commensal algae.

Sponges range in size from 1 to 2 cm (0.2 to 0.8 inches) to giants more than 1 m (3.3 feet) in diameter. Their age range is considerable. They reproduce by budding, which is an asexual process, and by gamete production, which results in new, sexually generated individuals. The poriferans are quite undifferentiated; new individuals have grown almost from individual single cells produced by homogenizing an adult and straining the homogenate. An old method of reseeding sponge beds was to cut up sponges and strew the pieces around the ocean bottom.

The Porifera are a very ancient phylum that dates certainly from the early Cambrian period. They do not fossilize well and have therefore left a very patchy record.

Since many poriferans live a very long time and cannot move to escape predation, they must defend themselves by manufacturing chemicals that are toxic to possible predators. These chemicals are of great interest to researchers who are looking for new compounds to use as pharmaceuticals. See CHEMICAL DEFENSES, PHARMACEUTICAL RESEARCH, SPICULE, SPONGE.

porpoise A small, gregarious whale of the order Cetacea, family Phocaenidae. Dolphins are frequently called porpoises, but the true porpoise is smaller than a dolphin. It is generally about 2 m (7 feet) long, has a rounded nose rather than a pointed one, and a dorsal fin.

The common porpoises are the following:

Harbor or common porpoise (*Phocaena phocaena*), a cold water animal of the Northern Hemisphere.
Dall (*P. dalli*), the largest porpoise, which is found most frequently in the North Pacific.
Finless porpoise (*Neomeris phocaenoidis*), the individual found most often in South Asia and East Africa.
Spring finned porpoise (*P. spinipinnus*), found near South America.
Spectacled porpoises (*P. dioptrica*), which are found in Antarctica.
California Gulf porpoise (*P. sinus*), sometimes placed in a separate category and sometimes lumped together with the spring fins. It is highly endangered.

The common porpoise plays before ships, as do dolphins. It has been known and trusted by man for centuries. The explosive sound porpoises make when emerging from the water is a well-known characteristic. Although they will eat fish and crustaceans, the standard meal of the porpoise is squid, which they prefer when they can find it. See CETACEA, DOLPHIN, WHALE.

port (1) The side of a vessel that is to the left of an observer standing on the vessel and facing the bow or front end. (2) A harbor. (3) An opening in the side of a ship for loading or unloading cargo, for ventilation, or for the display and use of gunnery. See STARBOARD.

portolan In Renaissance mapmaking, this was a very detailed coastal outline showing seaports, navigational hazards, and river mouths with channels, shoals, and sandbars. Distances from one focal point to other points of interest are essential. Latitude is often absent, and longitude was not yet a concept. The interior areas, since they were of no navigational interest, are blank.

Portuguese man-of-war A delicate, transparent, floating hydrozoan of the genus *Physalia*, phylum Cnidaria, found in warm waters in and near the Caribbean Sea and in the Pacific and Indian Oceans. The Portuguese man-of-war is known for its gas-filled, floating body, which ranges from 10 to 30 cm (3 to 12 inches) long and stands about 15 cm (6 inches) above the water. The float acts as a sail and is most often pale blue but may also be pink or lavender.

The Portuguese man-of-war lives and travels in spectacular colonies, which trail a mass of polyps which can be several meters long, with some over 50 m (160 feet) long. The organism is differentiated into structures that capture prey, others that eat it, and still others that reproduce. The tentacles bear nematocysts that give a powerful, painful sting and can cause fatal shock in a human victim. The only known predator of the Portuguese man-of-war is the sea turtle. *See* CNIDARIA, TOXIN.

posterior The rear or tail end of an animal. The word also denotes position. Thus, the tail of a fish is posterior to its pelvic fins. *See* ANTERIOR, DORSAL, VENTRAL.

potassium A chemical element of the alkali metal group occurring abundantly in nature. Potassium is an element necessary to animal and plant life. It is present in seawater as potassium ion, K^+, in roughly the same amounts as is calcium—about 0.04 grams per liter. The potassium argon dating of sediments is a fairly standard technique based on the degradation of radioactive potassium to inert argon in a process that takes thousands of years. *See* DATING, ELEMENT, RADIOACTIVITY.

prawn A large shrimp (genera *Pandalus* and *Peneus*), found in fresh, brackish, or salt water. Shrimp that are about 8 cm

(3 inches) or larger are called prawns. The tropical freshwater species are often much larger, and individuals of 18 to 20 cm (7 to 8 inches) are not uncommon. *See* DECAPODA, SHRIMP.

precession Rotation of the plane of orbit of the Earth (or the moon). This has an immediate effect on the Earth, causing variation in the tidal range. *See* LUNAR TIDE, TIDAL RANGE, TIDE.

predation The consumption of one animal by another. The patterns of predation describe the food chain, and this pattern is affected by several factors, the most obvious being the ratio of predators to prey. Temperature is also a factor, and with global warming, formerly normal patterns show changes. For example, slight warming temperatures affect the predation of *Pisaster ochraceus*, a sea star found on the Oregon coast. This animal that preys on mussels is rendered inactive when the temperature of the water in the intertidal areas drops. This occurs during upwellings, and in these periods the mussels become the dominant organism in the tidal pools. Once the water warms and the sea star is active again, the invertebrate population of the tidal areas becomes much more varied as the mussel population drops. This variation is significant since an El Niño year will result in significant change in water temperature. The nature and position of upwellings will also be affected by global warming.

Predators may also change their diets. If the normal prey is removed, a predator that is flexible in its feeding may switch to another food supply. This seems to have happened in the North Pacific. Commercial whaling in the area removed a large number of baleen and sperm whales. Since these animals are preyed on by orcas—killer whales—the animals moved on to other food supplies: harbor seals, fur seals, sea lions, and sea otters. All of these animals, newly preyed upon by killer whales, have shown greatly diminished populations, and the ecosystem as a whole has been disrupted. The removal of the whales has led to unintended and unexpected disruptions in all of the food web in the North Pacific. *See* FOOD WEB, GREENHOUSE EFFECT, SEA STARS.

predator-prey relationship The relationship between a hunting animal and its animal food. The food web consists of a large number of plant organisms, which serve as food for a somewhat smaller number of small animals that in turn supply the food for a still smaller number of larger, carnivorous animals. The usual ratio of predator to prey is that 100 kilograms of autotrophe (phytoplankton, plant, or bacteria) will feed 10 kilograms of herbivore, and that will in turn feed 1 gram of carnivore. There may be secondary and tertiary carnivores, and again the ratio of 10:1 holds. There is frequently more than one predator for each small, herbivorous (plant-eating) animal. Furthermore, some very large animals are herbivores, and some very small ones are carnivores. The very largest carnivores may even harbor plant populations on their bodies, such as the polar bear, which harbors algae in the hollow-stemmed hairs of its outer fur. Thus, the eater-to-eaten relationships are quite complex.

The food web of the Antarctic was assumed to be a fairly simple one to study; the largest population is that of krill. It was assumed that diatoms are the base of the food chain; they in turn are eaten by the krill, and krill are eaten by whales. The vast population of krill is also preyed upon by penguins and other birds, which are then preyed upon by seals. The food web is much more complex than it was originally thought to be. Krill are not the only food source. Copepods and salps are also preyed on by the bird population. All of the faunal populations are affected by weather, the nature of the sea ice that harbors the algae that feeds the diatoms, and seasonal fluctuation. The effect of whale hunting adds another variable to this already complex mix.

The relationship of a predator population to that of the prey shows that the predator population peaks after that of the prey. This in turn is echoed by the effect on the plant population fed on by the herbivores that form the bottom layer of the food web (or pyramid).

While there are organisms that do die of old age, those that are part of the regular food supply for others have nervous lives. Most prey organisms, however, do have some defenses, such as spines (with or without poisoned tips), a bad taste, mimicry or the ability to imitate other organisms so as to hide or appear potentially threatening, powerful partners (symbionts or commensals, such as in the case of the remora fish and shark), or chemical defenses. One example of the last type of protection is the toxic substances. A number of marine organisms produce toxins to discourage attack. For example, the Pacific sole renders itself distasteful to sharks, several species of puffer manufacture tetrodotoxin, and many sponges produce toxic materials as protective armor.

Predation is well established in the fossil record where, according to some researchers, it is a determining factor in the evolution of a number of organisms. The time span in the Paleozoic era saw an explosion of life-forms (the Paleozoic marine revolution produced a wide variety of shell-crushing predators). At the same time the shells of the surviving mollusks became much more protective. The mollusks that survived had shells that were more tightly coiled, or had spines. Brachiopods were extremely successful in the Paleozoic—550 million years ago until about 250 million years ago. Now they are a very small component of the ocean's life forms. However, drill holes appear in the earliest Paleozoic fossils; about 1% of the brachiopod shells show the marks of a predator, and by the end of the era most of the fossil brachiopod shells show the characteristic hole. Those animals were killed by drilling whelks or moon snails. The surviving brachipods today have much heavier shells and are attacked by drilling predators far less often than are ordinary snails and clams.

Another fossil of the same period that shows evidence of predation is the crinoid. The shell-crushing predators may not have broken crinoid arms to eat them. The damage may have been accidental as the predator sought prey that was hiding in the crinoid. Arm regeneration in crinoids leaves scars; the greatly increased number of such scars is evidence of damage that occurred during the Paleozoic marine revolution. Surviving crinoids show increased spines and thicker calyxes. *See* CARNIVORE, CHEMICAL DEFENSES, COMMENSAL RELATIONSHIPS, FOOD CHAIN, FOOD WEB, HERBIVORE, POPULATION, SYMBIOSIS, TOXINS.

pressure A force exerted on a surface. It is expressed in terms of force per unit area. The atmospheric pressure at sea level is that of a column of air exerting a force on a defined area. It is 101.325 kPa (kilopascals), where a pascal is defined as 1 newton per square meter, and a newton is defined as 1 kilogram-meter per second squared (kg.m/second^2). The atmospheric pressure is sometimes given as 14.7 pounds/inch^2, or 1 torricelli, or 760 millimeters of mercury (mm Hg).

As an organism descends in water, it has to deal with both the air pressure on the water and the pressure exerted downward on itself by the column of water through which it has already descended. At a depth of 10 m (32.8 feet), the pressure is an additional atmosphere (101.325 kPa). At 20 m (65 feet) it is 2 atm more than at sea level. The pressure increases by 1 atm per 10 m.

Most surface- or near-surface-dwelling marine organisms deal with the pressure they encounter by using a swim bladder. Some hadal species also have swim bladders. Other organisms deal with the need to equilibrate their internal pressure and the external pressure on them by absorbing water into their tissues, increasing their internal pressure. This renders the organism almost transparent. Other mechanisms include the storage of low-salt fat, or oil, or air. *See* DECOMPRESSION, DIVING, OCEAN-ATMOSPHERE RELATION.

Priapulida A phylum of wormlike, unsegmented animals. The giants in this group are 30 cm (12 inches) long; however, the average adult is about 12–15 cm (5–6 inches) long. They are found almost totally buried in sediment, at almost all depths, in cool to cold waters. Priapulids have tiny hooks on the proboscis (the head end that catches tiny prey). They have a chitinous cuticle that is shed as the animal grows. A midventral nerve cord runs the length of the body and terminates in a large ganglion that encircles the pharynx. Sexes are separate. Priapulids are likely related to the Kinorhyncha and Loricifera, all of which molt.

proboscis The food-obtaining structure on the "head end" in invertebrates. It may be extendable, and is frequently equipped with suctorial or hooking structures.

The term *proboscis* also refers to the nose or nasal structures of an organism. Thus, the elephant seals' large nose sacs are called probosces. *See* SEAL.

Procellariformes The order of marine birds normally found far offshore. Albatrosses, fulmars, petrels, and shearwaters, are all members of this order.

Prochlorococcus The smallest known photosynthesizing cyanobacterium. It lives in a wide band of open ocean ranging from the equator into warm (temperate) latitudes and is present in such numbers that it provides at least 30% of all primary production in the ocean and significantly contributes to the movement of carbon through the biosphere.

The range of this organism extends through the euphotic zone down to the point of almost no dissolved oxygen—the oxycline. Several relatively unusual pigments in addition to chlorophyll—

carotenes and phycoerythrins—enable the bacterium to make use of the very limited light available at depths of 150 to 200 m (500–670 feet). A related species is *Synechococcus*, and both seem to have viruses, phages that rearrange their single-strand DNA. A startling difference between the two bacteria is that *Synechococcus* can swim without any flagella or tail, while the *Prochlorococcus* cannot.

prokaryote According to some taxonomists, this is the superkingdom of organisms whose cells have no nucleus. Genetic material is dispersed throughout the cytoplasm. The bacteria are in this group.

protein A polymer of amino acids. *See* AMINO ACID, PEPTIDE FORMATION, POLYMER.

Proteobacteria A phylum in the superkingdom Bacteria. The most recent taxonomic classification of Proteobacteria dates to the 1980s and is subject to modification. These gram-negative bacteria are a very diverse group that is identified and separated into five lineages by their RNA sequences. They include organisms ranging from aerobes to obligate anerobes; many are pathogens. The phylum includes the enteric organisms that live in digestive tracts. *E.coli* and salmonella are proteobacteria, as are nitrogen-fixing bacteria on legume roots, rickettsiae, purple photosynthesizers, and the chemoautotrophic and chemoheterotrophic bacteria of hydrothermal vents and cold seeps. Proteobacteria are found on the shells of shrimp in hydrothermal vent communities. This is probably a commensal relationship. *See* EXTREMOPHILES, RIMICARIS, VENT COMMUNITIES.

Protista The simplest eukaryotes. They probably evolved from prokaryotes about 1.7 billion years ago. In the Lineaean system, this group would be divided between plants and animals. In the five-kingdom taxonomic scheme, the Protista are a kingdom unto themselves. This very diverse group of organisms have several characteristics in common; they all have DNA strands in nuclei that are enclosed within a nuclear membrane, and they have many organelles—structures within the cells that perform metabolic functions. Many are motile and have flagella or cilia, and most are aerobic although a few mud-dwelling species are facultative anerobes.

Their differences are varied; protists may be unicellular, colonial, or multicellular and are divided into four large groups: unicellular algae, multicellular algae, protozoa, and slime molds. Each of these may be further subdivided into phyla that in turn are the ancestors of multicellular organisms. They are Chromobionts, the yellow and brown algae; Chlorobionts, the green algae; Rhodophytes, the red and purple algae; Euglenozoa, algae that are mixotrophes; Dinozoa, the dinoflagellates; several parasitic phyla, Microsporidians and Myxozoa; Ciliata, ciliated organisms; and Arasiomycota and Myxomycota, the slime molds that were once thought related to fungi. RNA analysis shows that they are not similar to fungi at all. *See* ALGAE, CILIA, DINOFLAGELLATE, MIXOTROPHE, PROTOZOA, RHODOPHYTA.

Protoperidinium A colorless or yellow dinoflagellate that does not photosynthesize. It is present in most open ocean water and is one of the prey organisms that support the copepod population. *See* DINOFLAGELLATE.

protoplasm The substance of the living cell, enclosed by the cell membrane (i.e., the nucleus and cytoplasm of the cell). Chemically, the protoplasm is an aqueous solution of nutrients, sugars, lipids, starches, oils, and proteins. The protoplasm is distinctive for each different kind of cell.

Protozoa A phylum of minute eukaryotic unicellular organisms living in marine, freshwater, and damp terrestrial environments. There are an estimated 30,000 living species of protozoans. They range from

Acorn worm with elongated proboscis

plantlike forms to animal types. The animal members of the Protozoa include ciliates, foraminiferans, radiolarians, amebae, some flagellates, and sporozoans. Some protozoans—the euglenoids—contain chlorophyll, are green, and conduct photosynthesis in sunlight. They become colorless and predatory in the dark. The protozoa feed on detritus, bacteria, algae, and on each other. Populations of marine protozoa are greatest in polluted waters, although they are present in nutrient-poor environments as well. Ecologically they play a role in energy transfer. Protozoans are infrequently autotrophes; they feed on algae, bacteria, and microfungi. In an ecosystem they are primary consumers or decomposers that in turn provide food for other microinvertebrates. *See* PROTISTA.

Pseudocalanus A genus of copepods that are important in the food web of the polar regions. These organisms are the basic food for larger and commercially significant animals such as the dominant fish species. *See* CALONOIDA, COPEPODA, FOOD WEB.

Pterobranchia A very small class of Hemichordata. Unlike the other living class, the acorn worms, these tiny organisms (less than 1 mm or 0.04 inches), called zooids, form interconnected colo-

nies. They spend their lives within tubes made of collagen, a protein that is secreted from their mouthparts. They are filter feeders and live in benthic environments. *See* HEMICHORDATA.

Pteropoda A group of small gastropod mollusks characterized by winglike organs with which they swim. The sea butterfly, actually a small snail, is a pteropod. These animals live on or near the surface. Their shell is a delicate, transparent one about 0.5 cm (0.02 inches) long. There are significant deposits of such shells in calcareous sediments. The shell, because it is so light and thin, is particularly susceptible to dissolution in cold, carbon dioxiderich water and there are more pteropod remains in the waters of the North Atlantic and Mediterranean than in the Pacific. *See* GASTROPODA, SEA BUTTERFLY, SNAIL.

Ptolemy, Claudius (90?–168) An Alexandrian Greek mathematician, astronomer, and cartographer. He also collected earlier geographical works. Ptolemy's best known book was his *Geographical Treatise*—an eight-volume work and an atlas of the then-known world.

A series of Ptolemy's maps showed curved meridians with only the central one, which he called the "prime meridian," depicted as a straight line. Coincidentally, this Ptolemaic line is reasonably close to the present line of 0° longitude. Ptolemy drew his prime (or first) meridian as far west as he could and placed it near the Canary Islands.

Ptolemaic maps were lost in the general decline of science and scientific interest in the West that followed the decline of the Roman Empire. The maps reappeared about 1400 C.E., when they were reintroduced by Arab scientists and Byzantine Greeks. They had become widely known by the time of the great voyages of the 15th and 16th centuries. The error in Ptolemy's maps resulted from his use of Poseidonius's (135–219, Syrian-Roman traveler, natural philosopher and geometer) figure for the circumference of the Earth, rather than Eratosthenes' figure, which was more accurate and larger. Like almost all other ancient cartographers, Ptolemy knew that the world was spherical.

Ptolemy was a much better geographer than an astronomer. His view of the Earth in relation to other extraterrestrial bodies was mathematically correct but inaccurate. He believed that the Earth, not the Sun, was the center of the universe. This geocentric depiction of the Sun and the known stars and planets was of extremely long duration. It became codified and was accepted by astronomers and geographers in many places, and it remained for Copernicus, Brahe, Kepler, and Newton to disprove it.

The geocentric cosmos can very elegantly explain all observed astronomical phenomena. With it, stars can still be used to fix navigational points; however, the Spanish king Alfonso the Wise is supposed to have said, upon hearing the Ptolemaic exposition of the phenomenon of the eclipse: "If God had asked me how to design the world, I would have made it simpler." The story is probably apocryphal, but the statement points to the much later discovered reality of how the solar system works. *See* CHART; COPERNICUS, NICOLAUS; MAP; NAVIGATION.

Puerto Rico trench The deepest point in the North Atlantic, at 8,385 m (31,330 feet), north of the island of Puerto Rico and almost on the 20° north latitude line. The island is part of the Puerto Rico ridge, and the trench marks the edge of the Caribbean Plate.

The trench is a slip strike zone and shows continuations of the slip strikes seen on the island around it. It has been a feature on the surface of the Earth for more than 70 million years. It is partially filled with terrigenous sediment. *See* CARIBBEAN SEA, SLIP STRIKE, TRENCH.

puffer Also known as the globefish, a spiny-skinned, mostly warm-water marine fish with the ability to puff itself up. Puffers are of the family Tetradontidae, a name indicative of their fused teeth. There are about 90 known species.

Puffers have characteristic tough, spiny skins. They are usually about 30 cm (12 inches) long, although some species are much larger. The most obvious trait of the puffer is the ability to inflate to soccer ball proportions.

The sharp-nosed puffer, *Canthigaster*, is smaller and more colorful than the usual gray-brown color of the puffers. Porcupine fish, another group of puffers, are thick-bodied animals that live in shallow water. Their spines become prominent when their body is inflated.

The spines of puffer fish are a defense mechanism; any predator has to be willing to put with a mouthful of sharp spines if it wishes to dine on these fish. Some species, however, have poisonous spines, and others such as the fugu have poison-containing organs. When properly cleaned and prepared, however, the fugu is a delicacy consumed by Japanese gourmets. Because its organs are very poisonous, however, any flesh that has been touched by either the liver or the reproductive organs is extremely toxic, and several fatalities occur every year in Japan from eating contaminated fugu flesh. *See* FISH, TOXIN.

puffin A diving seabird related to the auk. It is somberly feathered in white and black or gray, but dramatically colored red, yellow, or blue on its legs and beak. The birds are about 30 cm (1 foot) long, and nest in large colonies on rocky cliffs or islands. Both parents raise their single chick in a burrow. The diet of puffins is fish. Some representative species are the following:

COMMON PUFFINS		
Common Name	Scientific Name	Habitat
Common puffin	Fratercula arctica	North Atlantic
Horned puffin	F. corniculata	North Pacific
Tufted puffin	Lunda cirrhata	Temperate to warm Pacific area

See BIRD.

purse seine A type of net used for trapping open-sea or pelagic fish. These nets are usually set by two boats and can be drawn together like a purse around a school of fish. The size of the mesh determines the target population. Purse seines with very fine mesh are used in biological sampling. Purse seines for tuna and other open ocean fish have recently been redesigned so that dolphins accidentally caught in these nets can escape, thereby helping preserve the dolphin population. See FISHING, NETS.

pycnocline A water layer that exhibits a relatively rapid increase in density with increasing depth. See ANTARCTIC BOTTOM WATER.

Pycnogonida An arthropod subphylum. These organisms live on ocean bottoms. The common name is sea spiders. The shallow-water species are less than 1 cm (0.4 inches) in diameter, while hadal species may be larger than 1 m (40 inches).

Pyrmnesiophytes (Pyrmnesiphyceae) Also known as Haptophyta because these tiny organisms possess a structure unique to them—the haptonema. It was once thought to be the remnant of a third flagellum, but this is not the case; its function is unknown. The best-known Pyrmnesiophytes are the coccolithphorids.

The 500 or so living species are collected in 75 genera. Their presence on Earth is well documented back to the Jurassic period, when they first appeared in great numbers indicating a prior existence in the Pennsylvanian. The greatest diversity of the prymnesiophytes occurred in the late Cretaceous period, and with the end of that era, a mass extinction removed more than 65% of the genera.

As a group, these organisms all photosynthesize, and with this activity they may contribute to massive algal bloom. Most species live in warm, tropical seas, although there are a few terrestrial and freshwater species. The prymnesiods contain pigments in addition to chlorophylls, notably those that give them their characteristic yellow color. See COCCOLITHOPHORADS, EMILIANA HUXLEYI, PHOTOSYNTHESIS.

Pyrrophyta A phylum of flagellated algae. The dinoflagellates are in this group. See DINOFLAGELLATES, FLAGELLA.

Quaternary period The part of the Cenozoic era that began about 4 million years ago and extends to the present. It is in turn divided into the Pleistocene epoch and the most recent epoch, the Holocene. The major event of the Quaternary is the appearance and then disappearance of the vast sheet of ice that covered most of the Northern Hemisphere from the North Pole to about 40° north latitude. The cooling trend that precipitated the glaciation started in the late Tertiary period. The Quaternary is also marked by the appearance of humans. *See* Cenozoic era, Pleistocene epoch, Tertiary period, *Appendix: Geologic Timescale.*

radar (RADAR) Radio Direction and Ranging: a device that uses ultra-high-frequency radio waves to measure the distance to an object by determining the time needed for the wave reflected by the object to return to some detection device. A directional antenna is used to determine the direction of the object, or the antenna can scan in continuous circles to locate any object in the path of its beam.

During World War II it was noted that rain and snow produced a particular signal on radar recorders. Using the Doppler effect, in which the frequency of a wave is shifted to a higher frequency if the source of the wave is moving toward the receiver, and to a lower frequency if the source is moving away from the receiver, it is possible to determine the amount of precipitation and air velocity of a storm. It is also possible to pinpoint the location of the precipitation and its intensity. This means that cyclonic storms can be mapped and tracked. Radar can also be used to track clear air turbulences (i.e., storms not associated with precipitation). A variation of radar, sonar, which employs sound waves, is used extensively in exploration, by measuring the time needed for the return of an echo. Given the medium in which sound waves travel, the speed of sound is constant, and the time it takes to travel to and from some object is therefore easily translated into distance. *See* NAVIGATION, WAVE, SONAR, SOUNDING, STORM, WEATHER.

radiata Another older name for coelenterates. *See* COELENTERATE, CNIDARIA.

radiation The movement of energy away from a source. The energy from the Sun that actually arrives at the Earth's surface is less than half of the possible total. The rest of the energy is reflected by the atmosphere back into space. The sea acts as a great heat sink for the Sun's radiation, since the water in it is heated directly. This causes both convection currents, which bring warmer water to cold areas, and evaporation. The water that evaporates then enters into the hydrologic cycle, in which it goes from water vapor to cloud to precipitation (rain or snow) to ice melt to runoff to the sea.

Warm water also warms the air above it, and the air, like water, moves from regions of higher temperature to regions of lower temperature. Moving air is wind. Thus, water air currents and weather changes are dependent on the radiant energy received on earth from the Sun. *See* ATMOSPHERE, EVAPORATION, METEOROLOGY, OCEAN, SOLAR ENERGY.

radioactivity The spontaneous disintegration of an isotope of an element, accompanied by the release of alpha, beta, and gamma radiation. Radioactive isotopes occur both on land and in the world's oceans. Some radioactivity comes from the interplanetary bombardment of atoms, causing them to break up and release particles and energy from the Earth itself or from the human production of nuclear materials. The cosmic (extraterrestrial) contribution to the radioactivity of the oceans comes from the isotopes H^3, Be^7, Be^{10}, C^{14}, and Si^{32}. These are "light atom" (meaning that they have smaller nuclei) radioisotopes. The principal radionuclides that date from the primordial earth are K^{40}, U^{235}, U^{238}, and Th^{232}.

In 1971 the Geochemical Ocean Sections Study began to inventory the contents of the oceans in order to establish a

Radiolaria

baseline, or "normal level" of radioactivity in seawater. The study measured, among other items, the deepwater concentration of carbon 14 in different oceans. The rate of decay of C^{14} is 1% per 83 years, and with this as a basic assumption, the turnover time in the various ocean deeps can be calculated. At depths greater than 1,500 m (4,800 feet), it is 510 years for Pacific waters, 250 years in the Indian Ocean, and 275 years in the deep Atlantic. Some deep water has stayed in the depths for more than 1,500 years, but the average turnover of all deep water is about 500 years. The outer limit of carbon 14 dating is about 40,000 years and possibly less.

It is also possible, using ratios of isotopes (forms of a single chemical element that have different atomic mass), to determine the ocean of origin of certain bodies of water. The ratio of neodymium 143 to neodymium 144 is the means for determining the quantities of Atlantic versus Pacific water going through the Drake Passage.

Potassium 40 decays to argon 40 very slowly. It is assumed that all of the original argon that was present in the primitive earth is now in the atmosphere. Thus any argon found in rocks—either marine or terrestrial, is the product of the decay of K^{40}. Since some of this also escapes into the atmosphere, corrections have to be made for this. If it is not corrected for, rocks appear "younger" than they actually are. The potassium-argon system is accurate for rocks 400 million years old and, along with the rubidium-87 to strontium-87 (Rb^{87}–Sr^{87}) system, is applicable to the dating of most sediments and of continental rock as well.

Uranium decays to protactinium (U^{238}–Pa). This process is useful in dating sediments less than 60,000 years old. Another decay process that is used for assigning age to recent sediments is that of aluminum-26 to beryllium-10 (Al^{26}–Be^{10}). See ALPHA RAY, BETA RAYS, ELEMENT, ISOTOPE.

Radiolaria A group of symmetrically shaped marine protozoans with radiating threadlike pseudopodia. They date back to the Precambrian. Their rapid diversification has made these protists a means of differentiating geological layers. The radiolarians flourished until the late Paleozoic, when they experienced a rapid diversification that coincides with the rise in dinoflagellates—a radiolarian food supply. The evolution of diatoms in the Cretaceous may have represented competition for silica, and the result was the construction of smaller and finer radiolarian tests. Radiolarians are found in the photic sunlit zones of all oceans. Their characteristic feature is a test, or external wall, of siliceous (silicon-containing) material. Numerous pseudopodia extend beyond the perforated test. The pseudopodia do not seem to propel the organism. Radiolarians drift with the current as part of the holoplankton. Radiolarians reproduce largely asexually. There is an occasional production of gametes, but the usual reproductive mode involves the test splitting in half, with one segment becoming part of each daughter cell. Upon the death of the animal, the tiny test (from 50 to 500 micrometers to 10^{-6} m) settles to the ocean bottom. Extensive deposits of tests are a sign of large populations of radiolarians and are found in regions of upwellings. The rising of nutrient material from the depths of the ocean supports the large populations of radiolarians that mark the Equatorial Divergence, the Antarctic Convergence, and the coast of South America. See PROTISTA, PROTOZOA.

Rajiformes Fish with dorsoventrally flattened bodies; this group includes guitarfishes, rays, sawfishes, and skates.

rat-tail fish These deep-sea fish have stout bodies and whiplike tails. See CHIMEARA, MACRURIDAE.

ray A cartilaginous, winglike fish of the order Batoide (or Rajiformes). They live in almost every ocean, and some species even move up rivers. Most are sluggish bottom dwellers living on mollusks. The manta ray, however, will occasionally break the surface. It is also one of the largest of the sea's creatures, with a 6 m (20 foot) "wingspan."

Other members of this group are the stingrays, electric rays, and guitarfish. All are characterized by a long, spiny, and sometimes venomous tail and highly modified pectoral fins that resemble wings. The gill slits on the underside are part of the very "streamlined" bodies of the rays. Water is brought into the body by a series of spiracles on the upper surface. Most rays bear live young. *See* CHONDRICHTHYES, OVOVIVIPARITY.

ray-finned fish Primitive fish. *See* ACTINOPTERYGII.

red algae *See* RHODOPHYTA.

red clay An oceanic sediment found at great depths. It consists of windblown particles of continental origin. Red clays dominate the sediments of the southwest Pacific Ocean. *See* OCEAN FLOOR, SEDIMENTS.

Red Sea A narrow, marginal sea of the Indian Ocean separating Africa from the Arabian peninsula. It is about 300 km (190 miles) wide at its widest point and 1,900 km (1,200 miles) long. The Sinai peninsula juts into the northern end of the sea, dividing it into the Gulf of Suez on the west and the Gulf of Aqaba on the east. The eastern portion is very deep (over 1,800 m or about 6,000 feet), while the Gulf of Suez is shallow. The entire region is geologically interesting, since it is part of a rift valley system extending from the Dead Sea through East Africa into Kenya. The northern end of the Red Sea is a downfault. The central and southern portions show complex rift activity, with new magma emerging and rising to spread the African and Arabian plates even further apart.

Water enters the Red Sea from the Gulf of Aden over the fairly shallow sill (125 m; 412 feet) at Bab el Mandeb. Practically no water enters from the surrounding lands except during the spring torrents, which bring sediment. The high rate of evaporation due to the hot, dry climate makes the Red Sea one of the saltiest bodies of ocean water on Earth, with a salinity of over 4.0% in summer. It is also the warmest. Some anomalous water, even hotter and saltier than the very warm, salty water to be expected in the Red Sea, has been retrieved from the bottom. Samples of water taken in the central channel from depths of 2,200 m (7,260 feet) and below have shown temperatures above 40°C and have more than a 20% salt content. The brine and the sediment associated with it are very rich in minerals, and the deposition of iron, manganese, zinc, lead, copper, silver, and gold is thought parallel to the processes of a billion years ago. The anomalous, warm deep water is thought by some to be connate or fossil water that contains dissolved minerals that originated elsewhere and are now appearing in the Red Sea.

The winds associated with the Red Sea are northerlies in the northwestern portion of the waterway and southerlies in the southeastern region. This makes sailing on the sea difficult. The technological answer was the dhow—a small, lateen-rigged ship. The triangular sails of this vessel are hung on yards that are suspended from the mast, making it possible to beat almost directly into the wind. *See* DHOW, HOT SPRINGS.

red tide Seawater covered or discolored by a large population of dinoflagellates; often fatal to many forms of marine life. Red tides are produced by the sudden growth, or bloom, of a dinoflagellate or other flagellate species. The Gulf of Mexico has red tides that are caused by *Gymnodinium brevis*. The result is often a gigantic fish kill. Wind and wave action move the "bloom" along on the water. Walvis Bay on the Atlantic side of South Africa is similarly affected. The red tides caused by the genus *Gonyaulax* result in human deaths because these dinoflagellates are ingested by clams and mussels, which then concentrate the paralyzing neurotoxin produced by these dinoflagellates. *See* DINOFLAGELLATE, TOXIN.

reef A mass or ridge of rock in either a freshwater or marine waterway. A reef is unlike a bar, which consists of unconsolidated sand and is thus subject to removal by wind, current, or both. A reef is either submerged coast, and therefore structurally like the rock of the nearby landmass, or a biogenic creation. A biogenic reef is constructed by corals and coralline algae. The remains of ancient reefs have been found on what is now land. An example of this is a mountain formation in New Mexico and the location of Carlsbad Cavern. When the North American continent west of the Mississippi was a vast oceanic embayment, El Capitan was a tropical reef. It is now both inland and at considerable elevation. *See* BARRIER REEF, CORAL.

reflection The return of a ray or wave of energy toward its source. When water waves (or light, radiation, or sound waves) strike a surface, the surface acts as a barrier and redirects the waves in the direction of their origin. *See* DIFFUSION, REFRACTION.

refraction The apparent bending of an energy wave on its passage from one medium (e.g., air) into another (e.g., water). Refraction is caused by the differing densities of the two media through which the wave passes, and the resulting change in velocity as the wave moves from one medium to the other. Sound, as well as light, can be refracted. This is important in the use of listening devices such as loran and sonar. The movement of sound waves through water is affected by temperature. In warm ocean levels, sound is refracted toward the surface. As the sound waves move deeper into sunless cold water, they are refracted slowly toward the ocean bottom. This creates a "dead zone," where a sound-producing object (i.e., a submarine) could be inaudible. *See* LORAN, SONAR.

relative humidity The ratio of the quantity of water vapor actually present in a sample of air to the quantity that the atmosphere can hold at that temperature and pressure.

The relative humidity is measured using two thermometers, one wet, the other dry. The temperature shown by the wet bulb will be some number of degrees lower than that of the dry one because the evaporation of water from the wet bulb requires energy (heat) and therefore the bulb will be cooled and show a temperature lower than that of the dry bulb. If the atmosphere is saturated with water vapor, no more water can evaporate and the two thermometers will have the same temperature reading. The relationship between the wet-bulb and dry-bulb readings is not a simple one and complex charts are used to determine the relative humidity on the basis of the two temperature readings and the barometric pressure. Electric hygrometers measure the water content of the air directly by measuring the electrical conductivity of the air. *See* AIR, FOG.

relict beach An archaic beachfront that is no longer at the seashore. A rising sea level has submerged some beaches, which are discernible on continental shelves. *See* BEACH, BLUE HOLES, CONTINENTAL SHELF, SUBSIDENCE.

Remipedia A class of primitive crustaceans first described in 1980. They are segmented and have walking legs on each segment. Unlike millipedes and centipedes, the segments are differentiated—as they are in crayfish. The biology of these animals is as yet unexplored.

remora Any member of a small group of specialized fish that can attach themselves to large fish or ships by means of a suction disk on the top of the head. They are anywhere from 30 to 90 cm (1 to 3 feet) long and dark gray or green in color. The remoras are most often associated as commensals with sharks. Like sharks, they live in warm waters worldwide. Their usual diet is detritus or anything that falls in their direction.

Remoras seem to feed on the stray bits that the shark neglects to snap up, although some seem to clean the larger animals, thus "paying" for their free ride. Remoras do not depend on larger fish for movement; they can and do swim well on their own. *See* COMMENSAL RELATIONSHIP, FISH, SHARK, SYMBIOSIS.

reproduction The great division in marine environments is between those organisms that reproduce asexually and those that produce eggs and sperm. The asexually reproducing organisms, including bryozoa, coelenterates, flatworms, sponges, and tunicates, will also have an occasional sexual phase. Many organisms will exhibit both sexual and asexual reproductive patterns. The Portuguese man-of-war, a colonial cnidarian, is an assemblage of fairly differentiated cells. It begins with a zygote, the product of an egg-and-sperm union, and as it grows, it asexually produces the different units that compose the mature organism. The organisms that reproduce sexually may be dioecious (distinct males and females) or monoecious hermaphrodites such as most snails and sea hares. Some, like the wrasse, can change sex with either adverse condi-

tions or age. Most dioecious animals just release eggs and sperm into the water and let the larvae fend for themselves. Others are viviparous (producing live young), or ovoviviparous. Some sharks are ovoviviparous; the young develop from eggs within the female's body. Some invertebrates are also brooders, notably, some corals. As a group, corals reproduce sexually, and the fertilized eggs can be widely dispersed to form new colonies. Fragmentation will also produce new colonies; if a reef is damaged by either human activity or a natural catastrophe such as a landslide, broken fragments can each be the start of a new colony. *See* CLAM, PENGUIN, SALMON, SHARK, RAY, WRASSE.

reptile Any member of the class reptilia, comprising air-breathing vertebrates characterized by scaly skin, internal fertiliza-

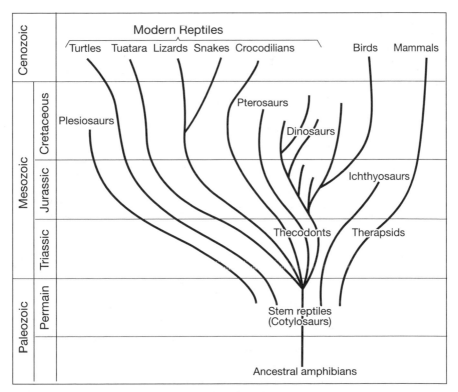

Evolution of reptiles

tion followed by the production of shelled but not always calcareous eggs, and in most species a three-chambered heart. The reptiles include the snakes, lizards, turtles, alligators, crocodiles, and their near relatives. Sea snakes and turtles are the most prominent marine reptiles, although some crocodilians live in brackish waters. The fossil record shows many extinct aquatic reptilian forms, such as the plesiosaurs and ichthyosaurs. *See* ICHTHYOSAUR, PLESIOSAUR, SNAKES, SEA TURTLE, VERTEBRATE.

research The systematic scientific study of a particular problem within a scientific discipline. *See* AIR, CHEMISTRY, GEOLOGY, METEOROLOGY, OCEANOGRAPHIC VESSELS, OCEANOGRAPHY, PHARMACEUTICAL RESEARCH, WEATHER, ZOOLOGY.

***Research,* HMS** A ship used in 1827 by the English sea captain Peter Dillon, to search for survivors of La Pérouse's expedition to Australia in 1787–88 on the *Boussole.*

The name *Research* was also given to a British surveying ship commissioned in 1889. Another vessel named *Research* was a nonmagnetic ship built to avoid mines, and launched during World War II. It was scrapped after just six years at sea. *See* LA PÉROUSE, JEAN-FRANÇOIS DE GALOUP COMTE DE.

respiration The use of a food (energy) source by some organisms. The reaction produces carbon dioxide CO_2 as a waste product. Since CO_2 is a primary greenhouse gas, its generation in the ocean must be of major consequence to the carbon cycle. The rate of respiration in the ocean is an ongoing research topic. Organisms produce compounds that utilize carbon, others decompose those compounds and release carbon. Current thinking on the relationship of the ocean to carbon levels points in the direction of the ocean as a vast carbon sink. The water can and does dissolve vast quantities of carbon dioxide. This is the vast store of carbon that is relatively unavailable for absorption by organisms or for escape

into the atmosphere. Thus cold and/or deep water that dissolves and holds more carbon dioxide than warm water is the reservoir of this greenhouse gas. *See* CARBON CYCLE, CARBON DIOXIDE, PHOTOSYNTHESIS.

Rhine River This western European river rises in central Switzerland and flows roughly northwest along a 1,300-km (820-mile) course to the North Sea. More than half of this course is navigable, which means that the Rhine is a very important river commercially.

The Rhine is one of Europe's significant historic, legendary, geographic, and political features. Its many falls and attractive mountainous surroundings have made it a center for tourism and vineyards. In other parts of its course, such as near Bonn, Germany, it is an industrial and therefore often polluted waterway. The city of Cologne, Germany, is one location that has been known for its "smelly river" since Coleridge's visit in the last years of the 18th century.

The delta region of the Rhine is one of fragmented streams. The Neder Rijn, or Lower Rhine, and the Waal are smaller rivers that rise from the breakup of the Rhine proper. These in turn divide again, forming the network of small streams that thread through the delta, which constitutes most of the land area of The Netherlands.

Current engineering works are changing the course and bottom topography of the Rhine throughout most of its length. The objectives are to control flow, improve navigation, and control pollution. The major mouths of the Rhine will all eventually be protected with a series of "superdikes" designed to control storm surges. Many of these have already been completed in the region around Eindhoven, The Netherlands. *See* NORTH SEA, STORM SURGE.

Rhizocephala An order of Cirripedia (barnacles). They are all parasites on other arthropods, usually decapods. *See* CIRRIPEDIA.

Rhodophyta The red algae, a division of the kingdom Plantae in some taxonomies; in the five-kingdom taxonomic structure, the rhodophytes are grouped with the other protists. Their chlorophyll is hidden by the pigments phycoerythrin and phycocyanins. Rhodophyta are usually found at depths greater than those at which green or brown algae flourish. It is believed that their red pigment enhances the feeble light that reaches these plants and enables them to photosynthesize despite the low-intensity light at these greater depths. See ALGAE.

Rhône River A great river 800 km (500 miles) long in southeastern France, that rises in Switzerland, flows through Lake Geneva, and then flows south through France into the Mediterranean Sea. The delta of the Rhône extends over 40 km (25 miles) from Arles to the Mediterranean, and is the major undersea feature of the western portion of this sea. The river is divided in its lower reaches into the Grand and Petit Rhône, with the marshy Camargue between them. See DELTA, MEDITERRANEAN.

ribbon fish Mesopelagic, long, thin, and silvery predators that belong to the Trachipteridae, a family in the order Lampriformes. Their maximum length is about 1.7 m (68 inches). They are characterized by a protruding upper jaw, a long dorsal fin, and the apparent absence of a swim bladder. Their usual prey is other fish, squid, and crustacea.

ribbon worm Any of a group of worms in the phylum Nemertea; they are marine forms of the free-living flatworms. These animals may be as long as 30 m (100 feet). The phylum contains both benthic (bottom dwelling) and pelagic (open-sea) forms. They are not to be confused with ribbon fish or Venus's girdle. See BENTHIC ECOSYSTEM, CTENOPHORA, LAMPRIDIFORMES, NEMERTEA, PELAGIC ENVIRONMENT, VENUS'S GIRDLE.

ribosymes Catalytic RNA molecules. Such structures were unknown before the 1980s, when the "RNA world" was hypothesized. Up to that point it was assumed that all enzymes were proteins. Once it was established that some RNA molecules can catalyze changes in the structure of other RNA molecules, a number of unexplained sequences in the chemistry of life's origins became clearer. Once researchers created an atmosphere that is thought to be that of the primitive Earth, these conditions could lead to the formation of cytosine, which is present in DNA and RNA, and of uracil, the nucleotide found only in RNA.

Ribosymes are frequently larger than the compounds that they cause to be created. The ribosymes are usually dimers (larger molecules) and almost always include a protein. Ribonuclease P, a ribosyme that is present in almost everything that lives, is a dimer built of a molecule of RNA and one of protein. The RNA part of the ribosyme is chemically active; it can break (cleave) some chemical bonds. However, the dimer is much more reactive and therefore a much better catalyst of essential-for-life reactions. The protein portion of a ribosyme cannot initiate bond cleavage. See ENZYME, LIFE, RNA.

Richelia A genus of cyanobacteria that forms commensal relationships with or is an epiphyte on diatoms. Only one species of this genus is known, R. *intracellularis* and this organism can exhibit several variations in form and morphology. It is widely distributed in warm to warm-temperate seas and apparently capable of living on the tests of several diatoms. It is now believed that this organism affects nitrogen fixation, but the process is unknown. Researchers are attempting to determine if there is a link between the "blooms" of diatoms and the possible source of the toxins produced. Some have hypothesized that the toxic material or its precursors are produced by the *Richelia*. See CYANOBACTERIA, COMMENSAL RELATIONSHIPS, DIATOMS, TOXINS.

rift valley A depression in the earth resulting from the collapse of a large section of the crust. Rift valleys are often part of the mid-ocean ridge system. They occur where crustal segments split and move away from the upwelling magma.

In some regions, a rift valley is the obvious separation of two plates. The African Rift Valley is an extension on land of the division of the African from the Arabian plates. It runs down the Red Sea and continues on what is the eastern portion of Africa, through Ethiopia and into Tanzania and Kenya. *See* FRACTURE ZONE, MID-OCEAN RIDGES.

Riftia pachyptila   The genus and species of the spectacular tube worms found on hydrothermal vents. This animal is encased in a chitinous tube that is, on average, 1.5 m (5 feet) long. Some are almost 3 m long. The bright red worm extends above the top of the tube, absorbing carbon dioxide and sulfides from the vent. The worm has no eyes, mouth, or gut. It is red because it has hemoglobin-based blood, and that is the medium that carries the noxious materials it absorbs to the bacteria in its interior. Those organisms are the worm's source of food, and without the commensal bacteria the worms die. The source of the bacteria is unknown. However, a larval worm has a mouth and a gut. Once the bacteria are established, those worm structures disappear. *See* COMMENSAL RELATIONSHIP, TUBE WORM, VENT COMMUNITY.

right whale   A large baleen whale with an extraordinarily large head. Adult right whales are 15 to 18 m (50 to 60 feet) long, about 30% of which represents the head of the animal. The name was originally applied to the Greenland or bowhead whale (*Balaena mysticetus*) because it was "right" for whalers. The resulting overfishing almost exterminated these animals, and they are still an endangered species.

The right whale is an attractive species, with blue-black superior surfaces and light undersides. It feeds on tiny crustaceans. Related species are the black right whale of the North Atlantic (*B. glaciales*), the pygmy whale (*Caperea marginate*) of South America, and *B. sieboldi* of the North Pacific. *See* BALEEN, CETACEA, WHALE.

Rimicaris   These are the vent shrimp. Before the identification of the Pompeii worm, this was the most extreme extremophile known. This shrimp genus is remarkable since it appears at hydrothermal vents in both the Atlantic and the Indian oceans. While the shrimp appear in more than one fairly unstable environment, the commensal proteobacteria that are associated with this arthropod genus vary considerably. The manner of dispersal of both is under study. *See* COMMENSAL RELATIONSHIPS, EXTREMEOPHILES, POMPEII WORM, PROTEOBACTERIA, VENT COMMUNITIES.

Rio Grande Rise   An aseismic ridge in the South Atlantic off the southeastern coast of South America. It continues the line of the Rio de la Plata and divides the Argentine and Brazil basins. This ridge is analogous to the Walvis Ridge, also in the South Atlantic and off the coast of Africa. *See* ASEISMIC RIDGE.

rip tide or rip current   Usually a rapid, narrow, surface current moving from shore out to sea at an abnormally rapid speed. These are not tides at all. While their rate is unusual, their direction away from the beach is normal. They occur on long, unbroken coasts, where waves approach parallel to the shore. If there is a sandbar or a groin, the water is constricted and may form a rip current. Because each wave pushes water shoreward, periods of high wave activity lead to the buildup of a mass of water near the shore, raising the level of the sea. This condition continues until the weight of the water breaks through the nearshore water at a low point and surges seaward for some period. Once the pressure is released, the wave activity and seaward flow of water returns to normal. A rip current may be noticeable from shore, manifesting itself as a line of choppy water; a corridor of water that is a somewhat different color; a line of foam,

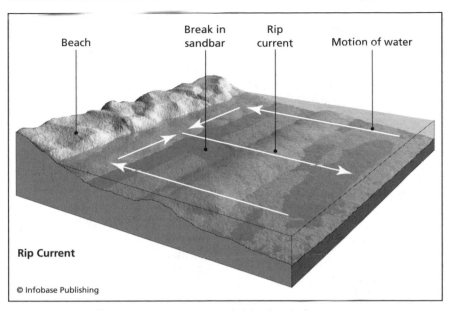

Beach | Break in sandbar | Rip current | Motion of water

Rip Current

© Infobase Publishing

Rip currents are an unseen and violent beach phenomenon.

debris, or seaweed that moves seaward; or an unusual breaking wave. *See* CURRENTS.

ripple marks Characteristic wave patterns on the ocean bottom. They are the result of wave action. They separate the bottom sediment into crest-and-trough patterns that mirror the crest-and-trough parts of waves. These bottom features are found in areas that have a bottom of sand or other finely divided sediment. Ripple marks may fossilize to later reveal a seafloor that bears the marks of ancient wave patterns. *See* TURBIDITY CURRENTS.

river bore A river tidal wave with a high, well-defined crest. It occurs at spring high tides (high tides at a new or full moon) if the river channel leading to the sea is constricted. The constriction concentrates the body of water rushing in with the force of the tide, and as this water sweeps upstream it has a well-defined, abrupt crest. The height of the crest varies; it is greater at the banks than at midstream. It also varies from one river to another and from one spot in a

given river to other locations. *See* BAY OF FUNDY, TIDE.

RNA Ribose nucleic acid, the substance that transcribes the information encoded by DNA into proteins in the living cell. The sugar unit in RNA is ribose, and the bases are adenine, guanine, cytosine, and uracil; the uracil replaces the thymine that would be the fourth base in DNA. RNA is a single-stranded chain. It is formed as a complement to DNA, which acts as a template for it. Where there is a guanine (G) on the DNA chain, the RNA base that pairs with it is a cytosine (C). The complement of adenine (A) is uracil (U). While DNA is a very large, double-stranded, much more complex molecule than RNA, there are more RNA molecules in cells.

RNA is present in several forms. Messenger, or m-RNA, carries the encoded message of the DNA from the nucleus of a cell to the ribosomes—the organelles that manufacture the species-specific protein encoded by the DNA and whose amino-acid sequence is established by the order of the base arrangement in the DNA.

Messenger RNA is constructed as a complement to only a part of one strand of a particular DNA (i.e., the segment of the DNA that is the instruction, or specific sequence of bases, for a particular series of amino acids). The m-RNA then moves out of the nucleus to the ribosome, where its own sequence of bases, complementary to those on the DNA, is translated into ribosomal RNA (r-RNA). This last complement of the DNA's complement is therefore identical to the structure encoded by the original DNA section being modelled. The ribosomal RNA is acted upon by transfer, or t-RNA, which brings the appropriate amino acids for the particular protein sequence being assembled, and lines them up on the ribosomal RNA to form the protein chain.

RNA synthesis is key to development of life. Current theory about the origins of life on Earth depends on a mechanism for the production of cytosine and uracil, two compounds that are present in RNA. Scientists at the University of California in San Diego have attempted to generate the necessary compounds from the ammonia, methane, and water that is thought to have been present in the primordial atmosphere. This is possible, and if the researchers are correct, the original genetic material was something that resembled RNA. *See* DNA, PROTEIN, RIBOSYME.

roaring forties An old sailing term referring to the prevailing westerly winds—winds moving from west to east in the 40 to 50° latitude range. Since there is less landmass in the Southern Hemisphere, the effect of these strong winds, which range up to gale force (thus giving them their name) is normally encountered there. Sailing ships such as clippers moved very quickly before such winds. *See* CLIPPER, WESTERLIES, WIND.

rock louse An isopod member of the class Crustacea that lives on the shore above the splash line (highest point of the high tide) and is not a louse at all. *See* ISOPOD, LITTORAL.

Romanche Deep *See* FRACTURE ZONE, PACIFIC OCEAN.

Rondelet, Guillaume (1507–1566) A French physician, traveler, professor of anatomy, and chancellor of the University of Montpellier (France). Rondelet was one of the "Encyclopedic Naturalists," and is a significant figure in the history of marine science because of his book on marine fish published in 1554. It was translated from the Latin four years later with the title *Complete History of Fish*. In it, Rondelet discusses freshwater as well as marine life; thus, he includes everything in aquatic environments from beavers to fish to dolphins. Almost all of his 300-odd entries are illustrated.

Rondelet analyzed the various systems of aquatic creatures in much the same manner that Aristotle did: with primary consideration of why and how a structure would work. He noted the swim bladder in freshwater organisms and then looked for it in marine specimens. He dissected dolphins, noting their relationship to both pigs and humans. His description of the sea urchin is excellent and a worthy complement to that of Aristotle. Rondelet's drawing of this creature is the earliest extant depiction of an invertebrate. *See* ANATOMY; ARISTOTLE; BELON, PIERRE; GESNER, KONRAD.

Ross, James Clark (1800–1862) A Scottish explorer who discovered the Ross Sea in the Antarctic in 1841. He is also noted for magnetic survey studies of the polar regions. He was present on an earlier expedition led by his uncle, Sir John Ross, during which he located the North Magnetic Pole in 1831. Subsequently, in the 1839–43 expedition, he sailed with the ships *Erebus* and *Terror* to find the South Magnetic Pole. In the Antarctic, he charted Graham Land, the Weddell Ice Sheet, and Victoria Land. This expedition established him as a notable scientist-explorer. *See* ANTARCTICA, EXPLORERS AND EXPLORATION, MAGNETIC POLE, NORTHWEST PASSAGE.

Ross, John (1777–1856) A Scottish explorer of the Arctic. Ross sought the Northwest Passage in the Arctic islands of the Canadian Northwest Territory. During his second voyage his nephew, James Clark Ross, located the North Magnetic Pole. A lasting body of work produced by Ross was his extensive charting of the islands of the arctic. *See* MAGNETIC POLE, NORTHWEST PASSAGE, ROSS, JAMES CLARK.

Ross Sea A sea arm of the South Pacific below the Antarctic Circle, approximately due south of New Zealand. It is bordered by Victoria Land on the east, and it merges into the Ross Ice Shelf south of 77° south latitude. The closest landmass (other than Antarctica) to the Ross Sea is New Zealand. The water of the Ross Sea is fairly constant in temperature (-2° to 0°C) and salinity (3.35 to 3.47%). The pattern of circulation is part of the East Wind Drift.

The sea covers a continental shelf that is about 300 to 450 m (1,200 to 1,800 feet) deep. Its depth is attributed to pressure from the weight of the ice shelf to the south. Ridges parallel to Victoria Land's coast show signs of Tertiary volcanic activity. The Ross Sea is relatively free of ice compared with other Antarctic waters. Admiral Byrd's *Little America* base is on its western coast. *See* ANTARCTICA, ICE SHELF.

Ross seal *Ommatophoca rossii* is named for the British Antarctic explorer, Sir James Ross, who discovered them in 1840. They are a small population and the least studied of the four Antarctic seals because they are generally inaccessible. The diet of these seals is largely cephalopods, which are in short supply in polar waters. An unusual feature of this species is that the female is larger than the male. *See* PINNIPED.

Rotifera Microscopic worms of the phylum Aschelminthes. Most are freshwater animals, but there are about 15 species that are marine. Rotifers have a "head" with a ciliated "mouth," and many are parasitic. *Seison* is a parasitic species that lives on the gills of crustaceans. *See* CILIA, PARASITE.

roundworm *See* ASCHELMINTHES, ANNELIDA.

ROV The acronym for remotely operated vehicles, now used extensively in marine research. These machines are manipulated by operators onboard surface vessels. These vessels are simpler to operate and safer than manned submersibles. These machines were first developed for commercial uses such as inspection of pipelines and offshore drilling platforms. They can range in size and complexity from a small box with a camera to something like a small truck that is loaded with recording equipment. The pressure that an ROV is to withstand depends on its depth range, and this is a major factor in its design. Some ROVs are equipped with movable arms that can perform a variety of tasks. Operating a submersible is an expensive and complex operation. NOAA is coordinator for a number of these exploratory tools, but not all are operated under its aegis.

The *Alvin* is probably the best-known submersible. It is owned by Woods Hole Oceanographic Institution and was launched in 1964 as a manned submersible. It has been rebuilt several times and can now plunge to depths of more than 4,400 m (14,764 feet) and stay submerged for about 10 hours. *Alvin* is capable of maneuvering and hovering; both are necessary for video recording.

Clelia—a three-person craft—is operated by Harbor Branch Oceanographic Institution. First launched in 1976, it was refitted in 1992 and is devoted to shallow-water research. Its range extends down to about 1,000 feet, and its advantage is that the crew can come close to organisms in areas of low visibility.

Both *Alvin* and *Clelia* are more than 20 feet long and both are tethered to a ship that supplies basic life support and light. *DeepWorker* is a small, very maneuverable submersible that is large enough for only one pilot-navigator-photographer. Its range extends about 2,000 feet down. The working arms on this craft can perform a number of tasks.

There are two submersibles named *Johnson Sea-Link;* Both are operated by Harbor Branch Oceanographic Institution. One was built in 1971, the other in 1975, and both are more than 20 feet long. For air supply and light, these submersibles use tethers to a mother ship, and both are capable of dives to depths of 3,000 feet. Marine archeology is one of the uses these vessels are put to; they can use hooks or claws or vacuum up samples of rock, bottom sediment, and organisms in the water. The *Sea-Links* have been involved in a number of explorations and salvage of historical importance. These submersibles were used to survey and recover artifacts from the wreck of the Civil War vessel the U.S.S. *Monitor,* which sank near Cape Hatteras, North Carolina. They were also used in the recovery of the space shuttle *Challenger.*

Mir I and *Mir II* are Russian research submersibles that are 7.8 m (about 24 feet) long and battery operated. Their maximum diving depth is 6,000 m (20,000 feet), which makes them the second-deepest submersible. The Japanese *Shinkai* can safely descend another 500 m. The Mirs are excellent vessels for photography and have been used in a number of documentaries and in the production of IMAX films.

The Hawaii Undersea Research Laboratory (HURL) uses two free, battery-powered submersibles—*Pisces IV* and *Pisces*—that are large enough for three people. They have explored the Hawaiian coasts and nearby seamounts down to depths of over 2,000 m (6,200 feet). These submersibles have excellent cameras and can function in low-light conditions. Colors and animal behaviors are very different in the natural light of a place than they are when the researchers turn on the lights.

The *Autonomous Benthic Explorer* (ABE) is a robot built in the mid-1990s. It can explore an underwater terrain for long periods and has self-contained power sources. The advantage of the ABE is that it needs no pilot or tether. This vessel can function down to depths of 5,000 m (17,000 feet). ABE is a good tool for exploring physical conditions on the sea

floor. It can relatively easily provide a continuous record of temperature, or water current, that would be difficult to produce even with multiple manned dives. This is not the only robot used in commercial exploration. ROPOS is the remotely operated platform for ocean science. This is a large cage that can either work in shallow water or, if encased in a steel cover, be lowered to depths of 350–500 m (450–650 feet), where it can roll out of its cage and function using a tether. This robot can lift and retrieve large objects as well as take photographs and employ sonar. Information is collected and transmitted to computers on the mother ship.

Hercules was specifically built to be a scientific instrument. It was intended by its creators at the Institute for Exploration to descend to depths of 4,000 m (2.5 miles) to study the geology and biology of the sea. It can also function as an archeological tool and recover objects from ancient shipwrecks. The ROV is handled by "pilots" on the tethered mother ship, and information is transmitted from the ROV to the ship by cable. *Hercules* operates with another ROV, *Argus* (the two are tethered together). The camera equipment is in the *Argus.*

Tiburon is another ROV designed to be a scientific instrument. It was built by the Monterey Bay Aquarium in California in 1997. Its depth range extends to about 4,000 m (13,200 feet), and it has seen service in the exploration and sampling of the biology and geology of the Pacific and specifically the Monterey Canyon area. *See* EXPLORATION, OCEANOGRAPHIC RESEARCH.

Royal Society The Royal Society of London for the Promotion of Natural Knowledge. Founded in 1660, it is the oldest scientific society in Great Britain and one of the oldest worldwide. Current membership in the society numbers about 1,000 and includes some members who are not British. The society started with an informal group of people interested in scientific subjects. Some of the founding members included John Dryden, Robert Hooke,

Christopher Wren, and John Evelyn. The group was given a charter by Charles II in 1662 but no financial support.

Beginning in 1665 the society published its *Philosophical Transactions,* a periodical that survives today. Notably, the Society actively encouraged interest in a variety of fields and recognized achievement by both British citizens and foreigners. Members of this period included Joseph Banks, Isaac Newton, and Edmund Halley. Benjamin Franklin was a colonial member. The enterprises the society sponsored ranged from voyages of exploration, the first of which was the 1768 voyage of Captain James Cook, to the 1919 expedition to photograph the solar eclipse, a voyage that was intended to and did prove Einstein's theory of relativity.

S

Sagitta The arrow worms, members of the phylum Chaetognatha. They are about 0.5 cm (0.2 inches) long and form an important part of the zooplankton. They are carnivores, feeding on smaller organisms. The arrow worms feed primarily on copepods and migrate vertically with them. The mass of sagittal population is staggering. It may reach 30% of the wet mass of the total oceanic copepod population. These carnivores are an essential component of the food web in every environment.

The arrow worms are exclusively marine animals and have been used as indicator organisms. The species of Sagitta in a particular area are specific to that region; thus, seawater entering the Atlantic from the Mediterranean will bring with it the arrow worms of the Mediterranean. The movement of the water can be traced by the species of sagitta in it. *See* CARNIVORE, CHAETOGNATHA.

Sahul Shelf A continental shelf off the Australian shore, under the shallow Timor Sea. It lies along the northwest coast of Australia, running from the arbitrary divide with the Arafura Shelf to the southwest. The Sahul Shelf extends to the Rowley Shelf and then almost to the Sunda Shelf, whose edge is marked by the Timor Trough.

The Sahul Shelf exhibits evidence of ancient erosion and geologically recent subsidence. The coral islands on the shelf are not oceanic, volcanic atolls but were once part of the Australian continental mass, and the land between them and Australia was once above sea level. The canyons in the Sahul Shelf are analogous to the ["Sunda River"] system. *See* ISLAND, SUBSIDENCE.

sail A cloth set on masts or vertical supports, and yards or horizontal supports that crosses these structures and is intended to catch the wind and propel a ship. In general, square sails are set on yards, triangular ones directly on the masts themselves. Ships carrying only triangular sails are also referred to as fore-and-aft rigged ships. However, some large square-rigged ships also carry triangular sails. Such sails on the bowsprit or leading sail support are jibs. If the triangular sail is hung on the mizzen (rearmost mast), it is a stay or spanker.

sailfish A tropical fish (*Istiophorus*) that is a relative of the swordfish and the marlin; it is sought by commercial and sport fishermen. The latter prefer it because it is both a fighting fish and a beautiful one. The sailfish is dark blue above, with a silvery ventral surface. It has a long, pointed snout and a distinctive "sail," its large dorsal fin. The average size of the sailfish is 3.4 to 3.5 m (11 to 12 feet), and it weighs about 90 kg (200 pounds). The diet of the sailfish consists largely of smaller, schooling fish.

Saint Lawrence, Gulf of A large, roughly triangular area, about 150,000 km^2 (90,000 square miles), of the North Atlantic Ocean along eastern Canada between Quebec, Newfoundland, New Brunswick, and Nova Scotia. The gulf is connected to the North Atlantic by the Straits of Belle Isle between Newfoundland and Labrador and the Cabot Strait between Newfoundland and Nova Scotia. The southerly portion is fairly shallow, with depths of about 50 to 80 m (180 to

250 feet). The dividing line is the Laurentian Channel, which runs northwest to southwest with an average depth of 300 to 450 m (1,200 to 1,500 feet.)

Cold water of the North Atlantic enters the area via the Strait of Belle Isle and then moves west, where it encounters St. Lawrence River water, which flows southeast between the Gaspé Peninsula and Anticosti Island. The salinity in the gulf is low, particularly in summer, when it is about 2.6%.

The Appalachian Ridge line bends at the Gaspé and then wanders through the gulf. The current shape of the gulf and of the St. Lawrence River channel was created during the last Ice Age. The sediments in it are due to current and recent geological glacial activity. *See* ATLANTIC OCEAN, LABRADOR CURRENT.

St. Peter and St. Paul A cluster of rocks protruding above the surface of the South Atlantic Ocean between Brazil and West Africa, about 1,000 km (600 miles) northeast of Natal, Brazil. These rocks are seamounts east of the Ceara Plain and the Mid-Atlantic Ridge. They were landmarks in the day of sailing ships. *See* ATLANTIC OCEAN, SEAMOUNTS.

salinity The amount of dissolved inorganic minerals (salts) in seawater. The standard measure of salinity is in grams per kilogram of water. The saltiness of the oceans has been exploited for millennia. Neolithic man harvested salt in salt pans and used the product as a trade good.

Explanations for the salt in the ocean's waters have been put forth for almost as long as salt has been used.

By the 17th century, careful experiments to determine the physical and chemical composition of seawater had begun. Robert Boyle, an English scientist well known for his work on air pressure, compared the specific gravities of seawater samples taken at different depths. He concluded that the salt content at all depths was essentially the same. Exceptions were the water of bays and of areas near undersea springs. Upon Boyle's suggestion for further research, others measured the specific gravity of seawater at different latitudes and established that there were variations in its saltiness.

It remained for the chemists Antoine Lavoisier of France and Tobern Bergman of Sweden to separately do successful analyses of seawater. Lavoisier's samples came from Dieppe, France, accounting for their less-than-average concentration of salt. He identified as constituents of seawater, sodium, calcium, and magnesium chlorides; sodium, magnesium, and calcium sulfates; and calcium carbonate. Edmund Halley (of comet fame) realized that Lavoisier was right in assuming the terrestrial origin of the salt in the sea. He then theorized that the dissolution and transportation of these minerals could be used as a measure of the age of the Earth. Because the ocean was in a steady state of salinity, said Halley, it had to have achieved its present state of saltiness long ago in very early geologic times.

MOST COMMON IONS PRESENT IN SEAWATER			
Ion	% of total salt (by weight)	Ion	% of total salt
chloride Cl^-	55.04	sodium Na^+	30.61
sulfate SO_4^{2-}	7.68	magnesium Mg^{2+}	3.69
bicarbonate HCO_{3-}	0.41	calcium Ca^{2+}	1.16
bromide Br^-	0.19	potassium K^+	1.10
borate HBO_3	0.07		

VARIETIES OF SALMON		
Generic name	Common name	Habitat
Salmo salar	Atlantic (parr and grilse are juveniles)	Scandinavia, North America, British Isles, Iceland, etc.
Oncorhynchus tshavytsha	Chinook, king, Colorado River, Sacramento	Central California to the Yukon
O. garhusha	Pink or humpback	California to Alaska, Japan
O. nerka	Red, sockeye, blueback, Fraser River	Oregon—North Alaska
O. keta	Dog, chum	Alaska, Siberia

The work of William Dittmar on the H.M.S. *Challenger* voyage established the composition of the sea. (See table.)

The total weight represented by the ions in the table is 99.95% of that of all ions present in seawater. *See* BRINE POOLS; CHALLENGER, HMS; *individual ions and salts.*

salmon Any of several large, anadromous food and sport fish of the order Salmoniformes. The name salmon is used for the Atlantic Ocean species (*Salmo salar*) and also for the red, pink, coho, king, chum, and Pacific salmon of the genus *Oncorhynchus*.

The Atlantic species is present in both European and American waters. It is carnivorous, feeding on crustaceans. The average adult fish weighs between 5 and 9 kg (11 to 30 pounds), although larger ones have been recorded. The color of the fish varies with its age and sex, the season, locale, and availability of food. Most adults, however, are silvery below and speckled brown on their upper surface. They may have red splashes of color on the brown surface. The adults, having matured in the ocean, ascend their native rivers to mate. The timing of the "salmon run," or movement of these fish upstream in any particular river, is a function of the temperature. Thus, Atlantic salmon are found in Irish rivers at an earlier date than in Norwegian rivers because the latter are colder. The female burrows a shallow depression in the river bottom, and the

eggs are fertilized in the rudimentary nest. When they hatch, early in the spring, the young salmon feed on insects. They look like trout at this stage and are called parr.

The adolescent salmon is called a smolt. At this stage it is about two years old and ready to go to sea. There it feeds on small crustaceans and grows rapidly. The average adult is about five years old when it ascends its native river to mate. It is an exhausting trip during which the fish do not feed. Many of the salmon die after mating, others, in their weakened and exhausted state, are caught by predators; some return to the ocean and repeat the trip about two years later.

The largest species of salmon are king salmon of the Pacific Ocean. Specimens of more than 50 kg (110 pounds) are known, but not common. The pink salmon are the smallest salmon species, at about 2.5 kg (6 pounds). These species differ in their appearance and the distance they travel to spawn, but they are in general more alike in pattern and more colorful than is the Atlantic species.

When they are ascending their native river, the physiologic changes in Pacific salmon, and particularly the king salmon, make it impossible for the adults to feed as they move upriver. If a given fish does not turn back almost immediately, it is committed to the upriver direction.

The steelhead trout (*Salmo gairdneri*) is close to the salmon in color and behavior and is related to the salmon. There are also other species that are migratory and called salmon, such as the wall-eyed pike, squawfish, and yellowtail, but these are unrelated to the true salmon.

Commercially, salmon are caught in a variety of nets or weirs. Because of serious overfishing, the range of many varieties, particularly in the Atlantic, has been severely curtailed. Extensive restocking has been going on, and some species are becoming more plentiful. Salmon are now being farmed successfully. This has decreased the fishing pressure on wild stocks. Another possible development in the management of wild salmon stock is the possibility of removing some of the dams that prevent many fish from successful spawning runs.

Salmon are now being commercially raised in captivity. It seems as though this practice would solve the problem of dwindling wild salmon stocks; however, large captive populations create problems that did not exist before the fish farming practices were put in place. The large weirs of farmed fish create pollution problems; this is analogous to a large cattle feedlot on land. The farmed fish, because they are in close proximity, are also possible sources of disease, which can then spread to the wild population. Several salmon species are endangered and some are extinct in some parts of their range. Those that are threatened are varieties of chinook, chum, coho, and sockeye.

salps Planktonic organisms that are found in open ocean; these small organisms may have considerable environmental importance. Salps are in the phylum Chordata; at some point in their development they have a dorsal nerve cord, a notochord. The subphylum is Urochordata; it is also called Tunicata. Most of these organisms have a "tunic" tissue of protein and carbohydrate that totally envelops the organism. The salps are in the class Thaliacea; they have the characteristic tunic and are pelagic organisms.

Salps resemble ctenophores: They are gelatinous, almost transparent, and cylin-der-shaped. The tunic is a trap with which they collect whatever they can. Salps will feed on whatever the tunic traps, although their major food source is phytoplankton, notably, diatoms. Although they have no coelom, tube-within-a-tube body plan, the mouth is at one end of the body cylinder and the atrial opening at the other. The muscles of the organism pulse water through the body. This also serves to propel it. Salps also swim by means of cilial action. The organisms are generally small, about 10 cm (4 inches) long as adults, although some species can be meters long. Aggregates, chains of individuals, can be very large.

The reproductive cycle of salps is unusual. Asexual salps are solitary; the sexual forms are the colonies. Females discharge one or two eggs, and these are fertilized by male aggregates. The embryos are nurtured by the aggregate, which has a placental structure. When the embryo is mature and released into the water, the placental aggregate becomes a male. The tiny salp grows very rapidly into an asexual solitary that reproduces by budding. A salp will mature in two days in warm to warm-temperate water. The budded salps form an aggregate.

Salps were very much unknown because their delicate tunics were destroyed by the nets used in sampling. Not until diving in open ocean became a more usual method of exploration were the numbers and importance of these creatures studied. Salps are an important factor in the distribution of carbon in the water column. They feed on diatoms and in some areas are the dominant herbivore, but they feed inefficiently. Salp excreta containing diatom fragments rains down on other organisms below them in the water column. Several species of fish eat salps, but there is not enough material in salps—they are largely water—to feed birds or sea mammals. A salp predator must have additional prey. The salps would constitute only a small part of the diet of a large vertebrate.

When salps were first identified, they were thought to be just warm water organisms. Since then, two common species have been found in the Antarctic, *Salpa thompsoni* and *Ihlea racovitzai*. The relationship of salp population to that of krill in the southern oceans is an area of current research interest. In general, salps are more common in warmer water than are krill; however, the salp population in Antarctic waters is increasing. This is possibly the result of the overall warming trend that is evidenced everywhere on Earth. Now both salps and krill are the basic foodstock of the large Antarctic vertebrates—whales, seals, and birds, especially penguins. *See* TUNICATA.

salt The common substance called salt is sodium chloride (NaCl). It is found in seawater and also as the mineral rock salt. Pure salt consists almost entirely of chloride, colorless, cube-shaped crystals. Dissolved salt is the most prevalent substance in seawater. Salt is obtained commercially by a series of processes from underground mines, by evaporation from salt pans, and others. Salt is used in the preservation and preparation of food, in commercial refrigeration, and in manufacturing other chemicals.

There are, however, other kinds of salts than NaCl. Chemically, a salt is defined as a compound which on dissolving produces negative and positive ions. The solubility of salts varies enormously. *See* MINERAL, SALINITY, SOLUBILITY, WATER.

salt domes Geologic formations consisting of evaporites, largely or entirely composed of sodium chloride, salt domes are most often vertical and cylindrical

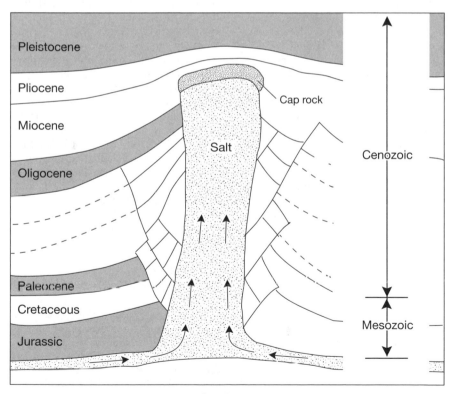

Salt domes

columns at least 1 km in diameter, which are thought to have developed in sedimentary basins and later been overlain by sediment. The entirety of the formation may then have been folded by upthrusting of the Earth's crust.

Salt domes are economically significant, since oil and gas deposits are often associated with them. The presence of a series of salt domes is often an indication of an oil or gas field in the same vicinity. When hollowed out, salt domes are also considered very safe storage units for "hot" radioactive waste material. *See* OIL.

salt flats Formed by the evaporation of a body of water. They usually occur in arid areas where an inflow of salt-containing water is dammed naturally and evaporates. This may result where isolated bays are separated from the sea by coral growth or the creation of barrier islands, or where a sedimentary basin or shallow sea subsides. A shallow shelf may also become a site for the deposition of more insoluble salts, sulfates being the first salts to precipitate. Nonmarine salt flats may result where a river deposits its load of dissolved material in a dry region. The Dead Sea is an example of a current, building salt flat. Some other such inland areas, such as the Bonneville Flats in the United States, were once marine. *See* COAST, ESTUARY.

salt marsh An area of brackish water usually built on a continental shelf. Only recently has the importance of the salt marsh ecosystem been recognized as vital to both esthetic and commercial interests. In earlier times a marsh was "waste land" and was considered to be an area fit only for outcasts. Civilization meant drainage and conversion of the area to farmland, such as was done centuries ago with the fen country of the east coast of England.

The typical salt marsh is the result of a silt-carrying river. The outflow to the sea is impeded either by a barrier island or by the slow subsidence of the region. The area of the marsh is a series of hillocks separated by water channels. It is not a permanent area, since the continued accretion of silt builds up the land and effectively dries it. This progression from marsh to dry land can be seen in the change in vegetation with time. The dominant plants are at first sea grasses, then salt-tolerating plants, then terrestrial ones. This in turn creates a similar change in the local fauna.

In general, the enclosed nature of a salt marsh means that the water temperature is somewhat higher than that of the nearby ocean. A salt marsh is also a region of shallow water. Both of these factors mean that there is increased photosynthesis, and the greater, resulting plant production in turn supports larger animal populations than those that would be found in a similar volume of ocean water. New soil brought by the river system increases the fertility of the region and so do the droppings of the animals that feed on the plants. The resulting quantities of shrimp and birds—to name just two groups of animals—are very large.

Many animals of the continental shelf region, especially some fish and crustaceans, spend some part of their lifespan in the marsh, as larvae, juveniles, feeding adults. Many marsh animals can spend some part of their lives out of the water or in water of low salt content. The clams of the littoral can exist in sand for hours awaiting the high tide that inundates them. Mussels and limpets cling to occasionally exposed rocks. Bird communities, both passarine and indigenous, use the grasses and local shellfish of the marsh as vital ingredients in their diets. Since a salt marsh, unlike a coral atoll, is not an isolated system, it is a more difficult ecosystem to study, and the interrelationship of many species in salt marshes is still unknown.

Marshes are ecologically fragile areas. They are sensitive to pollution, which may be chemical from industry or agriculture, and thermal pollution. The latter is frequently the result of river water being used as the coolant in industrial plants.

The tropical and semitropical version of a salt marsh is a mangrove swamp. Like any other salt marsh, it is vital to the ecosystem of the region. This positive effect is only recently entering public awareness. *See* DELTA, ESTUARY, POLLUTION.

salvage The saving, in navigable water, of a ship or its cargo. Salvage may involve towing a vessel that has mechanical problems, moving a stranded vessel to open water, or raising a sunken vessel. Salvage vessels are usually very large, powerful tugs equipped with various heavy-duty pumps, winches, and cranes. Salvage techniques may involve one or several methods of towing, lifting, and pumping out a damaged ship. In maritime law, the vessel to be salvaged and its goods belong to the owner unless abandoned. The owner of the vessel to be salvaged, who may pay by prearranged fee or a share of the material retrieved, contracts for the rescuer's service in salvage. The fee depends on the success of the salvage mission. *See* MARINE ARCHEOLOGY.

sampling The taking of specimens for testing. The subjects of the tests may be animal populations, nutrients, trace elements, or water. *See* RESEARCH.

San Andreas Fault A slip strike series of faults along the coast of California where the North American and the Pacific plates move past each other. It is analogous to other faults in the Pacific basin. The San Andreas is a line leading from the Gulf of California, where the East Pacific Rise dips under North America, northwest past San Francisco.

It has been theorized that all of the Pacific faults are rotating the land on their seaward sides in a counterclockwise direction. This would mean that those parts of California to the west of the fault are gradually sliding toward Alaska. *See* CALIFORNIA, GULF OF; CRUST; PACIFIC PLATE.

sand Granular mineral particles formed by the erosive activity of ocean or river currents, waves, wind, or glacial movement. Most sand is terrigenous and brought to

the ocean by rivers. The mineral most commonly found in sand is quartz, although it may contain volcanic rock, coral, or gypsum. These inclusions account for the variety of colors of sand and for its grittiness. Coral sand is powdery.

Sand is used commercially as an aggregate in construction. Concrete is sand in a lime matrix. Sand is also mined for use as an abrasive. *See* SEDIMENT, SILT.

sand crab A crab of the genus *Ocypoda* that lives on beaches in temperate to tropical habitats. There are about 200 species of these gray- or buff-colored animals, which are sometimes called ghost crabs. They live in burrows and are seen running on beaches above the high-tide line.

The American version of the sand crab is found on the Atlantic coast from about 40° north latitude to about 40° south latitude. This crab, *O. quadratus,* is about 4 to 5 cm (1.5 to 2 inches) in diameter. *O. ceratophthalmus* is found in the Indian Ocean and the eastern Pacific; *O. saraton* lives on the coast of East Africa and along the Arabian peninsula. The principal diet of the sand crab is insects. *O. quadratus* lives almost exclusively on sand fleas. *See* BACKSHORE, CRAB.

sand dollar A thin, disk-shaped echinoderm belonging to the class Asteroidea. The sand dollars are most often about 7 to 8 cm (3 inches) in diameter. The upper surface is marked with a characteristic five-petalled shape that is a modification of the water vascular system common to echinoderms. In some species, the petaloids are cut-outs, leaving the disk perforated.

The most often encountered species of sand dollar is *Echinoarachnius parma,* which is found in North American and Far Eastern waters. *See* ECHINODERMATA.

sand flea (sand hopper) *Orchestia agilis* is not an insect but an amphipod (crustacean) that lives in intertidal zones. The usual individual is less than 0.3 cm (0.125 inches). It looks like a seven-segment shrimp. Its tan-to-brown shell keeps it well-hidden on beaches. It is a detritus feeder and lives on decaying seaweed. It is notable for its ability to leap up to 50 cm (20 inches) in the air—hence its common name, sand hopper.

sand tiger shark (*Odontaspididae*)

Found on both sides of the Atlantic, the largest individual recorded was 3.6 m (22 feet). They have flattened heads, with short or bulbous snouts. These sharks are harmless, sluggish animals, inhabiting coastal waters on continental shelves, although some have been found at depths down to 1,000 m (3,300 feet). The usual prey of tiger sharks is fish, arthropods, and squid. Since they are sighted from beaches, they create human interest and excitement.

sandpiper

A common shorebird in the family Scolopacidae. They are also commonly called snipe or turnstones. They range in size but are characterized by fairly long legs and bills. They are usually gray, dark brown, black, or white with barred markings. Males and females look alike in general, but some species have more colorful breeding plumage. They feed on invertebrates and probe the mud flats in search of them. The average individual can run and walk as well as fly. Sandpipers congregate in large flocks and migrate as units. They are diminishing in population due to habitat destruction and hunting.

Sarcodina

A class of unicellular organisms in the kingdom Protista. There are over 11,000 species in the Sarcodina, including the Foraminifera and Radiolaria. *See* FORAMINIFERA, PROTISTA, RADIOLARIA.

Sarcomagistophora

Unicellular or colonial autotrophes or heterotropes. This is a subphylum of Protozoa. These organisms have flagella, pseudopodia, or both. The category is divided into three superclasses and upwards of 8,500 known species. One of the orders, Dinoflagellida, in the superclass, Masigophora, is a sig-

nificant part of the phytoplankton. This is then an important part of the food source for filter feeders, zooplankton, and other benthic animals.

sardine

A food fish of the herring family (Culpeidae) found in dense schools in almost all warm ocean waters. The adults are 25 to 30 cm (10 to 12 inches) long, dark green or blue on their upper surfaces, and silvery-white on the underside. Their backs are frequently marked with black. Sardines, like herring, are primitive bony fish with unscaled heads and no lateral line.

Sardines spawn in the spring, when the plankton population that they feed on is at a maximum. They rise and fall from more shallow to deeper waters on a daily basis following their prey, the zooplankton. Commercial fishermen take advantage of this behavior and net sardines at night, when both they and the plankton rise to the surface to feed.

The huge schools of sardines are a major factor in oceanic food webs. Since they are major feeders on the copepod and krill populations in the waters of the Pacific, the disruption of those populations in El Niño years has significant negative effects on the sardine population and in turn on the mammals and birds that feed on the fish. Fluctuations in sardine population are also cyclic and alternate with the anchovy population cycle. The sardine population is now rebounding.

The method used to preserve sardines as food depends on the locale. Some of the techniques in use have been used for a very long time. Sunken ships that were part of the commercial fleet of Rome carried pickled sardines in amphorae as cargo. Sardines preserved in salt can be found in the markets along the Mediterranean coast, as were those fish in the amphorae of long ago. Canning is today's most common process of preservation. The species of sardine caught depends on availability.

Some common sardine species and their locales are the following:

SARDINE SPECIES	
Species	*Range*
Pilchard (*Sardinia pilchardus*)	Southern and southwestern Europe and North Africa
Pacific (*Sardinops caerulea*)	Western North America
South American (*S. sagax*)	Western South America
Japanese (*S. melanosticta*)	Japan
S. neopilchardus	Australia
S. ocellata	South Africa

Besides being used as a food item, the sardine is also a source of oil. The oil is used in the manufacture of paints and varnishes, other chemical raw materials, and as an edible fat. Sardine meal, or what is left after the fat is extracted, is incorporated into animal feed.

The so-called Maine sardine is not a sardine but rather a small or juvenile herring. *See* FISH, FOOD WEB, HERRING, MARINE OIL.

Sargasso Sea A relatively still area in the North Atlantic, between 25° and 35° North Latitude and 40° to 70° West Longitude. It is the "eye" in the anticyclonic series of currents of which the Gulf Stream is one. Although eddy currents wander into it, the Sargasso is relatively warm and still. Its name comes from the large quantity of floating algae, or *Sargassum*, found in it. These branched algae support an ecosystem of small animals that use the "weed" as both a food supply and structural support.

The Sargasso Sea has been called an "aquatic desert," implying that little or nothing lives in it. It is a highly saline body of warm water floating on colder water. This thermal gradient keeps nutrient-rich water from welling up to the surface. Better collecting methods have proven that a vast number of fragile creatures do inhabit this region.

The Sargasso is the spawning ground for most of the European and American species of eels and for fabulous legends. The eels are real, the legends entertaining. *See* EEL, *FUCUS*, PLANKTON, SALPS.

Sargassum A tropical or subtropical alga. It looks like a vascular plant because it is branched. It is a truly floating alga that has pillowlike structures to buoy it up. It grows luxuriantly, sometimes forming dense mats. *Sargassum* supports an ecosystem of crustaceans, mollusks, small fish, and plankton that live on the plants or hide in the branches, feeding on the algae or on each other. *See* ALGAE, FUCALES, PHAEOPHYCEAE.

Sars, George (1837–1927) A Norwegian marine-life scientist known for his work on the Crustacea; the son of Michael Sars. Like his father, he was born in Bergen and later moved to Christiania (Oslo), where he spent the rest of his life. He was originally a doctor but eventually specialized in the Crustacea and wrote a nine-volume work on the crustaceans of Norway, which occupied him for almost 40 years (1890–1927). He was the outstanding expert of his time on the Crustacea and as such was the examiner of the *Challenger* expedition's crustacean samples.

In 1864 Sars was appointed by the government (at that time the government of Norway was Danish) to investigate the cod. He discovered that cod eggs float, rather than sinking, as had been theorized, and proved that this is true of almost all food-fish eggs. This fact is very important in the study of the life history and migration patterns of the oceanic fish that are to Denmark and Norway vital elements in the industry of the region. *See* CHALLENGER, HMS; SARS, MICHAEL.

Sars, Michael (1805–1869) A Norwegian naturalist and marine zoologist. He was born in Bergen and educated as a pastor; natural history was his hobby. As an amateur naturalist, he published a work on natural history in 1829. Another work, on the littoral fauna of Norway, appeared in 1846. By then Sars was no longer an amateur, and accepted an appointment as professor of zoology at the university in Christiania (now Oslo), Norway.

Sars made a number of important discoveries in marine zoology. The discovery of the sessile (stationary) stage in medusal development of the *Scyphozoa,* the development of mollusks from free-swimming ciliate larvae, and the first record of a living crinoid are among his contributions. In conjunction with his son George, he worked assiduously to disprove the prevailing theory of an abiotic zone in the ocean. *See* CRINOIDEA; FORBES, EDWARD; MEDUSA.

satellite In astronomy, a celestial object that revolves with or around a larger celestial object. Artificial satellites are man-made objects placed in orbit around the Earth for scientific, technological, and military uses such as reconnaissance, surveying, meteorological observation, navigational assistance, and communications, for relaying radio and television signals.

The space age began in 1957 with the Soviet Union's *Sputnik,* which made possible the first real look at Earth. It became quickly apparent that satellite sightings from several places on Earth at exactly the same time would make triangulation, and therefore exact mapping, possible.

The American *Landsat* spacecraft created a photographic record of Earth. This was the first proof of Columbus's accurate guess. He postulated that the Earth was egg (or pear) shaped. The northern end is more pointed; the southern is rounder. Used with computer enhancement, different terrestrial structures, soil conditions, land usage, and subsurface conditions became visible. In water, plankton concentrations and their concomitant fish predators produce a distinctive color when viewed by satellite. Although the individuals cannot be seen, the entire population can be mapped and migrations of fish can be tracked.

Satellite cartographers map volcanoes, currents, geothermal hot spots, and other energy resources. They also map other planets, using the data produced by the *Voyager* series of satellite photographs. *See* MAPS, MOON, NAVIGATION.

Sauropterygia An order of Mesozoic reptiles. These were the marine dinosaurs: the nothosaurs, plesiosaurs, and placodonts.

saw shark Two genera, the *Pliotrema* and *Pristiophorus,* of small sharks are found in the warm waters of the Southern Hemisphere. The long, thin, serrated snout is a distinguishing feature.

saxitoxin A paralytic poison produced by at least two species of dinoflagellate. About 20 derivatives of this molecule are known; all are paralytic poisons that are transmitted via shellfish to all other animals that ingest the shellfish. This is a considerable problem for commercial fish and shellfish industries in the Pacific. *See* DINOFLAGELLATE, TOXIN.

scales The outer covering of fish is a dermal bony structure. It is extremely flexible. The scales may overlap as in most modern (teleost) fish, these are elasmoid scales. They have thin, flexible fibrous bone. The two types of elasmoid scales are cycloid, found on salmon and related fish, and ctenoid, typical on the perches and sunfishes. The cycloid type will continue to grow larger in annular rings as the fish grows.

The scales on elasmobranchs are placoid; they do not overlap and have small barbs that point at the tail. These barblets are composed of dentine, enamel, and bone. Fossil fish show scale patterns, some of which resemble those of extant fish, others that are different. Cosmoid and ganoid scales are more typical of extinct fish or very primitive relicts. The gar and other primitive actinopterygians have ganoid scales that are similar to tooth enamel or bone.

scallop A bivalve mollusk of the family Pectinidae; a common genus is *Pectin.* There are over 400 species of scallop, and their range is worldwide, extending from the littoral regions of the coast to ocean depths of over 4,000 m (13,000 feet). Scallops are an important food and are taken commercially off the New England and Canadian coasts, usually from depths of about 40 to 120 m (130 to 390 feet).

The scallop has an almost hemispherical, ribbed shell with winglike projections at the hinges. The hinge is controlled by the animal's large adductor muscle. Scallops range in size from 2.5 to over 15 cm (1 to 6 inches); the smaller ones are found in bays or estuaries, and their color varies from white to yellow, red, or purple. The outer edge of the mantle has numerous blue eyes. These are visible just inside the partly open shell of the undisturbed scallop as it rests on the ocean bottom.

Scallop eggs are fertilized in the water. They hatch quickly to release free-swimming larvae, which may develop a byssal extrusion (a thread by which they attach themselves to a surface). Scallops are filter feeders and live in sand or on gravel in clear water.

While some scallops are sessile (stationary), the average individual can and does move by rapidly opening and closing its shell. This jet propulsion mechanism moves the scallop along.

The normal predator of the scallop is the sea star (starfish). Humans have known the scallop as food for thousands of years. The pretty shells have found their way into decorative motifs since the very beginnings of human history near the sea. During the Middle Ages, the scallop shell was adopted as a religious pilgrim's sign. *See* BYSSAL THREADS, FILTER FEEDERS, MANTLE, MOLLUSK, SHELL.

scaphopods Also known as tusk shells, Scaphopoda are a class that is part of the phylum Mollusca. The unique feature of

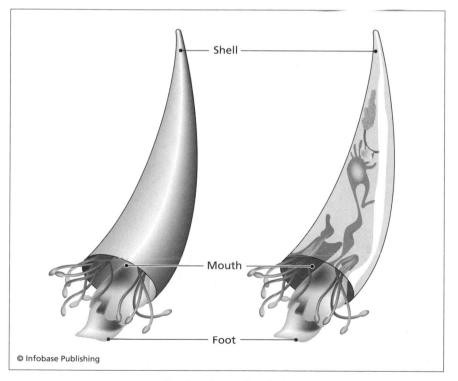

Scaphapoda, small mollusks

these marine burrowers is a shell that is cylindrical and slightly curved; it resembles a tiny tusk. This is the air exchange mechanism as well. The animal respires by using the mantle as a gas exchange medium; it has no gills. The usual scaphopod is about 5 mm (0.02 inches) long at maturity and spends its adult life with its head buried in the bottom sediment and only the top of its shell extended into the water. They are most often found in deep water ranging down to 4,000 m (13,000 feet). The estimated number of species ranges from 750 to 1,200. It is also assumed that about half this number are extinct. These organisms first appeared in the fossil record in the Devonian period. They may be older; there is some doubt about the burrows in Ordovician rocks, but those may also be pteropod remains. By the Cretaceous period the scaphopods are well defined.

These tiny creatures have a very insignificant eyeless head that is equipped with food-sensing organs, or statocysts. Cilia that resemble tentacles capture detritus, formaniferens, and an unlucky buried bivalve and bring these to the mouth. The mouthpart is a radula. The distinctive radula is snail-like, but the retractible foot is a bivalve's characteristic feature. Scaphods are either male or female; both release gametes into the water, and the female produces one egg at a time. The young hatchling floats as part of the plankton until it settles into the mud and finishes its transformation into an adult.

These animals are found in large numbers; wampum, a favored bead of American Indians, are scaphopod shells. See MOLLUSCA.

scarp A sharply falling undersea slope. It separates plateaus or regions of gentle slope. See CANYON, CONTINENTAL SLOPE, OCEAN FLOOR, PLATEAU.

scattering layers See DEEP SCATTERING LAYERS.

scavenger An animal that feeds on dead organisms. The decomposing bacteria are the ultimate and smallest scavengers. See BACTERIA, DECOMPOSERS, DETRITUS.

schooling A behavior of fish related to the herd behavior of many terrestrial organisms. A possible rationale for schooling is that it reduces predation. Since a predator can eat just so much, it cannot devour an entire school of fish. Some individuals are bound to survive. This ensures a breeding population in the school. Socially, this instinctive response to predation is involuntary, and it is thought by some biologists to be individual. Thus, as each individual fish moves to escape predators and maximize its feeding, all others do the same, resulting in a large number of individuals in one area at one time doing the same thing. See FISH, PELAGIC ENVIRONMENT.

schooner A sailing vessel with two or more masts carrying fore-and-aft sails. A number of vessels of this type were used in voyages of exploration. The *Fram*, a vessel famous in Arctic exploration and now in a museum in Oslo, is a schooner. See EXPLORERS AND EXPLORATIONS, *FRAM*, SAIL.

Scleratina The order of reef-building coral that contain zooxanthellae. The commensal algae give the coral its brilliant colors. These corals grow at a rate of 3–15 m (10–50 feet) per 1,000 years. A dead coral reef is obvious because it is white. See ALGAE, COMMENSAL RELATIONSHIPS, CORAL, RHYDOPLEYN.

Sclerospongiae A class of Poriphera. These organisms have large, complex skeletons composed of calcium carbonate, $CaCO_3$, in the form of aragonite or silicon dioxide, SiO_2, and spongin—a flexible proteinaceous material. These are the coralline sponges. There is a considerable fossil record of these organisms. There are a few extant species in the West Indies and in the Pacific. See PORIFERA, SPONGE, SPONGIN.

Scoresby, William (1789–1857) An English scientist and explorer of the Arctic. His father was a whaler, and Scoresby went to Greenland and its surrounding waters on numerous voyages with his father. The northernmost point he reached was 81°30', in 1806.

Scoresby retired from the sea in 1822 and became a clergyman. Like many other country vicars of the time, he was a natural scientist and author who wrote of his arctic experiences and on magnetism.

Scotia Ridge A part of the mid-ocean ridge that connects the southern end of South America to Antarctica by way of a long loop of raised seafloor that runs through the South Sandwich Islands. The ridge ends abruptly in the Scotia Trench east of the South Sandwich Islands. Its deepest point is the Meteor Deep, 8,260 m (more than 25,000 feet) below sea level. *See* MID-OCEAN RIDGES.

Scotia Sea The area of the southern Atlantic Ocean bounded by the Falkland Islands and South Georgia, South Sandwich, South Orkney, and South Shetland islands. It was first explored by Sir Francis Drake in 1578.

The conformation of the sea bottom, particularly at the western end of the Scotia Sea, channels the water pouring through the Drake Passage north and east across the Atlantic. There is a bend in the Antarctic Convergence that reflects this.

About 30% of the seafloor consists of foraminiferan calcareous remains, and about 30% consists of the siliceous remains of diatoms. The rest is sand and silted sediment, with a considerable occurrence of manganese nodules.

The warm, deep water of the Scotia Sea moves south, bringing nitrates and phosphates toward Antarctica. This accounts in part for the incredible plankton populations south of the Antarctica Convergence. *See* ANTARCTIC CONVERGENCE, DRAKE PASSAGE.

Scott, Robert Falcon (1868–1912) An explorer and officer in the British navy. Scott was given command of the British National Antarctic Expedition,

which was charged with the scientific exploration of Victoria Land (Antarctica). He sailed in the HMS *Discovery* in 1901 and spent two seasons (two southern summers) in Antarctica. During the first summer season (1902–03), Scott, Wilson, and Shackleton reached 82°16'33" south latitude, which was then the record for southernmost travel on Earth. He sailed for the Antarctic again in 1910 with the intention of reaching the South Pole before another explorer, the Norwegian Roald Amundsen.

Scott and his party set off in Antarctica in November 1911, and he and four others reached the South Pole on January 12, 1912, only to find Amundsen's markers already there. The Scott party's trek back to their supply depot was disastrous: One man died of injuries; another committed suicide, not wishing to be a burden; and Scott and the two remaining men were trapped by a blizzard and died. The tragic end of the Scott attempt to reach the South Pole caught the public eye, especially in England, where it was viewed as a national loss. *See* AMUNDSEN, ROALD; ANTARCTICA; EXPLORERS AND EXPLORATIONS.

Scripps Institution of Oceanography A research facility founded in 1903 by Edward Scripps and his half sister as the Marine Biological Station of San Diego, California. It is now part of the University of California at San Diego, and operates as a center for research in marine sciences. Its major subdivisions are marine geology, sedimentation, air-sea interactions, ecology, and paleontology. The institution maintains research vessels as well as satellite receiving equipment. Scripps Canyon, a submarine canyon off San Diego, was explored by the institution and its research ships.

scuba (SCUBA) An apparatus used for breathing while swimming underwater. Scuba diving is probably the best-known diving technique and is the one that made undersea exploration accessible to the nonprofessional diver. It has made it possible for man to enter into the marine environment. With scuba, both the biologist and the archeologist could for the first time observe an undisturbed site. Although nets, seines, and hooks have their uses in studies of undersea phenomena, accurate undersea observation and photography became available as adjuncts of the scuba diving method.

The Self-Contained Underwater Breathing Apparatus (SCUBA) was developed in the 1940s and rapidly became widely exploited. Almost any healthy person can use this apparatus, given reasonable precautions. The depth limit of scuba equipment is set, rather arbitrarily, at 50 m (160 feet), and for a short period. The length of stay at this or at any depth depends on the individual diver's physiology.

The scuba equipment most often used by amateur divers (all nonhelmeted divers) is of the "open-circuit" type. This consists of a compressed air tank that is carried by the diver. Its mouthpiece is a regulator supplying air at normal atmospheric pressure into a face mask. The diver exhales into the mask and the characteristic stream of bubbles is produced. Closed circuit scuba apparatus involves either pure oxygen or an oxygen-helium mixture. Neither technique is safe without extensive training and preparation. Semiclosed scuba circuits recycle the exhaled oxygen. *See* COUSTEAU, JACQUES-YVES; DIVING; MARINE ARCHEOLOGY.

Scyphozoa A class of coelenterates comprising true medusae. Their outstanding characteristic is their four-part radial symmetry. Both the hydroid (polyp) and medusal generations of the Scyphozoa can be quartered into four identical units.

All Scyphozoa have a bell- or umbrella-shaped body with tentacles in the subumbrellar area. Nematocysts or stinging cells cover the tentacles. The mesogleal jelly that forms the interstitial material of the jelly is fibrous. These fibers, which stiffen the umbrella, are unique to the class Scyphozoa. The interior of the jelly has a central, four-pouched stomach and primitive

Oral arm — Tentacle

Pelagia, a type of jelly

digestive system. Jellies are dioecious, with a distinct separation of the sexes.

Jellies may vary considerably in size. Some—the Stauromedusae—are nonswimmers and sessile. Some scyphozoans may live several years. They can be any color, and range from colorless to brown, green, pink, or violet. Several species are luminescent. The range of the class is from the ocean surface to almost any depth except the hadal zone, and from tropical waters to the subpolar regions.

The oldest fossil that bears evidence of being a scyphozoan is a Jurassic medusa, but some researchers believe that Cambrian rocks have traces of what may be medusa arms. *See* CNIDARIA; COELENTERATA; MEDUSA; NEMATOCYST; SARS, MICHAEL.

sea A term used to mean the ocean in general, or a large marginal part of a particular ocean, or a large landlocked body of water that may have an oceanic outlet. The Caribbean is a marginal sea, as is the Sea of Japan. The Aral and Caspian seas are really lakes.

sea anemone   A sea polyp that resembles a flower, of the phylum Cnidaria, class Anthozoa, and order Actiniaria. Sea anemones are common littoral animals whose general body plan is a ring of tentacles surrounding an oral disk. These animals' stinging cells or nematocysts, which are arranged in multiples of six, are on the tentacles and rarely on the columnar body. There is usually a pedal disk by which the animal attaches itself to a solid surface. Some species are floaters. The anemone has strong longitudinal muscles that can move it along the bottom. Even the most seemingly sessile (nonmoving) individuals can move, albeit slowly.

Most anemones live in rock pools near the surface, although some species are found at great depths. They are most abundant in tropical waters, where they are also more likely to be very large. The giants of the Great Barrier Reef can be up to 1 m (40 inches) in diameter. These animals are attractively colored—specimens have been found with colors ranging from delicate ivories to yellows, pinks, reds, and browns. Some species are hermaphroditic, with a single organism exhibiting both male and female sex characteristics, while others are dioecious. Like many other sessile animals, the anemones can be quite long-lived. Aquarium specimens have been known to live 50 years and longer. Anemones manufacture an array of toxic substances that are best described as avoidants. They defend themselves from predation by making compounds that taste bad, and therefore predators will not take more than one bite. These chemical defenses are also employed in territorial marking; even members of the same species will defend a rock or part of the sea floor from encroachers who would be competitors. Because these animals live a long time, the nature of their chemical defenses is of research interest to pharmaceutical companies.

Anemones will eat anything that they can trap with their tentacles. They spit out the nonfood components of a meal. Their central cavity is almost filled with a giant digestive gland.

Symbiotic and commensal relationships with other animals are well-documented phenomena in the anemone's life. Wrasses, clowns, and other fish live in the tangle of the anemone's tentacles, luring prey and catching crumbs from the host's meal. Hermit crabs "adopt" a particular anemone and attach it to their shell. When the crab outgrows the shell and moves into another one, it takes its anemone with it to the new shell. *See* CHEMICAL DEFENSES, CNIDARIA, COMMENSAL RELATIONSHIPS, LITTORAL, PHARMACEUTICAL RESEARCH, SESSILE.

sea bass   The common name of the perchlike, mostly marine fish of the family Serranidae. This family numbers over 400 species, which are found in tropical and semitropical waters. They are the groupers, coneys, grasbys, jewfish, and sea bass. The individuals range in size from 3 to 5 cm (1 to 2 inches) to very sizable fish over 2 m (7 feet) long.

Members of the subfamily Moronidae live in colder waters and include the familiar temperate-water fish, the striped bass (*Morone saxatilis*) and the white perch.

The entire group of sea bass and related fish are fished by both sport and commercial fishermen from tropical to temperate regions of the Earth. The group is largely carnivorous, feeding on crustaceans and mollusks. Their breeding patterns vary considerably. Some species are hermaphrodites, others are dioecious, and some have rather exotic behavior: The groupers, for example, change sex in maturity. *See* FISH, GROUPER, HERMAPHRODISM.

sea butterfly   A pteropod marine snail of the class Opisthobranchia. The sea butterfly is part of the plankton population in open ocean water. It lives in the water column but not at the surface, and it is therefore holoplanktonic. It is a gastropod that has lost most of its snail-like characteristics. The shell is absent in some; the gills have disappeared; and the foot, the structure that moves a snail forward, has evolved into two winglike structures, called parapodia. These structures flap and enable the organism to float or drift with the current. They also produce a mucus web that is used to trap smaller plankton or detritus. The tiny body is usually about 1 cm (0.3 inches) long and follows its plankton food supply in its diel vertical migration. Sea butterflies are usually passive feeders; however, at night they are active hunters at the ocean surface.

The taxonomy of these animals is being revised. Some taxonomists will not use the designation pteropod since some in the suborder Thecosomata have a shell and those in the Gymnosomata do not. *See* PTEROPODA.

sea cave A wave-carved cave in a cliff facing the sea.

sea cucumber A warty, stationary, marine animal of the class Holothuroidea (phylum Echinodermata), which resembles a squash or cucumber. Aristotle referred to an animal that he called a holothurian, but it is not known if he meant the sea cucumber. This common name comes from Pliny's *cucumis marinus*.

Holothurians are found in almost every marine environment. They have elongated bodies, most often 18 to 30 cm (7 to 12 inches) long. There is a tentacle-surrounded mouth at one end and an anus at the other. While the holothurians do have the five-part body that is typical of the echinoderms, the sea cucumber is not quite symmetric. It lies preferentially on one side, and that side is usually lighter in color than the other. It can be white, pink, salmon, or yellow, while the upper surface is frequently much more drab in shades of gray, olive, or brown. The exterior of the animal is a warty and knobby surface that produces a sticky mucus. This sticky material entraps the prey, which is then conveyed to the mouth by the tentacles. The spiny body wall of the sea cucumber contains calcareous ossicles (calcium-containing bonelike structures) that are unique to a species of these creatures and are used as a means of classification. Both the tentacles and the tiny podia operate by way of the water vascular system, which is a common feature of all the echinoderms. When the animal moves, it does so slowly and looks like a giant, sluggish caterpillar.

The reproductive pattern of the sea cucumber is also fairly typical of the rest of the echinoderms. Some are hermaph-

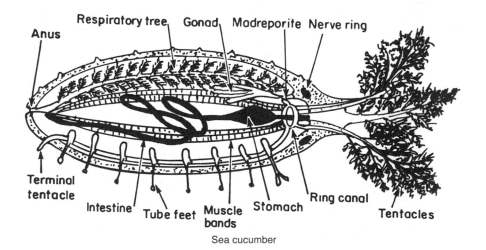

Sea cucumber

roditic (with both male and female characteristics on a single organism), while others are dioecious. Some species, particularly Arctic ones, brood their eggs.

Sea cucumbers manufacture a toxic substance with which they appear to kill small animals and stun larger ones such as fish. The toxin does not seem to interfere with the use of these animals as food for humans. Dried sea cucumbers are often used as basic ingredients in fish sauces and stews in the Far East.

As is the case with many long-lived and fairly sessile organisms, sea cucumbers have commensal organisms living on or near them. The sea cucumber, like other long-lived sessile organisms, produces toxic substances with which it defends itself from predation. These compounds are of interest as starting materials for pharmaceutical research. The sea cucumbers are also parasitized, particularly by marine snails. *See* CHEMICAL DEFENSES, COMMENSAL RELATIONSHIPS, ECHINODERM.

sea fan *See* GORGONIANS.

sea feather Another name for the echinoderm that is also known as sea lily or feather star.

sea floor *See* CONTINENTAL DRIFT, OCEAN BASINS, OCEAN FLOOR.

sea gooseberry A translucent ctenophore with a rounded body and longitudinal rows of cilia. *See* COMB JELLIES, CTENOPHORA.

sea hare A warty shell-less gastropod of the genus *Aplysia* and of the class Opisthobranchia. Some species have a rudimentary shell hidden by the mantle. The common name of these animals is based on the appearance of the anterior tentacles, which resemble the ears of a hare, and the arched back.

These hermaphrodites live in warm to tropical seas and produce strings of eggs that when dry look like seed cases. The diet of the sea hare is largely seaweed, but these animals occasionally capture small crustaceans, annelids, and other mollusks.

Aplysia produce a purple liquid as camouflage. This liquid is a bile protein and serves as a chemical defense for this rather slow-moving creature. The purple cloud of ink and the white viscous substance called opaline make a product that contains a protein called escapin. It and several related compounds exist only in the sea hare's exudates and the glands that produce them. Their primary function is as a deterrent, although since the exudates have antibacterial and cell-destroying properties, they may also protect a wound created during an attack (since one of the breakdown products of escapin is hydrogen peroxide). The pharmaceutical properties of the opaline and ink gland products are being examined. The sea hare's other passive defense mechanism also relies on the algae that it eats. Its skin is covered with chemicals that help it avoid being eaten. *See* CHEMICAL DEFENSE, GASTROPODA, HERMAPHRODITE, MANTLE.

sea horse One of a group of small marine fish of the family Syngnathidae, order Gasterosteiformes, that have a distinctive horselike head and neck and a prehensile tail. They are poor swimmers and therefore hold onto seaweed. The sea horse moves vertically using its swim bladder as a float. It feeds on tiny crustaceans and other zooplankton.

The singular structure of the sea horse is its brood pouch, which is beneath the male's tail. The female departs after depositing the fertilized eggs in the brood pouch, and the male tends the eggs. When they hatch, the young emerge from the male.

sea ice Any ice, floating or fast, that is formed from freezing ocean water. *See* FAST ICE, ICE, PACK ICE, PANCAKE ICE.

sea ice biome A habitat that had received no research attention before the last two decades of the 20th century. Although there are hundreds of hectares of sea ice, it was only recently noted that they are home to a large number of small organisms that play a significant role in the food webs of the polar regions. Sea ice harbors bacteria, algae, protists, and small invertebrates, all of which are adapted to the variations in temperature, water chemistry, sunlight, and nutrients.

The algae living in the ice are an important foodstock for young krill, which in turn support the largest of the marine mammals, the whales. These photosynthesizers also produce a range of compounds that absorb ultraviolet light, and the addition of these compounds to the krill are important to its survival. The sea ice algae and bacteria have provided another model of possible life forms that may exist on planets other than Earth. If these organisms can thrive here, can they exist on other planets or moons that are essentially large ice structures?

The loss of polar ice in the current period of global warming means that about one-quarter of the ice now extant will disappear and the primary food production of the area would increase about 10%. Open and warmer water means more phytoplankton. *See* POLAR BIOME.

seal A flippered marine mammal of the families Phocidae and Otariidae, of worldwide range but prevalent in cool to cold coastal waters. Seals vary in size from 1 to 8 m (3 to 25 feet); they have sleek, streamlined bodies and paddle with their flippers, which are modified forelimbs. Their bodies are well insulated with fat, which may serve as a pressure regulator as well as a thermal insulator. Both functions are necessary, since seals dive rapidly to depths of over 250 m (775 feet) and stay submerged for 20 to 30 minutes at a time.

Seals are divided into earless and eared groups. The ear, where it exists, is very small. The earless seals, or Phocidae, number 13 genera and 18 species. These are generally small, and their major means of propulsion is their tail. The eared seals constitute the family Otariidae. They consist of seven genera, which comprise 12 species. This group is the better known of the two and includes the sea lions and fur seals. They have longer flippers and propel themselves forward with these forelimbs. The eared seals also have much more mobility on land. While their motion on the shore is a comical waddle, it is much more directed than what the earless seal can manage. It is the eared seal that is the circus and aquarium clown.

The average seal has a dense, lustrous, brown pelt, the shade of which varies considerably, from almost black in some species to a sandy buff.

The bull male of the sea lion group grows much larger than the female, and these enormous animals collect large breeding harems if they can. The young males, or bachelors, live in groups on parts of the beach out of the territory of any bull and constantly try to effect "harem takeovers."

The young of seals are born after a gestation of about a year. They feed at first on a milk that is extremely rich in both fats and proteins, then progress to a regurgitated diet provided by the mother, and, in a few weeks, are swimming and catching their own fish and crustaceans. Infants and their mothers have strong scent markings—a factor that is vital for the survival of the pup, since it permits the pup and its mother to seek and find each other in the crowded rookeries.

The dynamics of a seal colony have been extensively studied using elephant seals as subjects. These animals are somewhat larger than the average seal, and also have huge noses, from which they derive their name. They were hunted for fur to near extinction in the 19th century. Effective protection by the United States and Mexico has reestablished a large population of elephant seals on the islands off California. A measure of the size of the population was attempted by banding the pups. The protection was effective, since there were six pups born in 1911 and 25,000 in 1982. The extent to which the area can support the population growth is still not known. *See* MAMMAL, PINNIPED.

Sealab A U.S. Navy project designed to determine the ability of humans to live at relatively great depths (180 m or 600 feet) for long periods. The Sealab habitat was occupied for over a month at one point in its use. The breathing mixture pumped into it was one of oxygen and helium. *See* DIVING.

sea lettuce Several species of green algae found in temperate to cold ocean water. *Ulva* is found in small sheets that are only two cells thick, which float in the open ocean. The tropical counterpart of *Ulva* is the spongeweed. *See* ALGAE, CHLORO-PHYLL, GREEN ALGAE, SPONGEWEED.

sea level The difference in the height of the sea from some arbitrary point on land; in reporting sea level, the value used is an average between the water level at high and low tide.

Changes in sea level depend on a number of factors, such as atmospheric pressure, salinity, and most noticeably, temperature. The volume of cold or high-salinity water is less than that of an equal weight of warmer or fresher water. A high atmospheric pressure tends to flatten the water surface; atmospheric lows have the opposite effect.

The effect of the Earth's rotation is evidenced in the Northern Hemisphere by an accumulation of water on the land to the right of a stream, whereas the accumulation is to the left in the Southern Hemisphere. Thus, oceanic gyres will create differing mean sea levels, a phenomenon that was observed by early explorers of the Caribbean. The mean sea level is highest for the North American continent at Nova Scotia.

Long-term effects of changes in sea level are being carefully studied. Previous advances and retreats of the continental ice sheets have altered the level of the sea to a mark about 100 m (330 feet) below the present mean level. The mean sea level has moved up about 12 cm (4.5 inches) since 1900. The change has been directly correlated with a global warming trend.

This in turn has been ascribed to the combination of the general warming that has been going on since the last glacial period and the greatly increased use of fossil fuels in the last 100 to 150 years. *See* GREEN-HOUSE EFFECT, SUBSIDENCE.

sea lily *See* CRINOIDEA.

sea lion A large, social marine mammal that lives in large herds. It is a feature of the North American Pacific coast. *See* PINNEPEDIA.

sea monsters Very large sea creatures that may have been the nucleus for some legends about sea monsters. There are giant pelagic (open sea) and benthic (bottom-dwelling) animals. A few that may have been the foundation for the monster theory could be whales, giant worms, ribbon worms, ribbon fish (Lampridiformes), giant clams, squid, crabs, and even the walrus. *See* GIANT CLAM, LAMPRIDIFORMES, RIBBON WORM, SQUID, SUNFISH, WHALE.

sea moth Sometimes called a sea dragon. It is an actinoptygian (primitive) fish that lives in the grasses near tropical shores. These ferocious-looking small fish (less than 10 cm or 4 inches) have been found in the waters of Hawaii, East Africa, India, Indonesia, and Japan.

seamount An undersea, relatively isolated mountain rising from the seafloor to a height of about 1,000 m (3,300 feet). If it has a pointed summit it is called a sea peak; if it is flat-topped, it is called a guyot.

The shape of most seamounts is elliptical. Only the very small ones have steep sides; large ones have gently sloping sides. These features are true for the seamounts present in every ocean. Hundreds of these mountains have been mapped, particularly in the Pacific, and it has been estimated by oceanographers that this number is only a small portion of the total. Although they are separate entities, seamounts occur in groups or chains, as do island arcs.

Seamounts are of volcanic origin, and may result from eruptions along crustal fissures or hot spots. In most cases the volcanic cone shows signs of weathering (erosion). There may be sedimentary accretion on a seamount, or a coral reef that has used the dormant or dead volcano as a base for growth.

Some seamounts, including a significant number with coral or limestone tops, are below the level of coral growth. This suggests subsidence by an unknown mechanism or a community that does not need to be near the surface. There now exists a growing body of evidence that chemosynthetic bacteria live in the basement basalts and extract energy from the rock itself.

Seamounts are isolated, and the result it that they, like islands, develop biota that is unique to that seamount. They are centers of species diversification. While there are communities of organisms at each seamount, on most of the seamounts in the Pacific the dominant organisms are gorgonians—sea fans. Ongoing research is attempting to identify the dispersal mechanisms of the characteristic corals. Other organisms that are found on Alaskan seamounts include bamboo corals and crinoids. These finds have completely set aside prior assumptions that corals are solely warm-water organisms. *See* CONTINENTAL DRIFT, CRUST, GORGONIANS, GUYOT, HOT SPOT, ISLAND.

sea mouse An annelid worm that is usually found buried in sediment in temperate zones. It may grow as long as 20 cm (8 inches), but most specimens are smaller. This detritus feeder has a bristly dorsal side that is frequently green and resembles fur.

sea pansy A soft coral cnidarian belonging to the class Anthozoa; it consists of a flat plate with polyps on one side. The genus *Renilla* is common to both coasts of North America. These cnidaria are frequently iridescent. *See* ANTHOZOA, CNIDARIA, CORAL.

sea pen A cnidarian of the genus *Pennatula,* belonging to the class Anthozoa, that looks like a feather. There are about 300 species of this colonial animal. They live in shallow to deep water, and their range is almost worldwide. The central stalk, or peduncle, of the sea pen bears the "branches." Many species are luminescent. *See* ANTHOZOA, BIOLUMINESCENCE, CNIDARIA, PENNATULACEA.

sea robin A marine fish of the family Triglidae, with an armor plated head. Some species have bony plates over their entire bodies. Sea robins have long, trailing feelers that explore the bottom, searching for mollusks and crustaceans.

The sea robins are often bright green or blue. They can, and frequently do, make audible sounds. The largest species is about 70 cm (28 inches) long.

These fish are occasionally used by humans as food. In the earlier years of the 20th century, sea robins were considered "trash fish" and were dumped overboard if they became entangled in lines set for other fish. *See* FISH.

sea slug A nudibranch. *See* NUDIBRANCH, Opisthobranchia.

sea snake A venomous, tropical marine serpent of the family Hydrophiidae. There are about fifty species. Their distinguishing feature is a flattened body; they swim using their tail as an oar. The usual individual is 1 to 1.3 m (3 to 4 feet) long, but the Japanese varieties may be twice that length.

Sea snakes may be found in tropical ocean water anywhere, but the largest number of species are found in Asiatic or Australian waters. Some species lay eggs on the beach, but most bear live young in the water. The sea snakes are fish eaters. They are most often solitary animals, although large groups of them are occasionally sighted. As a whole, sea snakes have especially lethal venom. The venoms are neurotoxins, which attack nerve centers that control breathing and heartbeat. *See* REPTILE, TOXIN.

sea spiders Small, long-legged arthropods that live on ocean bottoms. Also known as whip scorpions, they belong to the class Pycnogonida, and represent about 600 species. The individuals range in size from 2 to 3 mm (0.01 inches) to 50 cm (20 inches) or larger. The small ones are found in shallow water, the larger specimens are deep-sea dwellers.

Sea spiders have four to six pairs of walking legs. They are carnivorous animals, some feeding on other invertebrates by sucking them dry, while others tear food apart with powerful appendages and pass it onto a proboscis—a food-gathering mouth structure. While their reproductive mode is internal fertilization, the males incubate the eggs. *See* PROBOSCIS, PYCNOGONID.

sea squirt A sessile marine animal that looks like a lumpy potato; a member of the class Ascidacea, subphylum Tunicata. Sea squirts live in all seas, with a range from the intertidal flats to oceanic deeps. Their bodies are covered by a thick skin or tunic, which has two large openings, one of which sucks water in while the other expels it. The sea squirts feed by filtering debris, small animals, and plankton out of the water.

The sea squirts are on the whole small; individuals range in size from less than 1 cm to about 6 cm (0.4 to 3 inches) in diameter. Some species are colonial, others solitary. The animals are hermaphroditic, each individual having male and female gonads. The eggs are fertilized by sperm from another animal swept in by the water and retained within the ovary until ready to hatch. They hatch into free-swimming larvae that look like tadpoles. As the larvae mature, the notochord does not develop into a spinal column. Instead, the larval organism attaches itself to some likely spot and loses its vertebrate-like structures. In addition to their relatively common form of sexual reproduction, squirts also reproduce by budding.

In addition to their curious, almost vertebrate status, ascidians are peculiar because their blood and tissues contain vanadium oxides, compounds that no other animals concentrate in their systems. The peculiar circulatory system does pump blood; there is a heart that moves blood in one direction for a number of beats and then changes the direction. *See* NOTOCHORD, TUNICATA.

sea star Starfish are now officially sea stars, creatures of the class Asteroidea in the phylum Echinodermata. The stars are common inhabitants of all seas. The sea star most often encountered in a temperate zone tidal pool is yellow-orange and from 15 to 30 cm (6–12 inches) in diameter. Individuals that are much larger and brilliant blue or red are native to reef communities. The Great Barrier Reef in Australia is home to some of the most spectacular species.

The body plan of the sea star is a five-part arrangement. The typical tidal pool star has five arms; larger specimens have arms in multiples of five. The phylum name of these creatures implies a spiny skin; it is pebbly in texture and consists of horny plates. In many species these plates contain aragonite. This crystal form of calcium carbonate is much rarer than the calcite crystal. It has different light-refracting properties and is thought to function as the "light-sensitive" organ in sea stars.

The sea stars are quite mobile. The undersides of the arms are lengths of tube feet, a series of small and extremely flexible suction cups that operate on a hydraulic system. Water enters the animal through the madriporite—a porous structure on the aboral (upper) surface. Water channels connect this intake with the tube feet. The degree of suction that the tube feet can exert is a function of the quantity of water in the canals leading from the madriporite.

Sea stars will eat just about anything. They will ingest detritus if it is there, or they will attack mollusks, coral, or other echinoderms. The mode of ingestion may have been one of engulfing food in the primitive species, but modern echinoderms evert the stomach onto a meal and digest it on the spot. The stomach and

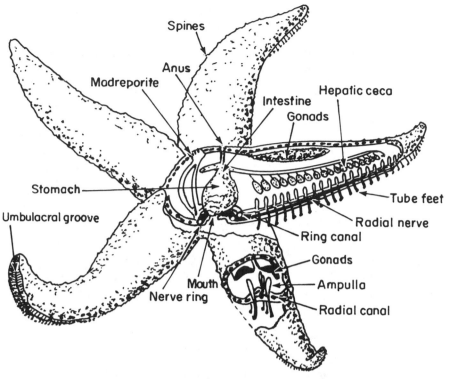

Spines

Anus

Madreporite

Hepatic ceca

Intestine

Gonads

Stomach

Umbulacral groove

Tube feet

Radial nerve

Ring canal

Gonads

Mouth

Nerve ring

Ampulla

Radial canal

Sea star

the digested meal is then sucked back into the animal. Sea stars have been a part of the fossil record since the Cambrian period; they are a very successful life-form. *See* ASTEROIDEA, BRITTLE STARS, CRINOIDEA, ECHINODERMATA.

sea turtle Any of several quadruped marine reptiles characterized by a typical shell made up of a series of plates, the shape and size of which are species specific. The shell covers both the dorsal and ventral surface of the animal. The head, limbs, and tail may be retracted completely into the shell. The sea turtles range in size and feeding habits, but all are strong swimmers that rarely come to shore. Females do this only to lay eggs on their native beaches. Thus, a male turtle may never have been on a beach after hatching. The sea turtles are classi-

fied as the leatherback (Dermochelyidae), green, loggerhead, hawksbill, and redleg (Cheloniidae).

The leatherback is the largest sea turtle. It is an omnivore whose shell is often 2 m (6.5 feet) long, and it weighs about 500 kg (1,100 pounds). The name refers to the a tough covering of floppy skin with the characteristic shell beneath it.

The green turtle is also omnivorous. It is an endangered species, and several attempts are under way to preserve it in the wild, restock it on native beaches, and keep it alive in zoos. The green turtle is about half the size of the leatherback. It is brownish green, but the name is probably due to its green fat. This species is most often hunted for its meat. It lives in tropical waters worldwide.

The loggerhead is about the same size as the green turtle. Its shell is red-brown, and its feeding pattern is also a carnivorous one.

The hawksbill is much smaller than the other sea turtles. The average individual weighs less than 50 kg (110 pounds). This species is also taken by hunters; its shell is used to make ornaments. It is a worldwide creature and an omnivore.

The redlegs—the Atlantic and somewhat larger Pacific species—are primarily tropical animals. They are 60 to 70 cm (25 to 30 inches) long and are omnivores.

The exploitation of the sea turtles for meat and jewelry has greatly decreased their number. Since reproduction is a very energy-intensive activity, these animals do not breed in lean years. The removal of eggs from beaches by humans (the eggs have long been considered as aphrodisiacs in the Caribbean) and the destruction of the beaches used for egg-laying have further decreased the likelihood of the turtles' survival without the active intervention of conservationists. One project for this purpose involves taking eggs from nests scooped out of the sand by female turtles and moving some of the clutch, which may contain as many as 100 eggs, to another beach where that species of turtle had appeared in the past. It is hoped that the hatchlings who make it back to the sea—despite the attempts of the shore birds to pick them off—will at maturity return to the new "native" beach.

While the sea turtles are primarily animals of tropical waters, some are blown off course and end up in cold or temperate-cold water. This puts them into thermal shock, since their reptilian circulatory system is not equipped to deal with temperature drops (which is why reptiles in the temperate zone hibernate). Since sea turtles are an endangered group, many locales post a "turtle watch." In New York waters, for example, sea turtles that come ashore are routinely brought to the New York Wildlife Conservation Society's aquarium, which arranges transport for them back to the

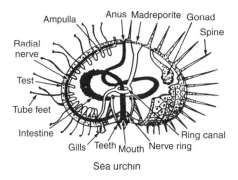

Sea urchin

waters around Florida. *See* EXTINCTION, POIKILOTHERM, REPTILE.

sea urchin An echinoderm (class Echinoidea), enclosed in a thin brittle shell covered with movable spines; it looks like a dark red, blue, or purple pincushion. The name is derived from the Greek for hedgehog. The sea urchin has been known for millennia and has left a fossil record since the Ordovician period, 500 million years ago.

The urchin is a regular echinoid (symmetric) with a roughly globular aboral (top) and a flattened oral (bottom) surface. Its mouth is on the oral surface and is equipped with hard teeth. Almost all of the urchin is covered with a calcareous, spiny test, or shell. In addition to the spines, five sets of podia (feet) protrude from the test. The urchin moves using both its feet and its spines.

Urchins are ubiquitous in tropical and warm temperate waters. They are used as a food by many people and have a place in the lexicon of both French and Chinese cuisines. The sea urchin's eggs, which develop rapidly after external fertilization, are primary tools for embryologists; the development of the eggs is a "textbook" illustration of the staging of cell division. *See* ARISTOTLE, ECHINODERMATA.

seawater Water in or from the sea. Seawater comprises about 97% of all the water on Earth. It is of remarkable worldwide consistency in terms of content. This was

SOME ELEMENTS IN SEAWATER	
Strontium (Sr^{2+}) ion	8 ppm (parts per million)
Oxygen (as O_2)	4.6–7.5 ppm (this varies with depth; the greatest concentration is at the surface)
Silicon (Si^{4+}) ion	3 ppm
Fluorine (F^-) ion	1.3 ppm
Nitrogen (N_2, NO_2^-, NO_3^-, NH_4^+)	0.5 ppm
Argon (A)	0.5 ppm
Lithium (Li^+) ion	0.17 ppm
Phosphorus (HPO_4^{2-}, $H_2PO_4^-$) ions	0.07 ppm
Iodine (I^-) ion	0.06 ppm
Carbon (in CO_2)	Traces

confirmed by the careful analyses done by William Dittmar, the chemist on the *Challenger* expedition of 1872–76. Dissolved materials, other than sodium and potassium chlorides, comprise only about 0.05% of all of the dissolved material in seawater. This small percentage includes traces of practically every element. But while they are present, most elements are present in amounts too small to be of commercial interest.

The carbonate and halite (chloride, bromide, iodide) deposits found on land are of marine origin and date from all geologic periods, from the Precambrian to the present. Since they are fairly consistent, it is assumed that the composition of seawater has been essentially the same through time. The composition of Precambrian water is not known but is assumed to be similar to the seawater of today. While the composition of seawater is relatively uniform, this does not mean there is no difference.

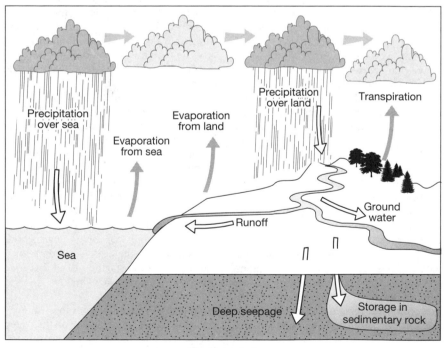

The water cycle

OCEAN AREA COVERED MILLIONS OF KM²(MI²)				
Sediment	*Atlantic*	*Pacific*	*Indian*	*Percent*
Calcareous Globigerina	40.1(15.5)	51.9(20)	34.4(13.3)	35
Pteropod	1.5(0.58)			
Siliceous Diatomaceous	4.1(1.58)	14.4(5.56)	12.6(4.86)	10
Radiolarian	uncertain	6.6(1.55)	0.3(0.116)	
Terrigenous Red Clay	15.9(6.14)	70.3(27.13)	16.0(6.17)	5

The patterns of water flow in various oceans have been established by core sampling. The study of foraminiferan remains shows that not only is there local variation in the temperature and salinity of seawater, because land formations hinder its circulation, but that these formations maintain the individual character of whole oceans. Thus, deep water flows from the Atlantic to the Pacific, but the water in the two bodies has slight but significant differences:

ATLANTIC DEEPS	SUBSTANCE	PACIFIC DEEPS
Relatively rich	$CaCO_3$	Relatively poor
Relatively poor	SiO_2	Relatively rich
Relatively rich	O_2	Relatively poor

The surface of the Atlantic is about 2° warmer and 1 ppm more saline than that of the Pacific. The difference in the waters dates roughly from the point about 12 million years ago when the Panamanian isthmus closed. Similar events isolated the Mediterranean. A comparison of the levels of the isotopes C^{13} and O^{18} was used to establish this point. The assumptions made were that there was unimpeded circulation prior to the change in landmass,

and that the rate of decay of the radioisotopes is known. *See* ANTARCTIC CURRENT, CARBON DIOXIDE; *CHALLENGER*, HMS; CIRCULATION; METEOROLOGY; MINERALS; RADIOACTIVITY; SALINITY.

seaweed The common name for marine protists that grow in salt water. *See* ALGAE, *FUCUS*, KELP, LAMINARIALES, *SARGASSUM*.

sea whip A gorgonian with polyps and extended tentacles that look like branches of small leafless trees. This animal manufactures a toxic chemical that protects it from predators. *See* GORGONIANS, PHARMACEUTICAL RESEARCH, TOXIN.

Sedgwick, Adam (1785–1873) An English geologist who studied and named some of the earliest segments of the Earth's history. Although he trained in mathematics, he was appointed professor of geology in Trinity College, Cambridge, in 1818 and proceeded to become a geologist. In 1827 he and Roderick Murchison investigated rock formations in western Scotland and then in southern England. They gave the Cambrian period its name, using an old name for Wales. This team also examined Devonshire and published their work in 1839.

A profound professional disagreement led to both the breakup of the team and the incorrect naming of some Paleozoic strata. This was not resolved until after the deaths of both Sedgwick and Murchison. *See* CAMBRIAN PERIOD, DEVONIAN PERIOD, ORDOVICIAN PERIOD.

sediment Matter that settles to the bottom of the sea. Sediments were first seriously sampled by Matthew Maury, and systematically studied by the *Challenger* expedition (1872–76). They are divided into two large groups, the terrigenous and the pelagic. The former is largely of terrestrial origin and consists of rock, gravel, sand, and mud—with volcanic intrusion, and coral debris. The pelagic sediments are those that "rain down" from all levels of the water and finally land on the bottom. These are the shells or skeletal remains of plankton.

The *Challenger* sampling showed the divisions of the types of sediment, with calcareous sediment disappearing at about 4,000 to 5,000 m (13,000 to 16,000 feet). The red clay that Murray described as soft, plastic, and greasy is of volcanic origin and terrigenous. At still greater depths radiolarians were found. Prior to the *Challenger* sampling, these were unknown. The sediment at these greater depths is a pale gray to yellow, claylike material.

Also found on the sea bottom are minerals precipitated from the water, and material of extraterrestrial origin. Although they constitute only a small part of the total sediment, cosmic dust, meteor fragments, and tektites are found in all old sediments and present-day bottom samples.

The location of sediments is a function of their deposition rates or their movement to the ocean by rivers, glaciers, and wind, and of the rate of generation of various materials by volcanic or biogenic activity. The sediments are then carried along and finally dropped in the ocean by the various currents that move solids horizontally and vertically.

Active transport of sediment tends to dominate sediment transport. This involves the movement of a great deal of solid material; a landslide near a volcanic site is an example. Passive sedimentation is the movement of fine sediment in the water column or drift of material in the path of bottom currents. The transportation of fine materials such as terrestrial volcanic ash is eolian, or wind-transported, passive sedimentation.

The remains of living organisms are found on the ocean bottom. They vary depending on their sources.

It is now known that there are diverse but sparse communities living beneath the sediment, subsisting on the basement rock. *See* CHALLENGER, HMS; FORAMINIFERA; OCEAN CURRENTS; OCEAN FLOOR; MAURY, MATTHEW F.; RADIOLARIA.

seeps Cracks in the undersea Earth's surface. Gases—methane and hydrogen sulfide in particular—emerge from these cracks, but they are not hot as are those that form black smokers. Seeps were discovered years after the finding of hydrothermal vents. While the hydrothermal vents are very dramatic, spewing hot, dissolved minerals into the cold water, slow leakings of oil and gas (almost always methane) also create suitable conditions for undersea communities.

OCEANIC SEDIMENTS			
Mineral	*Crystal Form*	*Organism*	*Genus*
$CaCO_3$	Calcite	Formaniferan (protozoan)	Globigerina
$CaCO_3$	Calcite	Algae	Coccolithophoridae
$CaCO_3$	Aragonite	Mollusks	Pteropod ooze
SiO_2	Opal (microcrystalline quartz)	Algae	Diatoms
SiO_2	Opal (microcrystalline quartz)	Protozoa	Radiolarians

These biomes are also dependent on chemical energy sources: They do not involve warmth or sunlight.

The Gulf of Mexico is the site of numerous seeps, but there are others in the Monterey Canyon and other sites in the Pacific Ocean. Very likely, seeps are a feature of areas near continents where fluid flows through the landmass and exits into the bottom water. Much of this fluid contains sulfur compounds and dissolved methane.

The biota that inhabits cold seeps is as varied as the vent biomes. Here numerous bacteria support an array of bivalves, worms, gastropods, crabs, anemones, and soft corals. See EXTREMOPHILES, ICE WORM, METHANE, SULFUR BACTERIA, VENT COMMUNITIES.

seiche The occasional, rhythmic rise and fall of water in enclosed lagoons or bays. It is not tidal. The phenomenon was carefully observed on Lake Constance (Switzerland) in the 16th century. The oscillation of the water is the result of a disturbance that sets up "waves." The time period of these "waves" is described by Merian's formula,

$$T = 2L \sqrt{gh}$$
$$L = length$$
$$\sqrt{gh} = depth$$
$$g = gravity$$

The so-called wave will bounce back and forth until its energy is dissipated through friction. The seiche is dependent on the shape of the basin in which it occurs.

Coastal seiches are derived from air-water interactions or long-period waves coming in from the open ocean. These are caused by strong winds or large atmospheric disturbances, or by earthquakes either on the coast or undersea. The "surfbeat"—a phenomenon of an embayed area—has also been named as the origin of seiches.

The effect of seiches can be good. They are the means of mixing water in a constricted area, and of distributing nutrients, dissolved gases, and heat. On the other hand, moored ships have been rammed into docks or even broken up when hit by a node of a strong seiche. See ESTUARY, TIDE, WAVE.

seine A net that hangs in the water with one edge held up by floats and the opposite edge held flush against the bottom by weights. When the sides of the net are brought together, they enclose the catch. The net and its contents are then drawn ashore or onto a fishing boat. A purse seine is one that is drawn shut by means of an interwoven rope. See FISHING INDUSTRY, NET, PURSE SEINE.

seismology The study of earthquakes and of artificially produced vibrations of the Earth, such as those caused by nuclear explosions. When earthquakes or landslides occur under the sea, they set long waves in motion. These waves are incorrectly called tidal waves; they are properly called tsunamis or seismic surges.

Whether naturally occurring or set off by explosives or by high-energy electrical discharges, these long waves have a measurable velocity. When this data is combined with gravity and magnetic measurements and with much higher frequency echo soundings, the result is a profile of the bottom. This profile is quite detailed since different rocks and sediments "reflect" signals differently, and this permits the composition of the bottom to be determined. See EARTHQUAKE, MOHO LAYER, OCEANOGRAPHY, SURVEYS, TSUNAMI.

semidiurnal tide Tides that occur in a cycle of two high tides and two low tides each tidal (lunar) day. See LUNAR DAY.

sensory perception See EYE, LATERAL LINE, NERVOUS SYSTEM.

Sepia A genus of cuttlefish. It also defines the black fluid or ink produced by some cephalopods as a protective device. See CUTTLEFISH, OCTOPUS, SQUID.

sessile A term used to describe an organism that is attached to another structure by its base (i.e., unable to move freely); a description of animals that do not move or whose movement is slow enough that they need other means of avoiding predation. Many species are sedentary, and these creatures range through almost every phylum. Some are motile for a part of their lives and sessile during the rest. The oyster is an example of a sessile adult that comes from a motile juvenile.

setae The small bristles either on arthropod (crab, lobster, etc.) exoskeletons or they are the paired projections on the segments of annelid worms. On polychaete worms, the setae are borne by the parapodia, the "false feet" that characterize this group. *See* ANNELIDA, ICE WORM, POLYCHAETA.

Seychelles Island group in the Indian Ocean off the eastern coast of Africa. The granitic islands are surrounded by a shallow-water coral reef. The area was studied by a multidisciplinary team from Cambridge (England) in 1998, the Southern Seychelles Atoll Research Programme or SSARP.

sferics Also known as atmospherics; the electromagnetic radiations that are end-products of atmospheric electrical discharges. They are the principal causes of radio static.

Since sferics are the result of discharges created by storms, they are used to track the paths of storms or to predict their activity. Different storms produce different discharge patterns. *See* ELECTROMAGNETIC SPECTRUM.

shad *See* HERRING.

shark Any of a widely diverse group of carnivorous, cartilaginous fish most prevalent in warm seas, where they are found at varying depths, with some found at great depths. The sharks belong to the families Carchariidae, or sand sharks; lamnidae or mackerel sharks; Carcharhinidae, or requiem sharks; Squalidae, or dogfishes; and many others. Most sharks kill their prey by biting chunks out of it. Some species have poisoned spines ante-rior to the dorsal spine, whose function may be connected with feeding and also may be a defense mechanism. The outstanding characteristics of the sharks are the five sets of exposed gill slits, an asymmetric tail (the top part is larger), numerous deciduous teeth (they fall out and are replaced), and placoid scales. These scales are not overlapping; each is small, uniform in size, and bears a tiny barb in it that points toward the tail.

Sharks vary in size, ranging from the small dogfish of about 1 m (3.3 feet) to the basking sharks and the great white sharks, which are among the largest fish and the largest of all sea creatures except the whales. The greatest number of shark species are relatively small, at less than 3 m (10 feet). The whale shark, the largest species, has been given an estimated maximum measurement of about 10 m (35 feet). It eats plankton and small fish. There is much conjecture (and hysteria) about the size and ferocity of sharks. There are, however, shark repellants. Some fish that sharks prey on manufacture toxic substances as protection.

Sharks have been intensively studied, putting some favorite myths to rest although perhaps only temporarily. Like other predators, they kill for food and do not eat for days or weeks afterward. They rest and sleep without moving.

Many sharks rise through the ocean in the daylight and descend at night in a diurnal pattern that mirrors the movement of their prey. Since they do not have a swim bladder, it is thought by some that the shark's enormous liver, whose volume consists largely of oil, may be an equilibrating device. Sharks also store urea (nitrogenous waste), which may also be a depth-equilibrating substance.

Sharks comprise a large number of ovoviviparous species with eggs that hatch within the female's body, as well as viviparous ones. The period of gestation is quite a long one: Dogfish pups may spend two years in the oviducts of the mother.

Sharks are caught commercially for food and for their livers. The liver is used as an oil source. Large quantities of squalene can be produced from shark liver oil; this compound, when purified, is used extensively in the preparation of pharmaceuticals and cosmetics.

Because there are so many gaps in what is known about sharks, the possibility of overfishing these animals is a very real one. *See* CHONDRICHTHYES, FISH, MARINE OIL.

shark repellent There is one group of chemical shark repellents that seem to work. They are based on a group of terpenoid compounds derived from the Moses sole, *Pardachirus marmoratus*, of the Red Sea, or from a related species, *P. Pavoninus*, a Japanese sole, but the actual formula is a secret.

shearwater A member of a group of oceanic, web-footed birds of the family Procellariidae, that glide along wave troughs. The puffins are typical members of this group. These birds are almost always flying or in the water. They come to land only to make nests and raise their chicks. They do so in dense colonies on offshore islands. Shearwater colonies are found from cold temperate to almost polar regions on the eastern side of the Americas. The colonies also are found off the coasts of Australia and New Zealand.

The single egg hatches in about seven weeks, and is fledged in about 10 to 12 weeks. The typical shearwater is a slender bird about 30 to 60 cm (12 to 25 inches) long. Its feathers are gray-brown to black on the upper surface; the underside is white. The body is long and the wingspan distinctive, being at least 1.5 times as long as the body. The hooked bill makes the shearwater an effective predator.

Widespread and typical species are the sooty shearwater (*Puffinus griseus*); the common shearwater, also known as the Manx (*P. puffinus*); and Newall's shearwater, which is found in Hawaii (*P. newelli*). The short-tailed shearwater (*P. tenuirostris*) is found in Australia or Alaska, and is hunted for its meat. *See* BIRD.

ship An oceangoing vessel, usually of some size. Using this somewhat arbitrary definition, the ocean voyages of the Polynesians and some of the Norse voyages were made in boats, not ships. The first significant ships were Phoenician galleys, which eventually became the ships used throughout the entire Mediterranean Sea.

Sails as a reliable means of propulsion appeared about 1000 C.E., and the original sailing ships had one mast, and later two and three. The rudder was developed about 1200 C.E. and allowed for greatly increased maneuverability.

Merchant vessels were usually not different from warships. There were a few exceptions: war galleys had rams with metal sheathings on them and defensive or offensive fortifications were constructed on these vessels to give the fighting crew a platform from which to operate. Gradually the difference between trading and fighting ships became more pronounced, and trading vessels became broader than warships and carried less sail. The breadth of the beam gave trading ships increased cargo space, and the decrease in sail meant that a smaller crew could work the ship; however, these factors meant that merchantmen were increasingly slower than warships. The size of the ships by the 15th and 16th centuries is significant: They were not large by our standards, the caravel averaging only about 250 tons and the largest carracks 1,500 to 1,600 tons.

The peak of success of the sailing merchant ship was the age of the clipper. These beautiful vessels were economically significant because they could handle small, expensive cargoes at great speed. However, they became uneconomical once steamships became reliable.

The first steamship, a merchant ship, crossed the Atlantic in 1838. The paddlewheeler came and disappeared within a decade, so rapid was the technical change that brought about a revolution in ocean travel. The largest ship of the time, the *Great Eastern* (19,000 tons), was a commercial failure but the beginning of a total change in transportation: Before this point ocean travel had not been commonplace, whereas by the end of the 19th century it was. *See* INDIVIDUAL SHIPS BY TYPE.

shipworm An elongated, boring marine clam that resembles a worm; also known as pileworm. There are about 14 genera of this family, the Teredinidae. They are of economic importance because they cause damage to wooden piers, pilings, and ships. *See* BORERS, CLAM, TEREDINIDAE.

shore The seacoast, from the lowest low-water line to the highest high-water line. *See* BEACH, COAST.

shrimp A group of important decapod crustaceans belonging to the order Natantia, with a semitransparent body and a flexible abdomen that ends in a fantail. There are roughly 2,000 species of shrimp, ranging from freshwater to deep-ocean dwellers, and of sizes from less than 7 cm to more than 20 cm (3 to 8 inches).

These delicate animals move rapidly in a backward direction by flexing their abdomen and tail. Most shrimp are omnivores; some are scavengers. They are rather typical of the decapods, living in almost every range of marine environment. They produce thousands of eggs upon a single mating, and these are carried on the swimmerets of the female as they develop. In some species the juveniles live as a separate, segregated group.

Shrimp are frequently hatched in the shallow waters of estuaries, and the continuing study of wetlands and marshes produces fresh information about the life cycle of this group of animals. This is of major interest not only to ecologists but also to commercial groups, since shrimp are an important part of the overall fish and shellfish catch. Not all shrimp are edible, but many, many tons of these crustaceans are taken annually in the Gulf of Mexico, the Mediterranean, the Baltic, the Sea of Japan, and other areas of the world's oceans.

Most shrimp species live as individuals; however, there are some species that live in family groups analogous to beehives or wasps' nests. These eusocial snapping shrimp live in the interstices of sponges. They are essentially one family as a beehive is one family, but in the shrimp colony there is a mating pair, the parents of all the nonmating sons. The adult nonmating sons' function is to defend the colony. It is not known if the sponge that provides a home to this synalpeid shrimp family derives some benefit from its tenants or is at least not harmed by them. *See* ARTHROPODA, DECAPODA, MANTIS SHRIMP, MARSH.

Sicily The largest Mediterranean island. It is situated off the coast of southern Italy, to which it was geographically connected in the Pleistocene epoch and possibly earlier. It has one of the world's most active volcanoes, Mount Etna. The whole island and the surrounding sea are subject to frequent seismic activity. *See* ISLAND, MEDITERRANEAN.

Sicily, Straits of A passage about 500 m (16,000 feet) deep between the island of Sicily and the Italian mainland. It separates the Mediterranean into two large areas. The cold Atlantic water that enters the Mediterranean at Gibraltar flows along the north shore of Africa and enters the eastern Mediterranean by passing through the Straits of Sicily. Geologically, this passage is old. It has certainly existed since the mid-Pleistocene and was probably a water channel before that time. *See* MEDITERRANEAN, STRAIT.

silicoflagellates Small phytoplankton that engage in photosynthesis and, like diatoms, have a glassy frustule or shell. They range in size from about 10 to 150 micrometers (1 micrometer = 1 millionth of a meter). More members of this group existed in earlier geologic times than do now. Their fossil remains are found in sedimentary rocks dating from the early Cretaceous period to the present. The tests of these organisms account for less than 3% of the marine sediment. While they have a long history on Earth, they were never a dominant group. However, they were common in the late Cretaceous and appear to have been more diverse in the early Cenozoic. There are fewer species now. *See* DIATOM, PHYTOPLANKTON, PLANKTON.

silicon A reactive, nonmetallic element, which in combined form is, after oxygen, the most abundant element in the Earth's crust. Its compounds are found in the sea, and are both biogenic and nonbiogenic in origin. Silicates predominate in the nonbiogenic sediments of the sea; quartz, feldspar, and other clay minerals are either carried into oceans by rivers, or are the result of undersea volcanic activity. Sand is silicon dioxide (SiO_2).

Biogenic silicates come from the tests (shells) of diatoms, silicoflagellates, and radiolarians. Differing depths have characteristic concentrations of silicates, as do different oceans. Deeps in the Indian Ocean and Antarctic deep water are particularly rich in silicates. There is obviously a rapid turnover in the use of silicon, but the exact mechanism by which this occurs is unclear. *See* DIATOM, ELEMENT, MINERALS, RADIOLARIA, SEDIMENT, SILICOFLAGELLATES.

sill An underwater ridge that separates ocean areas. It may be at the mouth of a fiord or between two oceanic basins. The Caribbean, for example, is separated into several basins by sills, and a sill separates the Mediterranean from the Atlantic Ocean at Gibraltar.

silt Finely divided particles that are the intermediate form of particulate matter between sand and clay. Particles of silt range from 0.01 to 0.05 mm in diameter. Mud is within this range. *See* SEDIMENT.

Silurian period The period following the Ordovician period of the Paleozoic era and preceding the Devonian. This time is usually given as 430 million to about 395 million years ago. The name Silurian was devised by Murchison, who chose it for the Silures, a tribe living in pre-Roman Wales.

The Silurian was an age of mountain building: The events of this time raised the Scottish Highlands in Britain and the Taconic Mountains in eastern North America, and closed the proto-Atlantic Ocean. Canada, Scandinavia, and South Africa were probably equatorial landmasses, Japan and the Philippines were near the North Pole, and South America and Australia were near the South Pole.

Generally speaking, the climate throughout this time was mild to tropical and the seas were shallow, a combination that encouraged the growth of calcareous invertebrates, which in turn produced vast limestone deposits. As the tiny animals died, their shells piled up in layers many meters thick. Niagara Falls is a prime example of a massive limestone formation of Silurian origin.

The first fish that had true jaws appeared in the Silurian, as did land plants. Fossil trilobites, graptolites, crinoids, brachiopods, and mollusks are also frequently found in the deposits of Silurian rock systems. These remains are of worldwide distribution and are used to date these rock formations. *See* FISH; INDEX FOSSIL; MURCHISON, RODERICK IMPEY; ORDOVICIAN PERIOD; PALEOZOIC ERA; *Appendix I: Geologic Timescale.*

Siphonophora An order of the Cnidaria that forms free-swimming or floating pelagic colonies of both polyp and medusal forms. *Physalia* (a genus of large oceanic Siphonophora, which includes the Portuguese man-of-war) is an example. *See* CNIDARIA, PELAGIC ENVIRONMENT, PORTUGUESE MAN-OF-WAR.

Siphunculida (**Sipuncula**) The peanut worms, a phylum of sausage-shaped worms. These animals are usually small, ranging from 1 to 2 cm (0.5 to 1 inches long). Their prominent physical characteristic is a long proboscis equipped with a circle of small tentacles. The proboscis collects plankton and debris. Siphunculids are usually burrowers in the ocean bottom; they live in all habitats, from intertidal zones to hadal depths down to 4,000 m (13,000 feet). Those that inhabit the depths are usually larger, some growing to lengths of 50 to 72 cm (18 to 30 inches).

Siphunculids live in commensal relationships with other animals. Their association with entoprocts and coral has been investigated. *See* BENTHOS, COMMENSAL RELATIONSHIPS, INTERTIDAL, PLANKTON, PROBOSCIS, TROCHOPHORE.

Sirenia The sea cows; a group of large, herbivorous, aquatic mammals that includes the manatee and the dugong. Most of them are extinct or on the endangered species list. The largest sirenian, Stellar's sea cow, was about 8 m (25 feet) long and weighed several hundred kilograms. It was hunted to extinction about 100 years after its discovery in 1741.

Dugongs live in warm, coastal Indo-Pacific waters; manatees live in the coastal waters of the Atlantic. They will enter estuaries, where they are frequently injured or killed by the propellers of motor boats.

The manatee is a fairly representative sirenian: It is 3 to 5 m (10 to 16 feet) long, rotund, slow-moving, and cumbersome. Manatees have strong flippers and are skillful swimmers. They live in herds, tend their single calves communally, and eat enormous quantities of *Zostera* and water hyacinths.

The sirenians are related to elephants. There are many more representatives of this group in the fossil record than exist now: 20 genera have been found from the Eocene to the present. *See* EXTINCTION, MAMMAL.

sirocco A hot, dry, sand-laden wind blowing north out of Africa. It is a regular feature in spring in the Mediterranean region. Since moisture is picked up by the wind as it passes the sea, the sirocco weather in Europe is humid. Sand is frequently precipitated with the rain brought by the wind.

In Egypt the same winds, blowing more easterly than northerly, are called the khamsin. *See* WIND.

Skaggerak A body of water in the northeastern part of the North Sea, between Denmark and Norway. It is the connecting waterway between the North Sea and the Kattegat, which leads to the Baltic. The two passages are ice-free year-round and are vital links in the shipping of the Scandinavian countries, the German northern ports, and all of the Baltic. Because of high volumes of river runoff, the salt level of both the Skaggerak and the Kattegat is low and the mean sea level of both is high. *See* BALTIC SEA, NORTH SEA.

skate A flat, cartilaginous fish with distinctive large pectoral fins which extend from nose to tail, giving the fish a diamond shape. The upper surface resembles sharkskin. Some skates have an electric organ in their tail and fend off predators with electric charges. There are 99 genera in this suborder, Rajoidea, of the order Batoidei. The skate has a worldwide range, and a very wide depth range as well. Skates have been found in depths beyond 2,700 m (9,000 feet).

Skates are bottom dwellers. They range in size from the Atlantic little skate (50 cm or 19 inches) to the big (2.5 to 8 m or 8 to 26 feet) Pacific skate. They are taken by both commercial and sport fishermen. *See* ELASMOBRANCHI, RAY.

skua A powerfully built brown seabird of the Arctic and Antarctic. The skua nests in either of the two polar regions. The great skua (*Catharacta skua*) is about 60 cm (24 inches) long. The three jaeger species, of the family Stercorariidae, are sometimes called the lesser Skuas. All skuas prey on other birds, taking eggs and young gulls, penguins, terns, and petrels. The skua, like gulls, will eat garbage and in dire need will eat its own young. *See* BIRD, CARNIVORE, JAEGER.

sloop A sailing vessel with one mast, rigged fore-and-aft. The name is also used today for smaller naval escort vessels.

smell To detect an odor. Sharks in general have been studied in attempts to understand their incredible tracking ability. Most sharks have an acute sense of smell; the olfactory sense of white sharks and hammerheads is most developed. The closely packed chemoreceptors in the shark's head allow it to detect the amino acid serine in a few parts per billion. White sharks may also detect taste, but that is less certain. The only evidence for that conjecture is sea otter carcasses that have been bitten by white sharks and not eaten. One explanation is that sea otters taste foul to sharks, but this is still only a hypothesis. *See* NERVOUS SYSTEM.

snail These animals are the largest class in the phylum *Mollusca*. The very large class includes organisms from the very smallest to large queen conchs, tritons, whelks, drills, cone shells, cowries, and others. Sea snails are present in every oceanic biome. Most are characterized by the single shell, the muscular foot that moves them along, and the specialized feeding structure—the radula. In some, the radula is highly modified and can function as a drill. Thus, some snails are carnivores and prey on other mollusks. While snails on land or in freshwater are almost exclusively herbivorous, sea snails are more omnivorous; they will ingest algae, sponges, tunicates, mollusks, and tiny crustaceans.

Most shelled marine snails have gills, which are absent in the pteropods and nudibranchs. In those animals the respiratory function is carried out by the "wings" in pteropods or the cerata in nudibranchs. They are also hermaphrodites. Some defend themselves using brilliant color (nudibranchs) or toxic substances, and some do both. The cone shells are notorious poisoners, and the chemical structures of their paralytic toxins are being explored in pharmaceutical research.

The current estimate of numbers of species is constantly being revised as more of the deep-sea fauna become known. One recently discovered organism is a hydro-thermal vent dweller. It is an armor-plated snail covered in iron sulfide crystals, iron sulfide being relatively abundant at hydrothermal vents. The armor resembles that on animals from the Cambrian period (540 to 500 million years ago). The snail was found at the base of an Indian Ocean black smoker in 2001 and is not yet named.

The discovery of deep-sea corals has opened several areas of study. Corals and snails form complex relationships: The corals provide living space for numbers of others. The current estimate is that there are about 150,000 species of mollusk, and that may easily rise to about 1 million. The preponderance of these mollusks are snails that live in a variety of food webs with the approximately 5,000 coral species. One of these relationships exists between deep-sea corals near Florida that are home to slit snails. These are a very ancient lineage that has long been thought of as a living fossil. They are large for snails and live in warm waters near the edge of the continental slope from South Carolina to the southern regions of Brazil. These snails are carnivores living on soft corals, sponges, and crinoids. Their different species live in very narrow ecological niches bounded by very small differences in depth. *See* COMMENSAL RELATIONSHIPS, CORAL, GASTROPODA, MOLLUSCA, NUDIBRANCH, OYSTER DRILL, PTEROPODA, VENT COMMUNITIES.

snorkel A breathing tube used by an underwater swimmer in conjunction with a mask in breath-hold or free diving. *See* DIVING, SCUBA.

sofar (SOFAR) Sound-fixing-and-ranging, a triangulation technique based on the mode of travel of sound waves underwater. Sounds created in deep ocean waters (e.g., by an explosion) move great distances with little attenuation. Land-based listening stations in several different locations that receive such sound signals can then pinpoint their origin. *See* NAVIGATION, SONAR, SOUNDING.

soft coral (Alcyonacea) Corals that, unlike the hard or true corals, do not surround themselves with a distinctively shaped calcareous (hard, calcified) skeleton. Instead, they are related to the gorgonians and with them constitute the subclass Octocorallia of the class Anthozoa, phylum Cnidaria, whereas the true corals constitute the subclass Hexacorallia. The names indicate the number of arms about the mouth of the animal and the number of symmetric segments of the body: six or multiples of six in the Hexacorallia and eight or multiples of eight in the soft corals.

Soft corals are the dominant corals in the Indian and Indo-Pacific oceans; the gorgonians are predominant in the Atlantic and eastern Pacific.

Alcyonaceans are usually found in symbiotic relationships with Zooxanthellae—the dinoflagellates. The dinoflagellates are photosynthesizing organisms, and the result of their pairing with soft corals is that the two animals can live in water that has no or little plankton. As long as there is light, the coral does not starve. *See* CHEMICAL DEFENSES, CORAL, DINOFLAGELLATE, GORGONIAN, SYMBIOSIS.

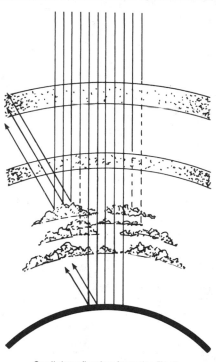

Sunlight reflection from the Earth

soft-shell crab A marine crab whose shell has yet to harden after molting. The edible blue crab, *Callinectes sapidus*, of eastern North America is one example.

solar energy Energy given off by the Sun in the form of electromagnetic radiation. It is the ultimate source of most of the energy on Earth. Solar energy is, overall, fairly constant, and the energy directed at the Earth, known as the solar constant, is 2 g cal/cm^2 minute or about 1.4 kW/m^2. About 35% of this energy is radiated immediately back into space by the top layer of the atmosphere. Only about 50 to 60% of the Sun's energy (i.e., only about 0.8 kW/m^2) reaches Earth either directly or indirectly via reflection from clouds.

The topmost layer of the ocean (top 10 m or 33 feet) absorbs heat and light. Some energy is radiated back into the atmosphere, and some is directed downward to a layer 10 to 20 times deeper (100 to 200 m or 330 to 650 feet). Some solar energy is also used to evaporate surface water. This energy—the latent heat of vaporization—is the driving force for oceanic winds, which in turn generate ocean currents and waves. The ocean currents are, in turn, the means of equilibrating water temperature. Through this mechanism, energy is transported from low latitudes to high ones at the rate of about 10^{15} (a quadrillion) calories per second.

Since there is considerable vertical temperature variation in the Earth's oceans, particularly in areas of high insolation (sunshine) and low latitude, large-scale experiments to produce energy by using thermal gradients in ocean waters (from 10° north latitude to about 10° south latitude) are being carried out. *See* ATMO-

SPHERE, ELECTROMAGNETIC SPECTRUM, INSOLATION, REFLECTION, TIDE, WIND.

solar tide   A partial tide caused by the gravitational pull of the Sun on the Earth. The Earth is closer to the Sun at perihelion because the Earth's orbit is an ellipse. Perihelion occurs in winter in the Northern Hemisphere, while aphelion (the farthest distance of the Earth from the Sun) occurs in summer. (The moon also moves closer to and farther from Earth, respectively, during these two seasons.) Because of both of these movements, the tidal ranges are greater in the Northern Hemisphere during the winter than they are in the summer.

The Chinese, by the second century B.C.E., recognized the correlation of both the Sun and the Moon in determining the tide. *See* APHELION, LUNAR TIDE, PERIHELION, TIDE.

soldier fish (Beryciformes)   These small animals (about 20 cm or 8 inches) live in caves or holes in reefs in warm-temperate waters. They are attractive red or red with black stripes. They are a worldwide group.

sole   An edible flatfish of the family Soleidae. There are about 100 species that live in temperate to tropical waters. The sole lies on the ocean bottom, on its white or blind side. Sole are reasonably large fish, averaging about 50 to 60 cm (1.5 to 2.0 feet) in length. The young sole swim upon hatching and have an eye on each side of the head. By the time the hatchling is a few days old, it has drifted to the bottom, and its left eye (usually) migrates to the right side of its head.

Sole are fished commercially, particularly in the Atlantic and Mediterranean. The Dover sole (*Solea solea*) is one of the most highly prized of these fish. *See* FLATFISH.

Solomon basin   A part of the Coral Sea in the southwestern Pacific Ocean, east of Papua New Guinea and west of the Solomon Islands. It lies at about 150 to 160° east latitude and 5 to 12° south latitude. The region was first approached by Europeans in 1567, when Alvaro de Mendoza entered the area as he traveled west from Peru. However, the Polynesians, Arabs, and Chinese had preceded him. Several other European voyagers also claimed to be the first on the scene. The *Challenger* expedition charted the area.

The northern reaches of the area, south of New Britain, constitute a geologically young geosyncline (depression) that is variously called the New Britain, Bougainville, or Solomons Trench. The deepest point of this depression is the Planet Deep, 9,140 m (29,988 feet) below the surface—a greater distance below sea level than Mount Everest is above it.

Seismic activity is frequent in the Solomon basin, as are undersea landslides. This, and the high rate of island weathering, produce a good deal of terrigenous sediment inshore. The sediment is largely from *Globigerina* in the southeast part of the basin. This part of the Pacific seems to have subsided in the late Tertiary, drowning terraces on some of the larger islands.

The climate in the area of the Solomon basin is hot and humid most of the year. There is a considerable cloud cover that keeps evaporation and therefore salinity down. The temperature of the water is about 27°C (80°F) year-round.

During World War II there were major battles in the Solomons, particularly on Guadalcanal. *See* EXPLORERS AND EXPLORATIONS, GLOBIGERINA, PACIFIC OCEAN, SUBSIDENCE, TRENCH.

solstice   The two days each year when the polar axis is inclined 23°30' toward the sun. The summer solstice in the Northern Hemisphere is on June 21 (except in leap years)—the day on which the noon Sun is directly overhead at the tropic of Cancer, 23°30' North Latitude. In the Southern Hemisphere, this day is the winter solstice.

On December 21 (except in leap years), the noon sun is directly overhead at 23°30' South Latitude—the tropic of Capricorn. This is the summer solstice in the Southern Hemisphere and the winter solstice in the Northern.

On the day of the winter solstice (in either hemisphere) the sun never rises above the horizon for an observer at the 66°30' parallel of latitude (the Polar Circle) or any point closer to the pole than the Polar Circle. See EQUINOX.

solubility The extent to which a substance will dissolve in a solvent. In the context of marine science, the solvent is water.

The extent to which a substance dissolves is a function of the properties of both the solvent and solute (the substance being dissolved). Solubility is determined by the difference in the intermolecular forces between the molecules of the pure solvent, the molecules of the pure solute (the initial state), and the molecules of the solvent and the solute in the solution (the final state).

Almost every substance has a measurable solubility in water, including those that are often thought of as insoluble. The solubility may be expressed either as milligrams per liter (mg/l) or millimoles per liter (mmoles/l). The solubility of any substance is variable, changing with changes in temperature and pressure. Calcium carbonate is a significant example of this: It is sparingly soluble in ocean water and with increasing depth precipitates out of the water as solid particles of salt; however, at still greater depth and pressure it redissolves. See MINERALS.

solution A homogeneous molecular mixture of two or more substances, of which one is usually a solid (the solute) and the other a liquid (the solvent). In the context of the marine environment, water is the solvent—the ingredient present in the greater quantity—and the other ingredient(s) are the solute(s) [the substance(s) dissolved].

Solutes may also be gases such as carbon dioxide, oxygen, or nitrogen, or they may be solids such as sodium chloride and various other salts, or metallic atoms. See MINERALS, SOLUBILITY, WATER.

Somali Current A strong, north-flowing current along the East African coast. It is part of the Indian Ocean gyre. The velocity of the Somali Current averages about 4 km/hour, and has been clocked as high as 7 km/hour. The current flows northeast for about 1,500 km (1,000 miles) and then turns east at the Horn of Africa and continues as the Monsoon Current. The stream exists only in summer. See CURRENT, GYRE, INDIAN OCEAN.

sonar (SONAR) The acronym for Sound Navigation And Ranging, a technique used at sea for detecting and determining the position of underwater objects (e.g., sunken ships, schools of fish) and for determining the depth of the water under a ship. See ECHO SOUNDING, LORAN, SOFAR, SOUNDING.

sounding An old nautical term for the measurement of the depth of a body of water. Many mechanical devices were proposed for this purpose, but none could replace the seaman's weighted line. Bonnycastle first attempted echo sounding in the 1830s, but the best readings were those of Sir James Clark Ross, who used a leaded line when he charted the Arctic in 1839–43.

Matthew Maury corrected depths erroneously reported by earlier surveyors. These were incorrect because it was not realized that the soft sediment through which the sounding weights fell constituted the sea bottom.

The *Challenger* expedition's work was meticulously careful. This group used weighted, thin piano wire for both sounding and sampling. See CHALLENGER, HMS; ECHO SOUNDING; MAURY, MATTHEW FONTAINE; ROSS, JAMES CLARK; SEISMIC SOUNDING; SONAR.

South American Plate One of the larger elements of the Earth's crust. It is approximately continuous with the western and Caribbean coastlines of the South American continent. On the Pacific side it abuts the overriding Nazca Plate. On the Atlantic side, the South American Plate spreads westward from the Mid-Atlantic Ridge. *See* ATLANTIC-TYPE MARGIN, CONTINENT, CRUST, MID-ATLANTIC RIDGE, MID-OCEAN RIDGES, NAZCA PLATE, PLATE TECTONICS.

South China Sea Part of the warm, complex ocean area at the joining of the Indian and Pacific oceans, bordered by the mainland of Asia, the Indonesian islands, Japan, and the Philippines. The area is pinched into a large gulf by Thailand. The South China Sea is a deep basin (over 5,000 m, or 15,000 feet); the circulation of its bottom water is impeded by the Bashi Sill between Taiwan and Luzon and the Karimata Sill between the China and Java seas and the Malacca Strait, the connection to the Indian Ocean. The sea is rich in sediment brought down the Red and Mekong rivers. The shoreline and near-shore undersea landforms of both of these exhibit the typical structures of drowned river valleys. The major climatic feature of the South China Sea is the monsoon. *See* MEKONG RIVER, MONSOON.

Southeastern Pacific Plateau The portion of the Pacific Ocean floor closest to South America. Except for some notable deeps, it is continuous with the Pacific Antarctic Ridge and rises to an average 2,000 to 3,000 m (6,650 to 9,900 feet) above the mean level of the ocean bottom. *See* MID-OCEAN RIDGES, PLATEAU.

Southeast Indian Rise A part of the mid-ocean ridge system that divides the Australian from the Antarctic Plate. *See* MID-OCEAN RIDGES, PLATE TECTONICS.

South Equatorial Current and Countercurrent Two fairly well-defined streams in the Atlantic; the current flows east to west and the countercurrent flows west to east. In the Pacific, however, island groups change the direction of the current, with the result that the continuous gyre is not as apparent. The South Equatorial Current enters the western Pacific and is turned southward through Melanesia into the Coral Sea and the Fiji Basin. *See* CURRENTS, EQUATORIAL CURRENTS, GYRE.

spanker A fore-and-aft sail on the mizzenmast (the mast behind the mainmast) on a sailing ship. *See* MIZZEN, SAIL.

spar A sail-supporting timber. It can be a mast, a yard, or a boom. *See* BOOM, MAST, SAIL, YARD.

spawning Either the production of eggs and sperm, or the production of young. Salmon travel upstream to their original hatchery streams to spawn, to release their eggs or sperm, and then die. Spawning also includes the appearance of horseshoe crabs on Atlantic beaches to lay their eggs, or of turtles that return to their native beaches to lay their eggs. *See* REPRODUCTION.

specific gravity The ratio of the density of a substance to the density of water, at 4°C and 760 torr, the normal atmospheric pressure at sea level.

sperm whale A large blunt-nosed whale (*Physter catodon*) noted for its enormous squared-off head containing a large closed cavity filled with a fluid mixture of spermaceti wax and sperm oil. The sperm whale, or cachalot, has small flippers, a lumpy back, and teeth. It is dark gray to brown. Males have an average length of about 20 m (65 feet); females are smaller. The sperm whale can dive rapidly to depths of 350 to 400 m (1,200 to 1,300 feet) and has been found at even lower depths. The whale lives in small herds in temperate to tropical waters. Its principal food is squid, including giant squid.

Spermaceti is a white, waxy substance obtained from the oil in the head of the sperm whale, while ambergris is a material obtained from the whale's digestive tract. Both substances were long used in the manufacture of cosmetics. Because of these valuable materials and its meat and blubber, the sperm whale was and still is hunted and is found today in ever-decreasing numbers.

Ambergris is a fatty substance that is supposed to protect the sperm whale's interior from the hard beaks of the squid it eats. It is, however, not found regularly in sperm whales and is therefore thought by some to be the result of an intestinal blockage. Its exact origin and function is quite uncertain.

Sperm whales have been found with scars of giant squid suckers on their hides. It was not until the 1970s that enough of a giant squid's body was found in a sperm whale to give credence to the existence of these squid. Up to that time many respectable zoologists believed that the huge squid scars were the result of battles between normal-sized squid and baby whales and that the scars grew in size as the whales grew.

The sperm whale has a pygmy relative, the kogia. It is a dolphinlike creature, 4 m (13 feet) long, and is not hunted. *See* AMBERGRIS, EXTINCTION, SQUID, WHALES.

spicule A small, calcareous, siliceous or protein support structure found in various invertebrates, such as sponges, echinoderms, chitons, and radiolarians. *See* SPONGE.

spider crab A thick-bodied scavenger crab with extremely long legs, of the family Majidae. The spider crab decorates its shell with bits of debris, which it uses as camouflage. The giant crab found near Japan, probably the world's largest anthropod, is a spider crab. Its maximum length, from claw to claw, frequently exceeds 4 m (13 feet). *See* CRAB, SEA MONSTER.

spit A narrow bank or small point of land usually composed of sand. Spits are the creations of longshore currents and are most often perpendicular to the mainland. *See* BEACH, COAST.

Spitsbergen The familiar former name of a group of islands west of northern Norway and in the Arctic. The formal name of the islands is Svalbard. Willem Barents mentions the islands in his log of 1596. The island group was annexed by Norway in 1925. *See* SVALBARD.

sponge A marine animal (phylum Porifera) characterized by a porous structure and a skeleton of interlocking, thornlike fibers that may be calcareous (Calcarea), or siliceous (Hexactinellida), or protein structures (Demospongiae). The last is the largest group of sponges. There are about 9,000 known species of sponge of which 20 or so live in fresh water. Marine sponges range from intertidal ones to hadal ones living at depths 8,500 m (26,000 feet) below sea level. Their color can vary from pale yellow to orange, red,

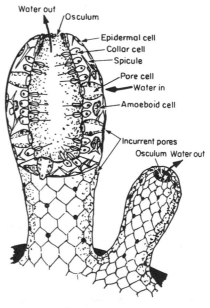

Colonial sponges

brown, green, and black. The glass sponges are found in deep water. These very pretty specimens, such as the Venus's flower baskets, are taken as collectors' items.

Some sponges are home to communities of mantis or snapping shrimp. These communities, like beehives, are all one family. In the case of the shrimp, they are a mated pair and their nonmating sons. The adult sons protect the colony that lives in the interstices of the sponge. The presence of the shrimp does not seem to harm the sponge. This may be a commensal relationship between shrimp and sponge, but that is not determined.

The sponge body is a series of many pores (ostia) and a large, single exit, the osculum. The exterior is usually a layer of flat cells called pinacocytes. Water enters the central cavity of the spongocoel because of the current created by the cells lining it, the choanocytes. The middle of the body wall is the mesoglea, a gelatinous material in which live the amebocytes—cells that move about and engulf and digest bacteria and debris brought in by the water current.

The amebocytes also produce motile egg cells which, when fertilized (most sponges are hermaphroditic), float around until they find a suitable spot for anchorage and a sessile adult life. Sponges also reproduce by budding and can regenerate completely: Sponge tissue can be ground up and passed through a sieve, and the tiny fragments will, if protected, grow into sponges.

As an industry, sponge fishing has all but disappeared.

While most sponges are relatively harmless, long-lived, sessile animals, they do have some chemical defenses. They apparently manufacture some toxic substances that kill bacteria. They can also produce offensive tastes and smells that discourage predators. Some sponges, such as the burrowing sponge, actually attack other organisms. These organisms dig into live coral or mollusk shells. Most burrowing sponges (*Cliona*) attack empty shells or dead coral. *Siphonodictyon* species attack live organisms and use toxic materials to stop coral growth. By this means, they prevent the coral from overgrowing the sponge's osculum. The result is a killed coral patch or a mollusk dead in its shell. *See* CHEMICAL DEFENSES, DENDROCERATIDA, HERMAPHRODITISM, PHARMACEUTICAL RESEARCH, PORIFERA, SPICULE, VENUS'S FLOWER BASKET.

spongin A hardened protein that forms the spicules, or support structures, of some sponges. *See* PORIFERA, SPICULE, SPONGE.

sponge weed A green alga that inhabits warm water. It grows in a branched colony, which may reach 2 m (6.2 feet) long. *See* ALGAE, CHLOROPHYTA, SEA LETTUCE.

spring bloom The sharp rise in the phytoplankton population as longer periods of light coincide with upwellings of nutrient-rich bottom water. *See* ALGAE, PHOTOSYNTHESIS, PHYTOPLANKTON.

spring tide A maximal tide—either the highest high tide or the lowest low tide. Spring tides correlate with the new and full moon of the lunar cycle. *See* LUNAR TIDE, SOLAR TIDE, TIDE.

squalene A hydrocarbon that is an intermediate compound in the biosynthesis of steroid hormones. Sharks accumulate squalene in liver tissue; they have been used extensively as a commercial source of a raw material for the pharmaceutical industry.

squall A wind of rapid onset. The winds in a squall may quickly reach velocities exceeding 70 km/hour (58 mph), and do not last long. They occur most often in connection with violent thunderstorms or cyclones.

In areas where there is cold mountain air that can quickly descend into a warmer valley, squally weather occurs frequently. The fiords of Norway and New Zealand are typical of such areas.

Mountain passes that channel the wind are another. The foehn in the Alps and the chinook in Alaska are examples of warm-air squalls.

A squall line is an approach line of a large system of unstable air. *See* METEOROLOGY, STORM, WIND.

square rigger A vessel bearing square sails that are set into the wind, as opposed to fore-and-aft sails. The latter are most often triangular. *See* CLIPPER, LATEEN, SAIL.

squid A member of the large order Teuthoidea of marine mollusks of the class Cephalopoda. Squid have a long body, a distinct head, and 10 arms, two of which are much longer than the others. These are used as tentacles. All the arms have undersides covered with suction disks. These in turn are controlled by giant neurons, and the whole is coordinated by a sizeable brain. The squid has exceptionally good eye-to-tentacle control. Like the octopus, it is teachable in an aquarium setting.

Squid range in size from 1 cm (0.2 inches) to the giants, 20 m (66 feet) long. The mantle totally covers the "pen," which is all that remains of the molluscan shell in these animals. The animal moves by drawing water into the mantle cavity and then forcibly ejecting it. The movement is a backward jet propulsion. Squid move quickly and capture fish and other prey easily by grasping them with their tentacles and tearing the prey apart with their beaky jaws. Camouflage is produced by ejecting either ink (a blue-black fluid) or a phospholuminescent material.

Squid are either male or female. In mating, the male inserts a sperm packet into the mantle cavity of the female. When the eggs are developed and fertilized, they are released into the water, where they develop into miniature squid. There is no separate juvenile form of this creature.

Squid have long been taken for human food in the Orient and the Mediterranean, and are now a commercial catch in the United States. They are also prey for whales, birds, and bony fish.

A live but dying giant squid caught near Bergen, Norway, in 1982 weighed more than 200 kg (480 pounds) and had tentacles that measured 8 m (25 feet). The blood of such an animal cannot carry enough oxygen in warm or shallow water to keep it alive. Such giant squid are rather sluggish animals and poor swimmers (for squid). *See* ARCHITEUTHIS DUX, MANTLE, MOLLUSK, SEPIA, SPERM WHALE.

starboard The side of a vessel to the right of an observer standing on it and facing the bow (front end). *See* PORT.

starch A high-molecular-weight polymer composed of glucose monomer units. Starch is a carbohydrate storage compound of many plants, particularly terrestrial plants. *See* SUGAR.

starfish This common name is no longer used for creatures of the class Asteroidea of the phylum Echinodermata that are inhabitants of all seas. They are not fish but rather ordinary inhabitants of littoral rock pools. Most are 15–30 cm (6 to 12 inches) in diameter, although some are larger and individuals 65 cm (26 inches) are known. *See* SEA STAR.

stay A rope that supports a mast or spar. Staysails, which are most often triangular, are hung on stays. Frequently the sails themselves are referred to as stays. *See* SAIL.

Stomatopoda The mantis shrimp, an order of the class Decapoda, that resemble the praying mantis, an insect. There are about 200 species of these arthropods living in tropical to warm seas. They are found from shallow water to depths of about 500 m (1,650 feet); however, most are creatures of the littoral. The mantis shrimp are about 4 to 35 cm (2 to 13 inches) long, but most species are in the 15- to 20-cm range. They are frequently brightly colored, with green and brown predominating, but red and blue are also seen.

These slender carnivores wait for prey (small worms, other anthropods) in rock crevices or burrows. They then pounce, in a move very similar to that of the praying mantis. The carapace (relatively hard shell) of this shrimp is most distinct, and serves to intensify its insectlike appearance. *See* DECAPODA, MANTIS SHRIMP, SHRIMP.

stone crab A large, edible, Caribbean crab (*Menippe mercenaria*). It is taken commercially by crabbers, who commonly harvest its large chela, or pinching claw, and leave the crab in the water to regenerate a new one. *See* CRAB.

stony coral A small marine invertebrates of the class Anthozoa, phylum Cnidaria, order Madreporaria or Scleractinia. This is the most familiar and widespread class of coral. There are over 1,000 species, only a few of which are solitary creatures. The range of the stony coral is from the intertidal regions to depths of 6,000 m (19,000 feet), and their color varies from the palest shades of yellow and pink to black. The color depends on the algae that live on the calcareous skeleton of the coral, which is white. The coral polyp is a typical cnidarian. It has a soft tubular body with a mouth at one end that is surrounded by a ring of tentacles. There are six tentacles, or some multiple of six, that trap plankton. The innermost layer of the three-layer body wall harbors the commensal zooxanthellae. These dinoflagellates are the photosynthesizers that supply both themselves and the coral polyp with a constant food supply. The presence of the commensals is important to the coral but obviously not the only food supply. Larger coral polyps can trap enough plankton to feed themselves, and the deep-sea corals that live below the euphotic zone do not harbor zooxanthellae.

The process by which the coral acquires calcium and deposits calcium carbonate is not understood, but it does hinge on the amount of energy (food) the coral has available. Thus, corals in warmer and shallower water, where there is more food available and more insolation for the photosynthesizing protists, grow faster. The branching corals are the fastest-growing, followed by finger corals and head or brain corals. Since the growth is an annual process, it can be measured by sampling a core and retrieving a section of several years' growth. This is a methodology similar to "reading" tree sections—dendrochronology.

Brain, mushroom, star, and staghorn are among the names of the different formations of specific colonial varieties of stony coral. The living coral grows about 0.5 m (20 inches) annually if it has optimum conditions of food, salinity, temperature, oxygen, and exposure to light. Stony corals are attacked by both animals that eat the live polyps and those that erode the reef they build. The grazers include some sea stars, notably the Australian crown-of-thorns, and the snails and fire worms in the Caribbean. A number of worms, bivalve borers, and burrowing sponges such as *Cliona* in the Caribbean are destructive animals that weaken the reef. *See* ANTHOZOA, CORAL, CNIDARIA, FIREWORM, POLYP, ZOOXANTHELLAE.

storm A meteorological disturbance involving a change in atmospheric pressure, high winds, high tides, and precipitation. If this condition moves over a large area and travels in a circular path it is called a cyclonic storm.

Extratropical cyclones start at the Polar Convergence, where there is cold air to the north and warmer air to the south (in the Northern Hemisphere, with the reverse holding true in the Southern Hemisphere). As warm air is lifted by an inrush of cold air, it is given a spin by the Earth's rotation. This is counterclockwise in the Northern Hemisphere and clockwise in the southern. As the air pressure decreases, the wind speed increases. This frequently leads to a warm front developing. If a cold front catches up with this warm front, a semistationary belt of bad weather occurs. This is very typical at the Polar Convergence.

Tropical cyclones develop in an analogous manner, usually between the 10 to 15° parallels of latitude. They are most likely in August and September in the Northern Hemisphere, and in January and February in the Southern. The most commonly observed path of these storms is west. This is a result of the trade winds, which peak and move from east to west. Tropical cyclones are hurricanes in the Atlantic and typhoons in the Pacific or Indian Ocean.

Storms of hurricane (cyclone) intensity are graded according to the Saffir-Simpson Hurricane Scale:

Category 1

- Winds of 74–95 mph;
- Storm surge of 4–5 feet above normal
- Minor pier damage
- Small craft in exposed anchorages torn from moorings

Category 2

- Winds of 96–110 mph
- Storm surge 6–8 feet above normal
- Coastal flooding
- Considerable pier damage
- Marinas flooded

Category 3

- Winds of 111–130 mph
- Storm surge 9–12 feet above normal
- Serious flooding, property destruction
- Flat terrain less than 5 feet above sea level flooded up to 8 miles inland

Category 4

- Winds of 131–155 mph
- Storm surge 13–18 feet above normal
- Flat terrain less than 10 feet above sea level flooded up to 6 miles inland
- Floating debris
- Major beach erosion

Category 5

- Winds greater than 155 mph
- Storm surge greater than 18 feet above normal

See BEAUFORT SCALE, METEOROLOGY, SIROCCO, SQUALL, WIND.

storm surge   A rise in the normal water level in a particular area of the ocean. Storm surges are caused by wind and by low pressure and are augmented by high tides. *See* STORM.

strait   A narrow, natural waterway between two large bodies of water.

stratiography   The study of layers of rock. This is a method for reading the geological history of a particular part of the Earth. *See Appendix: Geologic Timescale.*

stratosphere   Region of the atmosphere that is above the troposphere and tropopause. It is about 17 km (11 miles) above Earth's surface. Unlike the troposphere, the stratosphere becomes warmer in proportion to its distance from the Earth's surface; it is being warmed by the Sun. The air in the stratosphere is very thin, but it does circulate. The circulation is vertical as air enters from the equatorial tropopause, bringing nitrogen oxides, methane, and the chlorofluorocarbons. There is almost no water vapor in this air. It flows toward both poles before descending again. There are also waves in the stratosphere, driven by disturbances in lower air and by the Earth's rotation.

The lower layer of the stratosphere holds an aerosol layer of water with other droplets; some sulfuric acid is also found there. That is largely the product of airborne sulfur dioxide produced mainly by volcanoes. The rest of the sulfur dioxide is the product of burning fossil fuels. Other compounds, the chlorofluorocarbons, nitrogen oxides, and carbon dioxide (which is a refractor of solar radiation), are also present. These compounds all react with the vital ozone and deplete it. Ozone absorbs harmful ultraviolet solar radiation. Most of the ozone in the stratosphere is formed from oxygen reacting to sunlight. *See* ATMOSPHERE, OZONE, POLLUTION.

stromatolites Evidence of ancient bacteria. They are some of the oldest rocks on Earth and, on dissection, reveal that they are the fossilized remains of mat-forming, filament-forming bacteria. The accretions formed by these prokaryotes were reinforced by the inclusion of sediment. The structure grew larger because the bacteria continued to grow on its surface. Stromatolites can be flat, columnar, or round. Since they built up on the floors of very ancient seas, they were the bacterial analog and precursor of coral reefs.

Most of the stromatolites known date from the Late Archean to the Proterozoic eras of Earth's history. They were almost all gone by the late Cambrian period, probably because of the evolution of gastropod grazers. The few stromatolites that exist today are found in extremely saline environments such as Shark Bay in Australia. The water is too salty for snails to live in it. *See Appendix I: Geologic Timescale.*

sturgeon A group of large fish (genus Acipenser) widely distributed in northern Europe and North America. They live in the sea and spawn in rivers, to which they return more than once in a lifetime. Some species of sturgeon have abandoned the trip back to the sea and spend their entire lives in rivers. These fish are related to the paddlefish.

The sturgeon have several features of primitive fish, including an unequally lobed tail, laterally located bony plates (scutes), and no teeth in the adult. They are long-lived, individuals of more than 200 years having been tagged. Since they continue to grow, some individuals are very large.

Sturgeon are prized for both their flesh and for their eggs, which are considered the best caviar. They are therefore chronically overfished. The Caspian sturgeon are in danger of extinction because of serious overfishing and poaching. The collapse of the Soviet Union has exacerbated the problem.

Several anadromous species of sturgeon are listed on this page. The wholly fresh water animals, such as the shovelnose sturgeon, are omitted.

subduction A process that occurs when one edge of a crustal plate dives beneath the edge of a converging plate. Usually the plate subsiding is an oceanic one, and it dives beneath a continent-bearing plate.

COMMON STURGEONS			
Name	*Range*	*Length*	*Weight*
Common *(Acipenser sturio)*	Western Europe	3 m	225 kg
Atlantic *(A. oxyrhynchus)*	Eastern North America	3–4 m	50++ kg
Russian *(A. guldenstati)*	Black and Caspian Sea	3–4 m	250+ kg
Sterlet *(A. ruthenus)*	Black and Caspian Sea	1 m	–
White, Oregon, or Sacramento *(A. transmontanus)*	Western North America	3 m	130 kg
Chinese *(A. sinensis)*	China Sea, Yangtse	2 m	150 kg
Japanese *(A. kikuchi)*	Japan	2 m	130 kg

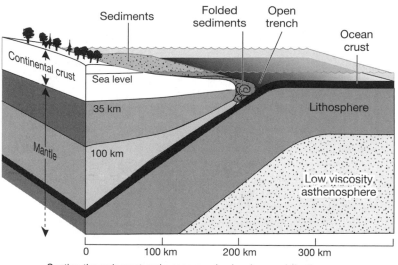

Section through crust and upper mantle showing a subduction zone
where a continental plate is advancing over an oceanic plate

The reason given for the sinking of oceanic plates is that they are denser than the granitic continental plates. *See* CRUST, PLATE TECTONICS, TRENCH.

submarine canyon A valley or system of valleys cut into the ocean floor. These may be of several types. They may be narrow, v-shaped clefts that are the visible edges of faults, broad valleys with flat floors that look like river valleys on land, or cuts in front of delta cones.

There is considerable discussion concerning the origin of the v-shaped canyons. They occur near the mouths, or at the sites of former mouths, of some large river systems. It is therefore very tempting to conclude that they are the drowned seaward ends of those river systems. Sediment is carried along the canyon's length, which can extend along the entire width of a continental shelf, then down the continental slope, and finally into fan-shaped patterns of valleys on the ocean floor.

Other gashes and seams in the ocean floor may be the result of glacial action or volcanic activity. Land and mudslides occur frequently. These change the sea-scape considerably, particularly in areas of high sediment buildup such as river deltas. *See* CONTINENTAL SLOPE, DELTA, OCEAN FLOOR, SEDIMENT.

submersibles Small vessels that can dive and remain submerged for long periods. They are designed to be easily moved and are battery-powered and capable of withstanding the great pressure of ocean depths. They are intended for the detailed study of small areas or the movement of equipment on the sea bottom. The submersible is usually moved from one location to another by tenders or mother ships.

The earliest submersible—the bathysphere—was used by Barton and Beebe. They descended to a depth of 1,000 m near Bermuda in 1934. The next real work on the use of the bathysphere at great depths had to await the 1950s.

Submersibles have been used in the inspection and repair of pipelines, cables, and drilling platforms. They are also used extensively in research. The *Alvin* and its tender *Lulu* have been involved in several very deep sea dives. One explored the East

Pacific Rise and its vent communities, and another, in 1974, the Mid-Atlantic Ridge. The latter was a joint Franco-American project that included the French submersibles *Archimede* and *Cyana*. The wreck site of the R.M.S. *Titanic,* which sank in 1912 in the North Atlantic, was explored by a submersible.

The trend in undersea exploration is now to use more unmanned submersibles. This is certainly safer, because sending people underwater is physiologically very stressful. Humans cannot remain submerged for long periods; they must have an air supply, and the ascent must be very slow to avoid nitrogen saturation. Instruments take up less room and have no requirements for comfort. Almost all current exploration is done with ROVs equipped with cameras that can operate in very low light, sound-recording devices, retrieval devices, and motion sensors.

See ALVIN; *ARCHIMEDE;* BEEBE, CHARLES WILLIAM; EAST PACIFIC RISE; MID-ATLANTIC RIDGE; ROV.

subsidence The dropping of dry land into the sea. This can have several causes. One might be a rise in the mean sea level, another the collapse of an isolated guyot, or coral island. The latter may be the result of volcanic activity. If a volcanic eruption empties an underwater chamber of liquid lava, the resulting cavity may be too fragile to withstand the surrounding water pressure. The collapse of the roof of such a chamber would also collapse any structure above it. *See* BLUE HOLE, VOLCANO.

subsurface current Deepwater current which is most often slower than a surface current. Its direction is not determined by that of the surface flow. *See* ANTARCTIC BOTTOM WATER, DEEP WATER.

Subtropical Convergence A not-quite-continuous line that divides subantarctic from subtropical waters. It hovers at 40 to 50° south latitude, about 10° north of the Antarctic Convergence. The temperature of the water north of the line is about

SUBMERSIBLES			
Name	*Site of Operation*	*Depth*	*Year*
Bathysphere—Barton-Beebe	Bermuda	1,000 m (3,228 feet)	1934
Kuroshio and *Oshoro Maru* (mothership)	Japan	205 m (680 feet)	1951–57
Galeazzi Diving Chamber—Elie Monnier	Mediterranean	300 m (1,000 feet)	1955–56
FNRS-3—Piccard	Dakar, Senegal	4,050 m (13,400 feet)	1954
Trieste—Piccard	Tyrrhennian	3,700 m (12,200 feet)	1956
Diving Saucer and *Calypso* (ship)— J.-Y. Cousteau	Mediterranean	300 m (1,000 feet)	1960
Archimède—French Naval Vessel	Mid-Atlantic Ridge		1961
Cyana—French Naval Vessel	Mid-Atlantic Ridge		
Alvin—U.S. (Woods Hole, Mass.) *Lulu* (tender)	Mid-Atlantic Ridge Pacific	3,000 m (10,000 feet)	1964

10°C warmer in winter than that to the south. The difference in summer is 12 to 18°C. The water that is north of the convergence is also saltier than that just to the south. *See* ANTARCTIC WATER.

sugar A simple chemical compound composed of carbon, hydrogen, and oxygen, also known as a carbohydrate. There are hundreds of known natural sugars; one of the simplest is glucose (see the diagram below), a structure that is one of the possible arrangements (configurations) of the molecular formula $C_6H_{12}O_6$. Glucose is important because it is a fundamental molecule in the metabolism of many sea and land organisms. Glucose can react with other molecules of glucose to form chains of varying length. Maltose, for example, is a disaccharide or a two-glucose unit. Larger chains of glucose (also called polysaccharides) are the glycogens, and still larger polymers are the starches and celluloses. Starch is actually a mixture of two polymers, one a straight-chain polymer of glucose called amylose, the other a branched polymer called amylopectin. Among other polysaccharides are algin (present in brown seaweeds) and chitin (the hard, outer covering, or carapace, of lobsters, crabs, shrimp, and insects).

Ribose is a five-carbon sugar ($C_5H_{10}O_5$) which is a vital part of the adenosine triphosphate (ATP) molecule—the so-called energy-storing compound in many metabolic systems. Ribose is the sugar that is built into the RNA molecule. It forms the polysaccharide skeleton. Ribose with one less –OH group, known as deoxyribose, is a constituent of DNA. *See* CARBOHYDRATES, POLYMER.

Sula Shelf A transverse ridge off the Norwegian coast. A raised region of ocean bottom perpendicular to the Mid-Atlantic Ridge, it is of biological interest because it is relatively warm due to the North Atlantic gyre. This area is known for its corals and other organisms, especially nudibranchs, that are ordinarily found in warmer regions. This is also a region of commercial fishing. The latter is damaging the corals and those organisms associated with it. *See* CORAL, GULF STREAM, MID-ATLANTIC RIDGE, NUDIBRANCH.

Sulawesi Sea Also known as the Celebes Sea; a body of water that separates Sulawesi (Celebes) from the Philippine islands. It is one of several ocean deeps with known volcanic activity. The flat bottom, which has a mean depth of about 4,000 m (13,000 feet), has deep, cold, Pacific bottom water entering the area by spilling over the sill near Mindanao. *See* BOTTOM WATER, PACIFIC OCEAN, SILL.

sulfur A nonmetallic element that occurs in nature either in its free form or combined with oxygen and/or other elements. Sulfur occurs in the sea both in offshore deposits in salt domes, and dissolved in water. The average concentration of sulfur in seawater is about 900 mg/liter. It is the third most available element dissolved in seawater. Sulfur is an essential element in proteins and is therefore present in all organisms. *See* ELEMENT, SULFUR BACTERIA, VENT COMMUNITIES.

sulfur bacteria Filamentous autotrophic (manufacturing their own food) bacteria that derive energy by oxidizing sulfides and building up carbohydrates from carbon dioxide. They occur in several marine habitats from shallow water to hadal depths. Large concentrations of sulfur bacteria were discovered living near hydrothermal vents in the seafloor where sulfurous gases escape from the submerged volcanic sites known as "black smokers." The bacteria thrive near the vents, where the water is at temperatures in excess of 100°C (212°F) and under very high pressure. Their unique metabolic system operates on the basis of the chemical equation:

$$ATP + SO_4{}^{2-} \rightleftarrows APS + energy$$

In ordinary cells, ATP (adenosine triphosphate) is an energy-storing compound. The phosphate bond of ATP breaks to form adenosine diphosphate and releases energy needed for the cell's metabolic functioning:

$$ATP + O_2 \rightleftarrows ADP + energy$$

In the sulfur bacteria, the sulfur compound functions in an analogous manner: The ATP reacts with the sulfur compound instead of oxygen, and produces a sulfur-containing adenosine compound with the release of energy. Since the bacteria utilize the energy in the bonds of these sulfur compounds, they function quite well in anerobic (oxygen-free) conditions. Sulfur bacteria exist wherever there are active hydrothermal vents and, subsequent to their discovery in deep seas, they had been found in other areas, including waters off the California coast near La Jolla, on the East Pacific Rise, and in the Bahamas. In addition to the sulfur bacteria, attendant vent colonies have been located near all black smokers.

Evidence of sulfur bacteria that lived on or near the Earth's surface was published in 2005. Fossilized evidence of pigments found in 1.6 billion-year-old rock in Australia indicate that photosynthesizing sulfur bacteria flourished in an atmosphere that would be toxic to life as we know it. These green and purple organisms needed both the hydrogen sulfide in the air and sunlight. Their habitat was most likely the ocean down to depths of perhaps 20 to 40 m (65–120 feet). Once the possibility of the green and purple sulfur bacteria became known, they were found in other places, notably at black smokers. There, the green sulfur bacteria do indeed photosynthesize at a depth of 2,400 m (7,875 feet). The light they use is very dim and emanates from the hydro-

thermal vent. Such data is significant to the scientists, exobiologists, who believe that life can be found on distant moons of Jupiter and Saturn despite their relative lack of sunlight.

The largest bacterium known is a sulfur bacterium, *Thiomargarita namibiensis*. It is large enough to be visible without magnification—0.0075 m. The organism was found in ocean sediments off the Namibian coast in an area that has a continental shelf and a strong ocean current that parallels it. The globular bacteria secrete sulfur inside their cell walls and store nitrate in a central structure. The latter makes them opaque, and they glow slightly. Since they grow in strands, the assemblage looks like a string of small pearls. These bacteria have melded the nitrate and sulfur cycles while living in nitrate and oxygen-poor, high-sulfide marine sediments. They produce mucus filaments that extend beyond the sediment into the open water, where the nitrates necessary for their metabolism can be captured. *See* AUTOTROPH, EXO-BIOLOGY, EXTREMOPHILE, RESPIRATION, VENT COMMUNITIES.

Sunda Shelf The largest shelf area in the ocean, almost 2 million km² (or 0.6 million square miles) in size. It underlies part of the Java Sea. The entire area, including Malaya, Sumatra, Java, and Kalimantan (Borneo) is a vast, drowned river system. The North Sunda River flowed northeast into the South China Sea from headlands in Malaya and Kalimantan. The South Sunda River also moved from Sumatra east into what is now the Makassar Strait between Borneo and Sulawesi (Celebes). The entire subsided river system is sometimes referred to as the Molengraaff in honor of the principal explorer of the region, G. A. F. Molengraaff, a Dutch geophysicist.

The Sunda Strait, which separates Sumatra from Java, is very shallow and probably the result of very recent seismic activity. Krakatau, in the Sunda Strait, is the most important volcano in the area. *See* CONTINENTAL SHELF, KRAKATAU, SUBSIDENCE.

sunfish The common name of this large sea fish is frequently associated with small freshwater fish related to trout; however, it gets its name from its habit of cruising near the surface, the reflection of the sun causes its body to glow. The *Mola mola* is about 2.5 m (8.5 feet) from mouth to the base of the tail and can weigh up to 2 tons. The food source of *Mola* is jellies. This harmless fish is frequently mistaken for a shark because of its prominent dorsal fin.

surface tension An intermolecular force that acts in liquids to hold the molecules at the surface more closely to each other than to the molecules within the body of the liquid. Surface tension minimizes the area of the surface. It also causes the curvature of the surface of a full glass of water, and the individual droplets or beads of water that form on a greasy surface. Soaps and detergents are called wetting agents, surface-active agents, or surfactants, and they reduce surface tension because they interfere with the intermolecular forces holding the water molecules tightly together at the surface.

Surtsey An island that formed off the coast of Iceland in 1963 as the result of a volcanic eruption. It quickly became both a geological laboratory and a tourist attraction. This island has suffered considerable wind and water erosion since its appearance. There is less of it now than there was five years after its emergence.

suspension feeders Organisms that strain food from the surrounding water. They can range in complexity from sponges and corals to the baleen whales. They live on plants, animals, or debris that drifts past them, or they may seek out small floating or swimming organisms. The feeding mechanisms of suspension feeders vary; some have large mouths that process large quantities of water, from which they extract their food. Such animals are sponges and baleen whales. Others use some organ to screen large quantities of water and pass on the entrapped food to small mouths. Clams, for example, screen the water by using their gills as strainers and pass the food particles they screen out into their relatively small mouths. *See* BALEEN, PORIFERA.

Svalbard The Arctic archipelago and part of Norway that is also known as Spitsbergen consists of four main islands and about 150 smaller ones, lying between 76.50 and 89° north latitude and between 10 and 34° east longitude. About 40% of the landmass is covered with glaciers. The southernmost island, Bjørnøya (Bear Island), was probably sighted by Jakob Heemskerck, a Dutch whaler, in 1596. Fridtjof Nansen used the islands as the base for his polar explorations. Svalbard is rich in coal deposits. It is an "arctic desert," with 200–300 mm (8–12 inches) of rainfall annually. It does, however, have fog, particularly in the summer when the average high temperature is about 6–8°C (43–47°F). A major industry is tourism—people come to observe the arctic midsummer light.

swell A long and massive wave or succession of long-period waves moving away from the cause of their generation. *See* WAVE.

swim bladder An organ regulating the buoyancy of most teleost (bony) fish. It originates as part of the digestive system. This gas-filled sac is used by the fish to equilibrate their internal pressure with that in the surrounding water as they rise or sink.

Some fish use the swim bladder as a sound-generating structure. In some primitive species it is part of the respiratory system and acts as an air (or oxygen) reservoir.

Fish that live very deep in the sea do not have swim bladders, nor do the cartilaginous fish. These organisms also experience great changes in the pressure of their surrounding environment, but the mechanism by which they are able to adjust to

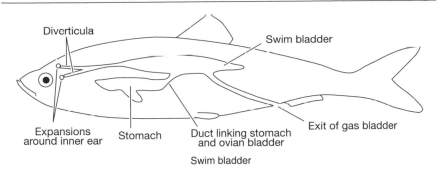

Diverticula

Swim bladder

Expansions
around inner ear

Stomach

Duct linking stomach
and ovian bladder

Exit of gas bladder

Swim bladder

these pressure changes safely is still largely unknown. *See* FISH, GRUNT.

swordfish A large oceanic game and food fish noted for its long, flat, sharp "sword," which projects forward for at least 60 cm (25 inches) from its head and is used to slash prey. Swordfish are impressive for their size: The average adult is longer than 4 meters (13 feet) and weighs between 400 and 500 kg (900 to 1,100 pounds).

These warm-water inhabitants are of the family Xiphidae in the order Perciformes. The most common species is *Xiphias gladius*. They are dark blue to blue-black on their upper surface and silver on the underside. Their body is long, scaleless, and slender; swordfish have neither pelvic fins nor teeth. The swordfish feeds mostly on other fish, particularly schooling fish. *See* FISH.

symbiosis An interrelationship between two organisms. *Mutualism* is a term applied to the relationship when both of the organisms derive some benefit from the arrangement. An example of mutualism is the presence of the tiny parrot fish that live in sea anemones and lure other fish into the mass of tentacles of the anemone, after which the parrot fish dine on the crumbs left by the anemone.

A more usual arrangement is a commensal one, in which organisms live together with only one of them having some advantages, although the arrangement is not detrimental to the other. Thus, the fungal colonies that reside on spider crabs camouflage the crabs, to their advantage, but the only evident benefit to the fungi is that in being transported from one locale to another, they may have a better chance at a food supply. Parasitism occurs when one species benefits at the expense of the other without giving anything in return. Many fish have both internal and external parasites. *See* COMMENSAL RELATIONSHIPS.

Synechococcus A tiny, recently discovered bacterial genus that is an important part of the food web of the open ocean. These photosynthesizing bacteria are primary producers on a large scale and one of the most numerous genomes. They are quite small, but there are so many of these tiny organisms that they constitute a significant part of the carbon-sequestering population. The are motile rather than drift organisms; although they lack a flagellum, they propel themselves through the water. These bacteria need manganese and calcium in particular and also a number of other metals, notably, iron. All are present in seawater. Their ability to sequester iron makes these organisms vital in the global iron cycle. *See* CYANOBACTERIA, FOOD WEBS, PHYTOPLANKTON.

T

tablemount *See* CORAL, GUYOT, ISLAND, SEAMOUNT.

tack A rope for controlling certain sails, or the sails themselves. When used as a verb, *tack* means resetting the sails in order to take advantage of a wind coming at an angle to the side of the ship (from offside). The direction in which the ship then moves, and the distance traversed while it is on a tacking course, is also called the tack. By tacking, or setting a series of zigzag courses, it is possible to sail against a prevailing wind. *See* SAIL.

tarpon A large fish found in warm-temperate to tropical waters. It lives inshore and can tolerate a wide variety of salt concentration. Juveniles can live in fresh water. All tarpons have characteristically large scales. They are the target of sport fishing and are diminishing in number. Several conservation groups are urging protective status.

Tasman, Abel Janszoon (1603?–1659) A Dutch explorer who sailed for the Dutch East India Company and after whom Tasmania (an island south of Australia) and the Tasman Sea are named. Van Diemen, the governor-general of the Dutch East Indies, sent Tasman to explore Australia in 1642. Several Dutch explorers had previously stopped at various points on the Australian coast, but Tasman circumnavigated the entire island and also explored Tasmania, which he named Van Diemen's Land for his patron. He sighted the north island of New Zealand—he called it Staten Landt—and thought it was part of South America. He mapped the Tongas, part of

the Fiji archipelago, part of the New Hebrides and Solomon islands, and New Guinea. Tasman then returned to Batavia (now Jakarta) after almost a year at sea. On a later expedition (1644) he mapped the Gulf of Carpentaria on the northern coast of Australia, south of New Guinea.

Tasman successfully led a commercial fleet from Batavia to Siam (now Thailand) in 1647. Later he was involved in a military action against the Spanish in the Philippines. He finally retired, rich and famous, to Batavia. *See* AUSTRALIA.

Tasman Sea An arm of the southwestern Pacific Ocean between Australia and New Zealand. It is named for Abel Tasman, the Dutch navigator who crossed it in 1642. He had previously landed on what he called Van Diemen's Land (Tasmania), believing it to be the Australian mainland.

The Tasman Sea roughly corresponds to a deep basin, with depths in excess of 4,000 m, that is known as the Tasman, or Thomson, or East Australian Basin. Its southern reach is Macquarie Island.

The East Australian Current dominates the circulation of surface water of the Tasman Sea. The water has a fairly high salt content near the surface and correspondingly less salt with increasing depth. The deep water contains traces of North Atlantic deep water.

The bottom sediments of the sea are calcereous, with considerable red clay. The Tasman Basin is yet another area assumed by some to have continental origin. According to one theory proposed around 1900 there was once a continent called Tasmantis, a Pacific Ocean analog of Atlantis. But while there are rises in the sea bottom such as the Lord Howe and Chatham, and many seamounts, the bottom of the Tasman Sea is oceanic in nature rather than continental. *See* AUSTRALIA, AUSTRALIAN PLATE, BOTTOM WATER, EVOLUTION OF OCEANS, LORD HOWE RISE, PACIFIC OCEAN.

taxonomy The systematic arrangement and naming of organisms. Although Aristotle (384–322 B.C.E.) and his pupil Theophrastus, (c. 371–287 B.C.E.) did botanical classifications, and there was no explanation for the variation in obviously related living things. Vesalius (1514–64) and then John Ray (1627–1705), an English botanist, linked organisms by a species concept. The Linnean (1707–78) innovation was a binomial nomenclature based on Aristotelean logic. The impact of this new classification scheme on Darwin's (1809–82) work was incredible. It meant that he could explain the rise of new variations that were still related but had eventually diverged. The families of finches were all still finches but had undergone modifications in their morphology and feeding habits—the various species had changed, but the genus remained constant.

The scheme still used to classify organisms—both those alive today and fossil finds—is a systematic designation of the organism in question by kingdom, phylum, class, order, family, genus, and species. All of these categories can have super- and subvariations. It is important to remember that these designations are man-made and are therefore subject to change or addition.

Present methods of classification also examine the protein sequences in organisms, since related species have certain biochemical features in common. Related organisms have similar enzyme systems and similar proteins that have the same amino-acid sequences. Thus, differences in the hemoglobins from one mammal to another can be minute, and the enzyme systems in humans and great apes are very similar, while those in humans and mice are less so. These changes occur with taxonomic divergence.

Cladistics is a method of classifying organisms according the evolutionary history and placing them on an evolutionary tree called a cladogram. The basis of this scheme is as follows:

a. groups of organisms that have an ancestor in common;

b. at each junction, there are two branching lines; and

c. evolution results in modification over time.

There is much argument about this approach and how to apply it. It does not reflect time and, in evolution, that is a major factor.

Collection of information about new organisms or relationships between known ones has always been problem. Establishment of online taxonomic schemes is aiding in the classification and tracking of species. One group, composed of the International Union of Biological Sciences (IUBS), the International Council for Science (ICSU), the Committee on Data for Science and Technology (CODATA), and the International Union of Microbiological Societies (IUMS), has collaborated to build a taxonomic database, called Species 2000.

CLASSIFICATION OF SMALL SHARKS	
Taxon	Dogfish
Kingdom	Animalia
Phylum	Chordata
Subphylum	Craniata
Class	Pisces
Subclass	Chondrichthyes
Order	Neoselachii
Family	Scyliorhinidae
Genus	*Scyliorhinus*
Species	*caniculus*

Example of a hierarchical classification scheme.

See ARISTOTLE; LINNAEUS, CAROLUS; RONDELET, GUILLAUME.

tectonics *See* OCEAN FLOOR, PLATE TECTONICS.

teeth Modifications of the dermal armor plates of primitive fish. The teeth in some fish and reptiles are the same three-layer structures of pulp, dentine, and enamel that are associated with mammalian teeth. They are, however, homodent teeth, meaning that they are all the same and are replaced throughout the life of the animal if they become worn away or are shed. Most mammals have heterodont teeth, consisting of incisors, canines, and molars in a number sequence that is species-specific. Some teeth are highly specialized: The saw of the sawfish, for example, has the only known set of external but real teeth. The tusks of the walrus are for display and for the defense of the territory of a large bull. The common notion that the walrus tusk is a digging tool is untrue. *See* FISH, NARWHAL, WALRUS.

teleost Any of the bony fish. They are also the most advanced in terms of evolution and the longest group of fish. Almost all sport, commercial, and ornamental fish are in this classification, which is not a taxonomic one. The one obvious uniform characteristic of the teleost fish, besides their calcified internal skeleton, is their tail, in which the upper and lower halves are of about equal size, whereas in the cartilaginous fish the tail has two lobes of unequal size.

The range of the teleosts is worldwide and goes from habitats in tidal pools to hadal depths. The size of adult telosts may range from 1 to 2 cm (less than 0.5 inches) to longer than 3 m (10 feet). Their reproductive systems vary from species that have distinct male and female members to others that consist of functional hermaphrodites (some of which can self-fertilize), while still others can change sex during their adult lifespan. Some teleosts are ovoviviparous, others viviparous. Most species fertilize eggs in the water, and these hatch to produce planktonic young.

Some teleosts move in schools, and others are bottom-dwellers; most are efficient swimmers because of their torpedo-shaped body and powerful tail. *See* FISH, HERMAPHRODISM, OVOVIVIPARITY.

temperature The degree of hotness or coldness of a body or an environment. Temperature was one of the properties of seawater that Robert Boyle—one of the original members of the Royal Society of London—thought should be studied. He made this suggestion as a part of a master plan for the advancement of science in 1666. Robert Hooke, a friend and later rival of Isaac Newton, designed several devices for the purpose of exploring Boyle's scientific program. Among these was a self-registering thermometer.

Marcet, in 1819, reported on the differences in temperature between bodies of water at the same degree of latitude and on the difference in density of seawater compared with fresh water.

Later in the 19th century, despite Humboldt's reports of a cold South American current in equatorial regions, the idea of a uniform temperature in the depths of the sea was accepted as truth. It was held to be almost axiomatic that deeps in the ocean were all at 4°C. Differences in the temperature of different bodies of water had to be disproved repeatedly by Lenz and others.

The temperature of any marine area depends on surface radiation, evaporation, and the movement of water. These are then subdivided according to seasonal variations and tidal oscillations. The temperature of the oceans is monitored constantly by a number of remote-sensing satellites operated by NOAA and SeaWiFS. These systems record small changes in the temperature of ocean water that is then translated into colored images. *See* CURRENTS; HUMBOLDT, FRIEDRICH HEINRICH ALEXANDER; INSOLATION; WEATHER; *specific bodies of water.*

temperature of animals Animals that maintain a constant body temperature are called homoiotherms (homeotherms) while animals that adjust their temperature to the ambient temperature are called poikilotherms. The designation of homoiotherm or poikilotherm is somewhat arbitrary.

Both forms of animals have a variety of mechanisms for maintaining a temperature that allows them to function at maximum efficiency and to achieve a balance between calorie requirements and temperature. The colder it is, the greater the need for food but the smaller the chance of finding it. Also as poikilotherms cool, they become sluggish and their hunting ability diminishes. The solution to maintaining a constant temperature for many organisms, both homoiotherms and poikilotherms, is hibernation or migration. Temperature is a factor in reptilian reproduction. Mammals and fish face the same temperature-control problem because some of their activities require a particular temperature. For example, swordfish have large fat deposits in and around their brains and eyes, and it is thought that this insulation makes it possible for them to dive deeply into much colder water and continue to function. *See* HOMOIOTHERM, POIKILOTHERM.

tentacle The arm of a mollusk such as a squid or octopus, a feeding structure surrounding the mouth of an invertebrate such as an anemone, or a group of organisms in a colonial cnidarian.

teratogenesis The damage to offspring before birth or reproductive failure. The case of ospreys and other fishing birds that are at the top of a food chain, which had accumulated a toxic concentration of chlorinated hydrocarbons, is well documented. This case was the basis of the ban on DDT as a pesticide. The persistence of DDT and related compounds in ocean waters has been investigated as a possible cause of reproductive failure and population decrease in oceanic mammals. *See* POLLUTION.

teredo The shipworm, a highly modified bivalve (mollusk) of the Teredinidae family that bores into wood. The free-swimming larvae settle in ships' bottoms or wharf pilings and continue to enlarge the burrows that they channel for their entire adult lives. Adult teredos may be 30 cm (1 foot) long or more. In these animals the characteristic shell of the bivalve is reduced to two digging scoops, which they bury in the wood they bore through. Only the very long siphons protrude from the burrow opening as evidence of the animal's presence. *See* MOLLUSCA.

tern Also known as the sea swallow, any of a group of about 40 species of gull-like birds of the order Charadriiformes, family Laridae, subfamily Sterninae, found almost everywhere in the world.

Terns are related to gulls, look similar, and also have webbed feet. They are generally thinner than gulls, however, and have longer, narrower wings, shorter legs, and forked tails. These birds are agile in the air and frequently migrate over very long distances to breed. Terns feed mostly on small fish and crustaceans, diving into the water while in flight. The lifespan of terns varies, but most are long-lived, and the arctic tern normally lives more than 20 years. Like many other long-lived animals, terns tend to be monogamous. The sound made by the tern is a characteristic "heh, heh," while gulls make a cackling sound. *See* BIRD, GULL.

Tertiary period The period of the Cenozoic era that followed the Cretaceous period and preceded the Quarternary. It started about 65 million years ago and lasted about 55 million years.

The Tertiary was the time when land bridges became submerged, cutting the connections between Siberia and North America, and those between the Indonesian islands. The Himalayas, Alps, Rockies, and Atlas mountains formed in the Tertiary.

The end of the Cretaceous was marked by the extinction of many species in a relatively short time. This left unfilled ecological niches to be filled by other species: snails, bony fish, and mollusks all proliferated. The climate at the start of this period was warm,

SOME REPRESENTATIVE SPECIES OF TERNS			
Name	Habitat	Color	Length
Atlantic fairy or white noddy (*Gygis alba*) or (*Anous albus*)	Tropical islands	White	30 cm
Pacific fairy (*Sterna nereis*)	Australia, New Zealand	White	20 cm
Inca (*Larosterna inca*)	Peru	Gray, white	10 cm
Black (*Sterna nigra* or *Chlidonias niger*) [this bird will fly from North to South America, annually]	Europe, Asia in summer; Africa in winter	Black head, white underside, gray wings	25 cm
Common (*S. hirundo*)	North temperate to south coastal zone	White, black head, red legs and bill	35 cm
Arctic (*S. paradisea*) [migrates longest distance]	North temperate to south temp.	White and black body, red legs and gray wings	38 cm
Sooty (*S. fuscata*)	Island colonies	White-fronted black body, gray wings	20 cm
Little (*S. albifrons*)	Europe, Asia, Australia *not* South America	White body with black wings	20 cm

then underwent a series of oscillations, and finally turned sharply colder. *See* CRETACEOUS PERIOD, QUARTERNARY PERIOD.

test The rigid covering of a number of widely divergent organisms. It is a product of the secretion of the animal that lives in it; it may also include compacted nonliving material such as sand grains. Among animals that live within tests as opposed to shells are some members of the Sarcodina, the sea urchins, and the tunicates. *See* SARCODINA, SEA URCHIN, TUNICATA.

Tethys Sea The great east-to-west waterway of the Mesozoic era that separated Laurasia from Gondwanaland. The shift of continental plates produced deformations and the uplifting of seabed sediments into the European Alps, the Caucasus, and the Himalayas. The remnants of the Tethys Sea are the landlocked Caspian and Aral seas. These were once linked with the Mediterranean, which also is a remnant of the once-larger Tethys. *See* ARAL, CASPIAN, GONDWANALAND, MEDITERRANEAN, MESOZOIC ERA, PANGAEA.

Tetraodontiformes An order of teleost (bony) fish that include boxfish, puffers, and trigger fish. These fish are also grouped as the Tetrodotoxia. They are characterized by two pairs of large protruding frontal teeth, tiny gills frequently missing opercula, a single posterior dorsal fin, and no pelvic fins. As a group, these fish contain a potent neurotoxin concentrated in the liver and gonads, although it is present in small amounts in other organs. The neurotoxin (tetrodotoxin) depresses heartbeat and can lead to coma and death if ingested by humans. It is now of medical interest as a lead compound in the development of pharmaceuticals. *See* CHEMICAL DEFENSES, PUFFER.

Thera Also known as Santorini, a Cycladean (Greek) island north of Crete. It is all that remains of a major volcanic island that exploded around 1500 B.C.E. The island is half of a volcanic cone composed of lava and pumice (a very light, porous stone resulting from frothy lava). Archaeological work has uncovered a Bronze Age culture on Thera with strong ties to that of Crete. This island is thought to be the basis of the Atlantis stories of Greek mythology and the biblical account of the drying-up of the bed of the Red Sea in Exodus. *See* CYCLADES, TSUNAMI, VOLCANO.

thermal equator An irregular and moving line that links temperature maxima. It is most often between 12° north latitude and 12° south latitude. It bends northward or southward over continents because land and the air above land heat up faster than does water or the air above water. Since there is more landmass in the Northern Hemisphere than in the Southern, the thermal equator is more northerly than southerly. *See* TEMPERATURE.

thermal spring A spring that produces water that has a higher temperature than the temperature of the surroundings. *See* VENT COMMUNITIES.

thermocline A temperature differential in the water. A decrease in water temperature with increasing depth is normal and expected. A thermocline exists where the decrease suddenly becomes sharper than in neighboring areas. This discontinuity at a specific point is either permanent or seasonal.

The permanent thermocline is the result of the descent of uniformly cold polar water. This moves toward the equator, sinking below warmer layers and preserving its own character for thousands of kilometers.

The seasonal thermocline is the result of the rapid warming of surface water in summer. The underlying water stays colder and does not warm up until later in the season. In some temperate regions the thermocline disappears in autumn. This is why coastal water is more agreeable for

swimming in the later summer and early autumn. *See* ANTARCTIC WATER.

Thomson, Charles Wyville (1830–1882) An English naturalist; organizer and leader of the *Challenger* expedition, the first oceanographic research voyage. Thomson's early interest in natural history led him away from a medical career at the University of Edinburgh to a lectureship in botany. He was first professor of natural history, then of zoology at several British and Irish universities.

By the 1850s Thomson was an established marine biologist, having studied fossil crinoids, trilobites, and coelenterates. On seeing Michael Sars's collection of organisms that were bagged at depths below the 300-fathom mark (500 m or 1,800 feet) in the Forbes's so-called "azoic zone," he was led to a sequence of dredging expeditions to see what really did live at great depths. This culminated in the organization of the *Challenger* expedition.

In 1868 Thomson and W. B. Carpenter, a vice president of the Royal Society, obtained Admiralty support for the project to dredge in the North Atlantic. They found an interesting collection of fauna. In addition, they observed that the temperature of the water at similar depths varied from one location to another. This led to an exhaustive analysis of the composition of the seawater. An extension of these projects in the North Atlantic led to an expedition of global scope.

The organization, writing, and editing of the results of the *Challenger* voyage took years. Thomson, the originator of this seminal project, was knighted in 1876 for his immense contribution to science. Unhappily, he did not live long enough to do more than just begin to arrange the *Challenger* notebooks. They were published by John Murray. *See* CHALLENGER, HMS; MURRAY, JOHN; SARS, MICHAEL.

thorium Isotopes that are used in the determination of the rate of sinking of organic matter. The decay of short-lived isotopes of thorium is an indication of the rate of sinking of organic materials. Longer-lived isotopes (Th^{230}) are also used to measure the rate of return to solution of materials that were used in tissue building in some organisms. The rate of return to solution is an indication of the turnover rate of materials in the oceanic biome.

thresher shark A shark of the genus *Alopias,* with a typically very long, scythelike tail that is about half the length of the animal and is the origin of its common name. *A. vulpinus* is a representative species of thresher. It is about 6 m (20 feet) long.

Thresher sharks are found in tropical and warm temperate waters of the Atlantic and eastern Pacific Oceans. Their principal food is cephalopods and schooling fish. It is thought that they use their long tail as a whip with which to stun prey, but not much is known about their feeding behavior. These sharks are not considered maneaters. *See* SHARK.

Thysanoessa A genus of krill. These are one of the several genera that are the primary food of several large polar mammals. *See* KRILL.

tidal bore *See* RIVER BORE.

tidal currents Local ocean currents that are important in the shallow waters off continental and island shores. The direction of these currents changes throughout the tidal cycle, making a clockwise gyre in the Northern Hemisphere and a counterclockwise one in the Southern. In open ocean the tidal currents become attenuated to the point of being undetected. *See* CURRENT.

tidal flat The area of shoreline above the normal high water mark that is a true beach. *See* BEACH.

tidal period The time between one high water mark and the next. *See* TIDE.

tidal range The difference in water height between high tide and the next low tide. *See* TIDE.

tide The regular movement of water vertically with respect to some point on land. Tides are classified as either diurnal, with one high tide occurring daily, or semidiurnal, with two high tides each day. There are measurable tides not only in the ocean but on enclosed bodies of water such as bays and large lakes.

The direct cause of the tide is the acceleration, as described by Newtonian mechanics, of the Earth as a response to the gravitational pull of the Sun and the Moon. Thus, the movement of water on Earth is a function of these celestial bodies, of the depth and shape of the coast, and of the Coriolis effect.

Tides are mapped for given areas; a chart that shows all points at which high tide occurs simultaneously is called a *cotidal* chart; another variety of chart connects points at which the high tide comes up to the same height above sea level. At some point in a body of water the lines of radiating tides—the cotidal lines—meet. This is the amphidromic or "notide" point. There may be more than one such point in a given area, particularly if that area is large, such as the Caribbean Sea.

The tide at any given location can be diurnal, semidiurnal, or unequal. The last means that the two high tides are not at the same height on any day. This unequal state is an intermediate state between diurnal and semidiurnal. The prediction of the time and height of a tide for a particular place can be done empirically on the basis of careful records of past tidal performance. Thus, the Moon has a specific effect on the tide, as does the Sun. This effect varies over the course of the lunar month. The greatest effect results when the Sun and Moon reinforce each other's gravitational effect. This occurs when the Sun and Moon are in line, on the same or opposite sides of the Earth, and the resulting tides are called *spring tides.* At the quadrature of the Moon, the Sun and Moon are out of alignment, the position of the Sun-Earth line being perpendicular to the Earth-Moon line. The result is called *neap tide.* Thus, there is a cycle each month ranging from high high tides (spring) to low high tides (neap). Any study of long-term tidal effects must take into account the variation in the Sun-Moon relationship over a 19-year period. This is called a saronic cycle.

The tide at any given point on the shore can therefore be predicted on the basis of past tidal activity. This point is extremely localized: The tide will exhibit variation from one spot to another on the shore of a single bay. Complicating factors are the movement with time of the amphidromic point, and inertia. It takes energy to move water, and as a result there is a distinct lag between the lunar high, which is the crossing of the meridian of the Moon, and high tide. This is the *intertidal interval,* which also depends on the arrangement of land and water bodies.

It is now possible to model tides with some success. Computer models are used for this and are a navigational asset, as are the older and still extremely useful tide tables. A good model must take into account all factors, and appropriately weigh each one. The calculations involved are complex and impossible without computers. *See* AMPHIDROMIC POINT, CORIOLIS EFFECT, DIURNAL TIDE, LUNAR TIDE, MOON, SEMIDIURNAL TIDE, SOLAR TIDE, SPRING TIDE.

tiger shark A large, banded, gray, brown, or black shark, *Galeocerdo cuvieri,* of the family Carcharhinidae found mainly in the Indian Ocean. These animals are about 5.5 m (18 feet) long. They live in warm tropical seas and near coasts. They are omnivorous carrion eaters, feeding on dead and dying animals, and are also known to tear fishing nets to get at the catch inside. While they are reputed to be maneaters, the usual diet of tiger sharks is other sharks, fish, birds, turtles, mollusks, and garbage. *See* SHARK.

Tigris and Euphrates Rivers Two major rivers of the Middle East (known in Arabic as Nahr Dijlah and Nahr al Furat, respectively). Both the Tigris and Euphra-

tes rise in Turkey and flow southeast into the Persian Gulf. Almost all of Iraq consists of the Mesopotamian Plain built of the alluvium carried downstream by these two rivers. Baghdad is on the Tigris, the easternmost of the two. The area between the two rivers has been intensively irrigated for milliennia. In this extremely hot, dry climate the rivers are the only water sources, and every civilization that has lived in the area has used them for irrigation. The canals built outward from the Tigris and Euphrates—the distribution system for the precious water—were the engineering triumphs of their day, and a good deal of the governmental bureaucracy dealt with their maintenance. They were finally abandoned after the devastation resulting from the Mongol invasion. *See* DELTA, INDIAN OCEAN.

tilefish A large, temperate-water marine food fish, *Lopholatilus chamaeleonticeps,* of the family Branchiostegidae, order Perciformes. Tilefish live off North America on the continental slope, where the continental shelf drops down toward the ocean floor. They are large, being over 1 m (3.3 feet) in length and weighing 15 kg (3 pounds) or more.

These fish are blue-green above with white undersides; their back has irregular yellow spots, while the head is frequently red or has red spots. Tilefish are burrowers: some live in solitary burrows, others in communal burrows that may be 3 m (10 feet) long. They hollow out areas in the continental slope and live in the hollows. There has been a noticeable change in the seascape of the continental edge as a result of tilefish activity. Their burrow-making has caused several collapses of the slope structure. *See* FISH.

Tintinnida An order of ciliated protozoans that are conical or trumped-shaped. Most are marine and are part of the plankton that supports the food web in the oceans. These tiny creatures (40–200 micrometers) have cilia surrounding an oral cavity and a "shell" (called a lorica) that is more like a tough membrane, with harder, grainy material embedded in it. The organism can retract completely into the lorica. These shells are so distinctive that they are used to differentiate among the hundreds of known genera. The organisms, found mostly in the Pacific, have a long history. About 12 genera have been found in limestones dating to the Jurassic period.

tissue A functionally specialized group of cells of an organism. Muscle tissue, for example, performs the function of moving an animal's body parts, while epidermal (skin) tissue protects the organism. In highly developed organisms, tissues are assembled into organs.

tombolo A sand or gravel bar connecting an island to another island or to the mainland. When larger, it may be a spit or a causeway. *See* BARRIER ISLAND.

Tonga trench A deep depression in the floor of the southwestern Pacific Ocean. Its average depth is almost 11,000 m (33,000 feet), and it is about 1,400 km (900 miles) long. This fairly narrow, steep-walled break in the Earth's crust lies south-southwest of Samoa. A belt of volcanic and seismic activity is west of the trench.

The Tonga trench and the Kermadec trench, with which it is almost continuous, form a line that, together with the Great Alpine Fault of New Zealand, creates a long slip-strike fault that is analogous to the fault in the Caribbean. *See* CRUST, KERMADEC TRENCH, PLATE TECTONICS, PUERTO RICO TRENCH, TRENCH.

toothed whales Whales of the suborder Odontoceti, with many conical teeth. This group is divided into five orders, commonly known as sperm whales, beaked whales, belugas and narwhals, dolphins and porpoises, and the freshwater group, river dolphins. There are about 65 species of Odontoceti. The toothed whales differ from the baleen whales, which feed by straining tiny organisms from the water. *See* WHALES.

torpedo fish A group of cartilaginous fish of the class Chondrichthyes, subclass Elasmobranchii. The torpedo fish are relatives of the sharks and rays. They are flattened fish whose habitat ranges from shallows to great depths, and from tropical to temperate waters. Their outstanding characteristic is the presence of electric organs on or near their head. These light-emitting fish form significant populations in abyssal ecosystems. They range in size from about 25 to 180 cm (10 to 70 inches). These sluggish detritus feeders have been known since antiquity. They appeared in the medicinal tracts of ancient Greece and Rome. Torpedo fish were to deliver electric shocks in an effort to cure a variety of ailments, particularly gout and migraine headaches. *See* ELASMOBRANCHII, LIGHT ORGANS.

toxin A poisonous substance produced by a spectrum of organisms, some of them marine. Bacteria, dinoflagellates, and algae produce poisonous chemicals, as do fungi (mycotoxins) and animals. Some toxins will pass through one species (the intermediary) to another without affecting the former. For example, mollusks may not be poisoned by the "red tides," but fish, birds, and humans who feed on the tainted shellfish will become ill. The distribution of the toxin-bearing dinoflagellates in any area is a function of the circulation of the water in that locale. One side of a bay may be relatively clear, and the fish and bivalves there are safe to eat, while other areas of the same bay may be deadly.

Animals that produce highly venomous toxins are usually tropical. They include lion fish, stone fish, and puffers. The most commonly reported marine toxin disease is ciguatera, which results from eating contaminated marine fish, grouper, barracuda, or snapper. Paralytic shellfish poisoning is associated with ingestion of contaminated shellfish that have ingested the appropriate dinoflagellates. For most of these toxins there are no antidotes.

There are several groups of compounds that cause neurologic effects in the organisms that are primary or secondary consumers. Some are produced by fish, notably the tetrodotoxins, but most are the products of algae or diatoms. Domoic acid is a neurotoxin produced by protists that acts on the glutamate receptors in the brain. The compound is produced by several red algae, notably *Pseudo-nitzchia* in response to stress. It is also produced by *Chondria armata*, a Japanese red alga known as domoi and an ingredient in traditional medicine. It is used to deworm mammals. Algae are ingested by bivalves or crustaceans, and the domic acid levels in those animals is concentrated. Therefore, after the invertebrates are eaten by mammals or birds, toxic effects are noted in the mammals or birds. This is a classic example of toxicity moving through a food chain.

The coasts of North America have more toxic algal species and more algal blooms than ever before. This is may be the result of more polluted waters or better testing methods or both. This increase in toxins in local waters has profound effects on human and animal health and on the commercial shellfish industry. Affected mammals such as dolphins and seals frequently exhibit pneumonia-like symptoms. The algae are not new, nor are their toxins. Sediments that are more than 1,000 years old contain both algae and toxins. The material is quite stable; it dissolves reasonably well in water and is undamaged by cooking or freezing.

Domoic acid, which some classify as a brevetoxin, initiates abdominal symptoms—vomiting, diarrhea, and cramps—within 24 hours. In severe cases, this is accompanied by headache, dizziness, and disorientation and, in the worst cases, seizures and cardiac arrhythmias and possible death. Short-term memory is frequently affected, and that condition is apparently permanent. There is no antidote. The Department of Health in Washington State maintains a Web site: www.doh.wa.gov/biotoxinmaps.htm. *See* DIA-TOMS, DINOFLAGELLATES, FOOD CHAINS, RED TIDES, TETRODOTOXIN, TETRODONTIFORMES.

trace metals Metals found in tiny quantities in seawater. Many have been detected using recently developed analytical equipment and techniques. While every metal dissolves to some extent, those present in quantities measured in parts per million are considered trace metals. The prevalent metals in seawater are the alkali and alkali earth metals. Others are present in tiny amounts, and they follow the biogenic patterns of upwellings and sinkings with the carbon, silicon, and nitrogen cycles. These are aluminum (Al), manganese (Mn), cobalt (Co), cerium (Ce), lead (Pb), thorium (Th), and bismuth (Bi). These are required by some organisms for their particular metabolic paths and are therefore regularly removed from the water and recycled. Some elements are present in the oceans in more than one valence state: chromium (Cr), arsenic (As), selenium (Se), iodine (I), and technicium (Te). Beryllium (Be), copper (Cu), gallium (Ga), and zirconium (Zr) are also present in seawater in minute amounts. To some extent, human activity has added to the metal load in seawater.

trade winds Systems of tropical winds that range from about 30° north latitude to a similar point south of the equator. They are northeasterlies, following a northeast-to-southwest direction in the Northern Hemisphere, and southeasterlies, with a southeast-to-northwest direction in the Southern Hemisphere. The trade winds meet at either the equator or at the Intertropical Convergence. In the western Pacific and the Indian Ocean, this pattern is overshadowed by the monsoons. *See* INTERTROPICAL CONVERGENCE, MONSOON, WIND.

transform fault A break in the crust of the earth. Transform faults are characteristic of the mid-ocean ridges, running perpendicular to the line of the ridge. The line of the ridge is displaced (offset) by the fault. *See* CRUST, MID-OCEAN RIDGES.

Transpolar Drift An Arctic current that moves very cold fresh water from north of the Laptev Sea across the Eurasian Basin to the Fram Strait and into the Atlantic. The Transpolar Drift enters the Atlantic by way of the East Greenland Current. *See* ARCTIC OCEAN, EAST GREENLAND CURRENT.

trawling The collection of material from the sea bottom by means of a trawler or gathering vessel, which pulls a conical net that entraps the target population in it. Trawling is used by both commercial fishermen and marine biologists, the former for gathering shrimp or bottom-dwelling fish such as flatfish or cod, the latter— using a net of much smaller mesh—for sampling specific depths for the population at that depth. *See* FISHING, NETS.

trematode (fluke) Numerous flatworms of the class Trematoda, including both external and internal parasites of animal hosts, that have a thick outer cuticle and one or more suckers or hooks for attaching to host tissue. *See* FLATWORM, PARASITE.

trench A deep depression in the Earth's crust, marking the boundary between plates. At a trench, one crustal plate dives beneath an adjacent one (subduction). The whole area is one of seismic instability, which is manifested by earthquakes. Examples of trenches are the Aleutian, Puerto Rico, and Tonga trenches. *See* CRUST, INDIVIDUAL TRENCHES, SUBDUCTION.

Triassic period Segment of the Mesozoic era that follows the Permian and precedes the Jurassic period. It lasted from about 225 million to 190 million years ago and was named for the three subdivisions of the period as recorded in rock formations: Lower, Middle, and Upper. The Middle corresponds to the notably marine part of the Triassic. The rock is sandstone and mudstone; it is found on all continents and is the residue of sedimentation in a hot climate. The landmass during the Triassic was a single super-continent, Pangaea. The Atlantic Ocean opened at the end of the Triassic.

The biota of the period included the first amphibians and gymnosperms (cone-bearing plants). Algal and coral reef builders and mollusks were present. On the whole, the Triassic did not leave an abundant marine record, possibly because the hot climate evaporated water from the seas and rendered them very salty.

trichocyst A rod-shaped organelle arranged to be ejected out of the body of an organism to stun, paralyze, or poison prey or predators, or to aid in the recovery of prey. Trichocysts are characteristic structures of ciliates. *See* CILIOPHORA.

triggerfish Belongs to the family Balistidae. They are found in tropical marine waters; most species live close to shore in fairly shallow waters. Their distinguishing characteristic is a dorsal spine that locks in place. This "trigger" makes it possible for the fish to wedge itself into crevices in a reef to protect itself from attack. These fish are found in an array of colors and are sought after by hobbyists for salt aquaria. They live on a diet of invertebrates. Although these fish are not commercially significant, they are caught and eaten. They have a significant relationship with Tetraodontoidei, a suborder, and are suspected of being poisonous.

trilobite An extinct marine arthropod. Trilobites were the dominant animals of the Cambrian period and finally disappeared by the Permian. They are used by paleontologists as index fossils for dating rock structures. These animals had a tripartite body similar to that of the arthropods of today. They may have looked like flattened centipedes.

Trilobites ranged in feeding habits from plankton and detritus feeders to carnivores. Their sizes varied from tiny species less than 1 cm (0.3 inches) long to those 40 to 50 cm (16 to 20 inches) long.

See ARTHROPODA, CAMBRIAN PERIOD, FOSSILS.

trochophore The free-swimming larvae of annelids and mollusks.

Tropic of Cancer and Tropic of Capricorn The 23.5° latitudes. The Tropic of Cancer is at 23.5° North Latitude, and the Tropic of Capricorn is at 23.5° South Latitude. The latitudes between these two lines mark the boundaries of the tropics.

troposphere The part of the atmosphere closest to the surface of the Earth. *See* ATMOSPHERE.

trough An undersea depression that is U-shaped rather than V-shaped like a trench or canyon, which it otherwise resembles; also, an atmospheric low about which winds circulate. *See* CANYON, TRENCH, WIND.

The same Indian Ocean area affected by two different tsunamis

tsunami A Japanese word for a wave system following a major sea disturbance such as an earthquake. It is not a tide; therefore the phrase "tidal wave" is an inaccurate one.

The tsunami is a gravity wave and as such is influenced by the water depth and the shape of the shoreline. Because of these variables, it is difficult to predict the outcome.

In this wave system, energy is transferred between the ocean and the atmosphere. These interchanges include an increased atmospheric pressure, which causes the ocean to become shallower. Thus, waves in the ocean can move faster than they normally would in deep ocean waters. This helps explain the very rapid movement of the water created by the Krakatau eruption in 1883. The undersea earthquake of December 2004 was very similar to that of Krakatau.

In a normal wave, the wave length (the distance from crest to crest) is between 30 and 60 m (100–200 feet) at most. In a tsunami this distance can be more than 50 km (100 miles). The speed of a wave's travel depends on its depth; the average depth of the Pacific is about 5 km (2.5 miles), and this means that the tsunami can move very quickly, as quickly as an airplane, in fact. In open water, a ship would not be disturbed by a tsunami moving under it, but unlike a normal wave, which produces pressure downward in the water column for hundreds of yards, the tsunami pushes downward to the ocean floor. This is the basis for the functioning of "tsunameters," instruments that measure the pressure of a tsunami passing over the surface. These instruments were developed by NOAA and placed at varying depths in waters 5 to 8 km (2.5 to 4 miles) deep as part of the Tsunami Warning System.

The relationship between earthquakes and tsunamis is not a direct one. Thus, a moderate quake (magnitude 7.0) such as the one in 1998 in New Guinea generated a massive tsunami, but one in the Aleutians in 2003 was a 7.8 magnitude quake, and the wave it generated was not dangerous.

The tsunameters could predict that, but they cannot predict the effect of this wall of water when it breaks on a shoreline. *See* ISLAND, KRAKATAU, VOLCANO, WAVES.

Tsushima (or Tashima) Current A western Pacific Ocean current which flows out of the Yellow Sea through the Korea Strait and into the Sea of Japan. It is a warm, low-salt surface current that branches into the East Korea Current and then again into the Tsugaru and Soya currents, all of which are warm-water flows.

The Tsushima Current is a spring and summer current; it disappears in the winter. *See* CURRENT, NORTH PACIFIC GYRE, SEA OF JAPAN.

tube worm *Lamellibrachia*, a species in the order Pogonophora. These giants live in vent communities. They grow to lengths of 3 m (10 feet) or more, enclosed in chitinous tubes, and live to great age. Some specimens from the Gulf of Mexico are more than 250 years old. The outstanding characteristic of these free-living creatures is that they have no digestive systems; they absorb materials from water and have a sulfur (not oxygen) metabolism. *See* POGONOPHORA, SULFUR, VENT COMMUNITIES.

tuna A large food and sport fish, of the family Scombridae, order Perciformes, that is related to the mackerel. Tuna are long, chunky fish characterized by a thin, shapely tail and a series of small fins on their dorsal and ventral surfaces, anterior to the tail. Like other long-distance swimmers, they have temperature-regulatory structures under their skin. These are energy conservers which can raise the temperature of the fish if it is in very cold water.

Tuna are generally dark blue or black on their upper surface, light or silvery below the lateral line, and frequently iridescent. They move in dense schools, a habit fully exploited by commercial fishermen. They are carnivores and feed mostly on herring.

Tuna are found in temperate to tropical water worldwide and are a food fish

COMMON SPECIES OF TUNA				
Common Name	Scientific Name	Length	Weight	Habitat
Bluefin	(*Thunnus thynnus*)	4–5 m	800 kg	Mediterranean or warm Atlantic
Albacore	(*T. alalunga*)	1 m	30 kg	Warm Pacific This fish migrates over great distances
Yellowfin	(*T. albacares*)	2.5 m	200+ kg	Tropical The name is derived from the yellow pectoral fins and gillcovers
Big eye	(*T. obesus*)	2.5 m	200+ kg	Tropical
Skipjack	(*Katsuwonus pelamus*)	85–95 cm	22–24 kg	North and Baltic Sea, Pacific This fish has stripes, not spots

almost everywhere. Attempts to farm tuna are under way in the western Pacific. The popularity of tuna is creating a problem; it is seriously overfished, resulting in declining populations wherein juveniles are overrepresented. This is not the only problem connected with tuna; it is also contaminated with mercury compounds. Public health officers in several countries, including the United States, caution people to restrict their intake of this fish. *See* FISH, SCHOOLING.

Tunicata A subphylum of marine chordates including the sessile (stationary) sea squirts. There are over 2,000 classified species of tunicates. Their range is worldwide, and they are found at every depth. Their notable characteristics are a flexible protective cover—the test or tunic—and a notochord, which is a feature of the larval state only. The lifestyles of tunicates vary from the pelagic to sessile, and they may vary in size from 1 to 2 mm (0.03 inches) to over 30 cm (1 foot). Colonies such as the Thaliaceae may be larger, some exceeding 1 m (40 inches) in diameter.

The tunicates are classified as Ascidiacea, sea squirts, which are sessile and either solitary animals or colonial species; the Thaliacea, which are pelagic and have a complex reproductive system that includes both sexual and asexual phases; and the Larvacea, which are tiny pelagic animals.

Tunicates have hearts and a network of blood vessels and nerves. The distributive function is, however, performed by cells called macrophages, which move within the animal. These cells are also the medium for gas exchange.

The question of the notochord has been debated for years. Tunicates were once described as "degenerate" organisms.

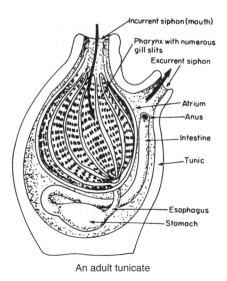

An adult tunicate

While this view is no longer held, there is still the problem of why this animal has a supporting skeletal unit as a larva, only to lose it in the adult form. Tunicates also use both iron and vanadium in their metabolic processes, and both metals are concentrated in their tissues. The need for either or both is unexplained. *See* LARVACEA, OIKOPLEURA, PELAGIC ENVIRONMENT, SEA SQUIRT, SESSILE.

Turbellaria A class of free-living flatworms. Most of these animals are aquatic. They are usually small, and many are barely visible to the unaided eye. Some are larger, and a few range in size up to about 60 cm (2 feet).

Many turbellarians are brilliantly colored. However, the best-known, the freshwater planarians are black. Turbellarians move by undulating the muscular layer that lies beneath their epidermis. The respiratory function is performed by the epidermis, through which gases move in and out by diffusion. The excretory system is very simple, and as with respiration, some of its function is undertaken by the epidermis. There is a "head" end and a distinct mouth on the ventral surface near the "head."

Turbellarians are hermaphrodites. They produce both eggs and sperm but can also reproduce by splitting laterally (down the middle from head to tail).

These organisms are frequently commensals or symbionts. They live on or near other marine invertebrates and sometimes harbor protists or nematodes. *See* COMMENSAL RELATIONSHIPS, FLATWORMS, SYMBIOSIS.

turbidity currents Currents in which there is suspended sediment. These currents occur on the continental slopes. Sediment in the water moves to a point where, due to gravity, it is deposited. Since most of the sediment in the ocean is the result of

river transport, this material is deposited in unconsolidated patterns at river outflows.

Turbidity currents arise quickly as the result of a flood, earthquake, or rapid deposit of river sediment. These trigger rapid water movements, which are the undersea equivalents of atmospheric storm surges. The rapid water flow will break up the unconsolidated deposits of sediment, causing rapid changes in the undersea landscape—usually landslides.

Turbidity currents are the cause of submarine cable breaks. The 1929 break in the trans-Atlantic telegraph cable was the result of an earthquake off New England. They are also thought to be the mechanism by which undersea canyons are originally carved or maintained free of sediment. Before the concept of plate tectonics was fully accepted, turbidity currents were considered by some geologists to be a factor in the explanation of continental drift. *See* EARTHQUAKE, PLATE TECTONICS, SEDIMENT, SUBMARINE CANYON.

Turkish Plate A small crustal plate that is moving between the African and Hellenic plates. It is headed for the Eurasian Plate. The Caucasus mountains are the seam of subduction along which this plate dips down under the Eurasian Plate. The buildup of pressure in this area makes it one of frequent seismic activity. *See* PLATE TECTONICS, SUBDUCTION.

turtle *See* SEA TURTLE.

Tyrrhenian Sea A roughly triangular part of the Mediterranean Sea between the Italian mainland, Sicily, and Sardinia. It is named for the mythical founder of the Etruscan kingdom. The deepest point, which is almost 3,600 m (12,000 feet) below sea level, is between the Gulf of Salerno and Sardinia. Atlantic water enters the Tyrrhenian Sea south of Sardinia and circulates counterclockwise. *See* MEDITERRANEAN.

ultraplankton *See* PLANKTON.

United Nations Conventions *See* LAW OF THE SEA CONFERENCE.

upwelling The rise of subsurface cold, dense water. It is most noticeable on the western coasts of continents. The deep water coming up does so to replace wind-displaced surface water. If the rising water is colder than the ambient air, fog results. The California coast near San Francisco is a prime example of a site where such colder water wells up. If the deep water is nutrient-rich, the result is a bloom of plankton at the surface. This in turn feeds a large fish and bird population, as evidenced on Antarctica and along the Chile-Peru coast of South America. *See* BOTTOM WATER, NUTRIENTS, PERU CURRENT, PLANKTON.

veliger The larva of a mollusk after it has developed its velum (swimming membrane). *See also* TROCHOPHORE.

Venezuelan Basin The second-largest portion of the Caribbean; it lies north of South America, west of Grenada, east of the Colombia Basin, and south of Santo Domingo and Puerto Rico. The average depth of the bottom of the Venezuelan Basin is about 4,400 m (14,000 feet). The Beata Rise separates the Venezuelan from the Colombia Basin.

The bottom topography of the basin is interesting, since it gives two reflections of sound. One explanation for this double echo is based on the theory of formation of the Caribbean Sea. If this sea was always oceanic, and then sank further still, it would have a different bottom composition than if it was once terrestrial and then sank. If the latter is the correct scenario, then an eroded landscape would have been overlain by sediment carried to it by wind and waves. *See* BEATA RIDGE, CARIBBEAN SEA.

vent communities Relatively recently discovered ecosystems, not based on photosynthesis, which occur at volcanic sites and at rifts where mineral-containing water emerges from breaks in the ocean floor. The ecosystem of the vent community is extraordinary because it exists in areas where the average depth is 2,500 m (8,200 feet) and the temperature in a hot vent is above 350°C (about 700°F) inside the vents and about 250°C (480°F) in the surrounding water. These communities are associated with black smokers. When its vent ceases to emit mineral oxides and sulfides, the vent community changes in character.

Ocean water, sulfur, manganese, and iron support the sulfur bacteria of the vent community, which depend on metabolic systems that utilize sulfur instead of oxygen. They also thrive in a high-temperature, high-pressure environment. The bacteria in turn support populations of tube worms, huge crabs, large oysters and clams, fish, and other biota that live in the constant dark of deep ocean at these seemingly poisonous sites.

Vent communities were first discovered in the 1970s, and are thought to be models of life on the primordial Earth. As such they are being intensively studied. So far, these habitats have been found in the Pacific Ocean at the East Pacific Rise, at the Galapagos Rift, near the Gulf of California, and on the Juan de Fuca Ridge. They have also been located in the Caribbean.

The theories about vent communities continue to arise. One recent idea is that the phototaxis—response to light shown by some vent organisms, may have been the precursor of photoreactions that enabled vent organisms to respond to food supply. Ultimately, this might have culminated in the development of photosynthesis. This theory meshes with the idea that the extremophiles found at vent communities and seeps are the ancestors of all the other organisms on Earth, but the path to the utilization of light to produce food is not simple, and the precursors are not yet apparent. *See* PHOTOSYNTHESIS, POGONOPHORA, POMPEII WORM, RIKTIA, SEEPS, SULFUR BACTERIA, TUBE WORMS.

ventral The lower part or underside of an animal. *See* DORSAL.

Venus comb *See* MUREX.

Venus's flower basket The skeleton of a tropical sponge. The "basket" is a graceful, J-shaped, tapered lattice about 25 cm (10 inches) long and tubular. The species *Euplectella* (class Calcispongiae, phylum Porifera), which give rise to the Venus's flower basket, attach themselves to sandy bottoms in tropical seas. They are most often found near the Philippines, and range as far north as Japan. They have also been collected in the Western Pacific and in the Indian Ocean. These sponges are prized by collectors. *See* PORIFERA, SPONGE.

Venus's girdle A member of the phylum Ctenophora. This beautiful undulating, luminous, transparent ribbon is actually a pelagic (open-sea) animal. *See* CTENOPHORA.

Verrazano, Giovanni da (1485–1528) Florentine navigator and explorer. He sailed to Newfoundland in 1508 and apparently made voyages to Carthage, Damascus, and the Eastern Mediterranean for the French. He met Magellan in 1517 in Seville, where both were looking for royal sponsors for voyages of discovery.

Verrazano was financed and given command of a voyage of discovery by François I of France. The Florentine bankers in Rouen also contributed to this, and were given shares of the possible profits. Four ships set sail for the New World, but only the largest, *La Dauphine* and *La Normande,* crossed the Atlantic. After a storm, the party made landfall on Cape Fear, in what is today North Carolina. When Verrazano sailed around the Outer Banks he thought he had found the Pacific. He then sailed north, looking for the passage to India and entered what is now called New York Bay on April 17, 1524. He continued up the coast to Rhode Island, Maine, and Newfoundland before turning east and back to France.

Verrazano made careful calculations of latitude for each of his anchorages. He kept logs of distance, and estimated longitude with reasonable success. Although he erred in some calculations, he realized that the landmass he found was nowhere near Asia and that it was no fringe of islands but a large continent interposed between the Atlantic and the Pacific oceans.

On a second voyage, again sponsored by France, Verrazano sailed to Brazil for wood that was in demand by the dyers. The profits made on this trip encouraged his backers, who then financed another attempt to find the strait to the Indies by sailing south. Verrazano was captured and killed by the inhabitants of what is now Guadaloupe in 1528, and his brother, Girolamo, brought the ships back to France.

vertebrate An animal with a backbone. These include the fish, amphibians, reptiles, birds, and mammals. As a group, the vertebrates have an internal skeleton with a separate head; the head has hinged jaws. The vertebral column is a series of bones, whose number is significant. Vertebrates also have two pairs of appendages that have been highly modified, and in some cases lost.

The bony vertebral column that protects the spinal cord—the great trunk of nervous tissue that extends from the brain down the back of vertebrates and gives off nerve branches to the body—is preceded in embryonic development by a notochord which develops ventral to the nerve cord. In many primitive animals this remains throughout life. The animals known as tunicates have a notochord as juveniles and then develop an adult body without the support of a bony or cartilaginous spinal column and skeleton. In more advanced forms it is lost and is replaced by a calcareous or cartilaginous skeleton.

The vertebrae of different species of vertebrates are of different shape, size, and number, depending on the animal. Fish have trunk and caudal vertebrae, snakes have the greatest number, but most of the bones of the spinal column are the same size and shape; in turtles several bones are fused together to form the carapace. Some of the vertebrae of birds also have been fused together for better aerodynamics. *See* BIRD, FISH, MAMMAL, NOTOCHORD, REPTILE, TUNICATA.

vertical migration *See* MIGRATION, PHYTOPLANKTON.

Vespucci, Amerigo (1454–1512) An explorer born in Florence to a family with Medici connections. As a member of this extended family he worked in their banking house and then in an affiliated bank and ship chandlery in Seville, Spain, where he outfitted the third of Columbus's voyages in 1498. In 1499 Vespucci sailed to the New World with Ojeda but left the voyage in Hispaniola, returning on his own. Based on that trip, he broadcast his own reputation as a mariner, which led to an invitation from King Manuel of Portugal to join Coelho's expedition. Vespucci accepted the royal invitation to accompany and chronicle the voyage which lasted from 1502 to 1504. His accounts, entitled *Lettera* and *Mundus Novus,* were published by a friend in Florence (1504–06). This led to the association of Vespucci's name with that of the new continent he explored, when Martin Waldseemuller, a publisher, produced a new edition of Ptolemy's *Geography* that showed the "new world" discovered by Americus (Amerigo), and the name was gradually adopted for the entire landmass, not just South America. Waldseemuller's map showed another ocean between America and India, because Vespucci—a keen observer of the geography, scenery, flora, fauna and indigenous population of the new lands—was one of the first to realize that the newly discovered lands were not a part of the Asian coast.

Vitiaz A Russian vessel that charted the Pacific Ocean trenches during the International Geophysical Year (IGY) of 1956–57. *See* IGY.

volcano A mound associated with a vent in the crust of the Earth, on land, or under water, from which molten or hot rock and steam are ejected. Undersea volcanic activity forms the Earth's undersea ridges. In an oceanic volcano, as in a terrestrial one, magma, the liquid rock from the Earth's interior, oozes out through long fissures or rises in a central channel, although the former is the more likely phenomenon at the crustal boundaries.

Fissure-flow volcanoes produce large areas of lava called pillows. These are similar to the gently sloped Hawaiian volcanoes and resemble gentle, rolling hills. Pillows are found in the Atlantic and are typical of slow, spreading ridges. The sheet flows of such areas as the East Pacific Rise are formed by very fluid lava with a glassy crust.

The volcano that has a well-defined central channel will form a cone whether it is on land or undersea. Seamounts have their origin in volcanoes of this central-

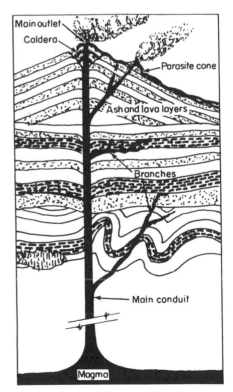

Cross section of a volcano

channel type. If there is a depression in the center of a seamount, it is called a crater if it is small and a caldera if it is large. Seamounts are usually bounded by fault lines; these sites of seismic and volcanic activity produce a fairly constant supply of molten rock. If it leaks into the rock at the top of a volcanic cone, this rock will eventually fill it, and the cone will become leveled off. This leveled structure is then called a guyot.

Volcanoes also occur within plates, at points of geologic stress. Such volcanoes tend to be in chains located obliquely in relation to the mid-ocean ridges.

Undersea volcanism is interesting and difficult to explore, but similarities to terrestrial volcanoes exist. The Marianas, the islands that include Guam and Saipan, are similar to the Cascade Range in the U.S. Northwest, and both are at subduction sites on major plates. Undersea volcanoes, like their counterparts on land, form sloping hills like the Hawaiian volcanoes. They can also form steep cones like Vesuvius from which gas and minerals are rapidly shot into the water. The list of communities of organisms that can survive in such extreme conditions continues to grow. *See* BLACK SMOKER, CRUST, EARTHQUAKE, GUYOT, HOT SPOT, LAVA, PLATE, PLATE TECTONICS, SEAMOUNT, THERA, VENT COMMUNITIES.

volvocales Largely freshwater, colonial green algae.

Wagenaer, Lucas Janszoon (1534?–1605) A Dutch sailor and cartographer who became a pilot and in 1582 began a compilation of pilots' charts for western and northern Europe. This collection eventually became *Spieghel des Zeevaerdt* (Mirror of Seafaring), which was published in 1584–85. It was a useful navigational aid and faithfully mapped the region from Cádiz, Spain, to the north of Holland. The Lord High Admiral of England was so impressed with it that he had it translated. The translation appeared in 1588, with the title *The Mariner's Mirror.*

Wallace, Alfred Russell (1823–1913) A British naturalist who discovered natural selection independently of Charles Darwin (1809–82). Wallace began his career as a surveyor and then became a self-taught botanist. He explored the Amazon River and survived a shipwreck in 1852. After a dramatic rescue, he transferred his interests to the Malay Archipelago, where he did intensive exploration (1854–62), and finally returned to England as an established naturalist. His book on the natural history of the East Indies, *The Malay Archipelago*, was first published in 1869. It was a continual best seller, and was reissued in several subsequent editions.

Wallace's *Species Notebook* was a scheme for grouping animals in terms of their geographical location and related species. This formed the basis of a scientific paper that appeared in 1855. The explanation given by Wallace for natural selection was formulated in 1858. He sent it—with a long explanation—to Darwin, who was already an authority on the subject. Darwin presented both his and Wallace's work

simultaneously in 1858 (Darwin's evolutionary views dated to the late 1830s, but he had not yet publicized them).

Wallace continued his research on evolution and the study of mimicry in both related and unrelated species. Toward the end of his life his ideas on the evolution of humans became entangled with his adherence to spiritualism, then a very popular pastime in England. *See* DARWIN, CHARLES; EVOLUTION; WALLACE LINE.

Wallace line A zoogeographical concept introduced by Alfred Wallace on the basis of his observations of the flora and fauna of South America and Malaysia. A Wallace line is said to exist between different but related species that are separated by a natural barrier such as a large river.

Wallace published a work on the zoogeography of Southeast Asia in 1860. He

Division of zoogeographic regions

introduced the concept of the line in 1863, and with it he divided Asian from Australian species. He drew the line between Bali and Lombok in the south, between Borneo and Sulawesi in the north, and in the east around the Philippines. As more information became available, the line shifted.

In 1902 Webber proposed a modification and expansion of the line. According to him, it would run from the Indian Ocean through Timor, then north through Molucca and into the Philippine Sea. *See* EVOLUTION; EVOLUTION OF OCEANS; MIGRATION; WALLACE, ALFRED RUSSELL.

walrus A large, tusked, seal-like mammal of Arctic waters, of the family Odobenidae. The walrus has many of the characteristics and the body plan of the seal, and no external ears. It has short red hairs on its wrinkled hide and a comical, brushlike moustache and whiskers.

The male is considerably larger than the female, with an average length in excess of 3 m (10 feet); walruses frequently weigh more than 1,200 kg (2,500 pounds). The tusks on the male are also large; these enlarged canine teeth are not digging tools but are used for both courtship displays and territorial defense.

The walrus is a colonial animal. It lives in large groups dominated by enormous harem-collecting males. The most usual habitat is near a stretch of reasonably shallow water. The walrus feeds almost exclusively on clams. The only serious danger to this fat and lazy animal comes from humans. *See* PINNIPED.

Walvis Ridge An undersea elevation near the western coast of Africa, which extends about 3,000 km (1,900 miles) into the southern Atlantic Ocean. It makes a rough line from the Tristan da Cunha island group on the Mid-Atlantic Ridge, northeast to the coast of Namibia. This aseismic ridge is very narrow and steep-sided, with drops of more than 1,000 m (3,300 feet) on its southern side. The expeditions that have mapped turbidity currents have shown that there is great

current activity in this area. Rapid currents scour the deeps near the ridge, and it is sediment-free.

On some maps the Walvis Ridge is called the Walfisch Ridge. *See* ASEISMIC RIDGE, MID-ATLANTIC RIDGE, TURBIDITY CURRENTS.

warm-blooded animals Animals that maintain a constant body temperature; also called homoiotherms (homeotherms). *See* HOMEOTHERMS, POIKILOTHERMS, TEMPERATURE IN ANIMALS.

water *See* SEAWATER.

water flea A small freshwater crustacean, of the genus *Daphnia,* that moves about on the surface of the water with sudden leaps.

waterspout A spinning, funnel-shaped column of water or spray pulled up by a whirling wind from the ocean surface. This rising wind-and-water funnel looks like a tornado. Waterspouts are formed in the same humid, hot, tropical air that builds thunderclouds. While they are most often tropical, they are sometimes found in middle latitudes, particularly on hot summer days. Waterspouts dissipate very rapidly if they move over land. *See* CYCLONIC STORM, HURRICANE.

wave A crest of water moving along the surface of the water in lines at right angle to the direction of motion. Waves are divided into two large groups: small waves, called ripples, with crests of equal size less than 2 cm apart, which result from surface tension, and longer waves, called gravity waves.

Gravity waves are then subdivided into shallow-water waves and deepwater waves, depending on the depth of the water through which they move. In shallow-water waves the water particles move forward and backward in a horizontal plane. The crests represent crowds of water particles; the troughs are sparse populations. A tidal or river bore is a classic shallow-water wave. A solitary wave is one that occurs when a

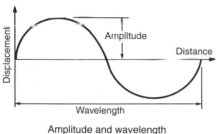

Amplitude and wavelength
for a simple sine wave motion

mass of water moves in one direction and an equal mass of water moves in the reverse direction. The water both in front of and behind the wave looks like it is standing still, but in fact is moving forward.

When one observes a shoreline, one sees a train of waves. They will decrease in velocity in shallow water. Thus, since the positive direction (i.e., the direction in which the wave train is going) moves faster than the negative bore, crests pile up and produce breakers. Breakers are the foaming delivery of water onto the beach. If the beach is one of gentle slope the breakers spill water onto it; if the beach has a steep slope, the breakers plunge onto it.

Deepwater waves are generated by storms, and can have crests miles apart. Wind generated waves start out short. There are conflicting theories about how these short waves become long. The velocity of fast, storm-generated waves is generally less than that of the generating wind—at most about 80% of it. Thus, long waves travelling from a storm center move more rapidly than the short surface waves. This is the "groundswell" that will precede and predict a coming storm.

By clocking the time between crests, it is also possible to determine where the storm is and where it was when the first waves arrived. This is extremely important in coastal areas, since violent storms at sea, or seismic activity undersea, will produce deepwater waves that will travel very rapidly and only slightly alter the condition of the surface. Enormous quantities of energy are expended in lateral motion in deep water; in shallow water this energy

is expended by smashing against the shore. The result of this impact with the shore is an instant and dramatic rise in sea level. Winds whip up the water, aggravating the situation. The effect on land is flooding of coastal areas to a variable degree depending on the height of the water and on the coastal area. *See* INTERNAL WAVE, RIP CURRENT, RIVER BORE, TSUNAMI.

wave height The vertical distance between the top or crest of a wave and the bottom of the preceding (or succeeding) trough. *See* WAVE.

wave period The time between the passage of one wave crest and the next crest through a specific point. *See* WAVE.

weather The state of the atmosphere. This results from the combined effect of clouds, precipitation, temperature, humidity, visibility, wind velocity, and barometric pressure. Weather is treated on two bases: geographic (global, regional, or local) and chronological (long- or short-term).

Much of the equipment of the meteorologist consists of the products of 17th-and 18th-century science. The 20th century contributed radar, dating techniques, very sensitive photography, and computer modeling, among other techniques. Today it is possible to track storms by radar and to use satellites to photograph developing weather patterns.

The study of seemingly local meterologic events in terms of global weather is now an unfolding area of research. The phenomenon of El Niño in the southern Pacific has a profound effect—a chilling one—on the weather in North America. This in turn affects the Atlantic Ocean, with an increase in iceberg formation as a direct result of the colder winter that comes with the Niño phenomenon. Another result of this weather disturbance is the increase in heavy rain in equatorial regions, which diminishes the velocity of the prevailing westerly winds.

Accurate meteorologic recordkeeping and modelbuilding by simulation both depend on large computers with extensive memory and calculating capability. The increasing accuracy of weather prediction depends on both such records and such models.

The use of satellites has made weather forecasting and long-range modeling possible. This has led to better communications and to the resulting decrease in deaths due to storms. The increasing construction in areas that are exposed to weather means that property damage from violent storms is increasing. The current trend is a warming one. That has happened in the past history of the Earth; considerable evidence that weather is cyclical comes from ice cores, coral cores, and dendrochronology. As warming continues, the incidence of violent cyclonic storms also increases. The production of greenhouse gases exacerbates a natural problem. *See* ATMOSPHERE, EL NIÑO, LA NIÑA, GLOBAL WARMING, GREEN VAST EFFECT, HURRICANE, METEOROLOGY, NOAA, OCEAN-ATMOSPHERE INTERCHANGE, STORM, WIND.

Weddell, James (1787–1834) An English sea captain in both the Royal Navy and the maritime trade. Weddell sailed in Antarctic waters, hunting seals in the post-Napoleonic period (after 1815). He visited the South Shetland Islands and named the South Orkney Islands in 1822. In 1822–24 he charted the waters near the South Shetlands and explored the sea named for him. His record mark for the most southerly point reached was 74°15' South Latitude before his path was blocked by ice. His memoir, *A Voyage Toward the South Pole*, was a popular success. It was published in 1822 and initiated efforts in several countries to explore the Earth's polar regions. *See* DUMONT D'URVILLE, JULES-SEBASTIEN CÉSAR; WILKES, CHARLES; ROSS, JOHN.

Weddell Current An Antarctic coastal flow of water at the ocean surface. It moves in an easterly direction from the Weddell Sea and is outside the West Wind Drift. *See* ANTARCTIC OCEAN, CURRENT, WEST WIND DRIFT.

Weddell Sea The southernmost part of the Atlantic Ocean; it is usually ice-filled. The area was explored by James Weddell in 1823. Coastal fossil-bearing limestone was found along the sea in 1902–04. The continental shelf of Antarctica runs under the sea for about 150 km (90 miles); the continental slope is relatively steep, and descends to depths of about 4,500 m (15,000 feet). This depth may be the result of the weight of the ice shelf. There is controversy about the origin of the Weddell Sea Basin.

The Weddell Sea is generally accepted as the source of most Antarctic bottom water. It is a cold sea, its water temperature generally being below -1°C, and is fairly salty for polar water at a salinity of about 3.41%, with higher percentages in some areas. There is little wind, which means that the huge icebergs produced in the sea tend to stay and fill the basin.

A recently discovered marsupial fossil located on the Antarctic Peninsula, western boundary of the Weddell Sea, is further evidence of the definite connections between South America, Australia, and Antarctica. The fossil was of South American origin; the link between this continent and the Antarctic mainland may therefore have continued into the Eocene. *See* ANTARCTIC WATER, EOCENE EPOCH, FOSSILS.

Wegener, Alfred Lothar (1880–1930) A German meteorologist and geophysicist best remembered for his theory of continental drift. He published this work in 1915 but was certainly not the first person to have noticed the coastal congruencies of the continents. That observation had been made early in the 17th century.

Wegener's theory was based not only on the similarities of the coastlines of various land masses, particularly in the Southern Hemisphere, but on similar fossil records on now widely separated landmasses as evidence of ancient climate conditions. This evidence had been rather inadequately explained away by the evocation of convenient "land bridges" linking these landmasses at some previous time. Wegener

called his primordial, warm, Permian continent Pangaea, from Gaea, the mythological
Greek "earth-mother." In 1937 Du Toit, a
French paleogeophysicist, divided this single, then hypothetical continent into two.
He called the northern portion Laurasia
and the southern portion Gondwana.

One of the great flaws of Wegener's
hypothesis was that he could not adequately explain the mechanisms by which
the fragments of his single continent had
separated, even though he made continuous and determined efforts to do so,
using paleoclimatology, balloon observations, and geodetic measurements.

The theory received considerable attention until 1928, when a panel of 14 geologists voted on it. Five were in favor of it,
and suggested future surveying to attempt
to prove it; two were uncertain; and seven
turned it down. From that time until the
work of Fred Vine, Drummond Matthews,
and others on paleomagnetism in the 1950s,
the Wegener theory was largely ignored.

Wegener was lost in a blizzard in central Greenland. He was there as a member
of a meteorological survey. See CONTINEN-
TAL DRIFT, MAGNETIC ANOMALY, PLATE
TECTONICS.

West Australian Current A portion of
the West wind drift, the current generated
in Antarctic waters, which turns north
along the western coast of Australia and
continues into tropical regions as a coastal
current. The West Australian Current is a
part of the Indian Ocean gyre. This surface stream is strongest in the summer. See
GYRE, INDIAN OCEAN, WEST WIND DRIFT.

westerlies Winds found in both the
Northern and Southern hemispheres,
between 30° and 60° of latitude. These
winds move from west to east. The westerlies arise in subtropical high-pressure
zones. The westerlies are very extensive and may reach up 15 km (9 miles)
into the stratosphere. They have velocities of about 10 km (6 miles) per hour
at the Earth's surface. Since wind velocity increases with altitude, at jet-stream

heights of over 10 km (10,000 m, 33,000
feet), the wind is moving at velocities
greater than 300 km per hour. See MON-
SOON, TRADE WINDS, WIND.

west-flowing currents Surface currents
that are propelled by the prevailing westerly winds. Most are at about the 45° parallel in both hemispheres, and they occur
in all oceans. The west-flowing current in
the North Atlantic is the Gulf Stream; in
the South Atlantic, the Brazil and Cape
Horn currents move west. In the Pacific
Ocean, the northern westflowing current
is the Kuroshio, while the Equatorial and
East Australian currents are in the Southern Hemisphere. In the Indian Ocean there
is no northern current; the landmasses are
in the way. The Agulhas Current in the
Southern Hemisphere is the end result of
the west-flowing Equatorial Current. See
INDIVIDUAL CURRENTS.

West Greenland Current An Atlantic Ocean current; a branch of the East
Greenland Current. It moves north to the
Davis Strait. The coastal water is cold and
of a low salt content, ranging from 3.1 to
3.4%, which is similar to that of the East
Greenland Current. The water farther
out to sea on the western side of Greenland—the North Atlantic gyre—is warmer
and saltier than that of the West Greenland Current, with an average salt level of
about 3.5%. The West Greenland Current
eventually loses its particular character by
dividing and merging with the Labrador
Current as the latter moves southward; the
rest disappears into Baffin Bay. See ATLAN-
TIC OCEAN, CURRENTS, EAST GREENLAND
CURRENT, GYRE.

West Indies (Antilles) Island groups
in the tropical Atlantic that stretch from
Cuba to the coast of South America. They
are volcanic in origin but are no longer
"hot." There is little volcanic activity now,
but Martinique was active in the early
20th century and Monserrat is now. The
rocks of these islands date from the Cenozoic. During the last ice age, much more of

the islands were above sea level. *See* Blue Holes, Caribbean, Island Arc.

West Spitsbergen Current A current that brings North Atlantic water through the Fram Strait between eastern Greenland and Svalbard (Spitsbergen). It makes a sweep around Jan Mayen Land before disappearing into the East Greenland Current. *See* Arctic Ocean, East Greenland Current, Fram Strait.

West Wind Drift The common name of the Antarctic Circumpolar Current. This is the tremendous volume of surface water that is moved by the prevailing westerly winds. *See* Antarctic Ocean.

whale oil Oil obtained by steamcooking the blubber or fat layers of captured whales. This was why whalers went to sea. The increasing rarity of whales, as a result of overhunting, has made them an unreliable source of oil. The problem of overhunting was first noted by Alexander Agassiz and underscored by Prince Albert of Monaco, two prominent, early 20th-century oceanographers. Some species may have been hunted to extinction. While some whale oils are quite specialized in use, such as spermaceti, the uses of whale oil could be filled by other substances. *See* Agassiz, Alexander; Albert I; extinction; marine oils.

whale shark A worldwide, tropical, and harmless giant shark. It is the largest living fish. Specimens of more than 9 m (30 feet) in length have been documented, and sightings of animals three times that size have been reported but not verified. The whale shark is gray or gray-brown on its upper surface, white below, and marked with small dark spots. This fish is a slow swimmer that ingests plankton and small fish in its path. *See* plankton, shark.

whales Large marine mammals of the order Cetacea, with highly modified forelimbs that have become flippers, no hindlimbs, a vestigial pelvis, and a flattened tail with flukes that act as stabilizers. There are 10 fossil forms and nine existing groups of whales, some of which are endangered.

The toothed whales are homodonts, each tooth looking like every other, and have a single nostril at the top of their head. This group includes river dolphins; beaked, sperm, and beluga whales; narwhals; porpoises; dolphins; and killer whales.

The baleen whales have a flanged lower jaw that acts as a giant strainer of the tiny animals that these whales feed on. The head is disproportionately large for the body. These animals, which include the gray, right, and humpback whales, have paired nostrils.

Most of the toothed whales school. The baleen whales tend to live in small groups called pods or are solitary creatures. Schools of whales are hierarchically arranged in the same way that elephant herds arrange themselves. Dominant males lead, others protect the herd's females and children, and dominant females organize the other females and juveniles. Play behavior among juveniles and adults is common, as is cooperative action and assistance. The latter may cross species lines—whales of one species will assist those of another.

Both the sight and hearing of whales are well developed. Most cetaceans make complicated and intelligible (to other cetaceans) sounds. The mechanism for the rapid swimming of whales is obvious. The body is extremely streamlined, sleek, and equipped with a massive and powerful tail. The explanation for the mechanism of whales diving rapidly to great depths and returning just as rapidly is not established. It is not known how a mammal can accomplish this without suffering severe brain damage, hypothermia, or the bends.

Biologists tend to correlate the longevity of whales (which tend to live as long as or longer than humans) with their long gestations of a year or more and long infancy of at least 18 months. Whales are born large and grow very rapidly on a diet of high-fat, high-protein milk.

Most whale species are well known for long migrations from feeding to breeding grounds. Some, however, do not migrate at all. Thus, lifestyle varies from the almost stationary to the baleens, which may travel 5,000 km (3,125 miles) to calving areas in warm water.

Toothed whales feed by capturing prey such as squid. The dentition of these whales varies; some have teeth in both jaws, others, such as the sperm whale, only on the lower jaw. The upper jaw of the baleen whales has plates suspended from it that are a curtain of bristling fringes of horn. This keratin material is chemically the same as hair or human fingernails. Baleens either skim the water or swim openmouthed, straining the water as they move forward, or gulp. Gulpers take mouthfuls of water and then expel it past the baleen. They then swallow whatever remains caught on the baleen.

Paleontologists believe that whales made the gradual transition from land to the sea in the early Eocene epoch, 55 million years ago. Fossil remains in what was the eastern Tethys Sea and is now in Pakistan, in addition to others in the Sahara, indicate a move to a shallow-water, intermediate form of these animals. The evidence for this theory is a particular ear development of the whales. *See* BALEEN, DOLPHIN, ECHOLOCATION, EXTINCTION, KRILL, PORPOISE, BLUE, SPERM, KILLER, RIGHT, AND HUMPBACK WHALES.

whelk A large, univalvular marine snail of the family Buccinidae, subclass Prosobronchia, class Gastropoda, that resembles the conch. Whelks feed on carrion but also eat other mollusks and fish, which they kill with their long proboscis. These animals have a worldwide range, living on sandy bottoms. The common North Atlantic species is *Buccinum undatum,* the waved whelk. Its shell is about 8 cm (under 3 inches) long, pale yellow to cream colored, and heavy for its size. The conchs are more often tropical or Pacific Ocean creatures. *Murex* or oyster drills are rock whelks. *See* GASTROPODA, MUREX.

White Sea Also known as the Byela More; an embayment of the Barents Sea south of the Kola peninsula in Russia. It is fairly shallow, with a maximum depth of less than 350 m (1,000 feet), and most of it is between 100 and 200 m (330 to 660 feet) deep. A sill prevents bottom water from mixing efficiently with the water of the Barents. The White Sea's water has a very low salt content, of 2.4 to 2.6%. Even the bottom water is of low salinity, at about 3.0 to 3.05%. This sea, like the Baltic, was an ice lake in the Pleistocene epoch, about 4 million years ago. *See* ARCTIC OCEAN; BARENTS SEA; DERYUGIN, KONSTANTIN.

white shark A dangerous, carnivorous shark, *Carcharodon carcharias,* of the family Isuridae, found in tropical to warm-temperate waters worldwide. This is the shark most likely to be a maneater. It is a handsome animal, gray-blue or gray-brown on its upper surface and creamy white on the underside. Old individuals tend to become whiter as they age. The reported maximum length of the white shark is greater than 10 m (33 feet), but the average individual is about 5 to 6 m (16 to 20 feet) long.

The white shark literally feeds on whatever it can catch. It will attack fish, birds, turtles, garbage, porpoises, and much else. Its voracious appetite and varied diet enhance its fearsome reputation.

In some classification schemes, the white shark is grouped with the mackerel sharks. *See* SHARK.

whiting A marine food fish and commercially important relative of the cod. The *Gadus* or *Merlangus merlangus* is the North Sea species of whiting. It is most often about 70 cm (28–29 inches) long and silvery-white with a black spot near the pectoral fins as well as the characteristic chin barbel of the codfish. The American species of whiting is *Menticirrhus.*

Other related species are the kingfish (*Menticirrhus saxatilis*), the rock whitings (*Odacidae*) of Australia and New Zealand, the *Sillaginidae* of the Indian and Pacific oceans, and the lake whitefish of North America. *See* COD.

Wiechert-Gutenberg discontinuity The layer that separates the Earth's mantle from the core. Its depth is estimated at about 2,900 km (1,800 miles). *See* CRUST, MANTLE, MOHO.

Wilkes, Charles (1798–1877) An American admiral and explorer who joined the U.S. Navy after service in the merchant marine. He worked with Ferdinand Hassler, the founder of the U.S. Coast and Geodetic Survey, and sailed on two surveying voyages between 1826 and 1833. In 1834 Wilkes headed the Navy's Department of Maps and Charts.

After much lobbying, Congress finally authorized and funded an expedition to the Antarctic. After buying instruments in Europe, Wilkes, in command of six ships, set sail in 1838. The first landfall on the voyage was Samoa. After refitting, the expedition sailed south and sighted land in January 1840. Taking advantage of the antarctic summer, Wilkes's group sailed along the coast of this land for 2,400 km (1,500 miles), and Wilkes named the territory Antarctica. The expedition then sailed north along the coast of the American continent, and finally ended its voyage in New York in June 1842. Their cargo of artifacts eventually formed the basis of the Smithsonian Institution's collection of Pacific Islands ethnology in Washington, D.C.

Willoughby, Hugh (?–1554) An English explorer remembered for his attempt to find the Northeast Passage. Willoughby had a military career of some note, but no sailing experience before he accepted an appointment as captain-general of a party of three ships which sailed from London in May 1553. The ships were separated by a storm in the far north. Two ships, the *Bona Esperanza* and the *Bona Confidentia,* were kept at sea by bad weather, and their crews, unprepared for the arctic winter, died of

exposure. The last entry in Willoughby's log was in early 1554. Willoughby's body and possessions were reclaimed by his pilot, Richard Chancellor. The log of the expedition was included by Hakluyt in his series. *See* CHANCELLOR, RICHARD; HAKLUYT, RICHARD; NORTHEAST PASSAGE.

wind A movement of air relative to the land or water surface. Winds are local or general. General winds move over great distances and have reasonably fixed locations and directions on a global scale. They are the Polar Easterlies and West Winds and the Trade Winds.

Winds are the result of several components; a principal one is that air heats more rapidly over land than over water. This warmed air is then moved toward the closest pole by convection currents in the atmosphere. As it moves, it is deflected by the rotatory motion of the Earth.

Local winds include coastal winds and winds moving down mountains or slopes. The coastal winds are subdivided into several categories, such as a sea breeze, which is a wind from cooler water moving onshore, and the bora, a cold mountain wind blowing down on a warm coastal area and the adjacent sea. The bora is a common phenomenon along the Mediterranean coast in the winter. *See* BEAUFORT SCALE, OCEAN-ATMOSPHERE, ROARING FORTIES, TRADE WINDS, WESTERLIES.

windward Facing the wind, or moving in the direction from which the wind originates. *See* LEEWARD.

Windward Islands *See* CARIBBEAN SEA.

Woods Hole Oceanographic Institution A private institution, founded in 1930, whose charge is the investigation of all aspects of oceanography. Interrelated research in physiology, microbiology, ecology, and animal and plant behavior is carried out by the permanent staff and visiting fellows. Woods Hole maintains several research vessels; one of them, *Alvin,*

BEAUFORT SCALE OF WIND FORCE

BEAUFORT NUMBER	GENERAL DESCRIPTION	MANIFESTATIONS	KILOMETERS PER HOUR	MILES PER HOUR	KNOTS
0	Calm	Smoke rises vertically	Less than 1	Less than 1	Less than 1
1	Light air	Wind direction shown by smoke but not by weathervanes	2–6	1–3	1–3
2	Slight breeze	Wind felt on face; leaves rustle; ordinary vane moved by wind	7–12	4–7	4–6
3	Gentle breeze	Leaves and small twigs in constant motion; wind extends light flags	13–18	8–11	7–10
4	Moderate breeze	Raises dust and loose paper; small branches are moved	19–26	12–16	11–14
5	Fresh breeze	Small trees in leaf begin to sway; crested wavelets form on in and waters	27–35	17–22	15–19
6	Strong breeze	Large branches in motion; whisling heard in wires; umbrellas used with difficulty	36–44	23–27	20–24
7	Moderate gale (high wind)	Whole trees in motion; difficulty in walking against wind	45–55	28–34	25–30
8	Fresh gale	Twigs broken off trees; progress generally impeded	56–66	35–41	31–35
9	Strong gale	Slight structural damage occurs; shingles and slate removed	67–77	42–48	36–42
10	Whole gale	Trees uprooted; considerable structural damage; seldom experienced inland	78–90	49–56	43–49
11	Storm	Very rarely experienced; widespread damage	91–104	57–67	50–56
12	Hurricane	Air at sea filled with foam and spray	Above 105	Above 68	Above 56

has been used on a number of deep dives in the Pacific to explore vent communities, and in the Atlantic to explore rift valleys and the wreck of the *Titanic*. *See* ALVIN, SUBMERSIBLES, VENT COMMUNITIES.

worm A term loosely applied to invertebrates that are longer than their cross sections. There are several phyla and many individuals of marine worms. They are Acanthocephala, Annelida, Aschelminthes, Chaetognatha, Echiuroidea, Nematoda, Nemertea, Pogonophora, Sipuncula, and Platyhelminthes. These include free-living and parasitic species. *See* INDIVIDUAL PHYLA.

wrack *See* EELGRASS.

Wrangell, Baron Ferdinand Petrovich von (1796–1870) A Russian diplomat-explorer who sailed to the Arctic on several voyages between 1820 and 1824, searching for islands north of Siberia. He failed to find what was eventually charted as the island group around 176° east longitude, the largest member of which is now called Wrangell Island. Between 1825 and 1827 Wrangell led a Russian circumnavigational expedition. From 1829 to 1835 Wrangell was the governor of Russian America—today known as Alaska—and opposed its sale to the United States.

wrasse An elongated, prickly-spined fish that is thin for its length and characterized by thick lips and sometimes protruding teeth. There are over 300 species of these members of the family Labridae, order Perciformes. They are usually inhabitants of tropical or warm temperate waters and range in size from 5 cm to 2 m (2 inches to 6.5 feet).

Wrasses are frequently part of the biota of coral reefs, where they establish commensal or mutually beneficial relationships with other fish. For example, they clean the external parasites from groupers, eels, and snappers. The relationship of the wrasse with the fish it cleans is not a peaceful one. The wrasse, in addition to picking off parasites, will attempt to pick off some of the fish's mucus layer, which apparently tastes good. Fish produce this mucus layer because it cuts down on friction and helps the fish move more easily through the water. Fish will object and harass the wrasse.

Most wrasses are brilliantly colored. The group also includes a number of species that change color with age. Wrasses can also change their sex as they change color. Interestingly, the sex change can occur more than once in the lifetime of an individual fish. *See* COMMENSAL RELATIONSHIPS, CORAL REEFS, SYMBIOSIS.

wrecks *See* MARINE ARCHAEOLOGY, SALVAGE.

Wüst, Georg Adolf Otto (1890–1977) A leading German oceanographer whose name is inexorably linked to the *Meteor* expedition. After the death of his mentor, Alfred Merz, Wüst was precipitated into the task of chief oceanographer of the German Atlantic expedition of 1925–27. This was the first study of all aspects of an entire ocean, and from the careful and extremely detailed measurements that were taken, an understanding of circulation, deep currents, temperature, and salinity strata was eventually achieved.

After World War II, Wüst reestablished the *Institut für Meereskunde* (Institute for Oceanography) at Kiel, Germany, and was its director. *See* ATLANTIC OCEAN; EKMAN, VAGN WALFRID; OCEAN CURRENTS.

xanthophyll   The yellow pigment present in yellow, orange, or brown algae. *See*
ALGAE, BROWN ALGAE, *FUCUS.*

Xiphosurida   *See* HORSESHOE CRAB.

Yangtze River Also known as the Ch'ang Chiang, Change Jiang, or Blue River; a river flowing east through central China from Tibet to the East China Sea. It is Asia's longest river, at over 5,000 km (3,100 miles) long, and in Chinese its name means "long river." The English name Blue River is peculiar since the water is most often a yellow-brown.

The Yangtze rises in Tibet and runs southeast, dropping in a series of waterfalls, then turns to the northeast, traveling through the North China Plain to the East China Sea and Yellow seas. While it is not as muddy as the Huang Ho (Yellow River), it still carries a great volume of alluvium, and extends its delta by about 20 m (about 65 feet) per year. Extensive irrigation and terracing work that has gone on for centuries has made efficient use of the great volume of the Yangtze's water. The valley of the river is an extremely important area agriculturally. The rapidly changing face of China is also a factor in the Yangtze River valley. As industrialization continues, the need for more electricity is propelling the massive dam constructions that will ultimately change the entire river valley. There is ongoing opposition to this massive project. Not only is it displacing millions of people, but environmentally, this massive hydroelectric project will alter the hydrology and climatology of the region. Geologically, the placement of such massive bodies of water on earthquake-prone land also poses problems. The earthquake incidence in such geologically unstable areas increases.

The outlet of the Yangtze is at Shanghai, which remains a principal port of China. The river is navigable for almost 2,000 km (1,250 miles) of its length by vessels of significant commercial size. *See* EAST CHINA SEA, YELLOW SEA.

yawl A fore-and-aft rigged sailing vessel with two masts. The main mast is much larger than the mizzen. *See* SAIL.

Yellow River *See* HUANG HO.

Yellow Sea Also known as Huang Hai; a marginal arm of the Pacific Ocean that lies between China north of the mouth of the Yangtze River and Korea. Its average depth is about 45 m (145 feet) because a great deal of the bottom is very shallow, particularly in the area of the Gulf of Chihli. The sea is postglacial in origin and is only about 11,000 years old. Except the Gulf of Chihli, the average depth of the Yellow Sea is 60 to 80 m (200 to 260 feet).

Both the Huang Ho (Yellow River) and the Yangtze empty into the Yellow Sea, accounting for its color. Its deep water is both cold and salty; the surface water, because of the great volume of river influx, is about 3.0 to 3.3% salt. This may be lower still in the summer.

The tidal range of the sea is high, at about 8 m or 26 feet. There are strong tidal currents in some areas.

The Tsushima Current flows through the Yellow Sea, bringing warm and more saline water into it. The climate is characterized by cold and dry weather in winter, warm and humid weather in summer. *See* MONSOON, TIDAL RANGE, TSUSHIMA CURRENT.

Yoldia Sea *See* BALTIC SEA; FINLAND, GULF OF.

Yucatán Current A surface stream in the Caribbean Sea, flowing from Honduras through the Yucatán Channel to the east of the Gulf of Mexico. In summer its maximum velocity is about 2 m per second (4 mph). *See* CARIBBEAN SEA, GULF OF MEXICO, GULF STREAM.

Yukon River A river in the Canadian northwest and in Alaska. It is one of the longest North American rivers, at 3,185 km (1,979 miles), and is named Yukon below the confluence of the Pelly and Lewes rivers. The merged stream flows from British Columbia (Canada) northward and then northwest. The White, Stewart, and then the Klondike rivers flow into the Yukon. This large volume of water from the northwest of Canada turns to enter Alaska at Fort Yukon. Only part of the Yukon is navigable; the channel is a shifting one. This factor and the cold weather make it unreliable for commercial shipping. The largest river town on the Yukon is Dawson in the Canadian Northwest Territories. The river flows through Alaska and empties into the Bering Sea. *See* BERING SEA.

Zoantharia *See* ANTHOZOA.

Zoanthidea Colonial polyps that resemble sea anemones. They are anthozoans of the phylum Cnidaria and are related to the corals. They inhabit sandy bottoms in tropical to temperate water, an ecological niche similar to that of the sponges.

The Zoanthidea collect prey by means of a mucus coating that they generate and to which plankton adhere. They tend to be small; the largest species are less than 20 cm (8 inches) in size. *See* ANTHOZOA, POLYP.

Zoochlorellae Green algae that are symbionts or commensals with the coral or coralline algae. *See* COMMENSAL RELATIONSHIPS, SYMBIOSIS, ZOOXANTHELLAE.

zoogeography A branch of biogeography concerned with the distribution of living animals in ecological space. Any species in any area is in an equilibrium with immigrants (i.e., new arrivals from other places), and with other species that are on the edge of extinction. Both of these processes are normal, but both are influenced by the activity of humans.

Continental drift has explained numerous finds of animal fossils in places where such animals could not exist today because of climate. It also accounts for the appearance of the same or similar species in areas that are now widely separated geographically but were once in close proximity. *See* CONTINENTAL DRIFT, ECOLOGY, HABITAT.

zooplankton Unicellular oceanic organisms that do not synthesize their own food but instead feed on those that do, either chemosynthetically or photosynthetically. The zooplankton may be carnivores, omnivores, or detritus feeders. *See* FOOD WEB, PLANKTON.

Zooxanthellae Symbiotic dinoflagellate algae. They derive their name from their flagellae and yellow pigments. These organisms conduct photosynthesis and live within other protists such as foraminiferans and radiolarians. They utilize the carbon dioxide produced metabolically by these hosts to manufacture sugars. These relationships are mutual or commensal, since the Zooxanthellae can leave and establish an independent existence.

The presence of the zooxanthellae within the polyps of the coral cannot be overemphasized. They are vital to the health of the coral. Since photosynthesis is the basis of this commensal relationship, the corals that derive the most benefit from their indwelling zooxanthellae are those in fairly shallow, warm water. All coral organisms and related cnidarians—the soft corals, sea anemones, gorgonians, and sea pens—are the most obvious hosts to one or another species of zooxanthellae. These dinoflagellates are also found in other, larger organisms. They are present in the giant clams (*Tridacna*) and in most nudibranchs, where they contribute to the stunning color patterns characteristic of nudibranch skins.

The presence of Zooxanthellae in coral is a stimulus to reef growth, since these protists increase the food supply available to the reef-building coral and coralline algae. *See* CORAL REEF, DINOFLAGELLATE, FORAMINIFERA, FLAGELLA, PROTOZOA, RADIOLARIA.

Zostera *See* EELGRASS.

APPENDIX I

GEOLOGIC TIMESCALE

CENOZOIC ERA			
Period	Subperiod	Years	Description
Quaternary period	Holocene (recent)	0 to 1.5 million years ago	—
	Pleistocene	1.5 to 4 million years ago	The ice ages occur in this period. In western Europe, the Alps are uplifted and the Mediterranean recedes. In the East, the collision of India with Asia closes the Tethys Sea.
Tertiary period	Pliocene	about 10 million years ago	—
	Miocene	25 million years ago	—
	Oligoncene	40 million years ago	—
	Eocene	55 million years ago	—
	Paleocene	65 million years ago	—
MESOZOIC ERA			
Period	Subperiod	Years	Description
Cretaceous	Upper	100 million years ago	—
	Lower	135 million years ago	The primitive continent of Gondwanaland breaks into the pieces now recognized as Africa, South America, and India. The western Alps are uplifted. Many species disappear during this period, an event referred to as the Cretaceous extinctions.
Jurassic	Upper	160 million years ago	—
	Middle	170 million years ago	—
	Lower	190 million years ago	The Tethys Sea extends to the Pyrenees. There are shallow seas east of the Rockies in North America.

Period	Subperiod	Years	Description
Triassic	Upper	205 million years ago	—
	Middle	215 million years ago	—
	Lower	225 million years ago	The Tethys Sea, which at its greatest extension reached from the Atlas Mountains in North Africa to the Himalayas, opens eastward. Similar plants and animals exist in Africa, South America, and India.
PALEOZOIC ERA			
Period	Subperiod	Years	Description
Permian	Upper	240 million years ago	The Paleozoic and particularly the Permian-Triassic transition is marked by extinctions of organisms, most noticeably the sessile ones, with the notable exception of the bivalves. The Permian epoch is particularly devastating, especially for organisms living near the surface of the sea. In general, the fish populations follow those of the corals, ammonites, and crinoids, and many genera disappear at the Permo–Triassic boundary.
	Lower	280 million years ago	—
Carboniferous	Upper (Pennsylvanian)	315 million years ago	—
	Lower	350 million years ago	Major plant growth on land, the basis of coal deposits. The ferns are the dominant plants.
Devonian	Upper	360 million years ago	—
	Middle	370 million years ago	—
	Lower	400 million years ago	The folding of land rocks, such as in Australia. Fish are the dominant animals.
Silurian	—	430 to 440 million years ago	The late Silurian is marked by the development of plants with cambium.

Period	Subperiod	Years	Description
Ordovician	Upper	445 million years ago	—
	Lower	500 million years ago	The major diversification of invertebrates occurs in the mid-Ordovician. Calcium carbonate shells appear on mollusks. Corals, bryozoans, brachiopods, crinoids, and nautiloids develop. Calcareous deposits are laid down over the Baltic Shield in Europe. Bone begins to form in some animals.
Cambrian	Upper	515 million years ago	—
	Middle	540 million years ago	—
	Lower	570 million years ago	At the division of the Cambrian and Precambrian, or somewhat later, the Caledonian geosyncline divides North America from Europe. Fossils and landforms that predate this event are identical. Terranes, inclusions of "foreign" territory, have been found in the Blue Ridge Mountains and the Piedmont Plateau in the southeastern United States.
PRECAMBRIAN ERA			
Period	Subperiod	Years	Description
—	—	4 to 1 billion years ago	Sponges with horny or calcareous spicules appear; carbonate plankton appeared about 1 billion years ago and filamentous green algae about 2 billion years ago. The first living system is very tentatively dated at about 4 billion years ago.

Science Online: Copyright © 2005 by Facts On File, Inc.

APPENDIX II

MARINE HISTORY:
A CHRONOLOGY OF SIGNIFICANT EVENTS

Year	Person	Event
7th c. B.C.E.	Kaleus	sails through the Pillars of Hercules (Gibraltar)
609	Necho II	commissions Phoenicians to circumnavigate Africa
6th c.	Necho II	starts canal to connect Red Sea to the Nile
600		Greek settlement at Marseille
5th c.	Anaximander	first idea of evolution; he believes that all land animals came from amphibians
4th c.	Hanno of Carthage	sails down the west African coast
350–300	Alexander the Great	commissions Admiral Nearchus to explore the Persian Gulf and the Indian Ocean
344	Aristotle	on Lesbos studying marine biology
110		mariculture (oysters) near Naples
81	Sujin	Japanese emperor encourages shipbuilding and the fishing industry
40 C.E.	Hippalus	Greek merchant sailor goes from Berenice (Egyptian Red Sea port) to India and back using monsoon direction change
79	Pliny	dies in Vesuvian eruption
140	Ptolemy	publishes map of the world—longitude
2nd c.		Chinese mariners sail to India
274	Ojin	Japanese long galley
450?	Hawaii Loa	Polynesians sail from Raiatea to Hawaiian Islands
7th c.	Brendan	sails the North Atlantic
700?		Arab sailors in Spice Islands (Moluccas)
850		astrolabe in common use
874		Iceland discovered
900		Greenland discovered
982		Greenland colonized
998	Biarni Hierulfson	sights North America
1000	Leif Ericsson	colonizes North America
1224	Abdullah ibn Ruml	geographical encyclopedia
1292		Venetian long galley developed

Year	Person	Event
1312		Canary Islands rediscovered
1337	William Merlee	at Oxford; attempts scientific weather forecasting
1405	Zheng He	sends a fleet of 63 junks on the first of six expeditions. Cheung Lo, admiral; they go to Philippines, Aden, East Africa
1415		Dutch fishing boats using drift nets
1419		Portuguese land on Madeiras
1421	Henry the Navigator	establishes a scientific observatory at Sangres
1432	Goncàlo Cabral	discovers the Azores, for Portugal
1434	João Dias	Portuguese, sails around Cape Bojador (Africa)
1455	Alvise da Cadamosto	sailing for Prince Henry discovers Cape Verde Islands
1470		Portuguese sail to the Gold Coast (Africa)
1486		Portuguese sail to Angola
1487	Bartholomeu Dias	sails around Cape of Good Hope
1492	Christopher Columbus	sails for Spain to the Spice Islands
1496	John Cabot	sails for Henry VII of England to North America
1497	Vasco da Gama	Portuguese, sails around Africa to India
1498	Columbus	discovers the Orinoco River
1500	Pedro Cabral	sails to South America and claims it for Portugal
1501	Amerigo Vespucci	sails to Brazil for Portugal. He realizes it is on a continent nowhere near Asia
1507	Martin Waldseemuller	map of the world with America as the name of the New World
1511		Portuguese reach Amboyna and Moluccas
1513	Juan Ponce de León	sights Florida
1517	Jorge Alvarez	Portuguese, sails to Canton
1519	Fernão Magellan	sails for Spain to the Orient by a westward route
1521	Francisco de Gordillo	sails up the coast of North America to the Carolinas—for Spain
1524	Giovanni da Verrazano	sails into New York Bay
1534	Jacques Cartier	sails to Labrador for France
1538	Gerardus Mercator	uses the name America to mean all of the lands of the western hemisphere
1539	Olares Magnus	map of the world
1542	Antonio da Mata	first European in Japan
1551	Pierre Belon	*Histoire naturelle des etranges poissons* (Natural history of foreign fish)
1553	Richard Chancellor	travels to Moscow via Archangel for England
1554	Guillaume Rondelet	*Livre des Poissons Marines* (Book of marine fish)

Year	Person	Event
1567	John Hawkins and Francis Drake	sail to the West Indies to harass the Spanish fleet
1569	Gerardus Mercator	publishes his atlas, cosmographic and navigational maps of the world
1570	Abraham Ortelius	first modern atlas
1576	Martin Frobisher	discovers Baffin Island and Frobisher Bay in the Canadian Arctic
1577	Francis Drake	begins his circumnavigation of the Earth
1581	William Borough	*Discourse on the Variation of the Compass*
1582	Richard Hakluyt	*Diverse Voyages Touching on the Discovery of America*
1584	Walter Raleigh	discovers and annexes Virginia (England)
1585	John Davis	sails into Davis Strait
	Lucas Waggenart, or Wagenar or Wagenaer	*Spiegel de Zeewart* (Mariner's mirror)
1586	Thomas Cavendish	leaves on voyage of circumnavigation
1587	Konrad Gesner	posthumous publication of *History of Animals*
1591	James Lancaster	first voyage to the East Indies
1592	Juan de Fuca	discovers British Columbia (Canada)
1595		Dutch expedition to the Orient
1596	Willem Barents	Dutch explorer to north, finds Spitsbergen and Barents Sea
1598	Olivier van Noort	Dutch, circumnavigation
	Visunsin	Korea, first ironclad ship
1599	Richard Hakluyt	map of North America
1600	William Gilbert	*De Magnete,* theoretical discussion on magnetism
1602	Galileo Galilei	laws of oscillation and gravitation
1603	Samuel de Champlain	explores and names New France (Canada)
1606	Luis Vaes de Torres	sails between Australia and New Guinea
1609	Henry Hudson	explores Delaware Bay and Hudson River
1616	William Baffin	explores Baffin Bay
1616	Willem Schouten and Jakob Le Maire	sail around the tip of South America and name the cape for Hoorn (Netherlands)
1617	Willibrod Snellius or Snell	uses trigonometric triangulation as a surveying technique
1622	Edmund Gunter	variation in the declination of the compass
1623		New Netherlands established
1626		Jardin des Plantes established in Paris
1627	Johannes Kepler	Rudolphine Tables of Fixed Stars

Year	Person	Event
1642	Abel Tasman	discovers Tasmania and New Zealand
1643	Evangelista Torricelli	designs the barometer
1661	Robert Boyle	*Skeptikal Chymist*
1662	Charles II	charters the Royal Society
1676	John Flamsteed	Royal Observatory (Greenwich) Astronomer
1683	Isaac Newton	mathematical theory of tides as response to gravitation of Sun and Moon
1699	William Dampier	sails to the Pacific for the Admiralty
1701	William Kidd	the pirate, hanged
1714	Gabriel Fahrenheit	mercury thermometer designed
		British government establishes a prize for an accurate measure of longitude
1728	Vitus Bering	sails through Bering Strait
	John Harrison	spring-driven clock wins Admiralty prize for longitude measure
1735	Carolus Linnaeus	*Systema Naturae* (Taxonomy)
1738	Daniel Bernoulli	*Hydrodynamica* explains pressure and velocity of fluids
1741	Alexei Cherikov	lands on California coast
1766	Louis de Bougainville	sails to the Pacific, discovers Tahiti
1767		American whaling ships in Antarctica
1768	James Cook	sail to establish a scientific station in Tahiti
1769	Benjamin Franklin	charts the Gulf Stream
1771	Bougainville	circumnavigation
1772	Joseph Priestley	discovers nitrogen
1777	Antoine Lavoisier	composition of air determined
1785	Comte de La Pérouse	voyage to the Pacific
1788	James Hutton	*New Theory of the Earth*
1789		mutiny on the HMS *Bounty*
1791	George Vancouver	explores the western coast of North Americ
1798	Baron Cuvier	publishes *Elementary Survey of Natural History of Animals*
1800	Alexander von Humboldt	explores the Orinoco
1801		metric system becomes the only system of weights and measures in France
1802	John Dalton	publishes his work on the atomic theory
	Nathaniel Bowditch	*Practical Navigator*
1804	John Stevens	of Hoboken, New Jersey, launches a steamship; the *Little Juliana* is a technical success and a financial failure
1805	Francis Beaufort	scale of wind velocities

Year	Person	Event
1807	Robert Fulton	commercial success of a steamboat, the *Clermont*
1809	Chevalier Lamarck	*Système des animaux sans vertèbres* (System of invertebrates)
1812	Baron Cuvier	*Recherche sur les ossements fossiles des quadrupeds* (Research on the bony fossils of quadropeds)
1815		volcanic explosion of Tamboro (Indonesia) and resultant tsunami
1818	John Ross	Northwest Passage attempt; finds cliffs in the Arctic near Baffin Bay
1822	Chevalier Lamarck	*Histoire naturelle des Animaux* (Natural history of animals)
1830	Charles Lyell	described geologic system of Earth's formations
1831	Charles Darwin	sails on the *Beagle*
	James Clark Ross	determines the magnetic North Pole
1832	Nils Nordenskjold	sails the entirety of the Northeast Passage
1839	James Clark Ross	sails to the Antarctic with the *Erebus* and *Terror*
1840	Louis Agassiz	*Etudes sur Glaciers* (Studies on glaciers)
	Charles Darwin	*Zoology of the Voyage of the Beagle*
1842	Matthew Maury	begins his research on sounding of the ocean bottom
1846		Smithsonian Institution founded
1858	C. Darwin and A. Wallace	in a joint paper at the Linnaean Society present the theory of the survival of the fittest
1859	Ferdinand de Lesseps	begins the Suez Canal
1865	Antonio Snider-Pelligrini	another proposal that the continents were once a single entity
1873	Julius Payer and Karl Weyprecht	discover Franz Josef Land—name it for the Austrian emperor
1876		HMS *Challenger* returns to England
1904	Robert Scott	to the Antarctic
1906	F. Nansen	publishes his account of the Norwegian North Pole Expedition
1909	Robert Peary	first man at the North Pole
1911	Roald Amundsen	first man at the South Pole
1912	R. F. Scott	attempts to reach the South Pole, dies
1916	Paul Langevin	designs submarine detection device
1919	Ernest Shackleton	publishes *South*, the account of the 1914–17 expedition to Antarctica
1923	K. M. Deryugin	USSR polar research station
1925		Meteor Expedition to study the Atlantic
1926	Amundsen, Ellsworth and Nobile	fly over the North Pole in an airship

Year	Person	Event
	Richard Byrd and Floyd Bennett	fly from Spitsbergen to the North Pole
		Scott Polar Research Institute established in Cambridge (England)
1935	William Ewing	seismic study of the sea floor
1947	R. E. Byrd	Little America, polar research station in the Antarctic
1957		International Geophysical Year (IGY) begins
1961	Robert Dietz	*Continent and Ocean Basin Evolution*
1977		discovery of vent communities
1984		discovery of cold seeps in the Atlantic Ocean
1993		volcanic eruptions on the Juan de Fuca Ridge followed by the discovery of many new chemosynthetic organisms
1995		submersible in the Challenger Deep finds another site of new-to-science chemosynthetic organisms unlike those at hydrothermal vents
2001		region west of Papua (Indonesia) identified as an "epicenter of biodiversity," where several new species of corals, shrimps, and vertebrates, notably sharks that seem to "walk" on their fins, are found
2004		massive undersea earthquake on December 26 occurs in the Indian Ocean west of Indonesia. The resulting tsunami devastates western Indonesia, Thailand, and the east coast of India. This quake occurs in roughly the same region where the 1883 quake struck
2006		*computer* projections of global warming trends predict that the fabled "Northwest Passage" sea route from Europe to Asia will be a reality in summertime before the end of the 21st century

APPENDIX III

TAXONOMIC CLASSIFICATION OF LIVING ORGANISMS FIVE-KINGDOM MODEL

Kingdom	Phylum	Class	Subclass
Monera (No nuclear membrane)	Bacteria		
	Cyanophyta (blue-green algae) ~230 species		
Fungi (none are aquatic)			
Protista (Nuclear membrane)	Chlorophyta (green algae) ~10,000 species		
	Pyrrophyta (dinoflagellates) 1,100 species		
	Phaeophyta (brown algae) 1,500 species (includes diatoms, watermolds and coccolithophorids)		
	Rhodophyta (red algae) ~4,000 species		
	Protozoa (no chlorophyll) ~27,000 species	Mastigophora (flagellates) Sarcodina (formaniferans) Ciliophora Spirozoa	
Metaphyta (plants)	Traceophyta (none are marine)	Gymnospermae	
		Angiospermae (flowering plants)	
Metazoa (animals)	Porifera (sponges) ~5,000 living species	Calcarea (calcium spines) Desmospongiae Sclerospongiae (coralline) Hexactinellida (siliceous spines)	
	Cnideria	Hydrozoa (polyps) Scyphozoa (jellyfish) Anthozoa (corals, sea anemones) Cubozoa	

Kingdom	Phylum	Class	Subclass
	Ctenophora (comb jellies)		
	Phatyhelminthes (flatworms)	Turbelleria (free-living)	
		Trematoda (flukes) *may be grouped with Turbelleria	
		Cestoda (tapeworms)	
	Nemertea (ribbon worms)		
	Aschelminthes (roundworms) [*this phylum may be considered differently and each class is called a separate "minor phylum" by some researchers.*]		
		Nematoda	
		Rotifera	
		Gastrotricha	
		Kinorhyncha	
		Pripulida	
		Acanthocephala	
		Ectoprocta (bryozoans)	
		Chaetognatha (arrow worms)	
		Cycliophora	
		Phoronida (horseshoe worms)	
		Echiuroida	
		Loricifera	
		Nematophora (horsehair worms)	
		Sipuncula (peanut worms)	
		Pogonophora (beard worms)	
		Brachiopoda (lampshells)	
	Mollusca (this classification is under review)		
		Monoplacophora	
		Amphineura (chitons)	
		Gastropoda (snails) ~65,000 species	
		Lamellibranchia or Pelecypoda ~7,500 species (oysters, clams)	
		Cephalopoda (squid, octopi)	
	Annelida (segmented worms)	Polychaeta ~90% of all annelids	
		Myzostomaria	
		Oligochaeta—largely freshwater	

Kingdom	Phylum	Class	Subclass
	Arthropoda (segmented legs) ~30,000 marine species	Chelicerata (horseshoe crab) Pycnogonida (sea spiders)	
		Crustacea (lobsters, crabs) ~26,000 species	Ostracoda Copepoda Branchiura Cirripedia (barnacles) Malacostracea
	Echinodermata (~6,000 species)	Crinoidea (sea lilies) Asteroida (starfish) Ophiuroidea (brittle stars) Echinoidea (sea urchins) Holothuroidea (sea cucumbers)	
	Hemichordata	Tunicata (sea squirts) Thaliacea (salps) Larvacea	Ascidiacea
		Cephalochordata	Lancelets
	Chordata	Pisces	
			Agnatha (jawless fish) Chondrichthyes (sharks, rays) Osteichthyles (bony fish)
		Amphibia—mostly freshwater Reptilia (snakes, turtles, crocodilians) Aves (birds) Mammals (seals, bears, whales)	

APPENDIX IV

RECENT OR ONGOING MARINE SCIENCE RESEARCH PROJECTS

IMMA   International Marine Mammal Association

BALTICSEAWEB   Baltic Marine Environment Information System

IFAR   International Fisheries and Aquatic Research

AWI   Alfred Wegener Institute for Polar and Marine Research

IIOE   International Indian Ocean Expedition 1990–1995

IOC   Intergovernmental Oceanographic Commission
                (branch of UNESCO; www.unesco.org/iocweb)

APPENDIX V

WEB SITES

GENERAL WEB SITES

National Oceanographic Data Center http://www.nodc.noaa.gov

Scripps Institution of Oceanography: University of California at San Diego http://scilib.
ucsd.edu/sio

Wind and Sea: Oceanic and Atmospheric Science Locator http://www.lib.noaa.gov/
docs/windandsea.html

OTHER USEFUL WEB SITES

Note: Some sites open using the www prefix; others do not. To access the sites listed below, readers should type URLs with or without the prefix. Most, but NOT all, are available without a subscription.

METASITES

This list of Web sites is a virtual supermarket; it has something for everyone, both on a particular subject and on a variety of levels. It is useful for seeing just how much information there might be on a topic—from newsletters for children to articles in professional journals. Another list of more subject-specific Web sites follows this one.

Biology

Biology Browser. URL: http://www.biologybrowser.org. This is the link to BIOSIS's teaching resources and to more than 24,000 biology Web sites. They range from details about new taxa to lesson plans for elementary school projects. Choose a location, a person, or an organism to search for.

British Library Online. URL: http://www.bl.uk/onlinegallery/ttp/ttpbooks.html. The digitizing process at the British Library is ongoing. They intend to make a number of their unique books and images available at this site.

Cold Spring Harbor Symposia. URL: http://library.cshl.edu/symposia. This laboratory site has collected a photo archive of participants.

The Molecular Level. URL: http://www.usrn.maine.edu/~rhodes/index.htm. This site covers basic biology and the chemistry needed for it, collected by the faculty of the University of Southern Maine, Portland.

Virtual Cell Animation Collection. URL: http://Vcell.ndsu.nodak.edu/animations. This collection is from North Dakota State University in Fargo. It provides information on anything from the action of transfer RNA to how mitochondria work and how photosynthesis occurs.

Definitions

The Dictionary of Algorithms and Data Structures. URL: http://www.nist.gov/dads. This scientific phrasebook concentrates on the mathematical end of science. It comes from the National Institute of Standards and Technology in Gaithersburg, Maryland.

Earth Sciences

Geo-guide. URL: http://www.geo-guide.de. This portal is maintained by German universities in English. It is a lead to more than 300 Web sites for earth sciences including geology, geography, meteorology, and paleontology.

Ecology

The Knowledge Network for Biocomplexity. URL: http://knb.ecoinformatics.org/home. html. This group has collected hundreds of data sets describing both short-term and long-term ecological functions. This database concentrates on biocomplexity—the big picture of how an ecosystem operates. Some data is available without charge.
National Biological Information Infrastructure. URL: http://159.189.176.5/portal/ server.pt. This is the home page for a government-supported clearinghouse for biological, environmental, and agricultural information. It is very detailed.

Environment

Environmental Protection Agency. URL: http://www.epa.gov. This is the EPA home page.
United Nations. Environment Programme: Global Data Portal. URL: http://geodata. grid.unep.ch. The site has collected information from every region on topics ranging from populations to freshwater availability to weather to natural disasters and pollution, among others.

Federal Science Register

Federal R & D Project Summaries. URL: http://ww.osti.gov/fedrnd. This site combines the workings of several U.S. agencies: Department of Energy, Environmental Protection Agency, National Science Foundation, and others. This is a site for professionals.

Films

Research Channel. URL: http://researchannel.org. This site is a compendium of lectures, interviews, forums, and films on a number of subjects. More than 25 universities, along with the National Institutes of Health, have collaborated to create the Research Channel, a reservoir of more than 1,700 programs.

Life

Library of Life. URL: http://bio-ditrl.sunsite.ualberta.ca. This is a good site for films, animation, and other teaching tools.
Martindale's The Reference Desk. URL: http://www.martindalecenter.com. This site is a bibliography of Web sites, each on a specific subject, such as chemistry, computers, and astronomy.
The Tree of Life. URL: http://tolweb.org. This site is an online glossary database that contains images showing the relationships of plants and animals. Contributed material is also used. The site is maintained by David Maddison of the University of Arizona in Tucson.

Museums

National Academy of Sciences Museum. URL: http://www.koshlandscienceniuseurn.org. The Marian Koshland Science Museum's portal is for students ages 13 and older.
National Biology Information Infrastructure. URL: http://usbiotechreg.nbii.gov. This site compiles data from the EPA, USDA, and FDA and covers topics such as drugs, genetically modified plants, and more.
National Institute of Standards and Technology. URL: http://www.nist.gov. This site provides information for mathematicians, computer scientists, chemists, and physicists.

National Oceanographic Data Center. URL: http://www.nodc.noaa.gov. This site makes over 12 databases accessible, each concentrating on oceanic information. It is part of the National Oceanic & Atmospheric Administration.

Natural History

American Museum of Natural History. URL: http://digitallibrary.amnh.org/dspace. This site is the archive of the periodical published by the American Museum of Natural History in New York, covering 1881 to the present. It includes the museum's other titles, as well as the Bulletin and Novitiate series.

Paleontology

The Paleobiology Database. URL: http://www.paleodb.org. This site is maintained by the University of California at Santa Barbara. It is searchable by species.

The Paleontology Portal. URL: http://www.paleoportal.org. This site is produced by the University of California Museum of Paleontology, the Paleontological Society, and others. Look for fossil sites here.

Research Expeditions

The Field Museum of Chicago. URL: http://www.fieldmuseum.org/expeditions. This site contains material about the Field Museum of Chicago's research interests in a number of areas, notably archeology and anthropology in the Americas.

Science Lectures

Vegas Science Trust. URL: http://www.vega.org.uk. This British site provides a number of lectures and interviews on science subjects. The lecturers are the researchers themselves.

Theories

Channel Thirteen. URL: http://www.thirteen.org/bigideas. This is part of a TV series called *Big Ideas*, and it explores a number of fields where there are no answers to questions. It includes questions about relativity, quantum theory, extraterrestrials, and more.

SUBJECT-SPECIFIC WEB SITES

Acid Rain

U.S. Geological Survey. URL: http://bqs.usgs.gov/acidrain. This site is a compilation of data on precipitation chemistry. It is sponsored by the U.S. Geological Survey.

Africa

Rockefeller Foundation. URL: http://www.africancrops.net. This site provides information collected by the Rockefeller Foundation concerning agriculture in Africa.

Algae

Partnership for Interdisciplinary Studies of Coastal Oceans. URL: http:// www.piscoweb.org/cgi-bin/qml/newalgaequery.qml. Information about the range, habitat, description, and abundance of algae from the western coast of North America has been collected by the Partnership for Interdisciplinary Studies of Coastal Oceans and put on this site. Photos are included.

Amber

Emporia State University. URL: http://www.emporia.edu/earthsci/amber/amber.htm. This site contains data provided by Susan Ward Aber of Emporia State University,

Kansas. It describes information she has collected about amber and the fossils found in it.

Amphibians

Global Amphibians. URL: http://www.globalamphibians.org. This is a population survey and assessment archive from The World Conservation Union, NatureServe, and Conservation International.

Snakes in the United States. URL: http://www.cnah.org/index.asp. This compilation of information focusing on snakes and amphibians is produced by Joseph Collins and others at the University of Kansas, Lawrence.

University of California. URL: http://www.amphibiaweb.org. This site covers taxonomy and species information.

Animals

Digital Library of Dolphin Embryonic Development. URL: http://www.neoucom.edu/DLDD. This site contains anatomical and morphological images of spotted dolphin embryos and fetuses.

The National Wildlife Health Center. URL: http://www.nwhc.usgs.gov. The National Wildlife Health Center is a clearinghouse for diseases that affect wild animals. Descriptions of this data are available on this site.

Natural History Museum in London, England. URL: http://www.nhm.ac.uk/visit-us/whats-on/temporary-exhibitions/wpy/OnlineGallery. This site contains contest-winning animal photographs that have been compiled by the Natural History Museum in London.

Anthozoa (*see also* Cnidarians, Coral)

Vreni Haussermann. URL: http://www.anthozoa.com. This site was created by Vreni Haussermann. It concentrates on corals and sea anemones.

Arctic (*see also* Climate, Global Warming)

NOAA. URL: http://www.arctic.noaa.gov/detect. This site, Arctic Change, is maintained by the U.S. National Oceanic and Atmospheric Administration. It provides information on the present state of Arctic ecosystems and on climate in historical context.

University of Waterloo, Ontario, Canada. URL: http://www.socc.ca. This site, The State of the Canadian Cryosphere, studies the snow, ice, polar ice caps, glaciers, and permafrost. it IS riosteci by the University of Waterloo, Ontario.

Astrobiology

San Francisco Exploratorium. URL: http://www.exploratorium.edu/astrobiology. This site looks at the possibility of life on other planets and moons.

Atmosphere (*see also* Meteorology)

European Geosciences Union. URL: http://www.copernicus.org/EGS/acp. This site leads to the Atmospheric Chemistry and Physics open-access journal.

Marion Koshland Science Museum. National Academy of Sciences, Washington, D.C. URL: http://www.koshlandsciencemuseum.org. This site looks at carbon dioxide and its implications for the atmosphere.

Bacteria

University of Birmingham, England. URL: http://colibase.bham.ac.uk. This database is devoted to the genetics of *E. coli*.

University of Nottingham, England. URL: http://www.nottingham.ac.uk/quorum. Bacteria can "sense" the presence of others. The chemical signals and their quantities are detected and monitored in this database of bacterial sensing.

Bacteriology

L'Ecole Nationale Veterinaire, Toulouse, France. URL: http://www.bacterio.cict.fr. This database contains current and previous names of bacteria.

Technical University, Munich, Germany. URL: http://www.microbial-ecology.net/probebase. This site covers bacteria that are difficult to culture in labs, which are studied by a group at the Technical University in Munich.

Biochemistry

The Department of Environment and Heritage, Canberra, Australia. URL: http://www.environment.gov.au/biodiversity/index.html. This is the government's Department of Environment, the sponsoring agency for Fauna Online and Flora Online. Both focus on biodiversity and concentrate on Australia's original biome.

European Bioinformatics Institute. URL: http://www.ebi.ac.uk/chebi. This site is a database providing information about the cellular reactions of the small molecules, ions and atoms, in a cell's environment.

Harvard Medical School. URL: http://chembank.broad.harvard.edu. This database examines a variety of biologically significant molecules.

Biology

American Association for the Advancement of Science. URL: http://www.biosciednet.org/portal. This is a site of teaching tools.

Center for Biodiversity Informatics. URL: http://www.ncbi.org.in. This site is devoted to the biome of India.

University of Arizona in Tucson. URL: http://www.biology.arizona.edu. This site is designed for undergraduates who are looking for more illustrative material and tutorials explaining concepts.

Birds

Bird Studies Canada. URL: http://www.bsc-eoc.org/avibase.jsp. This is a bird count site.

Ornithology Laboratory, Cornell and The Audubon Society. URL: http://www.ebird.org. This is a bird count site.

University of New Mexico. URL:http://elibrary.unm.edu/sora/index.php. This base is the Searchable Ornithological Research Archive (SORA).

Bryozoa

Natural History Museum, London, England. URL: http://www.nhm.ac.uk/hosted_sites/iba/bryozoa_home_page/default.html. This Web site concentrates on Permian specimens. It depicts both fossil and recent bryozoans.

Cell Membranes

Ion Channel Media Group, Montreal, Canada. URL: http://www.ebi.ac.uk/compneursrv/LGlCdb/LGICdb.php. This Web site shows and explains movement of materials cells.

University of Warwick, England. URL: http://www.chemsoc.org/exemplarchem/entries/2002/Tim_Smith. This tutorial, created by Tim Smith, shows and explains movement of materials cells.

Cell Nucleus

University of Alberta, Canada. URL: http://www.cellnucleus.com. This site depicts the workings of cells and has very good visual content.

Chemicals of Biological Interest

European Bioinformatics Institute (EBI). URL: http://www.ebi.ac.uk/chebi. Micronutrients, small molecular entities, alternate names, and other information has been collected in this database called ChEBI.

Chemistry

University of Oregon, Eugene. URL: http://greenchem.uoregon.edu/gems.html. This directory to environment-friendly chemistry, or green chemistry, is part of the movement to reduce the use of toxic materials. This directory is a compilation of lab procedures, tutorials, and information. It links to the EPA's software.

Chlamydomonas

Duke University, North Carolina. URL: http://www.chlamy.org. This is a Web site of everyone's favorite green glob.

Climate

Harvard University, Cambridge, Massachusetts. URL: http://www.snowballearth.org. Paul Hoffman posits that in Earth's history there were very cold spells. This site details that theory.

International Research Institute for Climate Prediction at Columbia University, New York. URL: http://iri.columbia.edu. This site contains climate history and predictions.

National Snow and Ice Data Center, Boulder, Colorado. URL: http://nsidc.org. This site discusses snow and ice.

NOAA. Washington, D.C. URL: http://www.beringclimate.noaa.gov/index.html. This site discusses Polar climate.

Cnidarians (*see also* Anthozoa, Coral)

Cnidarian Evolutionary Genomics Database or CnidBase at Boston University, Boston, Massachusetts. URL: http://cnidbase.bu.edu. This site details the history and morphology of sea anemones.

Colonial Organisms (*see* Bryozoa)

Commensal relationships

University of Connecticut, Storrs. URL: http://web.uconn.edu/mcbstaff/graf/Sym.html. Larger animals such as fish form commensal relationships with bacteria. The site details such interactions and relationships.

Cone Shells

Burke Museum of Natural History and Culture of Seattle, Washington. URL: http://biology.burke.washington.edu/conus/index.php. These molluscan carnivores produce a variety of toxic substances. This catalog, created by Trevor Anderson and Alan Kohn of the Burke Museum of Natural History and Culture of Seattle, is an attempt to sort the taxonomic details of these organisms.

Conversion Tables

University of Geneva, Switzerland. URL: http://mypage.bluewin.ch/berthod/vuc. This is a compiled means of converting units in one system into another (e.g., miles to kilometers).

Coral

NASA and the University of South Florida. URL: http://oceancolor.gsfc.nasa.gov/cgi/landsat.pl. This is a photographic archive.

Coral Reef

Reef Check. URL: http://www.reefcheck.org. This volunteer group is based in Pacific Palisades, California. It reports on reef health worldwide.

World Fish Center in Penang, Malaysia. URL: http://www.reefbase.org. This group collects information about the reef biome from many countries.

Darwin

American Museum of Natural History, New York. URL: http://Darwinlibrary.amnh.org. This Web site is the digital library of Darwin's work and that of others. It was collected by the museum to accompany the Darwin exhibit in 2006.

Devonian period

Academy of Natural Sciences, Philadelphia, Pennsylvania. URL: http://www.devonian-times.org. Dennis Murphy has collected this database. It includes basics on the flora and fauna of the period and is specifically about Red Hill, a Pennsylvania fossil site.

Diatoms

Academy of Natural Sciences, Philadelphia, Pennsylvania. URL: http://diatom.acnatsci.org/AlgaeImage. These tiny protists typify the water they live in. This site contains databases that provide a wealth of information about them. The databases were constructed by several people at the Academy of Natural Sciences in conjunction with an exhibit.

Dinosaurs (*see also* Paleontology)

Fred Bervoets of Rotterdam, the Netherlands. URL: http://www.dinodata.net. This database lists information about and provides pictures of hundreds of dinosaurs.

University of Chicago. URL: http://www.taxonsearch.org. This site has collected the newest names of dinosaurs and their relatives. Pictures of flying dinosaurs are also available.

DNA (*see also* Genetic Code)

Joint Genome Institute, Walnut Creek, California. URL: http://www.genornesonline.org. This Web site tracks the DNA sequencing projects at Genome's online database. It is maintained by Nikos Kyrpides.

Earth

American Geological Institute of Alexandria, Virginia. URL: http://www.earthscience-world.org/imagebank. This Web site contains pictures applicable to geology, oceanography, and climatology. They are from the Institute's collection.

Digital Library for Earth Systems. URL: http://www.dlese.org. This is a digital library system that has sites that cater to the full range of users, from elementary school students to professionals.

Encyclopedia of Earth and Space Science, University Corporation for Atmospheric Research, Boulder, Colorado. URL: http://www.windows.ucar.edu/windows.html. The Encyclopedia of Earth and Space Science Web site serves a range of grade levels with tutorials, photos of objects, and information about terrestrial phenomena.

NASA's Ames Research Center. URL: http://worldwind.arc.nasa.gov.

National Geophysical Data Center, Boulder, Colorado. URL: http://www.ngdc.noaa.gov/dmsp/download.html.

Universitat. Aachen, Germany. URL: http://www.blue-marble.de/night.php.

All three sites show Earth from space using software from NASA.

University of California in San Diego. URL: http://www.earthguide.ucsd.edu. This is a Web site bibliography.

Earthquakes (*see also* Tsunamis)

U.S. Geological Survey. URL: http://Earthquake.usgs.gov/image_glossary. This is the U.S. Geological Survey's visual glossary.

U.S. Geological Survey of Pasadena, California. URL: http://Pasadena.wr.usgs.gov/step. This local group has created a map illustrating the historical level of risk.

Ecology

Long Term Ecological Research (LTER). URL: http://TERnet.edu. This site has more than 2,000 data sets from sites in the United States and Antarctica.

Embryology

Cold Spring Harbor Laboratory. URL: http://www.gastrulation.org. This is a collection of videos from the CHL Press showing stages of embryonic development.

Environment

Columbia University. Center for International Earth Science. URL: http://wwf.ca/AboutWWF/WhatWeDo/The NatureAudit/InteractiveMap/Index.html. Or URL: http://sedac.ciesin.Columbia.edu/entri/index.jsp. These sites look at wildlife in North America and the effect of environment on wildlife populations. It is a compilation of material from the World Wildlife Fund Canada and Columbia University's Center for International Earth Science Information Network Law.

Environmental Protection Agency. URL: http://cfpub.epa.gov/si. This is the EPA public archive.

Harvard University. URL: http://www.ecolex.org/ecolex/index.php. This site is a global source for environmental legislation.

Environmental Management System

Environmental Protection Agency. URL: http://www.epa.gov/eims. This site has links to full-text reports and data from the EPA pertaining to environmental management.

Environmental Trends

World Resources Institute, Washington, D.C. URL: http://www.earthtrends.wri.org. This site includes figures from the World Bank, the U.S. Geological Survey, some UN agencies, and others.

Enzymes (*see* Peptidases)

Extremophiles

Marine Biology Laboratory of Woods Hole, Massachusetts, and Montana State University, Bozeman, Montana. URL: http://serc.carleton.edu/microbelife. The topics covered on this site range from tutorials on extremophiles to case studies.

Evolution

The National Academies of Science. URL: http://Nationalacademies.org/evolution. The National Academies of Science have produced a number of reports, position papers, and summaries of statements on evolution that can be found on this site.

University of Adelaide Library, Adelaide, Australia. URL: http://www.library.adelaide. edu.au/digitised/fisher. This archive was collected by R. A. Fisher, a British geneticist who combined evolutionary theory and statistical analysis of variance. His works are collected in a digital archive by the University of Adelaide Library.

University of California Museum of Paleontology. URL: http://evolution.berkeley.edu. This site is a simple discussion of evolutionary theory and appropriate responses to creationists. Age-appropriate material is available.

Eye (*see* Sight)

Fish

California Academy of Sciences. URL: http://www.calacademy.org/research/ichthyology/catalog/fishcatsearch.html. This comprehensive catalogue of fish describes and has photos of over 29,000 species.

Swedish Museum of Natural History. URL: http://www.nrm.seve/pisces/acara/welcome.shtml. Chiclids are the fish equivalent to Darwin's finches. They are represented in a database maintained by the Swedish Museum of Natural History.

Fossils

Roger Perkins, Jefferson, Arkansas. URL: http://www.fossilmuseum.net. This private collection, The Virtual Fossil Museum, is searchable by location or organisms.

Susan Ward Aber of Emporia State University, Kansas. URL: http://www.emporia.edu/earthsci/amber/amber.htm. This site contains data provided by Susan Ward Aber of Emporia State University. It describes information she has collected about amber and the fossils found in it.

Frogs

Midwest Frogs. URL: http://www.midwestfrogs.com. This commercial site collects films, soundtracks, and information about frogs. It concentrates on frogs of the midwestern United States and Canada.

Fuels

U.S. Geological Survey. URL: http://energy.cr.usgs.gov/oilgas/noga. This is a database about fuel reserves and exploration in the United States.

Genes and Genetic Testing

National Library of Medicine. Genetic Diseases. Bethesda, Maryland. URL: http://www.ghr.nlm.nih.gov/ghr. This site looks at the effects of genetic variation on human health.

Gene Therapy

Indiana University School of Medicine. URL: http://www.wellscenter.iupui.edu//MMIA/ htm/animations.htm. These animations were created by the medical school and are intended for use by high school students.

Genetic Code

National Library of Medicine, Bethesda, Maryland. URL: http://www.history.nih.gov/ exhibits/nirenberg. The history of DNA research is outlined here.

Genetics

Genetic Science Learning Center, University of Utah, Salt Lake City. URL: http://gslc. genetics.utah.edu. This is a simple tutorial.

Tech Museum of Innovation in San José, California. URL: http://www.thetech.org/ exhibits/online/ugenetics. The basics of genetics are available on this site from the Understanding Genetics exhibit at the Tech Museum.

Genome

Boston University. URL: http://www.stellabase.org. This site is devoted to one of the many sea anemones whose genome has been sequenced. John Finnerty and his coworkers at Boston University manage this site. It houses information enabling researchers to make gene comparisons.

Cold Spring Harbor Laboratory, Cold Spring Harbor, New York. URL: http://www. dnai.org. This is an interactive database aimed at students.

University of California, Riverside. URL: http://bioweb.ucr.edu/ChemMine/search. php?opt=2 or URL: http://www.ncbi.nlm.nih.gov/projects/CCDS. Chemicals are used to disable genes in order to study them. Lists of these chemicals are compiled by the National Institutes of Health and are found at this site, ChemMine. Here, molecules can be searched for by structure or chemical property.

Geology (*see also* specific periods)

Crystal Formation and Structure. Minerological Society of America. URL: http://www. geo.arizona.edu/AMS/amcsd.php. This site is a crystal structure database.

GEO-DATA Explorer, U.S. Geological Survey. URL: http://geode.usgs.gov. This is a portal for more than 100 geological and geographic data sets.

Geo-Images Project, University of California, Berkeley. URL: http://geoimages.berkeley. edu/GeoImages.html. These are all images of sites in the U.S. useful for teaching.

National Museum of Natural History. URL: http://www.mnh.si.edu/earth. This site explains the exhibit in the museum and supplements the displayed items. It will include more material as it expands.

Glaciers

National Snow and Ice Data Center, Boulder, Colorado. URL: http://nsidc.org/data/glacier_photo. The site has excellent glacier photographs.

Global Warming (*see also* Atmosphere-Carbon Dioxide, Climate)

Environmental Media Services. URL: http://www.realclimate.org. This is a blog on climate.

GreenHouse Gas Online

University of Edinburgh, Scotland. URL: http://www.ghgonline.org. This is a searchable database maintained by Dave Reay of the University of Edinburgh.

Hawaii

Flora. Smithsonian Institution. URL: http://rathbun.si.edu/botany/pacificislandbiodiversity/ hawaiianflora/index.htm or rathbun.si.edu/botany/pacificislandbiodiversity/marquesas-flora/index.htm. These are botany sites with up-to-date references to Hawaiian flora.

Histology

University of Iowa's Medical School, Ames. URL: http://www.siumed.edu/~dking2/ VSindex.htm. This site is a virtual slide collection of horseshoe crabs.

Hurricanes

NASA. URL: http://www.nasa.gov/vision/earth/lookingatearth/hurricane_2005.html. This site is part of NASA's home page. It details information about current storms and past events.

NASA Photographs. URL: http://msfc.nasa.gov/newsroom/camex/camphotos.html. This is a site of NASA's hurricane photographs.

Isotopes

Brookhaven National Laboratory, Upton, New York. URL: http://www.nndc.bnl.gov. This site is useful for information about radioisotopes and their decay products.

Jellies (formerly called jellyfish)

Harvard Museum of Comparative Zoology, Cambridge, Massachusetts. URL: http://jellieszone.com or URL: http://www.siphonophores.org. These two sites, compiled by a Harvard graduate student, provide everything you'd want to know about siphonophores, jellyfish, and "other gelatinous zooplankton."

Lava (see also Volcanoes)

U.S. Geological Survey. URL: http://volcanoes.usgs.gov/Products/Pglossary. This is an illustrated dictionary on volcanoes.

Lectins (see Proteins)

Limulus

University of Delaware. URL: http://www.ocean.udel.edu/horseshoecrab or URL: http:// wnw.horseshoecrab.org. This site provides information about the life, habitat, and evolution of horseshoe crabs.

Lipoproteins

Medical Research Council Laboratory of Molecular Biology, Cambridge, England. URL: http://www.mrc-lmb.cam.ac.uk/genomes/dolop. This site provides information about genomes. It also provides links to the database Swiss-Prot where one can analyze structure and sequence in these molecules.

Magnetic Field of Earth

U.S. Geological Survey. URL: http://geomag.usgs.gov. The Earth is changing its polarity. This site is a compilation of geomagnetic information from 14 observatories.

Mammals

American Society of Mammalogists. Editor, Virginia Hayssen, Smith College, Northampton, Massachusetts. URL: http://www.science.smith.edu/departments/Biology/VHAYSSEN/msi/msiacounts.html. This database covers taxonomy, anatomy, distribution, and other aspects of mammals worldwide.

Miocene Mammal Mapping Project, University of California, Berkeley. URL: http://www.ucmp.berkeley/edu/miomap. This site has maps of remains by species, age, and formation.

Maps

NASA and other U.S. agencies. URL: http://www.geodata.gov/wps/portal/gos. Not all maps are just geographical. The site includes information about wetlands, agricultural conditions, red tides, populations, and other materials.

Meteorology

University Corporation for Atmospheric Research, Boulder, Colorado. URL: http://www.meted.ucar.edu. This site consists of tutorials on the atmosphere.

Microbe Library

The American Society for Microbiology. URL: http://www.microbelihrary.org. The site contains photos, diagrams, tutorials, and other illustrative material. The course material is not free.

Microscopy

Dennis Kunkel, Kailua, Hawaii. URL: http://education.denniskunkel.com. This site is a personal collection of photos, diagrams, tutorials, and other illustrative material.

Migration (*see* **Turtles**)

Minerals

David Barthelmy of Houston, Texas. URL: http://webmineral.com. This site collects data on minerals: composition, crystal structure, and properties.

Molecular Biology

European Bioinformatics Institute (EBI). URL: http://www.ebi.ac.uk/chebi. This is another database that provides links between the growing number of medical and molecular biological sites. It was created by a multinational European group.

Josef Koenig of the Medical University, Vienna, Austria. URL: http://www.meddb.info. This database provides links between the growing number of medical and molecular biological sites.

Molecular Movement

Mark Gerstein, Yale University, New Haven, Connecticut. URL: http://bioinfo.mbb. yale.edu/molmovdb. How do molecules change their shapes to achieve different reactions? This is a database of macromolecular movements.

Mollusks (*see* **Cone Shells, Seaslugs**)

Nematodes

NEMBASE University of Edinburgh, Scotland. URL: http://www.nematodes.org/nematodeESTs/nembase.html. This is a genome database hosted by the University of Edinburgh that explores nematodes.

Ohio State University, Columbus. URL: http://www.oardc.ohio-state.edu/nematodes. This research effort is aimed at using nematodes to control other insect pests.

University of Nebraska, Lincoln. URL: http://nematode.unl.edu. This group studies these worms in relation to plant disease.

Nuclear Chemistry and Technology

Alsos Digital Library for Nuclear Issues, Washington and Lee University, Lexington, Virginia. URL: http://alsos.wlu.edu. This group collects information about the history and technology of fusion and fission.

Ocean Life

Census of Marine Life. URL: http://www.iobis.org. The Ocean Biogeographic Information System (OBIS) is a compendium of information about marine organisms.

Ocean Science

American Geological Institute, Alexandria, Virginia. URL: http://www.earthscience-world.org/imagebank. These pictures of Earth are applicable to geology, oceanography, and climatology.

European Geophysical Union (EGU). URL: http://www.cosis.net/rnernbers/journals/df/recent.php?j_id=9. This site contains general articles about oceans.

National Oceanic and Atmospheric Administration. URL: http://www.oceanexplorer. noaa.gov. This site is an educational adaptation of NOAA information.

U.S. Navy. URL: http://www.pmel.noaa.gov/vents/acoustics.html or URL: http://ocean-explorer.noaa.gov/explorations/sound01/sound01.html. These sites are compendia of recordings made by the U.S. Navy. The sounds recorded are naturally occurring, made by either marine vessels or marine organisms. They are maintained by NOAA.

Ozone

Ozone depletion (ozone hole). EPA. URL: http://www.epa.gov/ozone or URL: http:// daac.gsfc.nasa.gov/upperatm/CODI. These sites collect data and photographs of the ozone levels in the stratosphere.

Paleontology (*see also* specific geologic periods and Geology)

Life on Earth, University of California, Berkeley. URL: http://www.ucmp.berkeley.edu/historyoflife/histoflife.html. This exhibit is a teaching tool.

North America University of California Museum of Paleontology, the Paleontological Society et al. URL: http://www.paleoportal.org. This site is devoted to the biota of North America.

The Ryosuke Motani of the University of California, Davis. URL: http://www.geology. ucdavis.edu/~motani. Like the site directly above, this is where a similar collection is maintained.

Smithsonian Institution, National Museum of Natural History, Washington, D.C. URL: http:// www.nmnh.si.edu/paleo/PaleoArt. This site is a digital art archive.

Smithsonian Institution. National Museum of Natural History, Washington, D.C. URL: http://www.nmnh.si.edu/paleo/georime. This site lists periods, eras, and epochs and applies geologic time to modern-day places.

Sternberg Museum of Natural History, Hays, Kansas. URL: http://www.oceansofkansas. com. This museum has a collection of fossil ichthyosaurs, dolphinlike contemporaries of dinosaurs, in an exhibit called The Oceans of Kansas dating to the time when Kansas was part of the ocean. This is its online component.

University of California, Berkeley. URL: http://www.ucmp.berkeley.edu/collections/plants/clearedleaf.html. This paleobotany database examines fossil leaves as evidence of past climates, herbivore predation, growth patterns, and more.

Particulate Matter

EPA. URL: http://hei.aer.com/login.php. The measurements on this site are made by a private contractor and the results are not free, but the map showing what is where, is.

Peptidases

Sanger Institute in Hinxton, UK. URL: http://merops.sanger.ac.uk. This is the site of base MEROPS, which holds data on over 2,300 viruses, bacteria, and other organisms and their peptidases. These compounds are the enzymes that split bonds in proteins.

Plankton

Toby Tyrrell of Southhampton University, England. URL: http://www.noc.soton. ac.uk/soes/staff/tt/eh/index.html. This is a site for one organism, Emiliania huxleyi, or Ehux. Its prevalence and importance in the oceanic carbon cycle warrants the attention.

Plate Tectonics

UNAVCO, Boulder, Colorado. URL: http://Jules.unavco.org. Mapping software for earth movement and other material is mounted on this site showing water, vegetation, and more.

Pollution

Environmental Defense, an advocacy group. URL: http://www.scorecard.org. This site collects data from several sources, including the USDA, EPA, and state agencies.
National Library of Medicine, Bethesda, Maryland. URL: http://www.toxmap.nlm.nih. gov/toxmap/main/index.jsp. These maps were created using data from the EPA's Toxics Release Inventory.

Protista

University of Montreal, Canada. URL: http://megasun.bch.umontreal.ca/protists. This is a site for the taxonomy and classification of unicellular organisms.

Radioactivity

Oregon State University, Corvallis. URL: http://livingtextbook.oregonstate.edu. This site contains the chemistry, protocols, and more on radioactive elements.

Reproduction

Virginia Technical University, Blacksburg, Virginia, and Rockefeller University, New York. URL: http://www.mpf.biol.vt.edu/research/budding_yeast_model/pp. These two research groups have created a base of information about yeast, including its activity, mutations, and genetic drift.

Sea Anemone

John Finnerty et al. Boston University, Boston, Massachusetts. URL: http://www.stellabase.org. It is devoted to the one sea anemone whose genome has been sequenced. The site houses information enabling researchers to make gene comparisons. A companion site is the result of a North American and British cooperation: URL: http://www.nematostella.org.

Seamounts

NOAA. URL: http://www.oceanexplorer.noaa.gov. This site is a catalog of undersea volcanoes. Several additional sites are available:
URL: http://seamounts.sdsc.edu;
URL: http://earthref.org/databases/SC/main.htm;
URL: http://www.oceanexplorer.noaa.gov.

Seaslugs

Australian Museum, Sydney, Australia. URL: http://www.seaslugforum.net. This site contains information and photographs of seaslugs.

Species

The Catalog of Life Integrated Taxonomic Information System (IT IS) and Species 2000. URL: http://www.sp2000.org. This multinational group is attempting to categorize all living species. A search engine designed by Roderic Page of the University of Glasgow links images, protein and DNA sequences, recent literature and taxonomic data on species: URL: http://www.ispecies.org.

Statistics

Francis Galton. Gavan Tredoux, of Rochester, New York. URL: http://galton.org. This is a compendium of Galton's statistical writings. Another open site for free statistics software is URL: http://www.r-project.org.

University of California, Los Angeles. URL: http://socr.stat.ucla.edu. This site consists of Java applets for demonstrations and/or online experiments.

Sun

NASA. URL: http://sohowww.nascom.nasa.gov. This site consists of graphics from the Solar and Heliospheric Observatory (SOHO).

NOAA. URL: http://www.sec.noaa.gov/sxi. This is a site containing graphic data on the Sun's activities.

Tsunamis

Tsunami Information Center. URL: http://www.prh.noaa.gov/itic and University of Washington. URL: http://www.geophys.washington.edu/tsunami/welcome.html. Both of these sites list information about the earthquakes and earth movements that cause tsunamis.

Turtles

Seaturtle.org. URL: http://www.seaturtle.org. This site is completely devoted to sea turtles and includes satellite tracking, an image library, and more.

Vegetation

University of Alaska. URL: http://www.geobotany.uaf.edu/arcticgeobot. This site is a map of the polar regions.

Vent Communities (see Ocean Explorer)

Vertebrates

World Wildlife Fund. URL: http://www.worldwildlife.org/wildfinder. This site is a locator of over 30,000 species in 825 ecoregions.

Virus

Alexei Drummond and Andrew Ramaut, University of Oxford, England. URL: http://
beast.bio.ed.ac.uk/Main_Page. This database uses Bayesian analysis to plot the evolutionary trail of viruses. This is a software program called BEAST.

Scripps Research Institute, La Jolla, California. URL: http://viperdb.scripps.edu. This
site shows the distinctive outer shells of viruses.

Universite de Sherbrooke, Canada. URL: http://penelope.med.usherb.ca/subviral. This
subviral RNA database has hundreds of nucleotide sequences in its collection. It is
maintained by the viral research group.

The Viral Bioinformatics Research Center, University of Victoria, Canada. URL: http://
athena.bioc.uvic.ca. This site is a compilation of viral sequences for hundreds of
viruses to enable the comparison of viral genomes.

Volcanoes

U.S. Geological Survey. URL: http://hvo.wr.usgs.gov/kilauea/update. This site concentrates on one volcano, Kilauea.

Water

Rivers Heidelberg College, Tiffin, Ohio. URL: http://wqldata.heidelberg.edu. This
research group tracks water quality.

University of Arizona, Tucson. URL: http://www.sahra.arizona.edu/newswatch/index.
html. In this archive, the presence, absence, and function of water is studied.

Water Quality

U.S. Geological Survey for freshwater areas. URL: http://water.usgs.gov/nawqa. This is
the site for the U.S. Government freshwater sampling organization.

Weather

Passive Microwave Earth Science Information Partner. URL: http://pm-esip.msfc.nasa.
gov. This is a NASA-sponsored project for tracking storms, temperatures, and sea
levels. The site is run by The Passive Microwave Earth Science Information Partner.

Wetlands

U.S. Fish and Wildlife Commission. URL: http://wetlands.fws.gov. This federal organization looks at U.S. wetlands, flora, fauna, and health of biomes.

U.S. Geological Survey's National Wetlands Research Center. URL: http://www.nwrc.
usgs.gov. Like the site directly above, this site looks at U.S. wetlands, flora, fauna,
and health of biomes.

Wind

The Danish Wind Industry Association. URL: http://www.windpower.org/en/core.htm
The Danish Wind Industry maintains this site as a public service. It provides information about wind power and the future of energy.

Women in Science

University of California, Los Angeles. URL: http://cwp.library.ucla.edu. This is an
archive of contributions made before 1976 by 20th-century women. It was created
to accompany an exhibit.

Worms

Yeshiva and Columbia Universities, New York. URL: http://www.wormatlas.org. These two universities collaborated on this in-depth worm Web site that concentrates on the behavioral and structural anatomy of the worm.

BIBLIOGRAPHY

General

Ballard, Robert D. *The Lost Ships of Robert Ballard: An Unforgettable Underwater Tour*. San Diego, Calif.: Advantage Publishers Group, 2005.

Corfield, Richard. *The Silent Landscape: The Scientific Voyage of the HMS* Challenger. Washington, D.C.: National Academy Press, 2005.

Fonteneau, Gilles. *Murmurs from the Deep: Scientific Adventures in the Caribbean*. Berkeley, Calif.: Arcade Publishing, 2006.

Garrison, T. S. *Oceanography: Invitation to Marine Science*. 3rd ed. Pacific Grove, Calif.: Brooks/Cole, 1999.

Kennish, Michael J. *Practical Handbook of Marine Science*. 3rd ed. Boca Raton, La.: CRC Press, 2000.

MacInnis, Joe. *James Cameron's Aliens of the Deep: Voyages to the Strange World of the Deep Ocean*. Washington, D.C.: National Geographic Society, 2005.

Open University. *Case Studies in Oceanography and Marine Affairs*. Woburn, Mass.: Butterworth-Heinemann, 2000.

Pinet, Paul R. *Invitation to Oceanography*. Sudbury, Mass.: Jones & Bartlett Publishers, 2006.

Rozwadowski, Helen M. *Fathoming the Ocean: The Discovery and Exploration of the Deep Sea*. Cambridge, Mass.: Harvard University Press, 2005.

Theodorou, Rod. *To the Depths of the Ocean*. Barrington, Ill.: Heinemann Library, 2006.

Trefil, James S. *A Scientist at the Seashore*. Mineola, New York: Dover Publications, 2005.

Careers

Adams, Steve. *Marine Science Careers: A Sea Grant Guide to Ocean Opportunities*. Durham, N.H.: Maine-New Hampshire Sea Grant College Program, 2000.

Heitzman, William R. *Opportunities in Marine and Maritime Careers*. Lincolnwood, Ill.: NTC/Contemporary Publishing, 2000.

Reference

Allen, Gerald. *Marine Life of the Pacific and Indian Oceans*. Berkeley, Calif.: Periplus, 1997.

Harvey-Clark, Chris. *Eastern Tidepool and Reef: North Atlantic Marine Life*. Blaine, Wash.: Hancock House, 1997.

Lehr, Jay H., ed. *Water Encyclopedia, Oceanography, Meteorology, Physics and Chemistry*. New York: John Wiley & Sons, 2005.

Niesen, Thomas M. *Beachcomber's Guide to Marine Life of the Pacific Northwest*. Houston, Tex.: Gulf Publishing, 1997.

NOAA. *Atlantic Coast of North America*. New York: McGraw-Hill Professional Publishing, 2005.

———. *Central and Western Pacific Ocean and Indian Ocean*. New York: McGraw-Hill Professional Publishing, 2005.

Bibliography

———. *East Coast of North and South America Including Greenland.* New York: McGraw-Hill Professional Publishing, 2005.

———. *Pacific Coast of North America and Asia.* New York: McGraw-Hill Professional Publishing, 2005.

———. *Tide Current Tables 2007: Atlantic Coast of North America.* New York: McGraw-Hill Professional Publishing, 2006.

———. *Tide Tables 2007: Central and Western Pacific Ocean and Indian Ocean.* New York: McGraw-Hill Professional Publishing, 2006.

———. *Tide Tables 2007: East Coast of North And South America Including Greenland.* New York: McGraw-Hill Professional Publishing, 2006.

———. *Tide Tables 2007: Europe and West Coast of Africa.* New York: McGraw-Hill Professional Publishing, 2006.

———. *Tide Tables 2007: Pacific Coast of North America and Asia.* New York: McGraw-Hill Professional Publishing, 2006.

———. *Tide Tables 2007: West Coast of North and South America.* New York: McGraw-Hill Professional Publishing, 2006.

———. *West Coast of North and South America Including the Hawaiian Islands and the Alaska Supplement.* New York: McGraw-Hill Professional Publishing, 2005.

Snyderman, Marty. *California Marine Life: A Guide to Common Marine Species.* Niwot, Colo.: Roberts Rinehart, 1997.

Shwartz, M., Springer. *Encyclopedia of Coastal Science.* New York: 2006.

Snodgrass, Mary E. *World Shores and Beaches: A Descriptive and Historical Guide to 50 Coastal Treasures.* Jefferson, N.C.: McFarland & Company, 2005.

Stow, Dorrik. *Oceans: An Illustrated Reference.* Chicago, Ill.: University of Chicago Press, 2006.

Zilberberg, Louis-Jacques. *Elsevier's Dictionary of Marine Pollution: English-Spanish.* New York: Elsevier, 2000.

Biology

Bortone, Stephen A. *Seagrasses: Monitoring, Ecology, Physiology and Management.* Boca Raton, Fla.: CRC Press, 1999.

Boyle, Peter, and Paul G. Rodhouse. *Cephalopods: Ecology and Fisheries.* Oxford, UK: Blackwell Publishing Ltd., 2005.

Currents of Contrast: Life in Southern Africa's Two Oceans. Cape Town, South Africa: Stuik Publishers, 2006

Ellis, Richard. *The Search for the Giant Squid: An Authoritative Look at the Biology and Mythology of the World's Most Elusive Sea Creature.* New York: Viking, 1999.

Ellis, Richard. *Singing Whales and Flying Squid: The Discovery of Marine Life.* Guilford, Conn.: Globe Pequot Press, 2006.

Evans, David H. *The Physiology of Fishes.* Boca Raton, Fla.: CRC Press, 1997.

Haq, Bilal U. *Introduction to Marine Micropaleontology.* 2nd ed. New York: Elsevier, 2000.

Helvarg, David. *Blue Frontier: Dispatches from America's Ocean Wilderness.* San Francisco, Calif.: Sierra Club Books, 2006.

Kasyanov, V. L. *Larvae of Marine Bivalves and Echinoderms.* Enfield, Conn.: Science Publishers, 1998.

Kiehl, Mimi. *Wave-Swept Shore: The Rigors of Life on a Rocky Coast.* Berkeley, Calif.: University of California Press, 2006.

Lippson, Alice J. *Life in the Chesapeake Bay.* Baltimore, Md.: Johns Hopkins Press, 1997.

McLachlan, Anton. *The Ecology of Sandy Shores.* New York: Elsevier Science and Technology, 2006.

Probert, Keith et al.: *Southern Seas: Marine Life at 46° South Otago, New Zealand*: Otago University Press, 2005.

Rigsby, Michael. *Tide Pools of Southern California*. Tucson, Ariz.: Western National Parks Assoc., 2006.

Sokolova, M. N. *Feeding and Trophic Structure of the Deep-Sea Macrobenthos*. Enfield, Conn.: Science Publishers, 2000.

Thompson, Mary. *Endocrine and Reproductive Biology*. Enfield, Conn.: Science Publishers, 1997.

Biotechnology

Bates, Charles C., and G. L. Hanssen. *"Hydro" to "Navoceano": 175 Years of Ocean Survey and Prediction by the U.S. Navy, 1830–2005*. Green Valley, Ariz.: Corn Field Press, 2005.

Fingerman, Milton. *Recent Advances in Marine Biotechnology Aquaculture A. Seaweeds and Fishes*. Enfield, Conn.: Science Publishers, 2000.

———. *Recent Advances in Marine Biotechnology Aquaculture B. Fishes*. Enfield, Conn.: Science Publishers, 2000.

———. *Recent Advances in Marine Biotechnology: Biofilms, Bioadhesion, Corrosion and Befouling*. Enfield, Conn.: Science Publishers, 2000.

———. *Recent Advances in Marine Biotechnology: Environmental Marine Biotechnology*. Enfield, Conn.: Science Publishers, 1998.

Kjeldsen, Soren P. *Giant Waves: Origins, Simulation and Impacts on Ships and Platforms*. New York: Simon & Schuster, 2006.

Medwin, Herman. *Sounds in the Sea: From Ocean Acoustics to Acoustical Oceanography*. New York: Cambridge University Press, 2005.

Miller, Richard L. et al., eds. *Remote Sensing of Coastal Aquatic Environments: Technologies, Techniques and Applications*. New York: Springer, 2005.

Osinga, R., ed. *Marine Bioprocess Engineering*. New York: Elsevier, 1998.

Wozniak, Bogdan et al. *Light Absorption and Absorbents in Sea Water*. New York: Springer, 2006.

Chemistry

Brightwell, Christopher. *Marine Chemistry*. Neptune, N.J.: TFH Publications, 2006.

Gianguzza, Antonio. *Marine Chemistry: An Environmental Analytical Chemistry Approach*. Hingham, Mass.: Kluwer, 1997.

———. *Chemical Processes in Marine Environments: Risk Assessment of Delayed Non-Linear Responses*. New York: Springer-Verlag, 2000.

Hester, R. E. *Chemistry in the Marine Environment: Issues in Environmental Science and Technology*. London: Royal Society, 2000.

Ecology

Allen, Larry G. *Ecology of Marine Fishes: California and Adjacent Waters*. Berkeley, Calif.: University of California Press, 2006.

Alongi, D. M. *Coastal Ecosystem Processes*. Boca Raton, Fla.: CRC Press, 1997.

Barnes, R. S. K. *Introduction to Marine Ecology*. 3rd ed. Malden, Mass.: Blackwell Science, 1999.

Chang Chun Institute of Biology Staff. *Deep-Water Traction Current Deposits: A Study Of Internal Tides, Internal Waves, Countour Currents and Their Deposits*. Bethesda, Md.: International Scholars Publications, 1998.

Cote, Isabelle, et al., eds. *Coral Reef Conservation*. New York: Cambridge University Press, 2006.

Bibliography

Hall, Michelle K. et al. *Exploring Ocean Environments.* Florence, Ky.: Brooks/Cole, 2006.

Hall, Stephen J. *The Effects of Fisheries on Ecosystems and Communities.* Malden, Mass.: Blackwell Science, 1999.

Jennings, Simon. *Marine Fisheries Ecology.* Malden, Mass.: Blackwell Science, 2000.

Kennish, Michael J. *Estuary Restoration and Maintenance: The National Estuary Program.* Boca Raton, Fla.: CRC Press, 1999.

———. *Pollution Impacts on Marine Biotic Communities.* Boca Raton, Fla.: CRC Press, 1997.

Kingsford, Michael. *Studying Temperate Marine Environments: A Handbook for Ecologists.* Canterbury, Eng.: University Press, 1998.

Knox, George A. *The Ecology of Seashores.* Boca Raton, Fla.: CRC Press, 2000.

Kristensen et al., eds. *Interactions between Macro and Microorganisms in Marine Sediments.* Washington, D.C.: American Geophysical Union, 2005.

Livingston, Robert J. *Marine Eutrophication.* Boca Raton, Fla.: CRC Press, 2000.

Longhurst, Alan R. *Ecological Geography of the Sea.* New York: Elsevier Science and Technology, 2006.

McLachlan, Anton. *The Ecology of Sandy Shores.* Elsevier Science and Technology, New York, 2006.

Probert, Keith et al. *Southern Seas: Marine Life at 46° South.* Otago, New Zealand: Otago University Press, 2005.

Seaman, William. *Artificial Reef Evaluation with Application to Natural Marine Habitats.* Boca Raton, Fla.: CRC Press, 2000.

Sherman, Kenneth. *The Gulf of Mexico: Large Marine Ecosystem Assessment, Sustainability and Management.* Malden, Mass.: Blackwell Science, 1999.

Sherman, Kenneth. *Large Marine Ecosystems of the Indian Ocean: Assessment, Sustainability, and Management.* Malden, Mass.: Blackwell Science, 1998.

Souvorov, A. V. *Marine Ecologonomics.* New York: Elsevier, 1999.

Tang, Qisheng. *Large Marine Ecosystems of the Pacific Rim: Assessment, Sustainability and Management.* Malden, Mass.: Blackwell Science, 1999.

Thomas, Peggy. *Marine Mammal Preservation Bird Alert.* Brookfield, Conn.: Twenty-first Century Books, 2000.

Zedler, Joy B. *Handbook for Restoring Tidal Wetlands.* Boca Raton, Fla.: Auerbach, 2000.

Geology

Chough, Sung Kwun. *Marine Geology of Korean Seas.* 2nd ed. Enfield, Conn.: Science Publishers, 2000.

Eisma, D. *Intertidal Deposits, River Mouths, Tidal Flats, and Coastal Lagoons.* Boca Raton: CRC Press, 1998.

McAvoy, T. *Fault Detection II, Aerospace, Marine Systems.* New York: Elsevier, 1999.

Government

Baker, D. James. *Our Ocean Future: Themes and Issues Concerning the Nation.* Collingdale, Pa.: DIANE Publishing, 2000.

Hazardous Materials Transportation Act . . . 105 Congress. Washington, D.C.: U.S. Government Printing Office, 1997.

National Research Council. *Building Ocean Science Partnerships: The United States and Mexico Working Together.* Washington, D.C.: National Academy Press, 1999.

Meteorology

Benson, Keith R., and Helen M. Rozwadowski. *Ocean Weather Forecasting: An Integrated View of Oceanography.* New York: Springer, 2006.

Cap, Ferdinand. *Tsunamis and Hurricanes: A Mathematical Approach*. New York. Springer, 2006.

Extremes: Oceanography's Adventures at the Poles. Mission Viejo, Calif.: Watson Publishing, 2006.

Vallis, Geoffrey K. *Atmospheric and Oceanic Fluid Dynamics: Fundamentals and Large-Scale Circulation*. New York: Cambridge University Press, 2006.

Oceanography

Chamberlin, W. Sean, et al. *Exploring the World Ocean*. New York: McGraw-Hill, 2006.

Dinwiddie, Robert et al. *Ocean*. New York: Dorling Kindersley Publishing, 2006.

Garrison, Tom S. Brooks. *Enhanced Essentials of Oceanography*. Florence, Ky.: Cole Publishing, 2006.

Jochum, Markus, ed. *Physical Oceanography: Developments since 1950*. New York: Springer, 2006.

Robinson, A. R., et al. *The Global Coastal Ocean: Interdisciplinary Regional Studies and Syntheses*. Cambridge, Mass.: Harvard University Press, 2006.

Robinson, Ian S. *Understanding the Oceans from Space: The Unique Applications of Satellite Oceanography*. New York: Springer, 2006.

Savage, Stephen, et al. *Oceans*. Boston, Mass.: Houghton-Mifflin Company, 2006.

Sverdrup, Keith A., et al. *An Introduction to the World's Oceans*. New York: McGraw-Hill Higher Education, 2006.

Tjossem, Sara. *The Journey to PICES: Scientific Cooperation in the North Pacific*. Fairbanks, Alaska.: Alaska Sea Grant College Program, University of Alaska, 2005.

Juvenile

Berger, Melvin. *Dive! A Book of Deep-Sea Creatures*. New York: Scholastic, 2000.

Children's Press Staff. *Aquatic Life*. Danbury, Conn.: Children's Press, 1998.

Chanko, Pamela. *Sea Creatures*. New York: Scholastic, 1998.

Davis, Wendy. *Working at a Marine Institute*. Danbury, Conn.: Children's Press, 1998.

Esbensen, Barbara J. *Sponges are Skeletons*. New York: HarperCollins, 1998.

Fowler, Allan. *Shellfish Aren't Fish*. Danbury: Children's Press, 1999.

———. *Life in a Tide Pool*. Danbury, Conn.: Children's Press, 1997.

Jenkins, Priscilla B. *A Safe Home for Manatees*. New York: HarperCollins, 1997.

Kalman, Bobbie. *What is a Marine Mammal?* New York: Crabtree, 1999.

Littlefield, Cindy A. *Awesome Ocean Science*. Nashville, Tenn.: Ideals Publications, 2006.

Mara, Will. *The Four Oceans*. New York: Scholastic Library Publishing, 2005.

Milliken, Linda. *Ocean Life Activity Book*. San Clemente, Calif.: Edupress, 1998.

Moor, Jo Ellen. *Life in the Ocean: Grades 1–3*. Monterey, Calif.: Evan Moor Corp., 1998.

Strang, Craig. *Ocean Currents: Marine Science Activities*. Berkeley: University of California, Berkeley: Lawrence Hall of Science, 2000.

Tant, Carl. *Awesome Oceans: Advances in Marine Biotechnology*. Arlington, Tex.: Biotech Publishing, 1998.

Periodicals
General Interest

American Biology Teacher
Discover
Endeavor
National Geographic
Natural History

Bibliography

New Scientist
Wildlife Conservation

Professional periodicals
(cover a range of subjects and are readable by the general public)

America
American Naturalist
Darwin
Eos
Geotimes
Journal of Ecology
Nature
Proceedings of the National Academy of Sciences of the United States of America
Science
Scientific American